AF531089

The Saunders Series

Introduction to Topology 1983
Theodore W. Gamelin and Robert Everist Greene

Linear Algebra 1983
P. G. Kumpel and J. A. Thorpe

Differential Equations with Linear Algebra 1986
Zbigniew H. Nitecki and Martin M. Guterman

Principles of Topology 1989
Fred H. Croom

An Introduction to the History of Mathematics, Sixth Edition 1990
Howard Eves

Introduction to Proof in Abstract Mathematics 1990
Andrew Wohlgemuth

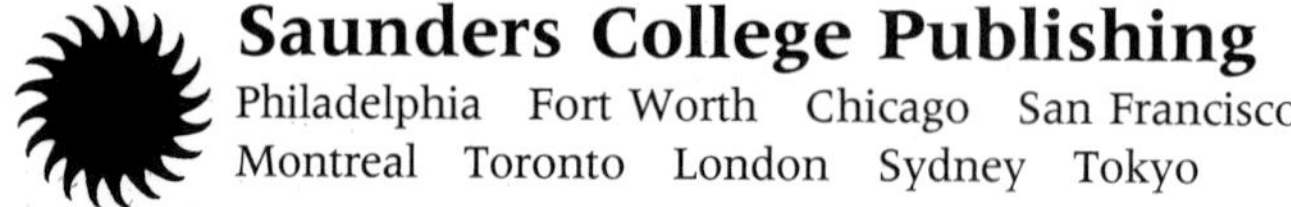
Saunders College Publishing
Philadelphia Fort Worth Chicago San Francisco
Montreal Toronto London Sydney Tokyo

Mathematical Logic:
Applications and Theory

Jean E. Rubin
Purdue University

Text typeface: Meridien
Compositor: Waldman Graphics, Inc.
Acquisitions Editor: Robert Stern
Managing Editor: Carol Field
Project Editor: Anne Gibby
Copy Editor: Diane Lamsback
Manager of Art and Design: Carol Bleistine
Art Director: Christine Schueler
Art and Design Coordinator: Doris Bruey
Text Designer: C.J. Petlich, Hunter Graphics
Cover Designer: Doris Bruey
Text Artwork: GRAFACON
Director of EDP: Tim Frelick
Production Manager: Bob Butler

Printed in the United States of America

ISBN 0-03-012808-0

Library of Congress Catalog Card Number: 89-043369

0123 0 016 987654321

Preface

This text is intended to be used in a first course on mathematical logic for undergraduate students of mathematics, science, computer science, and electrical engineering. Philosophy students and pre-law students may also find it useful. It is expected that most students taking the course will have had a course in calculus. A logic course for junior-level mathematics and computer science students has been taught successfully at Purdue University for the past five years using preliminary versions of this text.

The primary purposes of the text are to teach undergraduate students to use the English language correctly, to understand the notion of a formal language, to reason correctly, and to understand a mathematical proof. The text moves at a slow pace, introducing new concepts with many illustrative examples. Throughout the text, there are examples from mathematics. There are a large number of exercises, making it easy for instructors to assign homework problems, to choose problems for exams, and to do examples in class. Answers for selected exercises are included as part of the book, and there is a separate complete solution manual available for instructors.

In Chapter 1 we introduce a very simple formal language, the language of the Sentential Calculus. We study both its syntax and its semantics. Truth tables are used to describe the semantics. We provide many examples of translating English into this formal language thereby helping students to learn the structure of their own language a little better.

Chapter 2 covers the notion of a *proof* in the Sentential Calculus. We discuss various types of proofs (conditional proofs, indirect proofs, and so forth) and, in Section 2.6, give a set of general guidelines for constructing proofs. We also discuss the notions of valid and invalid arguments and introduce the notion of a counter-example to show an argument is invalid. This chapter serves to acquaint the students with various ideas which are more important in the context of the Predicate Calculus. (In the Sentential Calculus, there is an algorithm using truth tables to determine whether an argument is valid. "Proofs" are not necessary.)

The notions of *consistent* and *inconsistent* sets of sentences are discussed in Chapter 3. As an application of these ideas, we include some logical puzzles for the students to solve.

In Chapter 4, we discuss some applications of the Sentential Calculus primarily for computer science and electrical engineering students. These applications include Karnaugh maps, switching networks, and logic networks. However, we also include in this chapter, in Sections 4.1 and 4.2, the disjunctive normal form of a formula, a rule of inference for tautologically equivalent formulas, and the algebra of formulas. These latter notions are important for all logic students.

In Chapter 5, we extend the language of the Sentential Calculus to a language rich enough for mathematics, the language of the Predicate Calculus. To help students understand this new language, much time is spent in analyzing English and mathematical sentences and translating them into the language of the Predicate Calculus. Difficulties arise primarily because the English language, like any other spoken language, is ambiguous, whereas the language of the Predicate Calculus, like other formal languages, is not. Therefore, before translating an English sentence into the language of the Predicate Calculus, students must thoroughly understand the meaning and the syntax of that sentence. This type of analysis helps students understand complex sentences in English and in mathematics.

The semantics of the Predicate Calculus is discussed in Chapter 6. An interpretation structure for a formula in the Predicate Calculus is analogous to an assignment of truth values for the sentential variables in a formula in the Sentential Calculus. We discuss the notions of truth and falsity in an interpretation structure. An understanding of these notions will help students understand why the Rules of Inference introduced in the next two chapters preserve truth.

In Chapter 7, we start describing proofs in the Predicate Calculus, and this is continued in Chapter 8. In the beginning, the general procedure for proofs is to drop quantifiers, give a proof in the Sentential Calculus, and then add quantifiers. In these two chapters, we give the rules for dropping and adding quantifiers. We extend our guidelines for proofs in the Sentential Calculus (Section 2.6) to include guidelines for proofs in the Predicate Calculus (Section 8.4). The system of proof we use is a system of natural deduction, which formalizes informal mathematical proofs. Thus, in later chapters, when we discuss informal proofs, it is easy to see the relationship between formal and informal proofs. In Section 8.5, we compare the semantics and the syntax of arguments and indicate the role played by the Completeness Theorem and the Soundness Theorem.

Chapter 9 covers the concepts of invalid arguments and consistent sets of formulas. We show how interpretation structures are used to show an argument is invalid and a set of premises is consistent. In Section 9.3, we give a proof of the Compactness Theorem using the Completeness Theorem, and we include some applications of the Compactness Theorem. This section should be covered before Section 16.5, where we give a proof of the Completeness Theorem.

Chapter 10 introduces the notions of logical theorems and valid formulas. Additional rules of inference are added that allow us to use previously proved results to prove new results. Consequently, these rules allow us to shorten the length of proofs.

In Chapter 11, we add identity to the language of the Predicate Calculus. First we describe how identity is used in English. Then we add rules of inference for identity and describe the meaning of identity in interpretation structures. We give many examples of proofs in which identity occurs. Finally, in Section 11.6, we describe arguments in which obvious premises are omitted. For example, if John is in New York at a specific time, in ordinary discussions it would not be necessary to state that John cannot be in Chicago at that same time. However, if the fact that John is not in Chicago is needed to deduce some conclusion, it would have to be explicitly stated. This type of argument is called an *enthymeme.*

In Chapter 12, we summarize the important notions that were discussed in the preceding chapters for easy reference for the students.

The remaining chapters contain applications of logic. In Chapter 13, we discuss the relationship between logic and informal mathematical proofs. We use Group Theory as an example of a simple axiomatic system. Also in Chapter 13, we give other examples of theories and indicate some of the properties of theories that concern logicians—for example, consistency and characterization of models. This chapter may be used merely to show the students the relationship between formal and informal mathematical proofs.

Chapters 14 and 15 cover induction. In Chapter 14, we introduce mathematical induction in the setting of Peano Arithmetic. This chapter may be used merely to teach mathematical induction and/or it may be used to study another theory that is richer and has more applications than Group Theory. Chapter 15 continues with inductive definitions and complete induction. It describes various types of algorithms that are of interest to mathematics and computer science students. In the last section of Chapter 15, we give a brief discussion of automated theorem-proving. This topic becomes quite technical if one studies it in any depth, so this section is only intended to give the students the flavor of automated theorem-proving.

Finally, in Chapter 16, we give proofs of the main theorems that are used throughout the text, including proofs of the Soundness Theorem and the Completeness Theorem. This chapter is intended for those students who have a good mathematical background.

In the Table of Contents and in the text, we place an asterisk, (*), on chapters, sections, and exercises that are primarily theoretical.

This text is versatile. It can be used for several different types of logic courses. We indicate a few of these in the following list:

1. Core Course: 1.1 through 1.8; 2.1 through 2.6; 3.1; 4.1 through 4.2; 5.1 through 5.6; 6.1 through 6.2; 7.1 through 7.3; 8.1 through 8.5; 9.1 through 9.2; 10.1 through 10.4; 11.1 through 11.5.
2. Theoretical Course: Core Course; 9.3; 10.5; 14.1 through 14.4; 15.1 through 15.2; 16.1 through 16.5.
3. Course for Mathematics Majors: Core Course; 7.4; 13.1 through 13.5; 14.1 through 14.3; 15.1 through 15.2. (Sections marked with an asterisk are optional.)
4. Course for Computer Science Majors: Core Course; 3.2; 4.3 through 4.5; 14.1 through 14.3; 15.1 through 15.3. (Sections marked with an asterisk are optional.)

Normally, sections should be covered in numerical order except for the following: 3.2 is only a prerequisite for 15.3; 4.3 through 4.5, 7.4, 11.6, 13.4, 13.5, and 15.3 may be omitted without affecting continuity; all sections marked with an asterisk may be omitted unless Chapter 16 is covered; 10.5 is only a prerequisite for Chapter 16. Chapter 13 is not a prerequisite for later chapters except that the introduction to Chapter 13 introduces terminology that is mentioned in 14.1. Chapter 13 may be used to teach the relationship between formal and informal mathematical proofs, and Sections 14.2, 15.1, and 15.2 may be used to teach induction.

A text like this cannot be written without the help and encouragement of many people. In particular, I would like to thank Leonard Lipshitz, who has read and taught from many versions of the manuscript and has made innumerable helpful suggestions; my son Arthur, who read and corrected early versions of the manuscript; Andrea Casagranda, who helped me prepare the solution manual; Judy Snider, who typed most of the manuscript and the solution manual; Robert Stern, the senior mathematics editor at Saunders, who worked with me on this project; Anne Gibby, the project editor at Saunders, who coordinated the efforts of the copy editor, the typesetter, and the author; the students at Purdue who studied logic using the manuscript and who made helpful suggestions and/or found typographical errors; Bret Rhea, who proofread the solution manual; and, finally, the reviewers: Leonard Lipshitz, Purdue University; C. Ward Henson, University of Illinois at Urbana; Carol Wood, Wesleyan University; Mark E. Nadel, Mitre Corporation, Bedford, Mass.; Robert Fraenkel, MIT; Tom McLaughlin, Texas Tech University; Peter Hinman, University of Michigan; Don Pigozzi, Iowa State University; Michael P. O'Neil, Kansas State University; Thomas Jech, Penn State University; and Jacob Plotkin, Michigan State University.

Jean E. Rubin
West Lafayette, Indiana

Introduction

The American College Dictionary defines the word "logic" as "the science which investigates the principles governing correct or reliable inference" or "the system or principles of reasoning applicable to any branch of knowledge or study." This is a fairly accurate brief description of what we cover in this book. Our main objective is to present a system of rules of inference which allows us to reason correctly. Presenting these rules will not be very helpful unless we know how to apply them in our language, which is English. However, because any spoken language is very complicated, has exceptions to almost every rule, and, most important of all, is ambiguous, we approximate the English language with a formal language. In the process of studying rules of inference, we also study the structure of formal languages. In Chapter 1, we introduce a very simple formal language, the language of the Sentential Calculus. Later, in Chapter 5, our language is expanded to the language of the Predicate Calculus, thereby obtaining a language adequate for mathematics. The study of these languages will help you to understand quite complicated sentences in both English and mathematics. As a simple example, what is the relationship between the following five sentences?

(1) If I do not take an umbrella, then it rains.
(2) It rains if I do not take an umbrella.
(3) It rains unless I take an umbrella.
(4) If it doesn't rain, I take an umbrella.
(5) It doesn't rain only if I take an umbrella.

You should be able to convince yourself that all five sentences are different ways of saying exactly the same thing. On the other hand, the sentence

(6) If it rains, then I do not take an umbrella.

has a different meaning.

We shall also study how to negate a sentence. Consider the following six sentences:

(1) All cats are pets.
(2) No cats are pets.
(3) There does not exist a cat which is a pet.
(4) Some cats are not pets.
(5) Not all cats are pets.
(6) Some cats are pets.

An analysis of these sentences shows that (2) and (3) are equivalent, and each is the negation of (6). Also, (4) and (5) are equivalent, and each of these is the negation of (1). Frequently, it is important to understand the meaning of sentences like these. For example, in mathematics it is important to know what it means if some limit does *not* exist or if a function is *not* continuous or if a group is *not* commutative.

We shall also discuss how changing the order of words in a sentence may change the meaning of the sentence. For example, compare the two sentences:

(1) Only John loves Mary.

and

(2) John loves only Mary.

Also, compare the following sentences:

(3) Someone is hit by lightning every day.

and

(4) Every day someone is hit by lightning.

While most everyone would agree that (1) and (2) do not mean the same thing, many people would think that (3) and (4) do mean the same thing. Actually, with careful analysis, it can be seen that (3) implies that the *same* person is hit by lightning every day. While (3) and (4) may be used interchangeably in ordinary discourse without any misunderstanding, when English is replaced by a formal language, we have to carefully distinguish between these two types of sentences. In order to reason correctly, you have to understand the structure of the English language and eliminate ambiguity. Another striking example of ambiguity in English is the sentence "All that glitters is not gold." A literal translation of this sentence is "Nothing that glitters is gold." However, what we normally mean now by this sentence is "Not all that glitters is gold." In translating English into a formal language, you must abstract the content and then use the limited means that are available in a formal language to describe that content exactly. We hope these studies will help you to avoid ambiguous statements and help you to use the English language more precisely.

We frequently hear arguments in political speeches, in commercials, and in ordinary discourse, as well as in mathematics. For example, which of the following arguments are valid?

(1) All lions are mammals.
All tigers are mammals.
Therefore, all lions are tigers.

(2) Everyone wants white teeth.
John Jones, the famous athlete, has white teeth and brushes with Brand X toothpaste.
Therefore, you will have white teeth if you brush with Brand X toothpaste.

(3) If I am elected, I will improve the school system.
It is a good idea to improve the school system.
Therefore, it is a good idea to elect me.

Of course, none of these arguments is valid. The conclusions do not follow from the premises. In argument (1) above, both premises are true but the conclusion is false. Consequently, we know immediately that this argument is not valid. A valid argument must preserve truth. If the premises are true, the conclusion must also be true. In arguments (2) and (3), we do not know *a priori* whether the conclusion is true or false, but even without any knowledge of logic, most people would know from experience that these arguments are not valid. (Argument (2) is a form of *inductive* or *statistical* inference and will not be discussed here.)

To paraphrase a well-known poet: A proof is a thing of beauty and a joy forever. One of the aims of this book is to help you to understand proofs and to appreciate their beauty and structure and also help you to analyze arguments and decide whether the conclusions actually do follow from the premises. We give a system of "natural deduction," a set of rules of inference for deriving conclusions from a set of premises. We could have chosen many possible systems of natural deduction. Our purpose is to formalize the way a rational person reasons, and our rules were chosen specifically for this purpose. In other words, we have chosen rules which formalize the rules used implicitly in mathematical proofs.

There is dichotomy involved here. On the one hand, we have the notion of a valid argument. An *argument* is a set of premises and a conclusion; it is *valid* if whenever the premises are true, the conclusion is also true. On the other hand, we have the notion of a *proof* of an argument. There is a proof of a conclusion φ from a set of premises Γ if φ is derivable from Γ using the rules of inference. The notion of a "valid argument" is a semantical notion. It deals with meaning and truth. On the other hand, a "proof" is a grammatical or syntactical notion. It deals with the structure of the language. These two notions are related in the statement (*) below.

(*) An argument with a set of premises Γ and conclusion φ is valid if and only if φ is derivable from Γ using the rules of inference.

Statement (*) is a formulation of Gödel's Completeness Theorem. Thus, Gödel's Completeness Theorem serves as a bridge between the structure of an argument and its semantics.

Showing an argument is *not* valid is equivalent to showing that the premises are consistent with the negation of the conclusion. Proving that a set of premises is consistent is the same as showing that a contradiction cannot be derived from the premises. It follows from Gödel's Completeness Theorem that a set of premises is consistent if and only if it has a *model,* an interpretation structure in which all the premises are true. Gödel's Completeness Theorem is also a bridge between proof theory and model theory. Consequently, in this text, we not only discuss proof theory, which deals with the structure of the language, but also model theory, which deals with its semantics.

Another one of our aims is to apply our knowledge of a valid argument to mathematical theories. Chapter 13 is devoted to proofs and models for mathematical theories. It is in this chapter that we describe the relationship between formal and informal proofs. This is an important application of formal logic. Throughout the text, we give applications to mathematics by means of translations, proofs, and models. Additional applications of logic are covered in Chapter 4, switching and logic networks, Karnaugh (K) maps; Chapter 14, mathematical induction; and Chapter 15, inductive definitions, algorithms, complete induction, and automated theorem-proving.

We include proofs of most of the results we use, including a proof of Gödel's Completeness Theorem. The text is arranged so that the theory is there for those who want a theoretical approach, but the proofs and theory may be omitted by those who want an applied course.

Contents

Chapter 1

Language and Truth Tables for the Sentential Calculus

In the first four chapters, we develop a formal system called the *Sentential Calculus* and discuss some of its applications. We begin by studying the language of this system. The language is quite simple. It consists of sentences and sentential connectives. We shall use capital letters, $A, B, C, D, \ldots$, and these letters with subscripts, to represent sentences. The connectives we shall use are negation, conjunction, disjunction, the conditional, and the biconditional. The language is not very expressive (one would not write poetry in this language, for example), but, as we shall see, it does have important applications.

We shall be interested both in the *syntax* (structure, grammar) and the *semantics* (meaning) of this language. Both of these concepts are described below.

1.1 NEGATION

Negation is translated as "not" or "it is not the case that." The symbol we shall use is $\neg$. Other symbols that logicians use for negation are: $-$, $\sim$, and $^-$. Thus, if S is the sentence "It is snowing outside," $(\neg S)$ is "It is not snowing outside" or "It is not the case that it is snowing outside."

The effect of the symbol $\neg$ on the meaning of a sentence S is illustrated by the following table, which is called a *truth table for* $\neg$:

S	$\neg S$
T	F
F	T

Thus, if the sentence S is true, then $(\neg S)$ is false and if S is false, then $(\neg S)$ is true. (When there is no ambiguity, we shall frequently omit the parentheses from "$(\neg S)$.")

1.2 CONJUNCTION

The translation of "conjunction" is "and" or sometimes "but." The symbol we shall use is $\wedge$. Other symbols used for conjunction include & and $\cdot$. For example, if S is the sentence "It is snowing outside" and W is the sentence "It is warm inside," then $(S \wedge W)$ would be translated as "It is snowing outside and warm inside" or "It is snowing outside, but warm inside."

If, in addition, C is the sentence "It is cold outside," then the compound sentence "It is snowing and cold outside, but warm inside" would be translated into our language as $((S \wedge C) \wedge W)$.

The sentence "It is cold outside, but not snowing," is translated as $(C \wedge \neg S)$. The sentence "It is not the case that it is cold and snowing outside" is translated as $\neg(C \wedge S)$.

The sentence "It is not both cold and snowing outside" would be translated the same way: $\neg(C \wedge S)$.

The following is the truth table for $\wedge$:

A	B	$(A \wedge B)$
T	T	T
T	F	F
F	T	F
F	F	F

A conjunction $(A \wedge B)$ is *true* if and only if both conjuncts, A and B, are true. (We shall sometimes use "iff" as an abbreviation for "if and only if.")

1.3 DISJUNCTION

By "disjunction," in our language, we mean the "nonexclusive or" or "and/or." We use the symbol $\vee$. (Sometimes + is used as the symbol for disjunction.) For example,

Sentence	**Translation**
1. It is snowing or cold outside.	$(S \vee C)$
2. It is either snowing and cold outside, or warm inside.	$((S \wedge C) \vee W)$
3. It is neither snowing outside nor cold outside.	$\neg(S \vee C)$

Sentence	**Translation**
4. It is not snowing outside and it is not cold outside.	$(\neg S \wedge \neg C)$
5. It is snowing or cold outside, but not both.	$((S \vee C) \wedge \neg(S \wedge C))$

(Sentence 5 exemplifies how the "exclusive or" is translated in our language.)

The truth table for disjunction is as follows:

A	B	$(A \vee B)$
T	T	T
T	F	T
F	T	T
F	F	F

Thus, a disjunction is *false* iff both disjuncts are false.

The following are truth tables for the translations of sentences 3 and 4:

S	C	$(S \vee C)$	$\neg(S \vee C)$
T	T	T	F
T	F	T	F
F	T	T	F
F	F	F	T

S	C	$\neg S$	$\neg C$	$(\neg S \wedge \neg C)$
T	T	F	F	F
T	F	F	T	F
F	T	T	F	F
F	F	T	T	T

We see that the two formulas have the same truth table; that is, one formula is true if and only if the other is true. (See De Morgan's rules in Chapter 2.)

The truth tables for negation, conjunction, and disjunction show that these connectives are *truth functional*. Given the truth value, T or F, of each component part of a compound sentence containing these connectives, the truth value of the whole sentence is uniquely determined. An example of this is presented in the following truth table for the formula $S = ((A \wedge \neg B) \vee \neg(A \wedge C))$:

A	B	C	$\neg B$	$(A \wedge \neg B)$	$(A \wedge C)$	$\neg(A \wedge C)$	S
T	T	T	F	F	T	F	F
T	T	F	F	F	F	T	T
T	F	T	T	T	T	F	T
T	F	F	T	T	F	T	T
F	T	T	F	F	F	T	T
F	T	F	F	F	F	T	T
F	F	T	T	F	F	T	T
F	F	F	T	F	F	T	T

We see from the truth table that S is false if A, B, and C are each true, and S is true otherwise. (We also note that the truth table for a formula that contains n letters has 2^n lines. Why?)

The truth tables of the remaining connectives will also be defined so they are truth functional because we want truth tables to depend only on the structure of a sentence, not on its meaning.

1.4 THE CONDITIONAL

The normal translation of the conditional is "if · · · , then . . . ." We have various ways of expressing this in English, however, as we shall see below. The symbol for the conditional that we shall use is $\rightarrow$. (Some texts use the symbol $\supset$.) A sentence of the form "$(A \rightarrow B)$" is called a *conditional* statement; "$(B \rightarrow A)$" is called its *converse*; "$(\neg A \rightarrow \neg B)$" its *inverse*; and "$(\neg B \rightarrow \neg A)$" its *contrapositive*. Some of the ways "$(A \rightarrow B)$" is translated include "if A then B," "B if A," "A implies B," "B is a consequence of A," "B provided that A," "A is a sufficient condition for B," and "B is a necessary condition for A."

In the following examples, when we write a sentence followed by letters in parentheses, we mean that those are the letters that are to be used for the *atomic sentences*, those sentences that contain no connectives. We treat atomic sentences as our basic units. (We shall avoid using the letters "T" and "F" for atomic sentences because we use these letters for truth values.)

Sentence	Translation
1. A triangle is isosceles if it has two equal angles. (I, E)	$(E \rightarrow I)$
2. If $x^2 > 4$, then $x > 2$, provided that x is positive. (O, W, P)	$(P \rightarrow (O \rightarrow W))$
3. $x > 7$ implies $x + 1 > 8$, and conversely. (S, E)	$((S \rightarrow E) \wedge (E \rightarrow S))$
4. A necessary condition for two lines to be parallel is that they neither intersect nor coincide. (P, I, C)	$(P \rightarrow \neg(I \vee C))$
5. If certain numbers such as 5 or 6 are substituted for x in the inequality $x^2 - 6 < 2x + 4$, we obtain the false statements $19 < 14$ or $30 < 16$, respectively. (V, S, N, H)	$((V \rightarrow N) \wedge (S \rightarrow H))$

Notice that in some of the sentences given in the preceding table, we had to change the order of the atomic sentences before we could translate them. Sentence 1 was rewritten as "If a triangle has two equal angles, it is isosceles." (Sometimes, in conditional statements, "then" is omitted.) Similarly, sentence 2 was changed to "If x is positive, then if $x^2 > 4$, $x > 2$." Sentence 5 required more extensive changes: "If 5 is substituted for x in the inequality $x^2 - 6 < 2x + 4$, we obtain the false statement $19 < 14$, and if 6 is substituted for x in the inequality $x^2 - 6 < 2x + 4$, we obtain the false statement $30 < 16$."

In the conditional statement "$(A \rightarrow B)$," "A" is called the *antecedent* or *hypothesis*, and "B" is called the *consequent* or *conclusion*. Normally, when such a statement is used in English, there is some connection between the antecedent and consequent. We usually do not say "If $3 > 7$, then two is a prime." In order to make the conditional truth functional, however, we have to decide on a truth value for such a statement. It seems natural that if A and B are both true, then $(A \rightarrow B)$ is true, and if A is true and B is false, then $(A \rightarrow B)$ is false. We shall make the convention, which is used in mathematics, that if A is false, then $(A \rightarrow B)$ is true. Thus, the truth table for $\rightarrow$ is as follows:

A	B	$(A \rightarrow B)$
T	T	T
T	F	F
F	T	T
F	F	T

A conditional statement is false iff the antecedent is true and the consequent is false.

1.5 THE BICONDITIONAL

The symbol we shall use for the biconditional is $\leftrightarrow$. (Another symbol that is frequently used is $\equiv$.) The formula "$(A \leftrightarrow B)$" is translated as: A if and only if B, A is equivalent to B, or A is a necessary and sufficient condition for B. The natural meaning for $(A \leftrightarrow B)$ is that it is true if and only if A and B have the same truth value. That is,

A	B	$(A \leftrightarrow B)$
T	T	T
T	F	F
F	T	F
F	F	T

We would expect that "$(A \leftrightarrow B)$" and "$((A \rightarrow B) \wedge (B \rightarrow A))$" have the same meaning. This is indeed the case, as we see by the following truth table:

A	B	$(A \rightarrow B)$	$(B \rightarrow A)$	$((A \rightarrow B) \wedge (B \rightarrow A))$	$(A \leftrightarrow B)$
T	T	T	T	T	T
T	F	F	T	F	F
F	T	T	F	F	F
F	F	T	T	T	T

This gives additional support for the truth table we choose for the conditional.

Because "$((A \rightarrow B) \wedge (B \rightarrow A))$" and "$(A \leftrightarrow B)$" have the same meaning and "$(B \rightarrow A)$" is translated as "A if B," then another translation of "$(A \rightarrow B)$"

is "*A only if B*." The sentence "Virginia dates Oscar only if Walter is out of town" is interpreted as "If Virginia dates Oscar then Walter is out of town."

Similarly, the following three sentences have different shades of meaning, but are translated the same way in our language:

1. If an animal moos, then it is a cow.
2. An animal moos only if it is a cow.
3. The only animal that moos is a cow.

The connective *unless* is occasionally used in English. In our language, we shall translate "A unless B" as "$(\neg B \rightarrow A)$." The sentence "Oscar will come unless it rains" means the same as "If it does not rain, then Oscar will come." If we make a truth table for the formula "$(\neg B \rightarrow A)$" we see it has exactly the same truth table as "$(A \vee B)$." Thus, In our language, "A unless B" means the same as "A or B." Each of the following four sentences could be translated as "$\neg U \rightarrow R$" (or "$U \vee R$") in the language of the Sentential Calculus:

1. If I do not take an umbrella, then it rains.
2. It rains if I do not take an umbrella.
3. It rains unless I take an umbrella.
4. Either I take an umbrella or it rains.

The following table presents additional examples of translations:

Sentence	Translation
1. Oscar will flunk the logic course unless he studies and does his homework. (K, S, H)	$(\neg (S \wedge H) \rightarrow K)$ or $(K \vee (S \wedge H))$
2. A necessary and sufficient condition for two triangles to be similar is that they have equal angles (S, E)	$(S \leftrightarrow E)$
3. Assuming that either logic is difficult or that the text is not readable, Oscar will pass only if he concentrates. (D, R, P, C)	$((D \vee \neg R) \rightarrow (P \rightarrow C))$

Note that the sentence "The text is not readable" is not an atomic sentence, because it contains the connective "not." Also note that translations do not require any knowledge of underlying facts. For example, we do not have to know what sort of student Oscar is or what the definition of similar triangles might be to translate the preceding sentences.

1.6 LANGUAGE OF THE SENTENTIAL CALCULUS

The symbols of the language are as follows:

Sentential variables: $A, B, C, \ldots,$ and these letters with subscripts.

Connectives: $\neg$, $\wedge$, $\vee$, $\rightarrow$, $\leftrightarrow$.

Parentheses: (,) (Parentheses are used for grouping.)

Rules for Formula Construction

The following are rules for formula construction:

1. If A is a sentential variable, then A is a formula.
2. If P and Q are formulas, then each of the following are formulas:

$$(\neg P),\ (P \wedge Q),\ (P \wedge Q),\ (P \rightarrow Q), \text{ and } (P \leftrightarrow Q).$$

3. No other expressions are formulas.

(We shall have more to say about this type of definition, called an *inductive* definition, in Chapter 15. For now, it is sufficient to be able to distinguish a formula from a meaningless string of symbols.)

The definition of formulas given above describes how formulas are constructed from sentential variables. For each formula, we can construct a "tree" starting with sentential variables. The following tree for the formula $((A \wedge B) \rightarrow ((\neg C) \vee (A \leftrightarrow C)))$ is given as an example:

(4) $((A \wedge B) \rightarrow ((\neg C) \vee (A \leftrightarrow C)))$

(3) $(A \wedge B)$ $((\neg C) \vee (A \leftrightarrow C))$

(2) $(A \wedge B)$ $(\neg C)$ $(A \leftrightarrow C)$

(1) A B C A C

The first level of the tree consists of sentential variables. Each subsequent level is obtained by applying rule 2 to formulas on the preceding level or using formulas already constructed. (The formula "$(A \wedge B)$" on line (2) was repeated on line (3).)

1.7 SEMANTICS

The following truth table describes how we interpret the connectives in our language and thereby give meaning to arbitrary formulas:

P	Q	$\neg P$	$(P \wedge Q)$	$(P \vee Q)$	$(P \rightarrow Q)$	$(P \leftrightarrow Q)$
T	T	F	T	T	T	T
T	F	F	F	T	F	F
F	T	T	F	T	T	F
F	F	T	F	F	T	T

We mention the fact that there are connectives in English that are not truth functional. The connective "it is possible that" is an example. The symbol that is used is $\Diamond$. If S is the false statement "$2 + 2 = 5$," then $\Diamond S$ is false. However, if S is the false statement "Someone has lived to be 200 years old," $\Diamond S$ is true. It is possible that someone has lived to be 200 years old. Consequently, the truth value of $\Diamond S$ cannot be determined from the truth value of S alone.

The language we constructed in this chapter is much simpler than English (or any other spoken language). In our language, for example, "and" means the same as "but"; "only if" the same as "then"; and "unless" the same as "or." Also, there are many expressions in English that we do not have in our language. For example, the connectives "it is possible that" and "it is necessary that" are not expressible in the Sentential Calculus. In spite of all these limitations, we shall see in later chapters that the Sentential Calculus has many applications.

1.8 CONVENTIONS FOR PARENTHESES

To reduce the number of parentheses used in formulas, we shall use the following conventions:

1. The outermost parentheses around a formula will be omitted.
2. Negation binds more closely than any other connective.
3. Conjunction and disjunction bind more closely than the conditional and biconditional.

Without the Convention	**With the Convention**
1. $((\neg A) \wedge B)$	$\neg A \wedge B$
(Note the difference between "$\neg A \wedge B$" and "$\neg(A \wedge B)$.")	
2. $((A \wedge (\neg B)) \rightarrow C)$	$A \wedge \neg B \rightarrow C$
3. $((A \wedge B) \vee (\neg C))$	$(A \wedge B) \vee \neg C$
4. $(((\neg A) \vee B) \rightarrow (C \leftrightarrow D))$	$\neg A \vee B \rightarrow (C \leftrightarrow D)$
5. $(\neg(A \rightarrow B) \rightarrow (C \vee D))$	$\neg(A \rightarrow B) \rightarrow C \vee D$

Some sentences without parentheses are ambiguous. For example, "$(A \wedge B) \vee C$" has a different truth table from "$A \wedge (B \vee C)$." Thus, we cannot write "$A \wedge B \vee C$" without any parentheses (unless we introduce additional conventions for parentheses). Parentheses may always be added to a formula to clarify its meaning. In some instances, it might be advantageous to write "$A \wedge \neg B \rightarrow C$" as "$(A \wedge \neg B) \rightarrow C$" to emphasize that "$A \wedge \neg B$" is the antecedent. In most cases, we shall use the preceding conventions for reducing parentheses.

EXERCISES 1

A. Translate each of the following sentences into the language of the Sentential Calculus using the indicated letters for atomic sentences. (Answers for 5, 7, 11, 15, 17, 20, 22, and 24.)

(1) Oscar and Virginia both go to college. (O, V)
(2) Oscar registered for the logic course, but Virginia did not. (O, V)
(3) Neither Oscar nor Virginia like George. (O, V)
(4) If Oscar dates Virginia, then George does not. (O, G)
(5) Either Oscar dates Virginia or George dates Virginia, but not both. (O, G)
(6) Oscar will pass the logic course only if he studies. (P, S)
(7) Oscar will not pass the logic course unless he does his homework and studies. (P, H, S)
(8) Oscar will not pass the logic course if he neither does his homework nor studies. (P, H, S)
(9) It is not the case that Oscar will pass the logic course provided that he studies and does his homework. (P, S, H)
(10) A sufficient condition that Oscar pass the logic course is that he studies and does his homework. (P, S, H)
(11) If Oscar doesn't study and do his homework, then he will not pass the logic course. (S, H, P)
(12) If Oscar and Virginia work at a steady pace, then there is no gain or loss of efficiency when they work together. (O, V, G, L)
(13) If I miss my train, I will arrive 10 minutes late, assuming the next train is on time. (M, L, N)
(14) We will go to the park today provided that our car doesn't break down and it doesn't rain. (P, C, R)
(15) If logic is difficult, Oscar and Virginia will pass only if they study. (D, O, V, S)
(16) If two lines lie in a plane, a necessary and sufficient condition for them to be parallel is that they neither intersect nor coincide. (L, P, I, C)
(17) If Q is a quadrilateral, then Q is a parallelogram if and only if its opposite sides are both equal and parallel. (Q, P, E, L)
(18) If the function f is continuous on the interval (a, b), then either f has a maximum on $[a, b]$ or f is not continuous at both a and b. (C, M, A, B)
(19) A sufficient condition for the function f to have a maximum on

$[a, b]$ is that f is continuous on (a, b) and f is continuous at both a and b. (M, C, A, B)

(20) If f' is defined on the interval (a, b), a necessary condition for f to be increasing on (a, b) is that f' is positive on (a, b). (D, I, P)

(21) A necessary and sufficient condition for f' to be positive on (a, b) is that f' is defined on (a, b) and f is increasing on (a, b). (P, D, I)

(22) If $I = \int_a^b f(x)\,dx$ and A is the approximation of $\int_a^b f(x)\,dx$ using the trapezoidal rule, then $A \geq I$ if $f''(x) > 0$ for $x \in [a, b]$. (I, A, G, P)

(23) If Oscar's grades on his first two exams were 72 and 95, then his letter grades on these two exams were C and A, respectively. (S, N, C, A)

(24) If 3 and 4 are substituted for x and y, respectively, in the inequality $2x + y < x + 3y$, then we obtain the inequality $10 < 15$. (S_3, S_4, I)

(25) If $\vec{v}_1$, $\vec{v}_2$, and $\vec{v}_3$ are three vectors in $\mathbb{R}^3$ and their initial points are at the origin, then the set $S = \{\vec{v}_1, \vec{v}_2, \vec{v}_3\}$ is linearly dependent if and only if the three vectors lie in the same plane through the origin. (V, I, D, P)

B. Assign letters to the atomic sentences, and translate each of the following sentences into a logical formula. (Answers for 2, 5, 9, and 11.)

(1) If Virginia passed algebra or geometry, then she was admitted into college.

(2) If Virginia did not pass both algebra and geometry, then she would be admitted to neither Harvard nor Yale.

(3) It is not the case that if Oscar secures employment, then he will marry Virginia.

(4) Oscar will look for employment if he flunks out of college, unless he is admitted into a technical school.

(5) Oscar and Virginia will graduate from college only if he/she passes the logic course.

(6) A necessary and sufficient condition for three points A, B, and C to lie on a line is that the distance from A to C is the sum of the distance from A to B and from B to C.

(7) If f is continuous and differentiable on $[a, b]$, then either there is a horizontal tangent line to f between a and b or not both $f(a)$ and $f(b)$ are zero.

(8) A necessary condition for f to be continuous at $x = a$ is that $f(a)$ is defined and $\lim_{x \to a} f(x)$ exists.

(9) If x is an integer, then exactly one of x and $x + 1$ is an even integer.

(10) If f is continuous but not integrable on (a, b), then f is not defined at both a and b.

(11) If A is a false sentence, then the conditional sentence $A \to B$ is of the form $F \to T$ or $F \to F$, depending upon whether B is true or false, respectively.

(12) If f is a function and f has continuous derivatives, $f^{(n)}$, in a neighborhood N of zero, for $n = 0, 1, 2, \ldots$, then $f(x) = \sum_{k=0}^{\infty} \frac{f^{(k)}(0)}{k!} x^k$ for $x \in N$ if and only if the remainder,

$$R_{n+1}(x) = \frac{1}{n!}\int_0^x f^{(n+1)}(t)(x-t)^n\,dt,$$

approaches zero as n increases without limit.

C. Give the converse, inverse, and contrapositive of each of the following conditional sentences (Answers for 2 and 7.)

(1) If $\vec{v}$ is parallel to $\vec{w}$, then $|\vec{v}\cdot\vec{w}| = \|\vec{v}\|\,\|\vec{w}\|$.
(2) Two lines will intersect, if they are not parallel.
(3) If Oscar graduates from college, he will either look for employment or go to graduate school.
(4) If Virginia graduates from college and goes to graduate school, then she will not major in mathematics.
(5) If Virginia graduates from college and has an A average in mathematics, she will receive a graduate fellowship.
(6) Passing the algebra course is a necessary condition for Bill to graduate.
(7) A sufficient condition for a triangle to satisfy the Pythagorean Theorem is that it is a right triangle.
(8) A necessary condition for two triangles to be similar is that they have equal sides.
(9) A triangle is equilateral only if its three angles are equal or its three sides are equal.
(10) Three points lie on a circle only if they are not collinear.

D. Give a truth table for each of the following formulas. (Answer for 3.)

(1) $A \to (B \to \neg A)$
(2) $(A \wedge B \to C) \leftrightarrow (A \to C) \vee (B \to C)$
(3) $A \vee \neg B \to C \wedge \neg A$
(4) $A \wedge \neg A \to B \vee \neg C$
(5) $(A \wedge \neg B \wedge C \wedge D) \vee (A \wedge B \wedge \neg C \wedge \neg D)$

E. Show that each of the following pairs of formulas have the same truth table:

(1) $A \vee (A \wedge B)$, A
(2) $A \wedge (A \vee B)$, A
(3) $A \leftrightarrow B$, $(A \wedge B) \vee (\neg A \wedge \neg B)$
(4) $\neg(A \wedge B)$, $\neg A \vee \neg B$
(5) $A \to B$, $\neg A \vee B$
(6) $\neg(A \to B)$, $A \wedge \neg B$
(7) $A \to (B \to C)$, $A \wedge B \to C$
(8) $A \vee B \to C$, $(A \to C) \wedge (B \to C)$

F. Construct a tree starting from sentential variables for each of the following formulas:

(1) $A \to (B \to \neg A)$
(2) $\neg A \wedge B \leftrightarrow (B \to C \vee A)$
(3) $A \vee \neg B \to (\neg(C \wedge B) \leftrightarrow \neg A \wedge \neg B)$
(4) $A \wedge (B \vee \neg C) \to A \wedge (D \to C)$
(5) $\neg(A \wedge (B \leftrightarrow C)) \vee D \to C \wedge \neg D$

Chapter 2

Arguments, Valid and Invalid

In this chapter, we describe the important notion of a *proof* in the Sentential Calculus, but first we define some related concepts.

2.1 TAUTOLOGIES

Each formula in the Sentential Calculus has a truth table. If the last column of the truth table has all "T's," then that formula is called a *tautology*. A formula is a *tautology* if it is true for all possible truth values of its sentential variables. The following formulas are examples of tautologies: $A \vee \neg A$, $A \wedge \neg A \rightarrow (B \vee C \rightarrow D)$, $A \wedge B \leftrightarrow A \wedge B$, and $\neg(A \wedge B) \leftrightarrow \neg A \vee \neg B$. If a formula is false for all truth values of its sentential variables, it is called a *contradiction*. The negation of a tautology will always be a contradiction and vice versa. The formulas $A \wedge \neg A$ and $A \leftrightarrow \neg A$ are contradictions that frequently occur in practice.

If P and Q are formulas such that $P \rightarrow Q$ is a tautology, we say P *tautologically implies* Q, which is expressed symbolically as $P \Rightarrow Q$. If $P \leftrightarrow Q$ is a tautology, we say P is *tautologically equivalent* to Q, symbolically $P \Leftrightarrow Q$. For example, $A \Leftrightarrow A \wedge A$ because $A \leftrightarrow A \wedge A$ is a tautology, and $A \Rightarrow A \vee B$ because $A \rightarrow A \vee B$ is a tautology.

Usually, if we have a disjunction of formulas or a conjunction of formulas, we omit parentheses because the formulas are tautologically equivalent. So, for example, we normally will not distinguish between the two tautologically equivalent formulas $A \wedge (B \wedge C)$ and $(A \wedge B) \wedge C$, and we will write each as $A \wedge B \wedge C$.

If $P_1, P_2, P_3, \ldots, P_n$ and Q are formulas, we say $P_1, P_2, P_3, \ldots, P_n$ *tautologically imply* Q if $P_1 \wedge P_2 \wedge P_3 \wedge \cdots \wedge P_n \rightarrow Q$ is a tautology. As a consequence, we have the following theorem:

Theorem 2.1 *$P_1, P_2, P_3, \ldots, P_n$ tautologically imply Q if and only if whenever $P_1, P_2, P_3, \ldots, P_n$ are all true, Q is true also.*

Proof: Suppose $P_1, P_2, P_3, \ldots, P_n$ tautologically imply Q. Then $P_1 \land P_2 \land P_3 \land \cdots \land P_n \rightarrow Q$ is a tautology. If $P_1, P_2, P_3, \ldots$ and P_n are all true, to make the conditional statement true, Q must be true.

Conversely, suppose Q is true whenever $P_1, P_2, P_3, \ldots,$ and P_n are all true. Then $P_1 \land P_2 \land P_3 \land \cdots \land P_n \rightarrow Q$ is a tautology because if all the P_i's are true, by hypothesis, Q is true, so the conditional is of the form $T \rightarrow T$, which is true. If some P_i is false, then the conditional statement is of the form $F \rightarrow T$ or $F \rightarrow F$, depending on whether Q is true or false. In both cases, the conditional is true. Thus, the conditional statement is always true, therefore, it is a tautology. ■

Theorem 2.1 allows us to shorten the time it takes to determine whether a conditional statement is a tautology. For example, to determine whether the formulas $A \rightarrow B$, $\neg A \rightarrow C$, and $C \rightarrow B$ tautologically imply B, we prepare a truth table with eight lines. The only lines we have to look at, however, are those in which the three formulas $A \rightarrow B$, $\neg A \rightarrow C$, and $C \rightarrow B$ are all true.

A	B	C	$A \rightarrow B$	$\neg A \rightarrow C$	$C \rightarrow B$	B
T	T	T	T	T	T	T
T	T	F	T	T	T	T
T	F	T	F			
T	F	F	F			
F	T	T	T	T	T	T
F	T	F	T	F		
F	F	T	T	T	F	
F	F	F	T	F		

Lines 1, 2, and 5 are the only lines of the truth table where the three formulas are all true; on these lines, B is also true. Therefore, the three formulas $A \rightarrow B$, $\neg A \rightarrow C$, and $C \rightarrow B$ tautologically imply B.

On the other hand, suppose we want to show that the three formulas $A \rightarrow B$, $C \rightarrow \neg A$, and $C \rightarrow B$ do *not* tautologically imply B. All we have to do is find truth values for A, B, and C so that the three formulas are true but B is false. Starting with the conclusion, B, being false, we see that the first premise, $A \rightarrow B$, has a false consequent. In order for that to be true, A must be false. The second premise, $C \rightarrow \neg A$, now has a true consequent, so that is true whether C is true or false. The third premise, $C \rightarrow B$, however, has a false consequent, so C must be false to make it true. If we just construct the one line of the truth table

A	B	C	$A \to B$	$C \to \neg A$	$C \to B$	B
F	F	F	T	T	T	F

we see that it is possible to assign truth values to A, B, and C, so that the three formulas are true, but B is false. Thus, the three formulas do not tautologically imply B.

This discussion is leading up to the notion of a *valid argument*. An *argument* is a set of premises, $P_1, P_2, P_3, \ldots, P_n$, and a conclusion Q. An argument is *valid* if and only if whenever the premises are true, the conclusion is also true. That is,

$$P_1 \wedge P_2 \wedge P_3 \wedge \cdots \wedge P_n \Rightarrow Q \text{ or}$$

$$P_1 \wedge P_2 \wedge P_3 \wedge \cdots \wedge P_n \to Q \text{ is a tautology, or}$$

$$P_1, P_2, P_3, \ldots, \text{and } P_n \text{ tautologically imply } Q.$$

Example 2.1

Determine the validity of the following argument: If the dog is barking, then the dog is not in the house. If the dog is in the house, then someone is at the front door if the dog is barking. A necessary condition for the dog to be in the house is that the dog is barking. Therefore, someone is not at the front door unless the dog is barking.

Solution: Let the following letters represent the atomic sentences:

B: The dog is barking.

H: The dog is in the house.

D: Someone is at the front door.

Translating the argument into logical notation, we obtain the following:

Premises: $B \to \neg H$, $H \to (B \to D)$, $H \to B$

Conclusion: $\neg B \to \neg D$ (or $\neg D \vee B$)

We shall try to assign truth values to the atomic sentences so that the premises are true and the conclusion false. In order for $\neg B \to \neg D$ to be false, B must be false and D true. (A conditional statement is false if and only if the antecedent is true and the consequent false.) Moreover, we see that if H is false, all three premises are true. (The first premise is of the form $F \to T$; the second, $F \to (F \to T)$; and the third, $F \to F$.) Thus, the one line of the truth table

B	D	H	$B \to \neg H$	$H \to (B \to D)$	$H \to B$	$\neg B \to \neg D$
F	T	F	T	T	T	F

shows that it is possible to assign truth values to the atomic sentences in such a way that the premises are true but the conclusion is false. Such an assignment is sometimes called a *counter-example*. Consequently, the argument is *not* valid.

Example 2.2

Determine the validity of the following argument: If the dog is barking, then the dog is in the house. If the dog is in the house, then someone is at the front door unless the dog is not barking. Indeed, the dog is barking. Therefore, someone is at the front door.

Solution: First we translate the argument into logical notation. Using the same letters for atomic sentences as in Example 2.1, we obtain the following:

Premises: $B \rightarrow H$, $H \rightarrow (B \rightarrow D)$, B

Conclusion: D

If we start, as we did in the preceding example, to try to show that the argument is invalid, we would soon come to an impasse. (Try it!) Thus, we construct the whole truth table:

B	D	H	$B \rightarrow H$	$H \rightarrow (B \rightarrow D)$	B	D
T	T	T	T	T	T	T
T	T	F	F			
T	F	T	T	F		
T	F	F	F			
F	T	T	T	T	F	
F	T	F	T	T	F	
F	F	T	T	T	F	
F	F	F	T	T	F	

The only line of the truth table in which all the premises are true is the first line, and on this line, the conclusion is also true. Therefore, the argument is valid.

It is important to note that *the validity of an argument depends only on its structure*. The validity or nonvalidity of the arguments in Examples 2.1 and 2.2 did *not* depend on the interpretation or meaning of B, H, and D. It only depended on the *structure* of the premises and the conclusion.

For more complicated arguments, the truth table method to prove validity gets rather cumbersome. If there were ten atomic sentences, for example, the truth table would have $2^{10} = 1024$ lines. Although the problem can be done on a computer, it is still advantageous to have another method to show that an argument is valid. (The counter-example method, as illustrated in Example 2.1, is usually the best method to show that an argument is *not* valid.)

The method we shall describe to show that an argument is valid is called a *proof,* a *deduction,* or an *inference.* We shall give a few general rules, called *rules of inference,* that are sufficient to construct a "proof" of any valid argument in the Sentential Calculus. A rule of inference is a procedure for deducing a consequence from premises. The rules must be *valid* or *preserve truth;* that is, whenever the premises are true, the consequence must be true. Otherwise, there is no guarantee that the argument is valid.

2.2 RULE FOR TAUTOLOGIES

The first such rule we shall discuss is called the Rule for Tautologies, or Rule T.

Rule T (Rule for Tautologies) *A formula Q is deducible from formulas $P_1, P_2, \ldots, P_n$ if $P_1 \wedge P_2 \wedge \cdots \wedge P_n \Rightarrow Q$. (That is, $P_1 \wedge P_2 \wedge \cdots \wedge P_n \rightarrow Q$ is a tautology.) In particular, if Q itself is a tautology, then Q is deducible from the empty set of premises.*

It follows from Theorem 2.1 that Rule T is valid. Rules 1 through 4 listed below are four examples of Rule T:

Rule 1. From P and $P \rightarrow Q$ deduce Q.

Rule 2. From $\neg Q$ and $P \rightarrow Q$ deduce $\neg P$.

Rule 3. From $\neg P$ and $P \vee Q$ deduce Q.

Rule 4. From $\neg(\neg P)$ deduce P and from P deduce $\neg(\neg P)$.

Our preliminary definition of a *proof* is as follows:

By a (*formal*) *proof of Q from the premises $P_1, P_2, \ldots, P_n$,* we mean a finite sequence of formulas, $R_1, R_2, \ldots, R_m$, where R_m is Q, and each R_i is either one of the premises P_j or is deducible from preceding R_k's in the sequence using a valid rule of inference.

We use the notation "$P_1, P_2, \ldots, P_n \vdash Q$" if there is a proof of Q from $P_1, P_2, \ldots, P_n$. The symbol "$\vdash$" is sometimes called a "turnstile."

As an illustration of a formal proof, we shall prove that the following argument is valid.

Premises: $A \rightarrow \neg B,\ C \vee B,\ D \rightarrow \neg C,\ D$

Conclusion: $\neg A$

First, we shall describe informally how to deduce $\neg A$ from the premises:

Step 1: From D and $D \rightarrow \neg C$ deduce $\neg C$ using Rule 1.

Step 2: From $\neg C$ and $C \vee B$ deduce B using Rule 3.

Step 3: From B deduce $\neg(\neg B)$ using Rule 4.

Step 4: From $\neg(\neg B)$ and $A \rightarrow \neg B$ deduce $\neg A$ using Rule 2.

Somewhat more formally, the proof could be written as follows:

Proof	Reason
1. $A \to \neg B$	Premise
2. $C \vee B$	Premise
3. $D \to \neg C$	Premise
4. D	Premise
5. $\neg C$	T (Rule 1)
6. B	T (Rule 3)
7. $\neg(\neg B)$	T (Rule 4)
8. $\neg A$	T (Rule 2)

The proof is the finite sequence of formulas given on lines 1 through 8 above. Each formula in the proof is either a premise (1 through 4) or is derivable from preceding formulas in the sequence using a valid rule of inference (5 through 8). That is, line 5 is derived from lines 3 and 4 using Rule 1, line 6 is derived from lines 2 and 5 using Rule 3, and so forth. Normally, in a proof, we write the given premises at the beginning of the proof, but this is not required, because a premise can be written on any line of a proof.

2.3 RULE FOR PREMISES

Formally, our rule for premises, or Rule P, is as follows:

Rule P (Rule for Premises) *A premise may occur on any line of a proof.*

Moreover, in each proof, we shall indicate which premises each line depends upon. To do this, we shall associate with each premise a number. We shall use the number of the line where the premise first occurs. (Any other consistent numbering system would be just as good.) Thus, our formal proof of the preceding argument will look like this:

Premise Numbers	Line Numbers	Proof	Reasons
{1}	1.	$A \to \neg B$	P
{2}	2.	$C \vee B$	P
{3}	3.	$D \to \neg C$	P
{4}	4.	D	P
{3, 4}	5.	$\neg C$	3, 4 T
{2, 3, 4}	6.	B	2, 5 T
{2, 3, 4}	7.	$\neg(\neg B)$	6 T
{1, 2, 3, 4}	8.	$\neg A$	1, 7 T

Lines 1 through 4 are self-evident. To derive line 5, we used lines 3 and 4 and Rule T (Rule 1); thus, line 5 depends upon premises 3 and 4. To derive line 6, we used lines 2 and 5 and Rule T (Rule 3). Thus, line 6 depends upon premises 2, 3, and 4. Similarly, for the remaining part of the proof. In general, numbers preceding the line number in a proof are the numbers of the premises on which that line depends.

To construct proofs when the arguments are more complicated, we shall have to become more proficient in the use of Rule T, which means familiarizing ourselves with some more tautologies. For that purpose, we have made a list of some useful tautologies at the end of this chapter (see Section 2.7). In fact, this list of tautologies is *complete*; that is, any valid argument can be proved using tautologies from this list. Actually a much smaller list would be adequate. We shall delay the discussion of this type of completeness to Chapter 16 (see Section 16.3). In this list of tautologies, we have included a name for each tautology for historical reasons or, in some cases, as a mnemonic device. In most cases, there is no good reason for memorizing these names.

Clearly, every valid argument has a very simple proof. If $P_1, P_2, \ldots, P_n$ are the premises and Q is the conclusion of the valid argument, then

$$P_1 \wedge P_2 \wedge \cdots \wedge P_n \rightarrow Q$$

is a tautology. Therefore, the first n lines of a proof are the n premises and the $n + 1$st line is Q. That is not the point, however. The point is to proceed with small, obvious steps. We illustrate this in the next example.

Example 2.3

Show that the following argument is valid by giving a formal proof.

Premises: $A \rightarrow B \vee C$, $A \rightarrow \neg B$, $C \rightarrow \neg D$

Conclusion: $A \rightarrow \neg D$

Solution: First, we'll give an informal proof. (The numbers in parentheses refer to the tautologies given in Section 2.7. It will be useful for the reader to have a copy of this list.) We'll denote the premises by P_1: $(A \rightarrow B \vee C)$; P_2: $(A \rightarrow \neg B)$; and P_3: $(C \rightarrow \neg D)$.

Step 1: $P_2 \vdash B \rightarrow \neg A$ (20).

Step 2: $\vdash \neg A \rightarrow \neg A \vee \neg D$ (5).

Step 3: $\vdash \neg D \rightarrow \neg A \vee \neg D$ (5, 12).

Step 4: $B \rightarrow \neg A, \neg A \rightarrow \neg A \vee \neg D \vdash B \rightarrow \neg A \vee \neg D$ (7).

Step 5: $P_3, \neg D \rightarrow \neg A \vee \neg D \vdash C \rightarrow \neg A \vee \neg D$ (7).

Step 6: $B \rightarrow \neg A \vee \neg D, C \rightarrow \neg A \vee \neg D \vdash B \vee C \rightarrow \neg A \vee \neg D$ (8).

Step 7: $P_1, B \vee C \rightarrow \neg A \vee \neg D \vdash A \rightarrow \neg A \vee \neg D$ (7).

Step 8: $\vdash (\neg A \vee \neg D) \rightarrow (A \rightarrow \neg D)$ (17).

Step 9: $A \to \neg A \vee \neg D, \neg A \vee \neg D \to (A \to \neg D) \vdash A \to (A \to \neg D)$ (7).

Step 10: $A \to (A \to \neg D) \vdash A \wedge A \to \neg D$ (21).

Step 11: $\vdash A \to A \wedge A$ (19).

Step 12: $A \to A \wedge A, A \wedge A \to \neg D \vdash A \to \neg D$ (7).

(Later we shall introduce another rule, and give an easier proof of this argument.) The formal proof is as follows:

{1}	1.	$A \to B \vee C$	P
{2}	2.	$A \to \neg B$	P
{3}	3.	$C \to \neg D$	P
{2}	4.	$B \to \neg A$	2 T
	5.	$\neg A \to \neg A \vee \neg D$	T
	6.	$\neg D \to \neg A \vee \neg D$	T
{2}	7.	$B \to \neg A \vee \neg D$	4, 5 T
{3}	8.	$C \to \neg A \vee \neg D$	3, 6 T
{2, 3}	9.	$B \vee C \to \neg A \vee \neg D$	7, 8 T
{1, 2, 3}	10.	$A \to \neg A \vee \neg D$	1, 9 T
	11.	$\neg A \vee \neg D \to (A \to \neg D)$	T
{1, 2, 3}	12.	$A \to (A \to \neg D)$	10, 11 T
{1, 2, 3}	13.	$A \wedge A \to \neg D$	12 T
	14.	$A \to A \wedge A$	T
{1, 2, 3}	15.	$A \to \neg D$	13, 14 T

Notice that the formulas on lines 5, 6, 11, and 14 are tautologies, so no premise numbers are given. Frequently, such lines are omitted.

We are implicitly using a rule of substitution when we use the list of tautologies given at the end of the chapter. Each of the letters "P," "Q," and "R" represent a formula. So, for example, the valid rule

$$\neg(A \wedge B), (A \wedge B) \vee (B \to C) \vdash B \to C$$

is obtained from Modus Tollendo Ponens (3) by substituting "$A \wedge B$" for each occurrence of "P" and "$B \to C$" for each occurrence of "Q." Similarly, the following formula is a tautology:

$$(A \wedge \neg B \to C) \vee \neg(A \wedge \neg B \to C).$$

It is a form of the Rule of the Excluded Middle (10) with "$A \wedge \neg B \to C$" substituted for "P."

In the next example, we shall omit the numbers of the tautologies. (The reader should fill in the numbers for himself or herself.)

Example 2.4. Show that the following argument is valid by giving a formal proof:

Premises: $A \to B, \neg A \to \neg C \vee D, C \wedge \neg D$

Conclusion: B

Solution: The informal proof is as follows:

Step 1: $P_3 \vdash \neg(\neg C \vee D)$.

Step 2: $\neg(\neg C \vee D), P_2 \vdash A$.

(To obtain step 2, we used two rules. What are they?)

Step 3: $A, P_1 \vdash B$.

The formal proof is as follows:

{1}	1.	$A \to B$	P
{2}	2.	$\neg A \to \neg C \vee D$	P
{3}	3.	$C \wedge \neg D$	P
{3}	4.	$\neg(\neg C \vee D)$	3 T
{2, 3}	5.	A	2, 4 T
{1, 2, 3}	6.	B	1, 5 T

WARNING! Beginning logic students frequently make two common errors:

1. The fallacy of asserting the consequent: From Q and $P \to Q$, deduce P.
2. The fallacy of denying the antecedent: From $\neg P$ and $P \to Q$, deduce $\neg Q$.

Neither of these rules is valid, as is easily seen from truth tables. There are disastrous consequences from using invalid rules. A contradiction is deducible from an invalid rule and, by the Rule of Absurdity (9), every formula is deducible from a contradiction.

We shall show that we can derive a contradiction, $A \wedge \neg A$, using the fallacy of asserting the consequent, *FAC*:

1.	$A \wedge \neg A \to A \vee \neg A$	T
2.	$A \vee \neg A$	T
3.	$A \wedge \neg A$	1, 2 *FAC*

In *FAC*, we substituted "$A \wedge \neg A$" for "P" and "$A \vee \neg A$" for "Q." In general, if the rule

$$\text{from } P_1, P_2, \ldots, P_n \text{ deduce } Q$$

is invalid, then there is an assignment, s, of truth values for the sentential variables in the rule so that $P_1, P_2, \ldots, P_n$ are each true and Q is false. For each sentential variable A that occurs in the rule

$$\text{if } s(A) = T, \text{ replace } A \text{ by } B \vee \neg B,$$
$$\text{if } s(A) = F, \text{ replace } A \text{ by } B \wedge \neg B,$$

where B is a sentential variable that does not occur in the rule. Then, each of $P_1, P_2, \ldots, P_n$ becomes a tautology, and Q becomes a contradiction. Thus, a contradiction is deducible using the invalid rule.

If a contradiction is deducible in a system, the system is called *inconsistent*. Inconsistent systems are trivial and uninteresting because, in such a system, every formula is provable. It is, however, of no small interest to identify which systems are inconsistent, and a good deal of effort is spent on just such activity in more complicated settings.

2.4 RULE OF CONDITIONAL PROOF

Although the two rules of inference that we have, Rules P and T, are sufficient to prove any valid argument, it is convenient to introduce two additional rules in order to simplify proofs (see Theorems 16.3 and 16.4). The first is called the Rule of Conditional Proof, or Rule CP.

Rule *CP* (Rule of Conditional Proof)

If Q is deducible from P and a set of premises Γ then $P \to Q$ is deducible from Γ alone. In symbols, if $\Gamma, P \vdash Q$, then $\Gamma \vdash P \to Q$.

We shall first illustrate how Rule CP can be used to shorten the proof of the argument given in Example 2.3. We use Rule CP when we want to prove a conditional, $P \to Q$. The method is to take P as an additional premise, then deduce Q.

{1}	1.	$A \to B \vee C$	P
{2}	2.	$A \to \neg B$	P
{3}	3.	$C \to \neg D$	P
{4}	4.	A	P(for CP)
{1, 4}	5.	$B \vee C$	1, 4 T
{2, 4}	6.	$\neg B$	2, 4 T
{1, 2, 4}	7.	C	5, 6 T
{1, 2, 3, 4}	8.	$\neg D$	3, 7 T
{1, 2, 3}	9.	$A \to \neg D$	4, 8 CP

On line 4, we take A as an additional premise for the Rule CP. On lines 5 through 8, we deduce $\neg D$. At this point, we apply the Rule CP, so on line 9, we say that the conditional $A \rightarrow \neg D$ is deducible from our original set of premises. The numerals 4 and 8 on line 9 are the numbers of the line on which the temporary premise appeared, 4, and on which the conclusion appeared, 8. Under certain conditions, then, the Rule *CP* allows us to drop a premise. This is an important aspect of the rule. *The last line of a proof cannot depend upon any premises that were not in the original set of premises.* Otherwise, we're changing the question. It is trivial to prove any formula P if P is taken as an additional premise.

Let us examine the preceding derivation a little more closely. According to our preliminary definition of a proof, this is *not* a proof because, on line 4, we used a premise that was not in the original set of premises. On the last line, however, we do have the conclusion, and it only depends upon premises from the original set. We shall now give the "real" definition of a proof, to validate this type of phenomenon. It should be all right to introduce premises on any line of a proof as long as the last line only depends upon premises from the original set.

Definition 2.1 A *formal proof of Q from the premises* $P_1, P_2, \ldots, P_n$ is a finite sequence of ordered pairs $\langle \alpha_1, R_1 \rangle, \langle \alpha_2, R_2 \rangle, \ldots, \langle \alpha_m, R_m \rangle$ where each α_i is a set of formulas and each R_i is a formula such that

1. $\alpha_m \subseteq \{P_1, P_2, \ldots, P_n\}$ and $R_m = Q$.
2. Each R_i is either a premise or is deducible from preceding R_j's in the sequence using a valid rule of inference.
3. If R_i is a premise then $\alpha_i = \{R_i\}$.
4. If R_i is deducible from preceding R_j's in the sequence using a valid rule of inference then α_i is defined by the rule.

In Chapter 12 (see Section 12.2), we state each rule in such a way that its effect on the α_i's is clear. In this section, we shall explain the effect less formally.

In our proofs we abbreviate the α_i's by replacing each premise by the line number where the premise occurs. The symbol "$P_1, P_2, \ldots, P_n \vdash Q$" is an abbreviation for "there is a proof of Q from $P_1, P_2, \ldots, P_n$." Similarly, if Γ is a set of formulas and Q is a formula, "$\Gamma \vdash Q$" is an abbreviation for "there is a proof of Q from Γ."

Part 1 of Definition 2.1 states that the conclusion depends only on premises from the original set of premises; part 2 contains our preliminary definition of a proof; and parts 3 and 4 explain how to construct the set of premise numbers for each line of a proof. Part 3 says that if the formula on line i is a premise, then the set of premise numbers for line i is $\{i\}$. Thus, each α_i is the set of all formulas on which formula R_i depends. Auxiliary premises may be used in the course of a proof, but the conclusion must only depend on premises from the original set.

To illustrate part 4 of Definition 2.1, if the rule is Rule T and the tautology is $R_j \wedge R_k \rightarrow R_i$, then $\alpha_i = \alpha_j \cup \alpha_k$. See lines 5, 6, and 7 of the preceding proof, for example. Line 7 is obtained from lines 5 and 6 by using a tautology. Thus, the set of premise numbers for line 7 is {1, 2, 4} and {1, 2, 4} = {1, 4} ∪ {2, 4}, which is the union of the sets of premise numbers for lines 5 and 6.

On the other hand, if the rule is *CP* and R_j is the premise for *CP*, then $\alpha_j = \{j\}$. If we then deduce R_k ($\alpha_k = \Gamma \cup \alpha_j$) and if we use *CP* to get the next line, then R_{k+1} is $R_j \rightarrow R_k$ and $\alpha_{k+1} = \alpha_k - \{j\}$. On line 9 of the preceding proof, for example, we used Rule *CP*. The premise for Rule *CP*, *A*, was on line 4. We derived $\neg D$ on line 8. The set of premise numbers on line 8 is {1, 2, 3, 4}; thus, the formula on line 9 is $A \rightarrow \neg D$, and the set of premise numbers is {1, 2, 3, 4} − {4} = {1, 2, 3}.

Next, we shall show Rule *CP* is a valid rule. (Otherwise, we could derive a contradiction.)

Theorem 2.2 *Rule CP is valid.*

Proof: Suppose $P_1, P_2, \ldots, P_n$ are the premises in Γ. To show Rule *CP* is valid, we must show the following: If

(1) $P_1 \wedge P_2 \wedge \cdots \wedge P_n \wedge P \Rightarrow Q,$

then

(2) $P_1 \wedge P_2 \wedge \cdots \wedge P_n \Rightarrow (P \rightarrow Q).$

Suppose that (1) is true; that is,

(3) $P_1 \wedge P_2 \wedge \cdots \wedge P_n \wedge P \rightarrow Q$

is a tautology. Then we must show (2) is true; that is,

(4) $P_1 \wedge P_2 \wedge \cdots \wedge P_n \rightarrow (P \rightarrow Q)$

is a tautology. However, (3) ↔ (4) is a tautology (the Rule of Exportation-Importation, (21), with "$P_1 \wedge P_2 \wedge \cdots \wedge P_n$" substituted for "*P*," "*P*" substituted for "*Q*," and "*Q*" substituted for "*R*"). Thus, if (3) is a tautology, so is (4), so (1) implies (2).

■

2.5 RULE OF INDIRECT PROOF

The last rule we shall consider for the Sentential Calculus is the Rule of Indirect Proof, or Rule *IP*.

Rule *IP* (Rule of Indirect Proof) *If any contradiction, $Q \wedge \neg Q$, where Q is any formula, is deducible from $\neg P$ and a set of premises Γ, then P is deducible from Γ alone.* In symbols, if $\neg P, \Gamma \vdash Q \wedge \neg Q$, then $\Gamma \vdash P$.

Theorem 2.3 *Rule IP is valid.*

Proof: If $P_1, P_2, \ldots, P_n$ are the premises in Γ, we must show that

(1) $P_1 \wedge P_2 \wedge \cdots \wedge P_n \wedge \neg P \Rightarrow Q \wedge \neg Q$

implies

(2) $P_1 \wedge P_2 \wedge \cdots \wedge P_n \Rightarrow P$.

Suppose (1) is true. Then

(3) $P_1 \wedge P_2 \wedge \cdots \wedge P_n \wedge \neg P \rightarrow Q \wedge \neg Q$

is a tautology. The consequent, $Q \wedge \neg Q$, of (3) is always false. Thus, for (3) to be a tautology, one of the conjuncts in the antecedent must be false. If any of the P_i's are false, then

(4) $P_1 \wedge P_2 \wedge \cdots \wedge P_n \rightarrow P$

is true (because the antecedent is false). If $\neg P$ is false, then P is true. This also implies that (4) is true (because the consequent is true). Therefore, (4) is a tautology, so (2) is true. ■

Example 2.5 Show that the following argument is valid by giving a formal proof.

Premises: $A \rightarrow (B \rightarrow C)$, $C \rightarrow \neg D$, $\neg E \rightarrow D$, $A \wedge B$

Conclusion: E

Solution: We shall give an indirect proof and take $\neg E$ as an additional premise. The informal proof is as follows:

Step 1: $\neg E, P_3 \vdash D$.

Step 2: $D, P_2 \vdash \neg C$.

Step 3: $P_4 \vdash A$.

Step 4: $A, P_1 \vdash B \rightarrow C$.

Step 5: $P_4 \vdash B$.

Step 6: $B, B \rightarrow C \vdash C$.

Step 7: $C, \neg C \vdash C \wedge \neg C$.

Thus, it follows from the Rule of Indirect Proof that $P_1, P_2, P_3, P_4 \vdash E$. The formal proof is as follows:

{1}	1.	$A \rightarrow (B \rightarrow C)$	P
{2}	2.	$C \rightarrow \neg D$	P
{3}	3.	$\neg E \rightarrow D$	P
{4}	4.	$A \wedge B$	P

{5}	5.	$\neg E$	P(for IP)
{3, 5}	6.	D	3, 5 T
{2, 3, 5}	7.	$\neg C$	2, 6 T
{4}	8.	A	4 T
{1, 4}	9.	$B \rightarrow C$	1, 8 T
{4}	10.	B	4 T
{1, 4}	11.	C	9, 10 T
{1, 2, 3, 4, 5}	12.	$C \wedge \neg C$	7, 11 T
{1, 2, 3, 4}	13.	E	5, 12 IP

On line 12, we derived a contradiction using the four original premises and the premise given on line 5, $\neg E$. Thus, it follows from the Rule of Indirect Proof that E is deducible from the original four premises, and that is how we derived line 13 of the proof. Similar to the Rule CP, the Rule IP allows us to omit a temporary premise. The line numbers on line 13 are 5, the line number where the temporary premise occurred, and 12, the line number where the contradiction occurred.

In the following section, we give some guidelines for proofs in the Sentential Calculus and then give some more examples. The guidelines are possible strategies to try in a proof. Proofs are not always straightforward procedures.

2.6 GUIDELINES

1. To prove a conjunction, $P \wedge Q$, prove each of the conjuncts, and then use the valid rule, $P, Q \vdash P \wedge Q$.
2. To prove a conditional, $P \rightarrow Q$, take P as a premise, derive Q, and then use the Rule CP.
3. To prove a disjunction, $P \vee Q$,
   (a) Prove either P or Q, then use the rule $P \vdash P \vee Q$ or $Q \vdash P \vee Q$.
   or
   (b) Take $\neg P$ as a premise, derive Q, then use Rule CP to deduce $\neg P \rightarrow Q$. Then use the rule $\neg P \rightarrow Q \vdash P \vee Q$.
4. To prove a biconditional, $P \leftrightarrow Q$, prove $P \rightarrow Q$ and $Q \rightarrow P$. Then use the rule $P \rightarrow Q, Q \rightarrow P \vdash P \leftrightarrow Q$.

   If none of the preceding methods work, try
5. An indirect proof: To prove P, take $\neg P$ as a premise and derive a contradiction.
6. An alternative proof:
   (a) To prove P, find some sentence Q such that $Q \rightarrow P$ and $\neg Q \rightarrow P$. Then use the rule $Q \rightarrow P, \neg Q \rightarrow P \vdash P$. (See 8 on the list of tautologies.)

(b) To prove R from the premise $P \vee Q$, prove $P \to R$ and $Q \to R$, and then use tautology 8 to obtain $P \vee Q \to R$.

First, we'll give examples of alternative proofs.

Example 2.6 Show that the following arguments are valid by giving a formal proof:

(a) **Premises:** $A \to B$, $\neg A \to C$, $C \to \neg D$, $\neg B \to D$

Conclusion: B

(b) **Premises:** $A \vee B$, $C \to \neg A$, $\neg B \vee D$, $D \to \neg C$

Conclusion: $\neg C$

Solution:

(a)

{1}	1.	$A \to B$	P
{2}	2.	$\neg A \to C$	P
{3}	3.	$C \to \neg D$	P
{4}	4.	$\neg B \to D$	P

(We shall use lines 2, 3, and 4 to derive $\neg A \to B$, and then, with line 1, use an alternative proof to derive B.)

{2, 3}	5.	$\neg A \to \neg D$	2, 3 T
{4}	6.	$\neg D \to B$	4 T
{2, 3, 4}	7.	$\neg A \to B$	5, 6 T
{1, 2, 3, 4}	8.	B	1, 7 T

On line 8, we used an alternative proof.

(b)

{1}	1.	$A \vee B$	P
{2}	2.	$C \to \neg A$	P
{3}	3.	$\neg B \vee D$	P
{4}	4.	$D \to \neg C$	P

We shall prove $A \to \neg C$ and $B \to \neg C$.

{2}	5.	$A \to \neg C$	2 T
{6}	6.	B	P (for CP)
{3, 6}	7.	D	3, 6 T
{3, 4, 6}	8.	$\neg C$	4, 7 T
{3, 4}	9.	$B \to \neg C$	6, 8 CP
{2, 3, 4}	10.	$A \vee B \to \neg C$	5, 9 T
{1, 2, 3, 4}	11.	$\neg C$	1, 10 T

We obtained line 10 by using tautology 8.

There are other ways of proving parts (a) and (b) in Example 2.6, but an alternative proof is nonetheless often useful.

Example 2.7

For each of the following arguments, give a formal proof if it is valid; if not, give a counter-example—that is, assign truth values to the atomic sentences so that the premises are true and the conclusion is false.

(a) **Premises:** $A \rightarrow B \vee C,\ B \rightarrow \neg A,\ \neg D \rightarrow \neg C$

Conclusion: $\neg A \vee D$

(b) **Premises:** $A \rightarrow \neg B,\ (A \vee \neg B) \vee C,\ B$

Conclusion: $B \wedge C$

(c) **Premises:** $\neg A \rightarrow (B \vee C),\ \neg B \rightarrow (\neg A \wedge D),\ D \rightarrow (B \vee C)$

Conclusion: B

(d) **Premises:** $A \rightarrow C,\ \neg C \vee D,\ B \leftrightarrow D,\ B \rightarrow \neg(\neg A \wedge D)$

Conclusion: $A \leftrightarrow B$

Solution:

(a) We shall take A as a premise and try to derive D. (See 3(b) in the guidelines given above; $\neg A \vee D$ is tautologically equivalent to $A \rightarrow D$.) The following is an informal proof:

Step 1: $A, P_1 \vdash B \vee C$.

Step 2: $A, P_2 \vdash \neg B$.

Step 3: $B \vee C, \neg B \vdash C$.

Step 4: $C, P_3 \vdash D$.

Thus, by Rule CP, we obtain $A \rightarrow D$, which is tautologically equivalent to $\neg A \vee D$. The formal proof is as follows:

{1}	1.	$A \rightarrow B \vee C$	P
{2}	2.	$B \rightarrow \neg A$	P
{3}	3.	$\neg D \rightarrow \neg C$	P
{4}	4.	A	P(for CP)
{1, 4}	5.	$B \vee C$	1, 4 T
{2, 4}	6.	$\neg B$	2, 4 T
{1, 2, 4}	7.	C	5, 6 T
{1, 2, 3, 4}	8.	D	3, 7 T
{1, 2, 3}	9.	$A \rightarrow D$	4, 8 CP
{1, 2, 3}	10.	$\neg A \vee D$	9 T

(b) In this argument we are trying to prove a conjunction, $B \wedge C$. The method we shall use is described in Guideline 1, that is, deduce each of the conjuncts. One of the conjuncts is a premise, B, so that is easy to deduce. It remains to deduce C. For this, we shall use an indirect proof.

Informally, take $\neg C$ as a premise:

Step 1: $\neg C, P_2 \vdash A \vee \neg B$.

Step 2: $P_3, A \vee \neg B \vdash A$.

Step 3: $A, P_1 \vdash \neg B$.

Step 4: $\neg B, P_3 \vdash B \wedge \neg B$.

Thus, from Rule *IP*, we can deduce C.

Step 5: $C, P_3 \vdash B \wedge C$.

The following is a formal proof:

{1}	1.	$A \to \neg B$	P
{2}	2.	$(A \vee \neg B) \vee C$	P
{3}	3.	B	P
{4}	4.	$\neg C$	P(for *IP*)
{2, 4}	5.	$A \vee \neg B$	2, 4 T
{2, 3, 4}	6.	A	3, 5 T
{1, 2, 3, 4}	7.	$\neg B$	1, 6 T
{1, 2, 3, 4}	8.	$B \wedge \neg B$	3, 7 T
{1, 2, 3}	9.	C	4, 8 *IP*
{1, 2, 3}	10.	$B \wedge C$	3, 9 T

(c) To try to prove (c), we shall try an indirect proof; take $\neg B$ as a premise, and try to derive a contradiction.

Step 1: $\neg B, P_2 \vdash \neg A \wedge D$.

Step 2: $\neg A \wedge D \vdash \neg A$.

Step 3: $\neg A, P_1 \vdash B \vee C$.

Step 4: $\neg A \wedge D \vdash D$.

Step 5: $D, P_3 \vdash B \vee C$.

Step 6: $\neg B, B \vee C \vdash C$.

If C were false, we would have a contradiction, but we do not have any information to tell us that C is false. Let's try instead to find a counterexample.

Because B is the conclusion, B has to be false, and from the preceding discussion, C has to be true. Now we have to find truth assignments for A and D so that the premises are true. The following assignment works:

A	B	C	D	$\neg A \to B \vee C$	$\neg B \to (\neg A \wedge D)$	$D \to (B \vee C)$	B
F	F	T	T	T $(T \to T)$	T $(T \to T)$	T $(T \to T)$	F

Thus, if A and B are false, and C and D are true, the premises are true, and the conclusion is false, so the argument is not valid. Normally, to show an argument is not valid, try to prove it, and if you don't succeed, try to find a counter-example. One or the other must work.

(d) In this argument, we are trying to prove a biconditional, $A \leftrightarrow B$. The method we shall try is to deduce $A \rightarrow B$ and $B \rightarrow A$. (See Guideline 4.)

The formal proof is as follows:

{1}	1.	$A \rightarrow C$	P
{2}	2.	$\neg C \vee D$	P
{3}	3.	$B \leftrightarrow D$	P
{4}	4.	$B \rightarrow \neg(\neg A \wedge D)$	P
{5}	5.	A	P(for CP)
{1, 5}	6.	C	1, 5 T
{1, 2, 5}	7.	D	2, 6 T
{1, 2, 3, 5}	8.	B	3, 7 T
{1, 2, 3}	9.	$A \rightarrow B$	5, 8 CP
{10}	10.	B	P(for CP)
{4, 10}	11.	$\neg(\neg A \wedge D)$	4, 10 T
{4, 10}	12.	$A \vee \neg D$	11 T
{3, 10}	13.	D	3, 10 T
{3, 4, 10}	14.	A	12, 13 T
{3, 4}	15.	$B \rightarrow A$	10, 14 CP
{1, 2, 3, 4}	16.	$A \leftrightarrow B$	9, 15 T

Notice that to prove $A \rightarrow B$ (line 9), we used only premises 1, 2, and 3, and to prove $B \rightarrow A$ (line 15), we used only premises 3 and 4. To prove the biconditional, $A \leftrightarrow B$ (line 16), however, we used all four premises. Both in the assignments and in arguments occurring naturally, there may be redundant premises.

2.7 USEFUL TAUTOLOGIES

("P," "Q," and "R" denote arbitrary formulas.)

1. $P \wedge (P \rightarrow Q) \rightarrow Q$ (Rule of Detachment, Modus Ponens)
2. $\neg Q \wedge (P \rightarrow Q) \rightarrow \neg P$ (Modus Tollendo Tollens)
3. $\neg P \wedge (P \vee Q) \rightarrow Q$ (Modus Tollendo Ponens)
4. $P \wedge Q \rightarrow P$ (Rule of Simplification)
5. $P \rightarrow P \vee Q$ (Rule of Addition)
6. $P \rightarrow (Q \rightarrow P \wedge Q)$; $(P \rightarrow Q) \wedge (P \rightarrow R) \rightarrow (P \rightarrow Q \wedge R)$ (Rule of Adjunction)

7. $(P \to Q) \wedge (Q \to R) \to (P \to R)$ (Rule of Hypothetical Syllogism)
8. $(P \to Q) \wedge (\neg P \to Q) \to Q$
   $(P \vee Q \to R) \leftrightarrow (P \to R) \wedge (Q \to R)$ (Rules of Alternative Proof)
9. $(P \to Q \wedge \neg Q) \to \neg P$
   $P \wedge \neg P \to Q$ (Rule of Absurdity)
10. $P \vee \neg P$ (Rule of the Excluded Middle)
11. $\neg(P \wedge \neg P)$ (Rule of Contradiction)
12. $P \vee Q \leftrightarrow Q \vee P$
    $P \wedge Q \leftrightarrow Q \wedge P$ (Commutative Rules)
13. $P \vee (Q \vee R) \leftrightarrow (P \vee Q) \vee R$
    $P \wedge (Q \wedge R) \leftrightarrow (P \wedge Q) \wedge R$ (Associative Rules)
14. $P \vee (Q \wedge R) \leftrightarrow (P \vee Q) \wedge (P \vee R)$
    $P \wedge (Q \vee R) \leftrightarrow (P \wedge Q) \vee (P \wedge R)$ (Distributive Rules)
15. $\neg(P \vee Q) \leftrightarrow \neg P \wedge \neg Q$
    $\neg(P \wedge Q) \leftrightarrow \neg P \vee \neg Q$ (De Morgan's Rules)
16. $\neg(\neg P) \leftrightarrow P$ (Rule of Double Negation)
17. $(P \to Q) \leftrightarrow \neg P \vee Q$
    $\neg(P \to Q) \leftrightarrow P \wedge \neg Q$ (Rules for the Conditional)
18. $(P \leftrightarrow Q) \leftrightarrow (P \to Q) \wedge (Q \to P)$
    $(P \leftrightarrow Q) \leftrightarrow (P \wedge Q) \vee (\neg P \wedge \neg Q)$ (Rules for the Biconditional)
    $(P \leftrightarrow Q) \leftrightarrow (\neg P \vee Q) \wedge (P \vee \neg Q)$
19. $P \vee P \leftrightarrow P$
    $P \wedge P \leftrightarrow P$ (Idempotency Rules)
20. $(P \to Q) \leftrightarrow (\neg Q \to \neg P)$
    $(P \to \neg Q) \leftrightarrow (Q \to \neg P)$ (Rules of Contraposition)
    $(\neg P \to Q) \leftrightarrow (\neg Q \to P)$
21. $(P \to (Q \to R)) \leftrightarrow (P \wedge Q \to R)$ (Rule of Exportation-Importation)
22. $P \vee (P \wedge Q) \leftrightarrow P$
    $P \wedge (P \vee Q) \leftrightarrow P$ (Rules of Absorption)

EXERCISES 2

A. For each of the following pairs of sentences, use truth tables to determine whether

(i) $(a) \Rightarrow (b)$,

(ii) $(b) \Rightarrow (a)$,

(iii) $(a) \Leftrightarrow (b)$,

(iv) none of these.

(Answers for 1, 7, 9, and 10.)

(1) **(a)** $A \rightarrow (B \rightarrow \neg C)$, **(b)** $A \rightarrow (\neg B \rightarrow C)$

(2) **(a)** $A \wedge B \rightarrow \neg C$, **(b)** $A \rightarrow (C \rightarrow \neg B)$

(3) **(a)** $(A \wedge B) \wedge \neg C$, **(b)** $C \rightarrow (B \rightarrow A)$

(4) **(a)** A, **(b)** $A \vee (B \wedge C)$

(5) **(a)** A, **(b)** $(A \wedge B) \vee C$

(6) **(a)** $A \vee B \vee C$, **(b)** $A \vee (\neg A \wedge B) \vee (\neg B \wedge C) \vee \neg C$

(7) **(a)** $(A \vee B) \wedge C$, **(b)** $((A \wedge B) \vee (\neg A \wedge B) \vee (A \wedge \neg B)) \wedge C$

(8) **(a)** $(A \wedge B \wedge C) \vee (A \wedge B \wedge \neg C)$, **(b)** $C \wedge \neg C$

(9) **(a)** $A \wedge B \wedge C$, **(b)** $(A \vee \neg B) \wedge (B \vee \neg C) \wedge (C \vee \neg A)$

(10) **(a)** $(A \rightarrow B) \vee C$, **(b)** $(A \leftrightarrow B) \wedge C$

(11) **(a)** $(A \rightarrow B) \rightarrow C$, **(b)** $A \rightarrow (B \rightarrow C)$

(12) **(a)** $(A \leftrightarrow B) \leftrightarrow C$, **(b)** $A \leftrightarrow (B \leftrightarrow C)$

B. Each of the following formulas is a tautology that is obtained by a substitution in one of the tautologies given in the list of useful tautologies. In each case, state which tautology it is and what the substitution is. (Answers for 1, 5, and 9.)

(1) $(A \vee B) \vee (C \wedge D) \leftrightarrow (A \vee B \vee C) \wedge (A \vee B \vee D)$

(2) $\neg A \wedge (\neg A \rightarrow B \vee C) \rightarrow B \vee C$

(3) $(A \wedge B) \vee (A \wedge \neg B \wedge C) \leftrightarrow A \wedge (B \vee (\neg B \wedge C))$

(4) $(A \wedge B) \vee (A \wedge B \wedge C) \leftrightarrow A \wedge B$

(5) $(A \wedge B \wedge C \rightarrow D \wedge \neg D) \rightarrow \neg(A \wedge B \wedge C)$

(6) $A \wedge \neg A \rightarrow A \vee \neg A$

(7) $(A \wedge \neg A) \vee \neg(A \wedge \neg A)$

(8) $(A \wedge \neg B \wedge C) \vee (A \wedge \neg B \wedge \neg C) \leftrightarrow (A \wedge \neg B) \wedge (C \vee \neg C)$

(9) $\neg(A \vee \neg B \vee C) \leftrightarrow \neg A \wedge \neg(\neg B \vee C)$

(10) $(A \wedge B \rightarrow C \vee D) \leftrightarrow (A \rightarrow (B \rightarrow C \vee D))$

C. Translate each of the following arguments into logical notation using the given letters for atomic sentences. Also determine the validity of the argument using truth tables. (Answers for 2, 4, and 7.)

(1) If Oscar attends class, then either Miriam or Virginia attend class. Miriam is not attending class. Therefore, Virginia attends class if Oscar does. (O, M, V)

(2) If Oscar attends class, then so does Miriam, and if Miriam attends class, so does George. Oscar attends class unless George attends. Therefore, Miriam does not attend class. (O, M, G)

(3) If Oscar attends class, then Virginia attends class only if George attends. George does attend class. Therefore, if Oscar attends class, so does Virginia. (O, V, G)

(4) If both Oscar and George attend class, so does Virginia. George does attend class. Therefore, either Virginia attends class or Oscar does not. (O, G, V)

(5) If Oscar and George attend class, then either Miriam or Virginia attends. It is not the case that either George or Miriam attends class. Therefore, either Oscar or Virginia attends class. (O, G, M, V)

(6) If Oscar and George attend class, then so do Miriam and Virginia. It is not the case that George attends class only if Virginia attends class. Therefore, Oscar does not attend class. (O, G, M, V)

(7) A necessary condition for Oscar to attend class is that Miriam or Virginia attend. A sufficient condition for Virginia to attend class is that George attends. However, George does not attend unless Miriam attends, and Virginia attends only if Oscar attends. Therefore, Virginia attends class if and only if Oscar attends. (O, M, V, G)

(8) If Oscar attends class, then neither Miriam nor Virginia attend. If Virginia doesn't attend then neither does George, but if Miriam doesn't attend, then George does attend. Therefore, Oscar does not attend class. (O, M, V, G)

D. Construct a formal proof for each valid argument in part C. If the argument is not valid construct a counter-example. (Answers for 4 and 8.)

E. Determine the validity of each of the following arguments using truth tables. (Answers for 1, 4, and 8.)

(1) **Premises:** $O \rightarrow M \vee V,\ M,\ V \rightarrow O$

Conclusion: O

(2) **Premises:** $O \rightarrow (V \leftrightarrow G),\ \neg M \rightarrow (V \wedge \neg G)$

Conclusion: $\neg O \vee M$

(3) **Premises:** $O \rightarrow M,\ G \rightarrow V,\ \neg M \vee \neg V,\ G \vee \neg M$

Conclusion: $O \leftrightarrow \neg G$

(4) **Premises:** $O \rightarrow V,\ G \rightarrow M,\ G \rightarrow O \vee \neg M,\ G \vee \neg M,\ M \vee \neg V$

Conclusion: $O \leftrightarrow G$

(5) **Premises:** $O \wedge G \rightarrow V,\ V \rightarrow \neg M,\ \neg J \rightarrow M,\ M \rightarrow \neg J$

Conclusion: $G \rightarrow (O \rightarrow J)$

(6) **Premises:** $\neg O \rightarrow \neg V,\ O \rightarrow \neg G \vee M,\ \neg M$

Conclusion: $\neg G \vee \neg V$

(7) **Premises:** $(M \rightarrow O) \wedge (G \rightarrow V),\ M \vee G,\ O$

Conclusion: $O \wedge V$

(8) **Premises:** $M \vee V \rightarrow G,\ G \rightarrow V,\ O \rightarrow \neg J \vee \neg V$

Conclusion: $O \rightarrow (M \rightarrow \neg J)$

F. Construct a formal proof for each valid argument in part E. If the argument is not valid, construct a counter-example. (Answer for 4.)

G. Assign letters to the atomic sentences and translate each of the following arguments into logical notation. If the argument is valid, give a formal proof; otherwise give a counter-example. (Answers for 1, 5, 8(b), and 11.)

(1) If $a < b$ and $b < c$, then $a < c$. It is true that $a < b$, but it is not the case that $a < c$. Therefore, it is not the case that $b < c$.

(2) If $a > 0$, then a necessary and sufficient condition for $b > c$ is that $a \cdot b > a \cdot c$. It is true that $a \cdot b > a \cdot c$. Therefore, $b > c$.

In the remaining problems, we shall use the following abbreviations:

$x \ngtr y$ for it is not the case that $x > y$

$x \nless y$ for it is not the case that $x < y$

$x \neq y$ for it is not the case that $x = y$

(3) If $a > 0$ and $b > c$, then $a \cdot b > a \cdot c$, and if $a \ngtr 0$, but $b > c$, then $a \cdot b \ngtr a \cdot c$. Therefore, $a > 0$ if and only if $a \cdot b > a \cdot c$.

(4) If $a < 0$ and $0 < b$, then $a < b$. If $0 < c$ and $a < b$, then $a \cdot c < b \cdot c$. It is true that $a < 0$ and $0 < c$. Therefore, $0 < b$ only if $a \cdot c < b \cdot c$.

(5) A necessary and sufficient condition that $b < c$ is that $-c < -b$. Moreover, if $a + (-c) \nless a + (-b)$, then $b \nless c$. Therefore, $b < c$ if and only if $a + (-c) < a + (-b)$.

(6) If $a < 0$ and $b \nless 0$, then $a \cdot b \leq 0$. If $a \cdot b \leq 0$, then if $b < 0$, $a \nless 0$. If $a \cdot b \nleq 0$, then either $a < 0$ and $b < 0$, or $a \nless 0$ and $b \nless 0$. If $a \cdot b \leq 0$ and $b \nless 0$, then $a < 0$. Therefore, a necessary and sufficient condition for $a \cdot b \leq 0$ is that $a < 0$ if and only if $b \nless 0$.

(7) Either $a < 0$ or $b < 0$. Also, $a \cdot b \leq 0$ if either $a < 0$ and $b \nless 0$, or if $b < 0$ and $a \nless 0$. Therefore, if not both $a < 0$ and $b < 0$, then $a \cdot b \leq 0$.

(8) (a). Define a function f so that

$$f(x) = \begin{cases} x, & \text{if } x < 0 \\ 5, & \text{if } x \nless 0. \end{cases}$$

Therefore, $f(x) = x$ or $f(x) = 5$. (*Hint:* Let $Z : x < 0$, $X : f(x) = x$, $F : f(x) = 5$.)

(b). Define a function g so that

$$g(x) = \begin{cases} f(x), & \text{if } x < 2 \\ 6, & \text{if } x \nless 2, \end{cases}$$

where $f(x)$ is defined in part (a). Assume $x < 0$ or $x \nless 2$. Therefore, $g(x) = x$ or $g(x) = 6$. (*Hint*: If $Z : x < 0$ and $T : x < 2$, then $0 \leq x < 2$ is $\neg Z \wedge T$. Let $X : g(x) = x$; $F : g(x) = 5$; $S : g(x) = 6$.)

(c). Let g be the function defined in part (b). Assume that neither does $g(x) = 5$ nor does $g(x) = 6$. Therefore, $g(x) = x$.

(9) If f is continuous on $(a, b]$, then f has a maximum on $[a, b]$ if f is defined at a. If f is not defined at a, then there is a point c between a and b such that $f(c) \geq f(x)$ for all $x \in [a, b]$. If there is such a point c between a and b such that $f(c) \geq f(x)$ for all $x \in [a, b]$, then f has a maximum on $[a, b]$.

(a) Therefore, f has a maximum on $[a, b]$.

(b) Therefore, if f is continuous on $(a, b]$, then f has a maximum on $[a, b]$.

(Use the letters C, M, D, and B for the atomic sentences.)

(10) A necessary condition for f to have a maximum on $[a, b]$ is that f is continuous on $(a, b]$ and f is defined at a. A necessary and sufficient condition for f to have a maximum on $[a, b]$ is that there is a point c between a and b such that $f(c) \geq f(x)$ for all $x \in [a, b]$. In fact, there is a point c between a and b such that $f(c) \geq f(x)$ for all $x \in [a, b]$. Therefore, f is continuous on $(a, b]$ and f is defined at a. (Use the same letters for atomic sentences as in problem 9.)

(11) If one and two are substituted for x in the inequality $x^2 + 1 > x + 1$, we obtain $2 > 2$ and $5 > 3$, respectively. However, $2 \not> 2$, but $5 > 3$. Therefore, one is not substituted for x in the inequality $x^2 + 1 > x + 1$, but two is. (Use A, B, $C : 2 > 2$, and $D : 5 > 3$ as the letters for the atomic sentences.)

(12) If one and two are substituted for x in the inequality $x^2 + 1 > x + 1$, we obtain $2 > 2$ and $5 > 3$, respectively. However, it is not the case that $2 > 2$, but $5 > 3$. Therefore, not both one and two are substituted for x in the inequality $x^2 + 1 > x + 1$. (Use the same letters for atomic sentences as in problem 11.)

H. Give a formal proof for each of the following valid arguments. (Answers for 1, 4, 8, and 10.)

(1) **Premises:** $A \to (B \to C)$, $C \to \neg D$, $\neg E \to D$

Conclusion: $B \to (A \to E)$

(2) **Premises:** $A \lor B$, $C \to \neg A$, $B \to D$, $C \to \neg D$

Conclusion: $\neg C$

(3) **Premises:** $(A \lor B) \lor C$, $A \to (D \leftrightarrow \neg E)$, $B \to \neg(\neg D \lor E)$, $E \to \neg(C \lor D)$, $\neg E \to C \land D$

Conclusion: $D \leftrightarrow \neg E$

(4) **Premises:** $(A \to B) \lor (A \land \neg C)$, A, $C \to \neg B$, $\neg C \to B$

Conclusion: $\neg C$

(5) **Premises:** $(A \to B) \land (C \to \neg D)$, $(E \to \neg B) \land (\neg F \to D)$, $\neg E \lor F \to G$, $\neg B \to D$, $A \lor C$

Conclusion: $B \land G$

(6) **Premises:** $A \land C \to D$, $B \land C \to D$, $\neg A \land \neg B \to E \lor F$, $G \to \neg E$, $F \to H$, $C \land \neg D$

Conclusion: $\neg G \lor H$

(7) **Premises:** $A \land (B \lor C)$, $A \land B \to D \land \neg F$, $A \to (C \to \neg(D \lor \neg F))$

Conclusion: $D \leftrightarrow \neg F$

(8) **Premises:** $A \land B \to C$, $A \lor E$, $G \to (\neg C \land \neg D)$, $A \leftrightarrow B$, $A \to C$, $C \to \neg D$, $B \to D$

Conclusion: $\neg A \wedge (G \rightarrow E)$

(9) **Premises:** $\neg A \rightarrow C \vee D$, $B \rightarrow E \wedge F$, $E \rightarrow D$, $\neg D$

Conclusion: $(A \rightarrow B) \rightarrow C$

(10) **Premises:** $A \rightarrow B$, $C \rightarrow \neg B$, $\neg C \rightarrow D$, $A \wedge \neg D$

Conclusion: $E \rightarrow F$

(11) **Premises:** $B \rightarrow A$, $\neg A \vee C$, $C \rightarrow D \vee E$, $D \rightarrow F \wedge \neg B$, $E \rightarrow \neg A \wedge F$, $F \rightarrow A$, $\neg B \vee G \rightarrow H \wedge I$, $H \rightarrow B$

Conclusion: $H \wedge \neg H$

(12) **Premises:** $A \rightarrow (B \rightarrow C)$, $\neg A \rightarrow D \wedge \neg E$, $A \wedge B \rightarrow \neg C$, $D \rightarrow F \vee G$, $\neg B \rightarrow (G \rightarrow H)$, $E \vee (H \vee \neg G)$

Conclusion: $G \rightarrow H$

Chapter 3

Consistent and Inconsistent Premises

Which of the following sets of sentences are consistent?

1. If John killed Richard, then John will be prosecuted. John did not kill Richard, but John will still be prosecuted.
2. If the furnace is repaired, it will be turned on. The house will be heated if the furnace is turned on. The house is heated, but the furnace is not repaired.
3. Nebraska is north of Kansas, and Kansas is north of Oklahoma. However, Nebraska is not north of Oklahoma.

These are the types of questions we address in Section 3.1 (see Example 3.3). In Section 3.2, we apply logic to solve certain types of puzzles.

3.1 CONSISTENT AND INCONSISTENT PREMISES

A set of premises Γ is said to be *consistent* if it is not possible to deduce a contradiction from Γ; that is, $\Gamma \nvdash Q \land \neg Q$. (Equivalently, Γ is consistent if there is a formula P such that $\Gamma \nvdash P$. Why is this equivalent?) The method we use to show that $\Gamma \nvdash Q \land \neg Q$ is to show that the argument with premises Γ and conclusion $Q \land \neg Q$ is not valid—that is, find an assignment of truth values for the atomic sentences (sentential variables) in Γ such that all the premises in Γ are true, but $Q \land \neg Q$ is false. In this case, however, $Q \land \neg Q$ is a contradiction and is always false. Thus, to show that Γ is consistent, it is sufficient to find truth values for the atomic sentences in Γ such that all the formulas in Γ are simultaneously true. For example, consider the following premises:

If Oscar takes mathematics at college he will not have time to be on the basketball team. Oscar does not take mathematics at college and he does not have time to be on the basketball team.

Let the atomic sentences be M and B, where,

M: Oscar takes mathematics at college.

B: Oscar has time to be on the basketball team.

Then the premises will be

$$P_1 : M \to \neg B;\ P_2 : \neg M \wedge \neg B.$$

Premise 2 is true if and only if both M and B are false. If M is false, premise 1 is true. Therefore, the following one line of a truth table proves that this set of premises $\{P_1, P_2\}$ is consistent.

M	B	$M \to \neg B$	$\neg M \wedge \neg B$
F	F	T	T
		$F \to T$	$T \wedge T$

The consistency of a set of premises does not depend on whether premises are true or false; it only depends on their structure. Consider the following set of premises:

P_1: Two is greater than five. (G)

P_2: If two is greater than five, then six is less than one.

P_3: Six is less than one. (L)

Using the indicated letters for atomic sentences, we obtain

$$P_1 : G,\ P_2 : G \to L,\ P_3 : L.$$

Thus, if G and L are both true, all three premises are true, so this set of premises is consistent. It does not matter that, as stated, G and L are false. If it is *possible* to assign truth values to the atomic sentences in such a way that all the premises are true, then the premises are consistent. In fact, this is also a necessary condition for consistency, as is shown in the following theorem.

Theorem 3.1 *A set of premises is consistent if and only if it is possible to assign truth values to the atomic sentences in such a way that all the premises are true simultaneously.*

Proof: Let $\Gamma = \{P_1, P_2, \ldots, P_n\}$ be a set of premises. Suppose that it is not possible to assign truth values to the atomic sentences in Γ in such a way that all the premises are true. It follows that

$$(*) \qquad P_1 \wedge P_2 \wedge \cdots \wedge P_n \to Q \wedge \neg Q$$

is a tautology because its antecedent is always false. Thus, $\Gamma \vdash Q \wedge \neg Q$, so Γ is not consistent. Consequently, if Γ is consistent, then it is possible to assign truth values to the atomic sentences in Γ so that all the premises are true.

Conversely, if it is possible to assign truth values to the atomic sentences so that all the formulas in Γ are true, then (*) is not a tautology. Consequently, $\Gamma \nvdash Q \wedge \neg Q$, so Γ is consistent.

■

If a set of premises is not consistent, then the set is called *inconsistent*. Thus, it follows from Theorem 3.1 that a set of premises is inconsistent if, for every assignment of truth values to the atomic sentences, at least one of the premises is false. The premises cannot be true simultaneously.

Example 3.1. Show that the following set of premises is inconsistent:

$$A \to (B \to C),\ B \wedge \neg C,\ A.$$

Solution: We shall construct a truth table and show that, on each line of the truth table, at least one of the premises is false.

A	B	C	$A \to (B \to C)$	$B \wedge \neg C$	A
T	T	T	T	F	T
T	T	F	F	T	T
T	F	T	T	F	T
T	F	F	T	F	T
F	T	T	T	F	F
F	T	F	T	T	F
F	F	T	T	F	F
F	F	F	T	F	F

We see that, on each line of the table, one or more of the premises is false. Therefore, this set of premises is inconsistent.

As we saw earlier, the truth table method to show that an argument is valid is rather cumbersome if a large number of atomic sentences is involved. We found a better way to do it in the case of a valid argument (constructing a proof) and, indeed, there is a better way to show that a set of premises is inconsistent. Namely, give a proof of a contradiction from the premises. Γ is inconsistent iff $\Gamma \vdash Q \wedge \neg Q$ for some formula Q.

Example 3.2 For each of the following sets of premises, determine whether it is consistent or not. If the set is consistent, assign truth values to the atomic sentences in such a way that all the premises are true. If the set is inconsistent, give a formal proof of a contradiction.

(a) P_1: $C \to A$, P_2: C, P_3: $B \vee C \to \neg A$

(b) P_1: $A \to B \vee C$, P_2: A, P_3: $B \to \neg A$, P_4: $C \to D$

(c) P_1: $(A \to B)$, P_2: $(C \to D)$, P_3: $C \to A \vee \neg D$, P_4: $C \vee \neg D$, P_5: $D \vee \neg B$, P_6: $(A \wedge \neg C) \vee (\neg A \wedge C)$

Solution: (a) First we try to prove the premises consistent. If P_2 is to be true, then C must be true. If P_1 is to be true with C true, A must be true. On the other hand, if C is true, then $B \vee C$ is also true, so if P_3 is to be true, then A must be false. It appears that the premises are inconsistent. The preceding informal discussion is an informal proof of a contradiction. The following is a formal proof:

{1}	1.	$C \to A$	P
{2}	2.	C	P
{3}	3.	$B \vee C \to \neg A$	P
{1, 2}	4.	A	1, 2 T
{2}	5.	$B \vee C$	2 T
{2, 3}	6.	$\neg A$	3, 5 T
{1, 2, 3}	7.	$A \wedge \neg A$	4, 6 T

(b) Again, we shall try to prove that the premises are consistent. For P_2 to be true, A must be true. If A is true, to make P_3 true, B must be false. Since A is true and B is false, to make P_1 true, C must be true. Thus, for P_4 to be true, D must be true. Therefore, the following truth assignments to the atomic sentences make all the premises true:

A	B	C	D	$A \to B \vee C$	A	$B \to \neg A$	$C \to D$
T	F	T	T	T	T	T	T
				$T \to T$		$F \to F$	$T \to T$

This set of premises is therefore consistent.

(c) In this case, we shall also try to find truth assignments for the atomic sentences so that all the premises are true. The last premise, P_6, is a disjunction. To make that true, one of the disjuncts must be true. We'll consider two cases.

Case 1. $A \wedge \neg C$ is true. Then A is true, and C is false. If A is true, to make P_1 true, B must be true. If C is false, P_2 and P_3 are true, but to make P_4 true, D must be false. If D is false and B is true, however, P_5 is false. Thus, this truth assignment doesn't work, and we conclude $\neg(A \wedge \neg C)$.

Case 2. $\neg A \wedge C$ is true. Then A is false, and C is true. If A is false, P_1 is true. If C is true, to make P_2 true, D must be true. If C is true and A is false, to make P_3 true, D must be false. We derived the contradiction $D \wedge \neg D$, so this truth assignment doesn't work either. We deduce $\neg(\neg A \wedge C)$, and conclude that the premises are inconsistent.

Actually, the preceding informal argument is sufficient to prove inconsistency, but we give a formal proof below. We shall follow the outline of the informal argument in our formal proof.

{1}	1.	$A \to B$	P
{2}	2.	$C \to D$	P
{3}	3.	$C \to A \vee \neg D$	P
{4}	4.	$C \vee \neg D$	P
{5}	5.	$D \vee \neg B$	P
{6}	6.	$(A \wedge \neg C) \vee (\neg A \wedge C)$	P
{7}	7.	$A \wedge \neg C$	P (for IP)
{1, 7}	8.	B	1, 7 T
{4, 7}	9.	$\neg D$	4, 7 T
{1, 4, 7}	10.	$\neg(D \vee \neg B)$	8, 9 T
{1, 4, 5, 7}	11.	$(D \vee \neg B) \wedge \neg(D \vee \neg B)$	5, 10 T
{1, 4, 5}	12.	$\neg(A \wedge \neg C)$	7, 11 IP
{13}	13.	$\neg A \wedge C$	P (for IP)
{2, 13}	14.	D	2, 13 T
{3, 13}	15.	$A \vee \neg D$	3, 13 T
{3, 13}	16.	$\neg D$	13, 15 T
{2, 3, 13}	17.	$D \wedge \neg D$	14, 16 T
{2, 3}	18.	$\neg(\neg A \wedge C)$	13, 17 IP
{1, 2, 3, 4, 5}	19.	$\neg((A \wedge \neg C) \vee (\neg A \wedge C))$	12, 18 T
{1, 2, 3, 4, 5, 6}	20.	$(P_6) \wedge \neg(P_6)$	6, 19 T

(P_6 is $(A \wedge \neg C) \vee (\neg A \wedge C)$.) On line 7, we start case 1, taking $A \wedge \neg C$ as a premise, and on line 11, we contradicted P_5. We conclude, on line 12, $\neg(A \wedge \neg C)$. Then on line 13, we start case 2 and take $\neg A \wedge C$ as a premise. The contradiction is derived on line 17, from which we conclude $\neg(\neg A \wedge C)$. Consequently, we have contradicted P_6.

Example 3.3

Determine whether the three sets of premises given in the introduction are consistent.

Solution: In each case, we shall first translate the sentences into the language of the Sentential Calculus and then show each set is consistent.

1. $K \to P$, $\neg K$, P. If K is false and P is true, the three sentences are true.
2. $R \to O$, $O \to H$, $H \wedge \neg R$. The three sentences are true if H is true and R is false.
3. When we translate these sentences into the Sentential Calculus, we get formulas of the form: $N \wedge K$, $\neg O$. It is clear that if N and K are true and O is false, both sentences are true.

These are examples in which it appears from the meaning of the sentences that each set is inconsistent. After examining the structure of the sentences, however, we see that each is consistent.

3.2 LOGICAL PUZZLES

The type of argument illustrated in Example 3.2(c) can be used to solve certain types of problems by eliminating the cases that lead to contradictions.

Example 3.4 Four different people, Arthur (A), Betty (B), Charles (C), and Dorothy (D) have four different positions (not necessarily in the given order): lawyer (L), pilot (P), teacher (T), and veterinarian (V). Each person holds exactly one position, and each position is held by exactly one person. (In other words, there is a one-to-one correspondence between people and positions.) In each of the following cases, decide whether the conditions are consistent, and if they are consistent, decide whether each person's position is uniquely determined.

(a) (1) V is not A or C; (2) B is not V or P; (3) C is not L or P; (4) D is not L.

(b) (1) V is not A or C; (2) B is not P or T; (3) C is not L or P; (4) D is not P or V; (5) L is not B or D.

(c) (1) V is A or C; (2) B is not L or P; (3) C is T or V; (4) A is not L.

(d) (1) V is A or C; (2) B is not L or P; (3) C is not L or T; (4) D is not T or V; (5) A is not L.

Solution: (a) From (1) and (3), we deduce that C is not V or L or P. Because C must hold one of the four positions, C must be T. It follows from (1) that either B or D is V, but from (2), B is not V. Therefore, D must be V. From (2), B is not P. Because L and P are the only positions left, B must be L. The only thing left for A is P. It follows that the premises are consistent and each person's position is uniquely determined: Arthur is a pilot; Betty is a lawyer; Charles is a teacher; and Dorothy is a veterinarian. (Note that (4) is redundant.)

(b) We deduce from (2), (3), and (4) that neither B nor C nor D is P. Therefore, A must be P. It follows from (3) that C is T or V. But from (1), we see that C is not V. Thus, C must be T. From (2), we obtain that B is either L or V, but (5) implies that B must be V. The only position left for D is L, but by (5), D cannot be L.

In this case, the premises imply that A is P, C is T, and B is V, so the only thing left for D is L. But if D is L, (5) is false. Therefore, these premises are inconsistent.

(c) From (1), V is either A or C. Suppose A is V. It follows from (2) that B is either T or V. Because A is V, B must be T. Thus, T and V are taken by B and A, respectively, and this contradicts (3). Thus, A cannot be V.

Suppose, then, that C is V. We deduce from (2) that B must be T. Thus, A is either L or P, but from (4), A is not L, so A must be P. It follows that D

must be L. Now, all the premises are true, so we have shown that the premises are consistent. There is only one possibility: Arthur is a pilot, Betty is a teacher; Charles is a veterinarian, and Dorothy is a lawyer. (This argument is another illustration of an alternative proof. One of the alternatives, A is V, leads to a contradiction, so we drop that and try the other alternative, C is V.)

(d) Just as in part (c), it follows from (1) that there are two cases—either A is V or C is V. Suppose, first, that A is V. Then from (2), B is either T or V. But because A is V, B must be T. It follows from (3) that C is either P or V. But V is taken by A, so C must be P. Consequently, D is L, which is consistent with (4).

On the other hand, if C is V, then it follows from (2) that B is T or V. Because C is V, B must be T. If C is V, (3) is true. Now we have that A is L or P, and D is L or P. But it follows from (5) that A is P, so D must be L.

In this case, the premises are consistent, but there are two possible solutions. The first solution is Arthur is a veterinarian; Betty is a teacher; Charles is a pilot; and Dorothy is a lawyer. The other solution is Charles is a veterinarian; Betty is a teacher; Arthur is a pilot; and Dorothy is a lawyer.

Example 3.5

Four friends—Arthur, Betty, Charles, and Dorothy—are suspected of murder. They testify as follows:

Arthur: If Betty is guilty, so is Dorothy.

Betty: Arthur is guilty, but Dorothy is not.

Charles: I'm not guilty, but either Arthur or Dorothy is guilty.

Dorothy: If Arthur is not guilty, then Charles is guilty.

(a) Are the four statements consistent?

(b) If everyone is telling the truth, who is guilty?

(c) If the guilty people lie and the innocent (not guilty) tell the truth, who is guilty?

Solution: Let X represent the statement "X is innocent." Translating the testimony into the Sentential Calculus, we obtain the following.

Arthur: $\neg B \rightarrow \neg D$.

Betty: $\neg A \wedge D$.

Charles: $C \wedge (\neg A \vee \neg D)$.

Dorothy: $A \rightarrow \neg C$.

(a) Betty's statement will be true if A is false and D is true. To make Arthur's statement true with D true, B must be true. Dorothy's statement is true because A is false. Thus, to make Charles' statement true, because A is false, C must be true. It follows that there is an assignment of truth values that makes all the statements true.

A	B	C	D	$\neg B \to \neg D$	$\neg A \wedge D$	$C \wedge (\neg A \vee \neg D)$	$A \to \neg C$
F	T	T	T	T	T	T	T
				$T \to T$	$T \wedge T$	$T \wedge T$	$F \to F$

Thus, the statements are consistent.

(b) It follows from the preceding argument that the truth assignment given is the only one that makes the four statements true. Thus, if the four statements are true, Arthur is guilty and the others are innocent.

(c) In this case, we do not know, at first, which statements are true and which are false. We only know that the guilty people make false statements and the innocent make true statements. To decide who is guilty and who is innocent, we shall give an alternative proof.

First assume Arthur is guilty. Then A is false and $\neg B \to \neg D$ is false. This means B must be false and D true. But if B is false, then Betty's statement, $\neg A \wedge D$, must be false. However, we have a contradiction here because A is false and D is true, so Betty's statement is true. We deduce from this that Arthur is not guilty, so A is true.

If A is true, then Betty's statement is false, so B is false. Because Arthur is innocent, Arthur's statement, $\neg B \to \neg D$, is true. We have that both $\neg B$ and $\neg B \to \neg D$ are true, so $\neg D$ must be true. Thus, Dorothy is guilty, so her statement must be false. Because A is true and $A \to \neg C$ is false, C must be true. Thus, we have the following:

A	B	C	D	$\neg B \to \neg D$	$\neg A \wedge D$	$C \wedge (\neg A \vee \neg D)$	$A \to \neg C$
T	F	T	F	T	F	T	F
				$T \to T$	$F \wedge F$	$T \wedge T$	$T \to F$

In this case, we see that Betty and Dorothy are guilty and their statements are false, but Arthur and Charles are innocent and their statements are true.

There have been many stories about islands, usually isolated from the rest of the world, where some of the natives always tell the truth, and others never tell the truth. We have also discovered such an island. It is called TUFA Island. There are two races of people on TUFA Island: the Truth Tellers (called Tu's), whose statements are always true, and the Falsehood Tellers (called Fa's), whose statements are always false.

Example 3.6

On our first visit to TUFA Island, we met three natives—Ralph (R), Simon (S), and Tyrone (T). They each made the following statements:

Ralph: S and T do not belong to the same race.

Simon: R is a Tu.

Tyrone: R is a Fa.

Can it be determined to which race each person belongs?

Solution: It follows from the statements of Simon and Tyrone that one is telling the truth and the other is lying. Thus, Ralph's statement is true, so Ralph is a Tu. Consequently, because Simon is telling the truth, he is a Tu, and because Tyrone is lying, he is a Fa. Thus, each person's race is uniquely determined.

Example 3.7 On our next visit to TUFA Island, we came across four natives—Milo (M), Norbert (N), Opel (O), and Penelope (P)—who were seated at a round table as indicated below. (Each person is either a Tu or a Fa.) We asked each native the same question: "Do you and the person on your right belong to the same race?" Milo and Norbert answered "yes," and Opel and Penelope answered "no." Can it be determined to which race each of the four people belong?

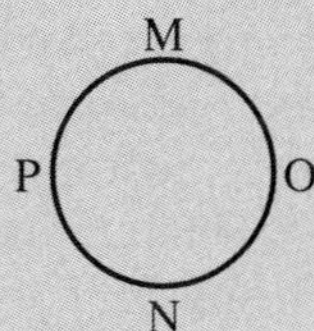

Solution: For each native, there are four possibilities. Each native is either

1. a Tu with a Tu on the right, or
2. a Tu with a Fa on the right, or
3. a Fa with a Tu on the right, or
4. a Fa with a Fa on the right.

The native asked the question in case 1 would answer "yes"; in case 2, "no"; in case 3, "yes"; and in case 4, "no." Thus, regardless of race, if there is a Tu on the right, the native answers "yes," and if there is a Fa on the right, the native answers "no." Because Milo and Norbert answered "yes," Penelope and Opel are Tu's, and because Opel and Penelope answered "no," Milo and Norbert must be Fa's. Again, each person's race is uniquely determined.

(References for additional logical puzzles include *What is the Name of this Book* by R. Smullyan, Prentice-Hall, 1978, and *New Puzzles in Logical Deduction* by G. J. Summers, Dover, 1968.)

EXERCISES 3

A. For each of the following sets of premises, assign letters to the atomic sentences and translate into the Sentential Calculus, and then determine whether or not each is consistent. If consistent, assign truth values to the atomic sentences so the premises are all true; if not consistent, give a formal proof of a contradiction. (Answers to 1, 6, 9, 11, 15, and 19.)

(1) If Oscar studies, he will pass his logic course. Also, if Oscar does not study, he will pass his logic course. However, Oscar does not pass his logic course.

(2) If Oscar studies, he will pass either his logic course or his mathematics course. However, Oscar does not study, but he will pass either his logic course or his mathematics course.

(3) If Miriam studies and Oscar does not distract her, Miriam will pass both her logic course and her mathematics course. Oscar does not distract Miriam. However, Miriam does not pass her mathematics course.

(4) If Oscar studies, he will either graduate this semester or next semester. If he graduates this semester, he will go to Law School; if he graduates next semester, he will not go to Law School. Indeed, Oscar does study.

(5) If Miriam studies, then she will graduate. If she graduates, then she will go to Law School and get married. If she goes to Law School, she will not have children this year, but if she gets married, she will have children this year. Miriam does not study, but she will graduate.

(6) Miriam will graduate unless she gets married. Miriam will go to work if and only if she gets married. If she goes to work, she cannot help Virginia with her homework. If Miriam does not help Virginia with her homework, Virginia will not graduate. Miriam will graduate if and only if Virginia does not graduate.

(7) If Oscar studies for his finals, then he will not go to the basketball game Friday night. If Oscar does not study for his finals, then he does not graduate. If Oscar does not graduate, then he will not get a job. If Oscar does not go to the basketball game Friday night, then he will not meet Virginia. However, Oscar does get a job and he does meet Virginia.

(8) If Oscar and Virginia are the same age, then George and Miriam are not the same age. Either Oscar and Virginia are the same age or Oscar is older than Virginia. If Oscar is older than Virginia, then Oscar is older than Miriam. It is not the case that if George and Miriam are the same age, then Oscar is older than Miriam.

(9) If Oscar and George both graduate, then either Oscar or George will get a job. If Oscar gets a job, then George will not; and if George graduates, then Oscar will not. Neither Oscar nor George graduates, but either Oscar or George does get a job.

(10) If Oscar graduates, then both George and Virginia graduate. If Oscar

does not graduate, then George graduates and gets a job. If George graduates, then if he gets a job, Virginia also gets a job. If Virginia graduates, then George gets a job. If Virginia gets a job, then George does not graduate.

We again use the following abbreviations:

$x \not< y$	for	$\neg(x < y)$
$x \not> y$	for	$\neg(x > y)$
$x \neq y$	for	$\neg(x = y)$

(11) If $a < b$ and $b < c$, then $a < c$. If $a < b$ and $b \not< c$, then $a \not< c$. If $a \not< b$ and $b < c$, then $a < c$. However, $a < c$ if and only if either $a < b$ and $b < c$, or a $\not< b$.

(12) If $a < b$, then $b < c$ only if $a < c$. If $a < b$ and $a < c$, then either $b < c$ or $b \not< c$. If $b < c$, then $a \not< b$. However, $b < c$ unless $a \not< c$.

(In Exercises 13 through 17, interpret a statement of the form "*P* is false" to mean the same as "*P* is not true." Take a statement of the form "*P* is true" to be an atomic sentence.)

(13) If the conditional statement $A \to B$ is false, then A is true and B is false. B is false if A is true or A is false. Indeed, the conditional statement $A \to B$ is false.

(14) A necessary and sufficient condition for the conditional statement $A \to B$ to be true is that both A and B are true. A sufficient condition for B to be false is that A is true. A necessary condition for the conditional statement $A \to B$ to be true is that A is true. However, the conditional statement $A \to B$ is false.

(15) If A is false, the conditional statement $A \to B$ is of the form $F \to T$ or $F \to F$, depending upon whether B is true or false, respectively. The conditional statement $A \to B$ is of the form $F \to T$ if B is true and A is false, but neither is B true nor is $A \to B$ of the form $F \to F$.

(16) If A is true, the conditional statement $A \to B$ is of the form $T \to T$ if B is true, and $T \to F$ if B is false. If B is false, $A \to B$ is false. However, $A \to B$ is not of the form $T \to T$, and it is true.

(17) If A is false, the conditional statement $A \to B$ is true, and if A is true, the conditional statement $A \to B$ is true, provided that B is true. A sufficient condition for $A \to B$ to be true is that A is true and B is false. However, the conditional statement $A \to B$ is false.

(18) If the function f is continuous on the interval (a, b), then f has both a maximum and a minimum on $[a, b]$ provided that f is continuous both at a and at b. The function f does not have a maximum on $[a, b]$ only if f is not continuous on (a, b); and f does not have a minimum on $[a, b]$ only if f is not continuous at a. If f has a maximum on $[a, b]$, then f is continuous at b, but if f has a minimum on $[a, b]$, then f is not continuous at b. In this case, f is continuous on (a, b).

(19) The function f is continuous on (a, b), f is continuous at both a and

b, and f has a maximum on $[a, b]$, but f has no minimum on $[a, b]$. If f is continuous on (a, b), then either f has both a maximum and a minimum on $[a, b]$ or f is not continuous at both a and b.

(20) If f is continuous on (a, b), then f has a maximum on $[a, b]$ if f is continuous at a, and f has a minimum on $[a, b]$ if f is continuous at b. If f has a maximum on $[a, b]$, then f is not continuous at b. If f is continuous on (a, b), then f is continuous at both a and at b. If either f has a maximum on $[a, b]$ or f does not have a minimum on $[a, b]$, then f is continuous on (a, b). If f is continuous at a or if f does not have a maximum on $[a, b]$, then f is continuous on (a, b).

B. For each of the following sets of premises, determine whether or not it is consistent. Give reasons for your answer. (That is, if the set is consistent, assign truth values to the atomic sentences so that the premises are all true; if not consistent give either a formal or an informal proof of a contradiction, or show by means of a truth table that the premises can never be true simultaneously.) (Answers to 3, 5, 9, and 13.)

(1) $A \wedge B \rightarrow C, A \wedge \neg C$

(2) $A \rightarrow B \wedge \neg C, A \wedge \neg B$

(3) $A \vee B \rightarrow \neg C, A \wedge C$

(4) $A \rightarrow B \vee C, A \wedge \neg B$

(5) $A \rightarrow B, C \rightarrow \neg B, \neg C \rightarrow D, A \wedge \neg D$

(6) $A \rightarrow (B \rightarrow C), \neg D \rightarrow B, C \rightarrow D, D \rightarrow \neg A$, A

(7) $A \rightarrow (B \leftrightarrow C), B \rightarrow D, B \vee C \rightarrow \neg D, A \wedge \neg D$

(8) $A \vee \neg(B \vee \neg C), A \rightarrow D, \neg(A \wedge D), B \rightarrow D, B \rightarrow C$

(9) $A \leftrightarrow B \vee C, A \vee \neg D, \neg B \vee D, C \vee \neg D, B \leftrightarrow \neg A$

(10) $A \leftrightarrow B \vee C, B \leftrightarrow \neg A, B \vee D, \neg C \vee \neg D$

(11) $(A \vee B) \vee C, A \rightarrow (D \leftrightarrow \neg E), \neg(\neg D \vee E)$,
$E \rightarrow \neg(C \vee D), \neg E \rightarrow C \wedge D, D \leftrightarrow \neg E$

(12) $A \wedge B \rightarrow C, A \vee E, G \rightarrow (\neg C \wedge \neg D), A \leftrightarrow B$,
$A \rightarrow C, C \rightarrow \neg D, B \rightarrow D, A \vee (G \wedge \neg E)$

(13) $A \rightarrow C, A \vee \neg B, C \rightarrow D \vee E, D \rightarrow F \wedge \neg B$,
$E \rightarrow \neg A \wedge F, A \vee \neg F, \neg B \vee G \rightarrow H \wedge I, H \rightarrow B$

(14) $A \wedge C \rightarrow D, B \wedge C \rightarrow D, \neg(A \vee B) \rightarrow E \vee F$,
$G \rightarrow \neg E, \neg F \vee H, C \wedge \neg D, G \rightarrow H$

(15) $A \rightarrow (B \rightarrow C), \neg A \rightarrow D \wedge \neg E, A \wedge B \rightarrow \neg C$,
$D \rightarrow F \vee G, B \vee (G \rightarrow H), G \rightarrow E \vee H, G \wedge \neg H$

C. At a famous murder trial, four witnesses—Arthur (A), Betty (B), Charles (C), and Dorothy (D)—testify under oath.

(i) Determine in each case whether the testimony is consistent.

(ii) If the innocent (not guilty) people tell the truth and the guilty lie, determine (if possible) who is guilty and who is innocent.

(Answers to 1 and 4.)

(1) Arthur: Either Betty or Charles is guilty, but I am innocent.

Betty: If Dorothy is guilty, then the rest of us are innocent.
Charles: If Betty is guilty, then I am innocent.
Dorothy: Either Arthur or Betty is innocent.

(2) Arthur: If either Betty or Charles is innocent, then so am I.
Betty: Arthur is guilty, and either Charles or Dorothy is guilty.
Charles: If Betty is innocent, then Dorothy is guilty.
Dorothy: If Betty is guilty, then Charles is innocent; however, Betty is innocent.

(3) Arthur: Either Betty is guilty and Charles innocent, or Betty is innocent and Charles guilty.
Betty: If Arthur or Dorothy is innocent, then so is Charles.
Charles: Either Arthur or Betty is guilty, but I am innocent.
Dorothy: Betty is innocent or Charles is guilty if and only if Betty is innocent and Charles is guilty.

(4) Arthur: Either I am innocent or Dorothy is guilty.
Betty: If Charles is guilty, then Dorothy is innocent.
Charles: If I am innocent, then Betty is guilty.
Dorothy: If Arthur is innocent, so is Betty.

(5) Arthur: Either Dorothy is innocent or Betty is guilty.
Betty: Either Charles is innocent or Dorothy is guilty.
Charles: Arthur is innocent or Dorothy is guilty, and Betty is innocent if and only if Arthur is guilty.
Dorothy: Arthur is innocent if and only if either Betty or Charles is innocent.

D. In this exercise, we have the same assumptions as in Example 3.4, and we ask the same questions. (Answers to 1 and 3.)

(i) Are the conditions consistent?, and, if so,

(ii) Is each person's position uniquely determined?

(1) (a) V is not B or C; (b) A is not P or V; (c) B is V or T; (d) C is V or L.
(2) (a) P is not A or B; (b) T is not A or D; (c) V is not A or C; (d) B is V or L.
(3) (a) T is not A or D; (b) V is not A or B; (c) B is L or P; (d) C is T or V.
(4) (a) A is not L; (b) B is not P; (c) C is not T; (d) D is not V; (e) V is not A or B; (f) L is not C or D; (g) T is not A or D.
(5) (a) L is not A or D; (b) P is not B or C; (c) T is not B or C; (d) V is not C or D.

E. We return once again to TUFA Island, where there are two races of people, the Tu's, who always tell the truth, and the Fa's who never tell the truth. (Answers to 1(a), 2(a), and 3(a,d).)

(1) We meet three natives R, S, and T, who make the following statements:
(a) R: We are all Tu's.
S: Exactly one of us is a Tu.
T: Exactly two of us are Tu's.

(b) *R*: None of us are Tu's.
S: Exactly one of us is a Tu.
(c) *R*: *S* and *T* belong to the same race.
S: Both *T* and I are Tu's.
T: *S* is a Fa.
(d) *R*: *S* and *T* are both Fa's.
S: *R* is a Tu.
T: *R* is a Tu.
(e) *R*: *S* and *T* do not belong to the same race.
S: *R* is a Tu.
T: *R* is a Tu.

Can it be determined, in each case, to which race each person belongs? Give reasons for your answer.

(2) In our travels in TUFA, we meet three more natives—*U*, *V*, and *W*—and we ask, "How many of you are Tu's?" The first person, *U*, responds in his native language, which we do not understand. However, *V* translates for us and says:
(a) *V*: *U* said, 'Exactly one of us is a Tu.'
W: Don't believe *V*. He is a Fa.
(b) *V*: *U* said, 'Exactly one of us is a Fa.'
W: *V*'s statement is true.
(c) *V*: *U* said, 'All of us are Tu's.'
W: That is false. *V* is a Fa.
(d) *V*: *U* said, 'None of us are Tu's.'
W: If *V*'s statement is false, then *U* is a Tu.
(e) *V*: *U* said, '*V* and *W* are Tu's.'
W: *V*'s statement is false.

In each case, decide whether the race of *V* and *W* is uniquely determined.

(3) It is traditional on the Island of TUFA that only Tu's marry Tu's and Fa's marry Fa's. However, the Smiths broke this tradition. One was a Fa and the other a Tu, but they married anyway and had a child, called Smith Jr., who alternately tells the truth and lies in any sequence of statements. But it is not known, a priori, whether the first statement is true or false.
(a) We meet three children on the island—*X*, *Y*, and *Z*. We know one is Smith Jr., another is a Tu, and the third is a Fa, but we do not know which is which. The first child, *X*, makes three statements:

X: I am Smith Jr. *Y* is a Fa. *Z* is a Tu.

Is the race of each child uniquely determined?
(b) The Smiths have three more children—ϵ_1, ϵ_2, and ϵ_3. In any sequence of statements made by any one of these children, there is exactly one false statement. However, it is not known in advance which is the false statement.

One day the Smiths discovered that cookies were missing from their cookie jar and their children were the only possible culprits.

They were questioned on the matter, and they each made two statements:

ϵ_1: (1) ϵ_2 and ϵ_3 both took cookies.
(2) Smith Jr. also took cookies, but I did not.

ϵ_2: (1) I did not take any cookies, but Smith Jr. did.
(2) ϵ_1 took some cookies.

ϵ_3: (1) ϵ_2 took cookies, but ϵ_1 did not.
(2) I did not take any cookies.

Can it be determined which children took the cookies?

(c) The Smiths were concerned about their older son Smith Jr. because the school principal told them that he was not attending all his classes. They asked their three younger children, ϵ_1, ϵ_2, and ϵ_3, which classes Smith Jr. had attended that day. They responded as follows:

ϵ_1: He attended mathematics. He did not attend history. He did not attend biology.

ϵ_2: He attended history. He attended English. He did not attend mathematics.

ϵ_3: He attended mathematics. He attended history. He did not attend English.

Assuming each child told exactly one lie, which classes did Smith Jr. attend?

(d) One day when the Smiths' three children, ϵ_1, ϵ_2, and ϵ_3, came home from school, their parents asked them how they did on their mathematics exam. They each made three statements, exactly one of which was false:

ϵ_1: My grade was lower than ϵ_2's grade. ϵ_2's grade was 70. ϵ_3's grade was 30 more than ϵ_2's grade.

ϵ_2: My grade was 60. That is 20 points less than ϵ_3's grade. It is also 10 more than ϵ_1's grade.

ϵ_3: My grade was higher than ϵ_2's grade. It was 30 more than ϵ_1's grade. ϵ_1's grade was 70.

Determine, if possible, each child's exam grade.

(e) One day, the Smiths' children, ϵ_1, ϵ_2, and ϵ_3, go to the store to buy some candy. When they come home, their parents ask, "How many pieces of candy do you have?" They respond as follows:

ϵ_1: ϵ_2 has seven pieces. I have fewer pieces than ϵ_2. ϵ_3 has three more pieces than ϵ_2.

ϵ_2: I have six pieces. I have two fewer pieces than ϵ_3. I have one more piece than ϵ_1.

ϵ_3: I do not have the smallest number of pieces. I have three more pieces than ϵ_1. ϵ_1 has nine pieces.

Assuming each child told exactly one lie, determine, if possible, how many pieces of candy each child has.

(f) A traveler in the land of TUFA comes to a fork in the road. One fork goes to Tufaria and the other to Fatusic. The traveler wants to go to Tufaria, but he does not know which fork to take. There are three children playing by the side of the road; one is Smith Jr., the second is a Tu, and the third is a Fa, but the traveler does not know which child is which. He may ask one of the children only one question. What question should he ask to find the correct way to Tufaria, and which child should he ask? (Remember that a Tu always tells the truth, a Fa always lies, and Smith Jr. alternately tells the truth and lies, but it is not known in advance whether his first statement is true or false.)

Chapter 4

Applications of the Sentential Calculus

The applications we discuss in this chapter include Karnaugh maps, switching networks, and logic networks. First, however, it is convenient to have a *normal form* for formulas in the Sentential Calculus.

4.1 DISJUNCTIVE NORMAL FORM

We have seen that every formula built up from sentential variables and the sentential connectives $\neg$, $\wedge$, $\vee$, $\rightarrow$, and $\leftrightarrow$ gives rise to a truth table. Conversely, we can ask whether every truth table comes from such a formula. In other words, given a truth table, can we always find a formula that has this truth table? Let's look at several examples. The formula we are seeking will be denoted by φ.

(1)

P	Q	φ
T	T	T
T	F	T
F	T	T
F	F	T

Take φ to be $P \vee \neg P$.

(2)

P	Q	φ
T	T	F
T	F	F
F	T	F
F	F	F

Take φ to be $P \wedge \neg P$.

(3)

P	Q	φ
T	T	T
T	F	T
F	T	F
F	F	T

Take φ to be $(P \wedge Q) \vee (P \wedge \neg Q) \vee (\neg P \wedge \neg Q)$.

(4)

P	Q	R	φ
T	T	T	T
T	T	F	F
T	F	T	F
T	F	F	F
F	T	T	T
F	T	F	F
F	F	T	T
F	F	F	F

Take φ to be $(P \wedge Q \wedge R) \vee (\neg P \wedge Q \wedge R) \vee (\neg P \wedge \neg Q \wedge R)$.

Theorem 4.1 *Given a truth table, there is a formula in the Sentential Calculus that has the given truth table.*

Proof: Suppose the sentential variables in the truth table are $P_1, P_2, \ldots, P_n$. (Then the truth table has 2^n lines.) If the last column of the truth table is all F's, take the formula to be $P_1 \wedge \neg P_1$. Hence, we may assume that a T appears somewhere in the last column. For each such line, write down the conjunction $X_1 \wedge X_2 \wedge \cdots \wedge X_n$, where X_i is P_i if P_i has the value T on that line and X_i is $\neg P_i$ if P_i has the value F on that line. The required formula φ is the disjunction of these conjunctions. It is clear that φ has the given truth table. ■

Notice that the formula φ has a special form. It is said to be in *disjunctive normal form* (DNF).

Definition 4.1 A formula φ with sentential variables $P_1, P_2, \ldots, P_n$ is said to be in *disjunctive normal form* if it is a disjunction of conjunctions of the form

$$X_1 \wedge X_2 \wedge \cdots \wedge X_n,$$

where each X_i is either P_i or $\neg P_i$. $i = 1, 2, \ldots, n$.

It is now easy to prove Theorem 4.2.

Theorem 4.2 *Every formula of the Sentential Calculus that is not a contradiction is tautologically equivalent to a formula in DNF.*

Proof: Write down a truth table for the formula. Because it is not a contradiction, there must be at least one T in the truth table. Use the procedure described in Theorem 4.1 to find the DNF.

■

Example 4.1 Find the DNF of the formula φ: $((P \vee \neg Q) \to R) \leftrightarrow (Q \wedge \neg R)$.

Solution: First make a truth table for the formula φ:

P	Q	R	$(P \vee \neg Q) \to R$	$Q \wedge \neg R$	φ
T	T	T	T	F	F
T	T	F	F	T	F
T	F	T	T	F	F
T	F	F	F	F	T
F	T	T	T	F	F
F	T	F	T	T	T
F	F	T	T	F	F
F	F	F	F	F	T

Thus, we see φ is only true on lines 4, 6, and 8 of its truth table. Using the procedure described in the proof of Theorem 4.1, the DNF of φ is

$$(P \wedge \neg Q \wedge \neg R) \vee (\neg P \wedge Q \wedge \neg R) \vee (\neg P \wedge \neg Q \wedge \neg R).$$

4.2 ALGEBRA OF FORMULAS

In this section, we discuss the simplification of formulas. In many applications of the Sentential Calculus, it is desirable to find a formula tautologically equivalent to a given formula, but containing fewer connectives and/or fewer occurrences of sentential variables. For example, consider the formulas

$$\varphi : (P \vee P) \to (P \wedge Q)$$

and

$$\psi : P \to (P \wedge Q).$$

We know that φ is tautologically equivalent to ψ, $\varphi \Leftrightarrow \psi$, because $P \vee P \Leftrightarrow P$. Moreover, ψ is simpler than φ because it contains one fewer occurrence of the sentential variable P.

In order to discuss this type of simplification, we shall use another rule of inference of the Sentential Calculus that we shall call the Rule for Tautologically Equivalent Formulas, or Rule *TE*.

Rule TE (Rule for Tautologically Equivalent Formulas)

Suppose φ_1, φ_2, ψ_1, and ψ_2, are formulas in the Sentential Calculus such that ψ_1 is a subformula of φ_1, φ_2 is obtained from φ_1 by substituting ψ_2 for some occurrence of ψ_1, and $\psi_1 \Leftrightarrow \psi_2$, then $\varphi_1 \Leftrightarrow \varphi_2$. (A *subformula* of a formula is a consecutive sequence of symbols from the formula that itself is a formula. The subformulas of a formula φ are all the formulas that appear on some level of the tree for φ. See Section 1.6.)

The rationale behind Rule *TE* is intuitively clear. We are replacing a subformula, ψ_1, of a formula φ_1, by another formula, ψ_2, which has exactly the same truth values as ψ_1, to obtain a formula φ_2. Thus, φ_1 and φ_2 should have exactly the same truth values. (Actually, Rule *TE* is a special case of the Rule *LE*, which we will study in Chapters 10 and 16.)

Example 4.2

Give a formal proof that each of the following formulas are tautologies:

(a) $\varphi_1 \leftrightarrow [P \rightarrow (P \wedge Q)]$, where φ_1 is $[(P \vee P) \rightarrow (P \wedge Q)]$.

(b) $\varphi_2 \leftrightarrow (\neg P \vee Q)$, where φ_2 is $[P \rightarrow (P \wedge Q)]$.

(c) $\varphi_3 \leftrightarrow [(P \vee Q) \wedge R]$, where φ_3 is $[[(P \wedge (Q \wedge R)) \vee (\neg P \wedge (Q \wedge R))] \vee (P \wedge (\neg Q \wedge R))]$.

Solution: Following each use of Rule *T* and *TE*, we shall give in parentheses the number of the tautology that was used. (See the list of useful tautologies at the end of Chapter 2.)

(a)		1.	$\varphi_1 \leftrightarrow [P \rightarrow (P \wedge Q)]$	*TE* (19)
(b)		1.	$\varphi_2 \leftrightarrow [\neg P \vee (P \wedge Q)]$	*TE* (17)
		2.	$\varphi_2 \leftrightarrow [(\neg P \vee P) \wedge (\neg P \vee Q)]$	1 *TE* (14)
		3.	$\varphi_2 \leftrightarrow [(\neg P \vee Q) \wedge (\neg P \vee P)]$	2 *TE* (12)
		4.	$[(\neg P \vee Q) \wedge (\neg P \vee P)] \rightarrow (\neg P \vee Q)$	*T* (4)
	{5}	5.	$\neg P \vee Q$	*P* (for *CP*)
		6.	$P \vee \neg P$	*T* (10)
		7.	$\neg P \vee P$	6 *T* (12)
	{5}	8.	$(\neg P \vee Q) \wedge (\neg P \vee P)$	5, 7 *T* (6)
		9.	$(\neg P \vee Q) \rightarrow [(\neg P \vee Q) \wedge (\neg P \vee P)]$	5, 8 *CP*
		10.	$[(\neg P \vee Q) \wedge (\neg P \vee P)] \leftrightarrow (\neg P \vee Q)$	4, 9 *T* (18)
		11.	$\varphi_2 \leftrightarrow (\neg P \vee Q)$	3, 10 *TE*

(c) In this derivation, we shall combine lines that use the commutative rules (12) and the associative rules (13) with other lines. (As an exercise, extend the proof so that each use of one of these rules appears on a separate line). In the proof, we shall first simplify the first two disjuncts.

1.	$[(P \wedge (Q \wedge R)) \vee (\neg P \wedge (Q \wedge R))]$ $\leftrightarrow [(P \vee \neg P) \wedge (Q \wedge R)]$	*TE* (12, 14)
2.	$[(P \wedge (Q \wedge R)) \vee (\neg P \wedge (Q \wedge R))]$ $\leftrightarrow (Q \wedge R)$	*TE* (*)
3.	$\varphi_3 \leftrightarrow [(Q \wedge R) \vee (P \wedge (\neg Q \wedge R))]$	2 *TE*
4.	$\varphi_3 \leftrightarrow [(Q \vee (P \wedge \neg Q)) \wedge R]$	3 *TE* (12, 13, 14)
5.	$\varphi_3 \leftrightarrow [((Q \vee P) \wedge (Q \vee \neg Q)) \wedge R]$	4 *TE* (14)
6.	$\varphi_3 \leftrightarrow [(Q \vee P) \wedge R]$	5 *TE* (*)
7.	$\varphi_3 \leftrightarrow [(P \vee Q) \wedge R]$	6 *TE* (12)

The tautology above that is indicated by "(*)" was actually proved in lines 4 through 10 of part (b). That is, $[(P \vee \neg P) \wedge Q] \Leftrightarrow Q$ (or equivalently, $[Q \wedge (\neg P \vee P)] \Leftrightarrow Q$), for all formulas P and Q.

We can see from the preceding examples that if our list of tautologies is extended to include tautologies like those indicated by a (*) in part (c) of Example 4.2, we can simplify logical expressions using Rule *TE* similarly to the way we simplify algebraic expressions using algebraic cancellation rules, distributive rules, and so forth. With this in mind, we list in Figure 4.1 those tautological equivalences that are most useful. In this list, we use the symbol $\top$ for a tautology and $\bot$ for a contradiction.

1. $(P \vee \neg P) \Leftrightarrow \top$
2. $(P \wedge \neg P) \Leftrightarrow \bot$
3. $\neg\top \Leftrightarrow \bot$
4. $\neg\bot \Leftrightarrow \top$
5. $(P \vee \top) \Leftrightarrow \top$
6. $(P \vee \bot) \Leftrightarrow P$
7. $(P \wedge \top) \Leftrightarrow P$
8. $(P \wedge \bot) \Leftrightarrow \bot$
9. $(P \vee P) \Leftrightarrow P$
10. $(P \wedge P) \Leftrightarrow P$
11. $(P \vee Q) \Leftrightarrow (Q \vee P)$
12. $(P \wedge Q) \Leftrightarrow (Q \wedge P)$
13. $\neg(\neg P) \Leftrightarrow P$
14. $[P \vee (P \wedge Q)] \Leftrightarrow P$
15. $[P \wedge (P \vee Q)] \Leftrightarrow P$
16. $[P \vee (Q \vee R)] \Leftrightarrow [(P \vee Q) \vee R]$
17. $[P \wedge (Q \wedge R)] \Leftrightarrow [(P \wedge Q) \wedge R]$
18. $[(P \vee Q) \wedge (P \vee R)] \Leftrightarrow [P \vee (Q \wedge R)]$
19. $[(P \wedge Q) \vee (P \wedge R)] \Leftrightarrow [P \wedge (Q \vee R)]$
20. $\neg(P \vee Q) \Leftrightarrow (\neg P \wedge \neg Q)$
21. $\neg(P \wedge Q) \Leftrightarrow (\neg P \vee \neg Q)$
22. $[P \vee (\neg P \wedge Q)] \Leftrightarrow (P \vee Q)$
23. $(P \rightarrow Q) \Leftrightarrow (\neg P \vee Q)$
24. $(P \leftrightarrow Q) \Leftrightarrow [(P \wedge Q) \vee (\neg P \wedge \neg Q)]$
25. $(P \leftrightarrow Q) \Leftrightarrow [(\neg P \vee Q) \wedge (P \vee \neg Q)]$

Figure 4.1

Remember that when a tautological equivalence given in Figure 4.1 is used to simplify a formula, any formula may be substituted for "P," "Q," and "R" as long as the same formula is substituted for each occurrence of the same letter. For example,

$$[(\neg A \wedge B) \wedge ((\neg A \wedge B) \vee (A \vee B \rightarrow \neg C))] \Leftrightarrow (\neg A \wedge B)$$

is a substitution instance of 15 in Figure 4.1.

In the next example, we shall give informal proofs of tautological equivalences using Rule *TE* and Figure 4.1.

Example 4.3

Give an informal proof of the following tautological equivalences:

(a) $[(A \wedge \neg B \wedge C) \vee (\neg A \wedge \neg B \wedge C) \vee (B \wedge C)] \Leftrightarrow C$

(b) $[(A \wedge (\neg B \vee C)) \vee (\neg A \wedge C) \vee ((A \vee \neg B) \wedge \neg A)]$
$\Leftrightarrow (\neg B \vee C)$

Solution: In parentheses following each line, we give the number(s) of the tautological equivalences from Figure 4.1 that were used. We shall usually omit mention of the commutative and associative rules (11, 12, 16, and 17).

(a) $[(A \wedge \neg B \wedge C) \vee (\neg A \wedge \neg B \wedge C) \vee (B \wedge C)]$

$\Leftrightarrow [((A \vee \neg A) \wedge (\neg B \wedge C)) \vee (B \wedge C)]$ (19)

$\Leftrightarrow [(\neg B \wedge C) \vee (B \wedge C)]$ (1, 7)

$\Leftrightarrow [(\neg B \vee B) \wedge C]$ (19)

$\Leftrightarrow C$ (1, 7)

(b) $[(A \wedge (\neg B \vee C)) \vee (\neg A \wedge C) \vee ((A \vee \neg B) \wedge \neg A)]$

$\Leftrightarrow [(A \wedge \neg B) \vee (A \wedge C) \vee (\neg A \wedge C) \vee$
$(A \wedge \neg A) \vee (\neg B \wedge \neg A)]$ (19)

$\Leftrightarrow [(A \wedge \neg B) \vee (A \wedge C) \vee (\neg A \wedge C) \vee$
$(\neg B \wedge \neg A)$ (2, 6)

$\Leftrightarrow [((A \wedge \neg B) \vee (A \wedge C)) \vee ((\neg A \wedge \neg B) \vee$
$(\neg A \wedge C))]$ (11, 12)

$\Leftrightarrow [(A \wedge (\neg B \vee C)) \vee (\neg A \wedge (\neg B \vee C))]$ (19)

$\Leftrightarrow [(A \vee \neg A) \wedge (\neg B \vee C)]$ (19)

$\Leftrightarrow (\neg B \vee C)$ (1, 7)

Normally, to simplify a formula that contains $\rightarrow$ or $\leftrightarrow$, first use one of the tautological equivalences from 23 through 25 to obtain a tautologically equivalent formula that only contains $\neg$, $\vee$, and $\wedge$.

4.3 KARNAUGH MAPS

A *Karnaugh map*, or *K-map*, is a pictorial method for simplifying formulas of the Sentential Calculus in disjunctive normal form. We shall only use it here for formulas that have at most four distinct sentential variables, but it can also be used if there are more. (See the references at the end of this section.)

Let us start with formulas with two sentential variables, say A and B. To save space and to be consistent with engineering notation, we shall use the following notation in this section and the next two sections:

$$\overline{P} \text{ for } \neg P,$$
$$PQ \text{ for } P \wedge Q,$$

and in general,

$$P_1 P_2 \ldots P_n \text{ for } P_1 \wedge P_2 \wedge \cdots \wedge P_n.$$

A K-map is a condensed truth table. As an example, let φ be the formula $AB \vee A\overline{B}$ and ψ, $A \vee \overline{A}B$. Then φ is represented in the K-map in Figure 4.2 and ψ in Figure 4.3.

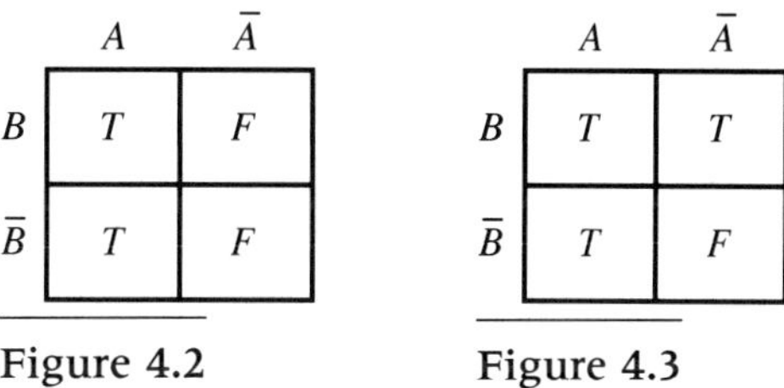

Figure 4.2 **Figure 4.3**

A K-map is read like a multiplication table. Formula φ is true iff either AB is true or $A\overline{B}$ is true, so we put T's in the squares that represent AB and $A\overline{B}$ and F's in the other squares (Fig. 4.2). Similarly, ψ is true iff A is true or $\overline{A}B$ is true, so in the K-map for ψ, we put T's in every square under the A and also in the square corresponding to $\overline{A}B$ (Fig. 4.3.)

Each square in a K-map represents a *fundamental conjunction*. A conjunction $X_1 X_2 \ldots X_n$ is called a *fundamental conjunction* if for each i, X_i is P_i or X_i is $\overline{P}_i$, where $P_1, P_2, \ldots, P_n$ are distinct sentential variables.

Figure 4.3 would be interpreted as the formula

$$AB \vee A\overline{B} \vee \overline{A}B. \tag{1}$$

However, (1) is tautologically equivalent to $A \vee \overline{A}B$.

By *adjacent* squares in a K-map, we mean squares whose fundamental conjunctions differ by one symbol. (We usually construct K-maps so that squares that have a side in common are adjacent, but, as we will see below, there may be additional adjacent squares.) For example, in Figure 4.3, the fundamental conjunctions AB and $\overline{A}B$ lie in adjacent squares and so do AB and $A\overline{B}$.

The following theorem is useful when using K-maps.

Theorem 4.3 *Suppose P and Q are each fundamental conjunctions of $n \geq 2$ symbols. Suppose, also, that P and Q lie in adjacent squares of a K-map. Then $P \vee Q$ is tautologically equivalent to a fundamental conjunction of $n - 1$ symbols. In particular, if*

(i) $P \Leftrightarrow P_1P_2 \ldots P_{n-1}R$

and

(ii) $Q \Leftrightarrow P_1P_2 \ldots P_{n-1}\overline{R}$,

then

(iii) $(P \vee Q) \Leftrightarrow P_1P_2 \ldots P_{n-1}$.

Proof: There is no loss of generality in assuming (i) and (ii). (Why?) Then (iii) follows by using the Rule *TE* and formulas 19, 1, and 7 of Figure 4.1. ■

Theorem 4.3 provides a means for simplifying formulas using *K*-maps. To aid in the simplification, we shall put loops around the *T*'s in adjacent squares of a *K*-map. The *K*-map for $AB \vee A\overline{B}$ is given in Figure 4.4, and that for $AB \vee A\overline{B} \vee \overline{A}B$ is given in Figure 4.5.

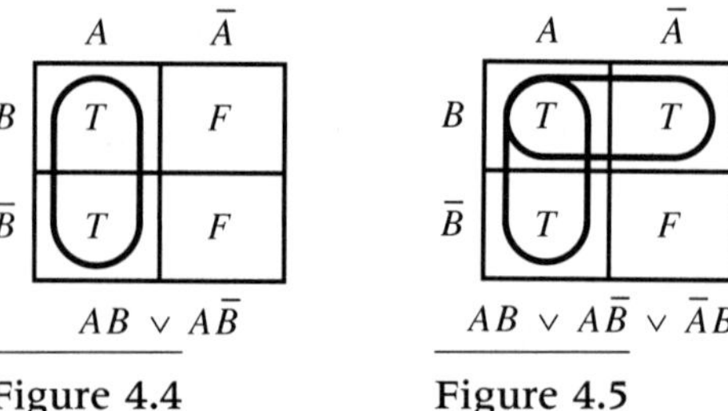

Figure 4.4 **Figure 4.5**

Using Figure 4.4 and Theorem 4.3, we see that

$$(AB \vee A\overline{B}) \Leftrightarrow A.$$

In Figure 4.5, we have a choice. If we use the loop around the first column, the simplification is

(*)
$$\begin{aligned}(AB \vee A\overline{B} \vee \overline{A}B) &\Leftrightarrow ((AB \vee A\overline{B}) \vee \overline{A}B)\\ &\Leftrightarrow (A \vee \overline{A}B).\end{aligned}$$

On the other hand, if we use the loop around the first row we obtain:

(**)
$$\begin{aligned}(AB \vee A\overline{B} \vee \overline{A}B) &\Leftrightarrow ((AB \vee \overline{A}B) \vee A\overline{B})\\ &\Leftrightarrow (B \vee A\overline{B}).\end{aligned}$$

Finally, if we use both loops:

(***)
$$\begin{aligned}(AB \vee A\overline{B} \vee \overline{A}B) &\Leftrightarrow ((AB \vee A\overline{B}) \vee (AB \vee \overline{A}B))\\ &\Leftrightarrow (A \vee B).\end{aligned}$$

It is clear that (***) is simpler than either of the other two. *The first rule in diagramming K-maps is that every T for which it is possible should be included in a loop.*

Next, we shall consider the case of three sentential variables A, B, and C. The form for the K-map is given in Figure 4.6.

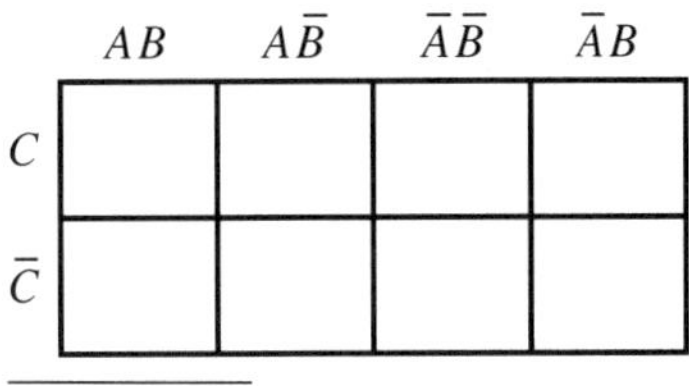

Figure 4.6

Similarly, the form for four sentential variables A, B, C, and D is given in Figure 4.7.

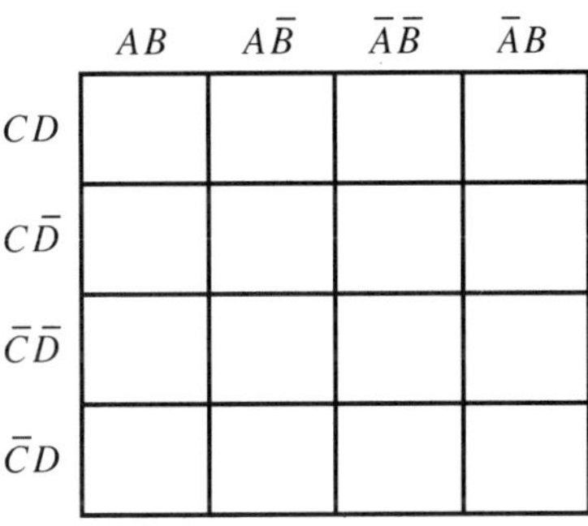

Figure 4.7

In Figure 4.6, each square represents a fundamental conjunction of three symbols, whereas in Figure 4.7, each square represents a fundamental conjunction of four symbols. For example, the formula

$$AB \vee \bar{A}C \vee BC$$

is diagrammed in Figure 4.8, and the formula

$$AD \vee ABCD \vee AB\bar{C}D \vee \bar{A}BCD \vee \bar{A}B\bar{C}D$$

is diagrammed in Figure 4.9. Note that to diagram the conjunction AB in Figure 4.8 we put a T in all squares where the fundamental conjunction contains A and B as conjuncts; similarly, for the conjunctions $\bar{A}C$ and BC. Also, in Figure 4.9, to diagram AD, we put a T in all squares that have an A and a D in their fundamental conjunctions. (These formulas will be simplified in Example 4.4(b) and (d).)

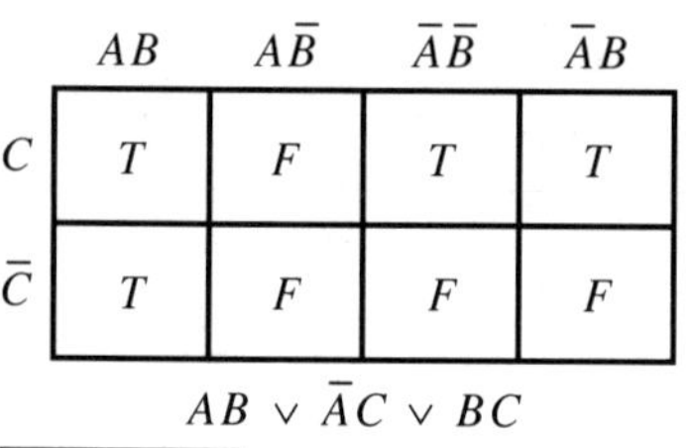

$AB \vee \bar{A}C \vee BC$

Figure 4.8

	AB	$A\bar{B}$	$\bar{A}\bar{B}$	$\bar{A}B$
CD	T	T	F	T
$C\bar{D}$	F	F	F	F
$\bar{C}\bar{D}$	F	F	F	F
$\bar{C}D$	T	T	F	T

$AD \vee ABCD \vee AB\bar{C}D \vee \bar{A}BCD \vee \bar{A}B\bar{C}D$

Figure 4.9

To aid us in simplifying K-maps like those given in Figures 4.8 and 4.9, the following theorem is useful.

Theorem 4.4 *Suppose P_1, P_2, P_3, and P_4 are each fundamental conjunctions of $n \geq 3$ symbols. Suppose also that either P_1, P_2, P_3, and P_4 lie in a row (or column) of a K-map or their squares form a square array of adjacent squares in a K-map. Then their disjunction, $P_1 \vee P_2 \vee P_3 \vee P_4$, is tautologically equivalent to a fundamental conjunction of $n - 2$ symbols. In particular, if*

(i) $P_1 \Leftrightarrow Q_1Q_2 \ldots Q_{n-2}RS,$

(ii) $P_2 \Leftrightarrow Q_1Q_2 \ldots Q_{n-2}R\bar{S},$

(iii) $P_3 \Leftrightarrow Q_1Q_2 \ldots Q_{n-2}\bar{R}\,\bar{S},$

(iv) $P_4 \Leftrightarrow Q_1Q_2 \ldots Q_{n-2}\bar{R}S,$

then

(v) $(P_1 \vee P_2 \vee P_3 \vee P_4) \Leftrightarrow Q_1Q_2 \ldots Q_{n-2}.$

Proof: There is no loss of generality in assuming (i) through (iv) because, by hypothesis, P_1, P_2, P_3, and P_4 lie in a row, column, or square. Let P be the fundamental conjunction $Q_1Q_2 \ldots Q_{n-2}$. Then

$$\begin{aligned}(P_1 \vee P_2 \vee P_3 \vee P_4) &\Leftrightarrow PR(S \vee \bar{S}) \vee P\bar{R}(S \vee \bar{S}) \\ &\Leftrightarrow PR \vee P\bar{R} \\ &\Leftrightarrow P(R \vee \bar{R}) \\ &\Leftrightarrow P.\end{aligned}$$

(The student should fill in the details.) ■

Example 4.4

Simplify each of the following formulas using K-maps:

(a) $\varphi_1 : ABC \vee AB\bar{C} \vee A\bar{B}C \vee A\bar{B}\bar{C}$

(b) $\varphi_2 : AB \vee \bar{A}C \vee BC$

(c) $\varphi_3 : A\bar{B}CD \vee A\bar{B}C\bar{D} \vee A\bar{B}\bar{C}\bar{D} \vee A\bar{B}\bar{C}D$

(d) $\varphi_4 : AD \vee ABCD \vee AB\bar{C}D \vee \bar{A}BCD \vee \bar{A}B\bar{C}D$

(e) $\varphi_5 : ABCD \vee \bar{A}\bar{B}CD \vee \bar{A}B\bar{C}\bar{D} \vee \bar{A}B\bar{C}D \vee AB\bar{C}D \vee \bar{A}\bar{B}\bar{C}D$

(f) $\varphi_6 : \bar{A}\bar{B}CD \vee ABC\bar{D} \vee A\bar{B}C\bar{D} \vee \bar{A}\bar{B}C\bar{D} \vee A\bar{B}\bar{C}\bar{D} \vee A\bar{B}\bar{C}D \vee \bar{A}\bar{B}\bar{C}\bar{D} \vee \bar{A}B\bar{C}\bar{D}$

(g) $\varphi_7 : ABCD \vee A\bar{B}CD \vee ABC\bar{D} \vee A\bar{B}C\bar{D} \vee \bar{A}\bar{B}C\bar{D} \vee A\bar{B}\bar{C}\bar{D} \vee \bar{A}\bar{B}\bar{C}\bar{D} \vee A\bar{B}\bar{C}D.$

Solution:

(a) We give the K-map for φ_1 in Figure 4.10:

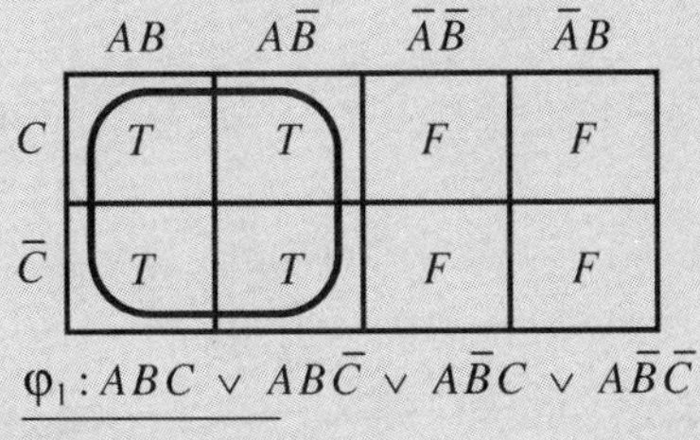

$\varphi_1 : ABC \vee AB\bar{C} \vee A\bar{B}C \vee A\bar{B}\bar{C}$

Figure 4.10

Simplifying using Theorem 4.4, we get

$$\varphi_1 \Leftrightarrow A.$$

(b) The *K*-map for φ_2 is given in Figure 4.11.

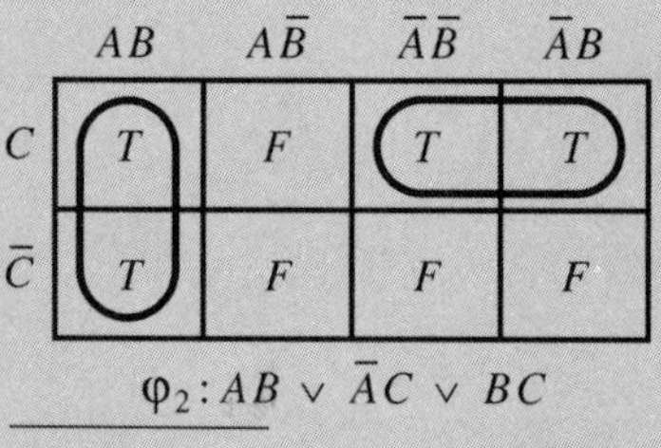

$\varphi_2 : AB \vee \bar{A}C \vee BC$

Figure 4.11

Using Theorem 4.3, we obtain

$$\varphi_2 \Leftrightarrow [(ABC \vee AB\bar{C}) \vee (\bar{A}\bar{B}C \vee \bar{A}BC)]$$
$$\Leftrightarrow (AB \vee \bar{A}C).$$

(c) The *K*-map for φ_3 is in Figure 4.12:

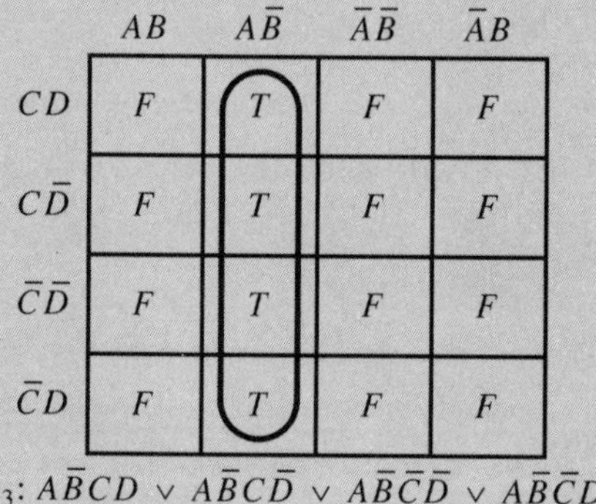

φ_3: $A\bar{B}CD \vee A\bar{B}C\bar{D} \vee A\bar{B}\bar{C}\bar{D} \vee A\bar{B}\bar{C}D$

Figure 4.12

It follows from Theorem 4.4 that

$$\varphi_3 \Leftrightarrow A\bar{B}.$$

(d) We map φ_4 in Figure 4.13. Remember that to map *AD* we put a *T* in all squares that have both an *A* and a *D* in their fundamental conjunctions. Also note that the squares $ABCD$, $A\bar{B}CD$, $AB\bar{C}D$, and $A\bar{B}\bar{C}D$ form a square array of adjacent squares; and that the squares $ABCD$, $\bar{A}BCD$, $AB\bar{C}D$, and $\bar{A}B\bar{C}D$ also form a square array. The corresponding loops are indicated in the *K*-map in Figure 4.13. Thus, using Theorem 4.4, we obtain

$$\varphi_4 \Leftrightarrow [(ABCD \vee A\bar{B}CD \vee AB\bar{C}D \vee A\bar{B}\bar{C}D)$$
$$\vee (ABCD \vee \bar{A}BCD \vee AB\bar{C}D \vee \bar{A}B\bar{C}D)]$$
$$\Leftrightarrow (AD \vee BD).$$

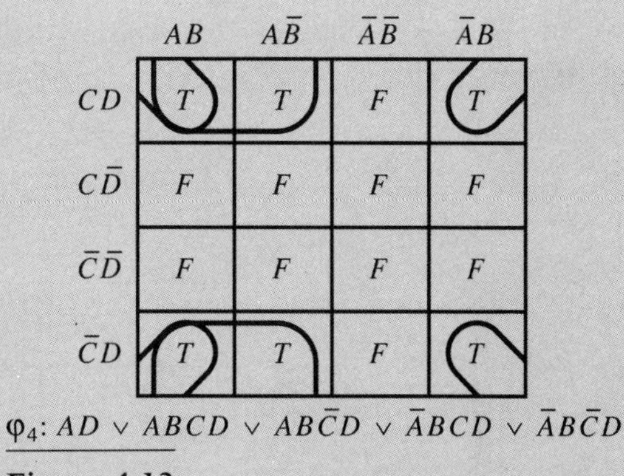

φ_4: $AD \vee ABCD \vee AB\bar{C}D \vee \bar{A}BCD \vee \bar{A}B\bar{C}D$

Figure 4.13

On the other hand, if we avoid overlapping loops and use the fact, as before, that $ABCD$, $A\bar{B}CD$, $AB\bar{C}D$, and $A\bar{B}\,\bar{C}D$ form a square array, but $\bar{A}BCD$ and $\bar{A}\,B\bar{C}D$ are adjacent, we obtain the simplification.

$$\varphi_4 \Leftrightarrow [(ABCD \vee A\bar{B}CD \vee AB\bar{C}D \vee A\bar{B}\bar{C}D) \vee (\bar{A}BCD \vee \bar{A}B\bar{C}D)]$$
$$\Leftrightarrow (AD \vee \bar{A}BD).$$

Clearly, $AD \vee \bar{A}BD$ is not as simple as $AD \vee BD$. Therefore, when using K-maps, *combine T's in as large a loop as possible, even if some loops overlap.*

(e) See Figure 4.14 for the K-map of φ_5:

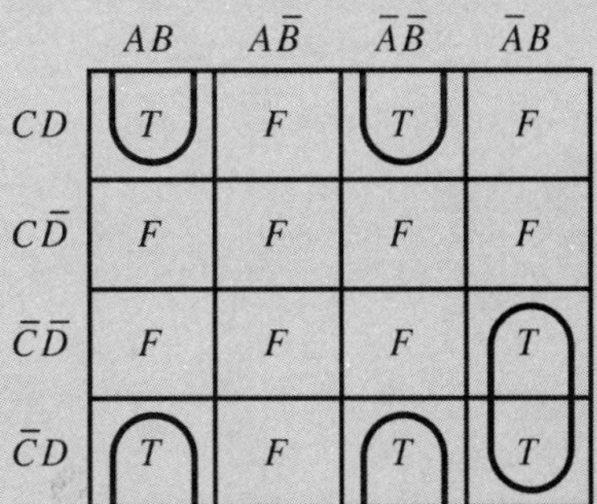

φ_5: $ABCD \vee \bar{A}\bar{B}CD \vee \bar{A}B\bar{C}\bar{D} \vee \bar{A}B\bar{C}D \vee AB\bar{C}D \vee \bar{A}\bar{B}\bar{C}D$

Figure 4.14

In Figure 4.14, there are other ways to put loops around adjacent pairs. For example, $\bar{A}\,\bar{B}\bar{C}D$ and $\bar{A}B\bar{C}D$ are adjacent. However, if we put a loop around these two squares, then we could not combine $\bar{A}B\bar{C}\bar{D}$ with anything. *Thus, another rule for using K-maps is to start by combining those fundamental conjunctions for which there is only one possibility.* Using Theorem 4.3 and Figure 4.14, we obtain

$$\varphi_5 \Leftrightarrow ABD \vee \bar{A}\bar{B}D \vee \bar{A}B\bar{C}.$$

(f) The formula φ_6 is mapped in Figure 4.15:

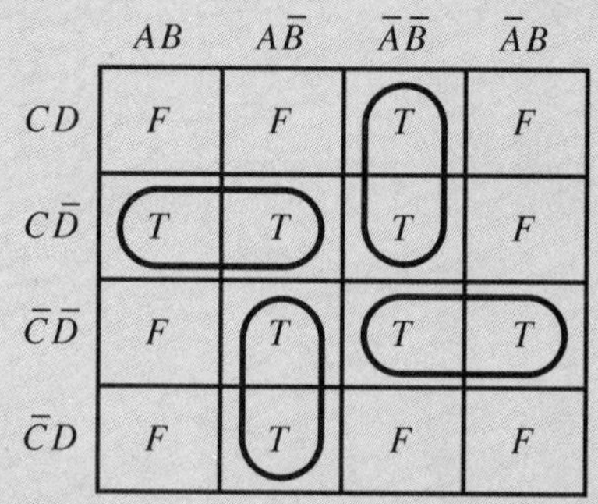

	AB	$A\bar{B}$	$\bar{A}\bar{B}$	$\bar{A}B$
CD	F	F	T	F
$C\bar{D}$	T	T	T	F
$\bar{C}\bar{D}$	F	T	T	T
$\bar{C}D$	F	T	F	F

φ_6: $\bar{A}\bar{B}CD \vee ABC\bar{D} \vee A\bar{B}C\bar{D} \vee \bar{A}\bar{B}C\bar{D} \vee A\bar{B}\bar{C}\bar{D} \vee A\bar{B}\bar{C}D \vee \bar{A}\bar{B}\bar{C}\bar{D} \vee \bar{A}B\bar{C}\bar{D}$

Figure 4.15

It is tempting to put a loop around the middle square array, but then there would be four fundamental conjunctions ($ABC\bar{D}$, $\bar{A}\bar{B}C\bar{D}$, $\bar{A}B\bar{C}\bar{D}$, and $A\bar{B}\bar{C}D$) with no loops around them. This would violate the rule stated in part (e). Notice that after we combine those fundamental conjunctions for which there is only one possibility, all the *T*'s are included in some loop, so there is no need to add the loop around the middle four *T*'s. Using Figure 4.15 and Theorem 4.3, we obtain

(*) $$\varphi_6 \Leftrightarrow AC\bar{D} \vee \bar{A}\bar{B}C \vee \bar{A}\bar{C}\bar{D} \vee A\bar{B}\bar{C}.$$

(On the other hand, if we did put a loop around the middle four *T*'s and used Theorem 4.4, we would obtain

$$\varphi_6 \Leftrightarrow \bar{B}\bar{D} \vee ABC\bar{D} \vee \bar{A}\bar{B}CD \vee \bar{A}B\bar{C}\bar{D} \vee A\bar{B}\bar{C}D,$$

and this is clearly not as simple as (*).)

(g) See Figure 4.16 for the *K*-map of φ_7:

We follow the rule of combining those fundamental conjunctions for which there are no other possibilities, and we combine them in as large a loop as possible. In this case, a loop of size four. We obtain the *K*-map in Figure 4.16. Thus, simplifying φ_7, we obtain

$$\begin{aligned}\varphi_7 \Leftrightarrow\ & [(ABCD \vee A\bar{B}CD \vee ABC\bar{D} \vee A\bar{B}C\bar{D}) \vee \\ & (A\bar{B}C\bar{D} \vee \bar{A}\bar{B}C\bar{D} \vee A\bar{B}\bar{C}\bar{D} \vee \bar{A}\bar{B}\bar{C}\bar{D}) \vee \\ & (A\bar{B}CD \vee A\bar{B}C\bar{D} \vee A\bar{B}\bar{C}\bar{D} \vee A\bar{B}\bar{C}D)] \\ \Leftrightarrow\ & (AC \vee \bar{B}\bar{D} \vee A\bar{B}).\end{aligned}$$

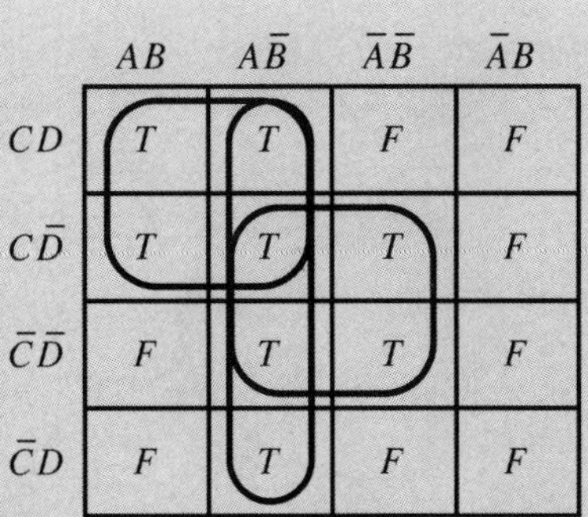

$\varphi_7: ABCD \vee A\bar{B}CD \vee ABC\bar{D} \vee A\bar{B}C\bar{D} \vee \bar{A}\bar{B}C\bar{D} \vee A\bar{B}\bar{C}\bar{D} \vee \bar{A}\bar{B}\bar{C}\bar{D} \vee A\bar{B}\bar{C}D$

Figure 4.16

Summarizing the rules for combining *T*'s in *K*-maps:

1. Start making loops around those *T*'s for which there is only one possibility, and make the loops as large as possible. (In our case, either size two, four, or eight. See Exercise 4 G.)
2. Continue making loops as large as possible around *T*'s until as many *T*'s as possible are included in some loop.

Let us look at one more example to illustrate these rules.

Example 4.5 Use a *K*-map to simplify

$$\varphi: B\bar{C}\bar{D} \vee \bar{A}B \vee A\bar{B} \vee \bar{B}D \vee \bar{B}C.$$

Solution: First, in Figure 4.17, we map φ.

	AB	$A\bar{B}$	$\bar{A}\bar{B}$	$\bar{A}B$
CD	*F*	*T*	*T*	*T*
$C\bar{D}$	*F*	*T*	*T*	*T*
$\bar{C}\bar{D}$	*T*	*T*	*F*	*T*
$\bar{C}D$	*F*	*T*	*T*	*T*

Figure 4.17

In the diagram in Figure 4.17, we see that $A\bar{B}\bar{C}\bar{D}$ can only be included in one loop of size four. Let us draw that first (Fig. 4.18):

	AB	$A\bar{B}$	$\bar{A}\bar{B}$	$\bar{A}B$
CD	F	T	T	T
$C\bar{D}$	F	T	T	T
$\bar{C}\bar{D}$	T	T	F	T
$\bar{C}D$	F	T	T	T

Figure 4.18

If we do not overlap with the T's inside the loop in Figure 4.18, we see that each of $\bar{A}\bar{B}C\bar{D}$ and $\bar{A}\bar{B}\bar{C}D$ can be included in only one loop of size four. We add these two loops to Figure 4.18 and obtain Figure 4.19:

	AB	$A\bar{B}$	$\bar{A}\bar{B}$	$\bar{A}B$
CD	F	T	T	T
$C\bar{D}$	F	T	T	T
$\bar{C}\bar{D}$	T	T	F	T
$\bar{C}D$	F	T	T	T

Figure 4.19

The largest loop that includes $AB\bar{C}\bar{D}$ is a loop of size two. We have two choices for such a loop. We can combine $AB\bar{C}\bar{D}$ with $\bar{A}B\bar{C}\bar{D}$ or with $A\bar{B}\bar{C}\bar{D}$. Because the latter is already included in a loop, we shall combine $AB\bar{C}\bar{D}$ with $\bar{A}B\bar{C}\bar{D}$. The completed K-map is shown in Figure 4.20:
Thus, using Figure 4.20, we have,

$$\varphi \Leftrightarrow [(AB\bar{C}\bar{D} \vee \bar{A}B\bar{C}\bar{D}) \vee (A\bar{B}CD \vee A\bar{B}C\bar{D} \vee A\bar{B}\bar{C}\bar{D} \vee A\bar{B}\bar{C}D) \vee$$
$$(\bar{A}\bar{B}CD \vee \bar{A}BCD \vee \bar{A}\bar{B}C\bar{D} \vee \bar{A}BC\bar{D}) \vee (\bar{A}\bar{B}CD \vee \bar{A}BCD \vee \bar{A}\bar{B}\bar{C}D \vee \bar{A}B\bar{C}D)]$$

(*) $$\Leftrightarrow (B\bar{C}\bar{D} \vee A\bar{B} \vee \bar{A}C \vee \bar{A}D).$$

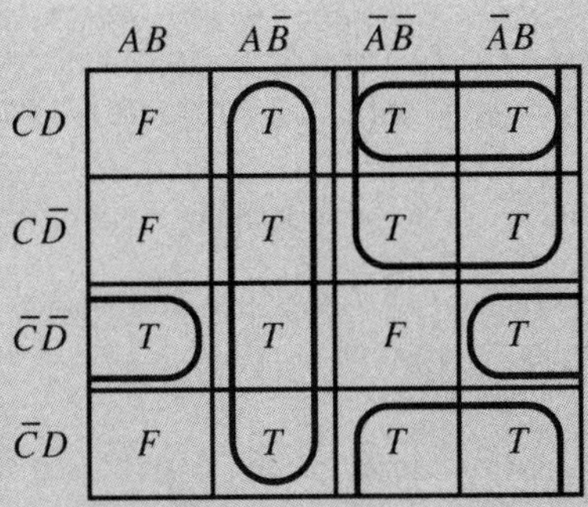

Figure 4.20

However, Figure 4.20 was not the only alternative. We could have started with Figure 4.17 by noting that $\bar{A}B\bar{C}\bar{D}$ can only be included in one loop of size four. Then, continuing in a similar manner, we would have obtained Figure 4.21:

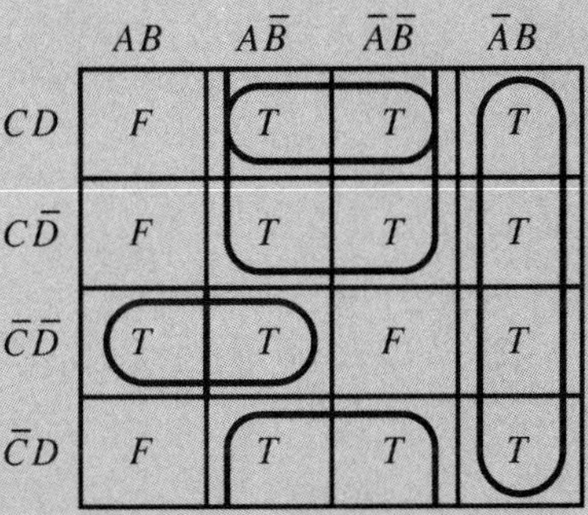

Figure 4.21

Simplifying φ using Figure 4.21,

(**) $$\varphi \Leftrightarrow A\bar{C}\bar{D} \vee \bar{A}B \vee \bar{B}C \vee \bar{B}D.$$

Comparing (*) and (**), we see that neither is simpler than the other. In this case, we have at least two equally good simplifications. The best *K*-map for a given formula need not be unique.

The advantages of a *K*-map are: first, it gives us the DNF of the formula (if we do not already have it); second, it tells us how to group the disjuncts (sometimes with repetitions when loops overlap) most efficiently for the simplification.

(For additional information about *K*-maps, see *Introduction to Switching Theory and Logical Design* by J. Hill and G. Peterson, John Wiley & Sons, New

York, 1968; *Boolean Algebra and Switching Circuits* by E. Mendelson, Schaum's Outline Series, McGraw Hill, 1970; *Little Bits of Digital Wisdom*, 10th ed., a Lecture-Workbook for EE 266, by David G. Meyer, Purdue University, 1984.)

4.4 SWITCHING NETWORKS

A switching network consists of a number of wires that are capable of carrying current along with a number of switches, each of which is capable of creating a gap in the wire. The simplest switching network is a single wire with no switches. Current can flow through the wire with no impediment. A little less simple is one wire and one switch.

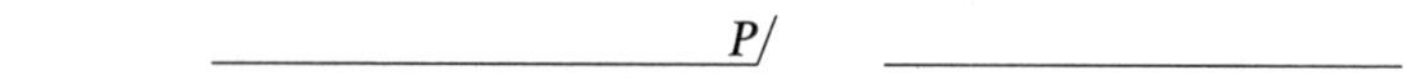

When P is closed, current can flow, but when P is open, it creates a gap, so current cannot flow. In general, current will flow through a switching network if there is a way to get through the network through closed switches.

In Figure 4.22, we show n switches *in series*, and in Figure 4.23, n switches *in parallel*.

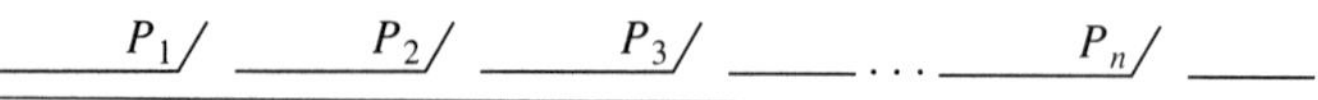

Figure 4.22. Switches in series.

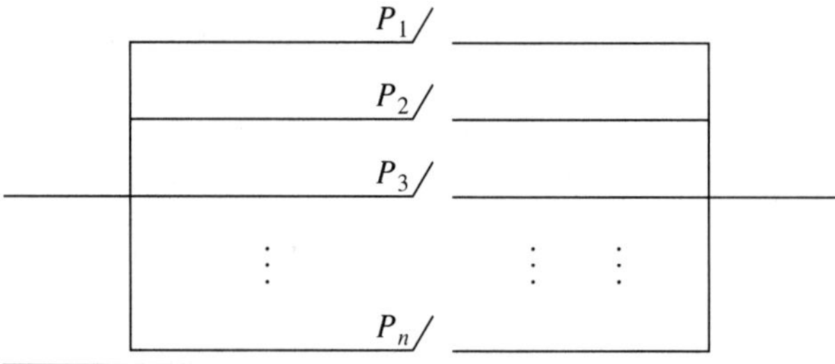

Figure 4.23. Switches in parallel.

Sometimes there is a relation between switches. They may close together and open together, in which case we shall use the same symbol for each switch. When one is closed, the other is open, and vice versa, we shall call one switch P and the other $\overline{P}$. In the network in Figure 4.24, for example, the two switches labeled with a "P" are either both open or both closed; if the switch labeled "Q" is open, the one labeled "$\overline{Q}$" is closed, and vice versa; similarly, if the switch labeled "R" is open, the one labeled "$\overline{R}$" is closed, and vice versa.

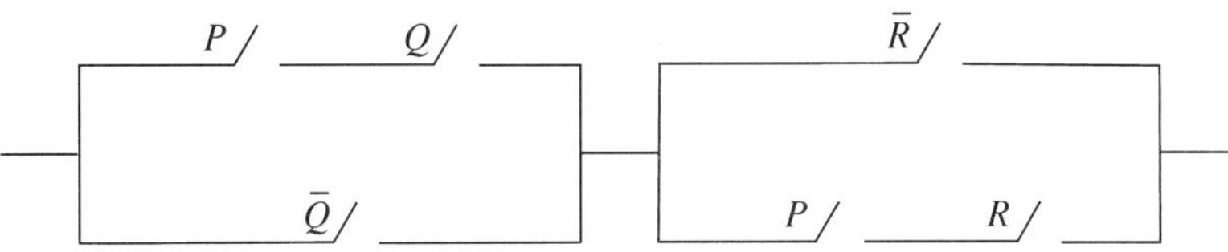

Figure 4.24

Switching networks can be represented by formulas in the Sentential Calculus. In Figure 4.22, for example, current will flow through the network iff all switches P_1, P_2, P_3, . . . , P_n are closed. If we match a closed switch P_i with a truth value of T to the sentential variable P_i and an open switch P_i to the truth value of F to P_i, we see that current flows through the network given in Figure 4.22 iff the formula $P_1P_2P_3 \ldots P_n$ has the truth value T; that is, all the switches are closed or, equivalently, all the sentential variables are assigned the value T.

Similarly, current will flow through the network given in Figure 4.23 iff at least one switch is closed. Equivalently, current will flow through this network iff the formula

$$P_1 \vee P_2 \vee P_3 \vee \cdots \vee P_n$$

is assigned the truth value T.

In this manner, we can associate a formula of the Sentential Calculus with any switching network. For example, current will flow through the network given in Figure 4.24 iff either

(i) P, Q, and $\bar{R}$ are closed,

(ii) P, Q, and R are closed,

(iii) $\bar{Q}$ and $\bar{R}$ are closed, or

(iv) P, $\bar{Q}$, and R are closed.

Therefore, we can represent this network by the formula

(*) $$PQ\bar{R} \vee PQR \vee \bar{Q}\bar{R} \vee P\bar{Q}R.$$

On the other hand, current will flow through the network in Figure 4.24 iff

(v) P and Q are closed, or $\bar{Q}$ is closed

and

(vi) $\bar{R}$ is closed, or P and R are closed.

The corresponding formula is

(**) $$(PQ \vee \bar{Q})(\bar{R} \vee PR).$$

Therefore, formulas (*) and (**) are tautologically equivalent. In fact, any formula φ that is tautologically equivalent to (*) (or to (**)) has the property that, for any assignment of truth values to its sentential variables, φ is assigned the truth value T iff current flows through the network, (again, under the assumption that a sentential variable is assigned the truth value T iff the corresponding switch is closed and a sentential variable is assigned the value F iff the corresponding switch is open).

We say two networks are *equivalent* iff the formulas that represent them are tautologically equivalent. Or, what is the same thing, two networks are equivalent if current flows through one iff it flows through the other. The first problem we shall address is how to find the simplest network that is equivalent to a given network. By the simplest network, we generally mean the one that is easiest and cheapest to build and that is the one that is represented by a formula in the Sentential Calculus that has a minimum number of sentential variables and binary connectives. Normally, negation symbols are just applied to sentential variables. (Thus, for example, we would use the formula $\neg P \neg Q$ instead of $\neg(P \vee Q)$.) These are just the types of simplified formulas that we discussed in the last two sections.

Example 4.6 Simplify the network given in Figure 4.24,

(a) using Figure 4.1 and tautological equivalences,

(b) using K-maps.

Then draw the simplified network.

Solution: In either case, the first thing to do is to find a formula of the Sentential Calculus that represents the network. We shall use formula (**) above and call it "φ". $\varphi : (PQ \vee \overline{Q})(\overline{R} \vee PR)$.

(a)

$$\begin{aligned} \varphi &\Leftrightarrow (P \vee \overline{Q})(Q \vee \overline{Q})(\overline{R} \vee P)(\overline{R} \vee R) && (18)\\ &\Leftrightarrow (P \vee \overline{Q})(\overline{R} \vee P) && (1, 7)\\ &\Leftrightarrow P \vee \overline{QR} && (18) \end{aligned}$$

(The numbers in parentheses following each formula refer to the tautological equivalences given in Figure 4.1. We have omitted references to the commutative and associative rules—11, 12, 16, and 17.)

(b) We give the K-map for the network in Figure 4.25 using the formula (*) above:

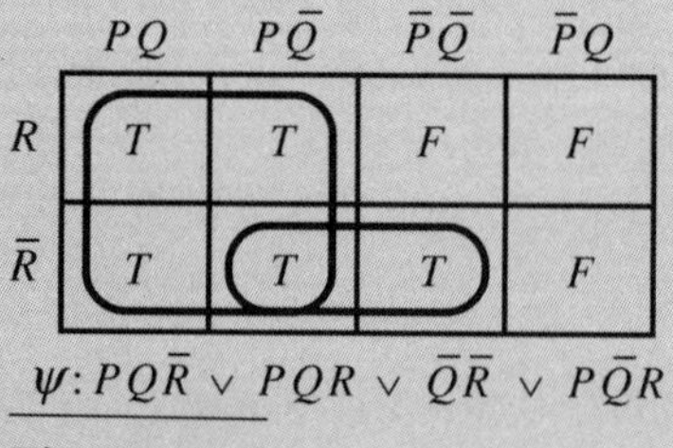

Figure 4.25

Thus, we see,

$$\begin{aligned}\psi &\Leftrightarrow (PQR \vee P\bar{Q}R \vee PQ\bar{R} \vee P\bar{Q}\bar{R}) \vee (P\bar{Q}\bar{R} \vee \bar{P}\bar{Q}\bar{R}) \\ &\Leftrightarrow P \vee \bar{Q}\bar{R}.\end{aligned}$$

Comparing (a) and (b), we see that (a) requires more ingenuity than (b). It requires more skill in manipulating tautological equivalences and, *a priori,* we do not know the most efficient way to simplify the formula. On the other hand, with K-maps, the best way to start is to write the formula as a disjunction of fundamental conjunctions. We can always do this because the DNF is such a formula and there is an algorithm for finding the DNF. The K-map tells us how to group the fundamental conjunctions in the DNF so that we can use the tautological equivalences more efficiently. Frequently, the best way to simplify a formula is to use some combination of the methods given in (a) and (b).

In either case, you should always *check your answer.* Check to see if the simplified formula has the same truth table as the original formula. A quick check in our case would be to show that both formulas are true if either P is true or if $\bar{Q}\bar{R}$ is true; and that both formulas are false if both P and $\bar{Q}\bar{R}$ are false.

Finally, to complete Example 4.6, we draw the network corresponding to the simplified formula in Figure 4.26:

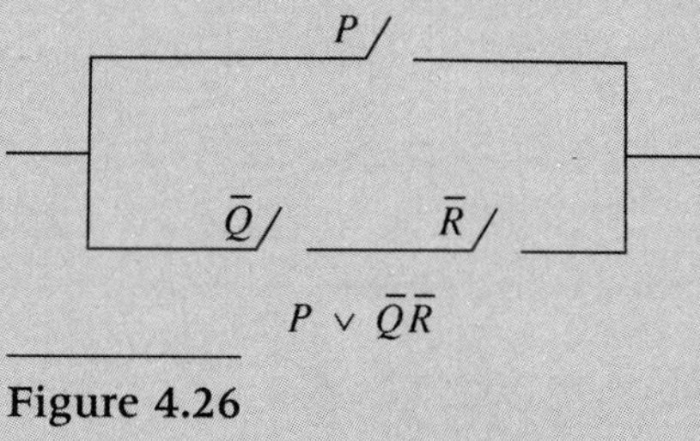

Figure 4.26

We have shown above how to simplify networks. Our next task is to design a switching network to serve a certain purpose.

Example 4.7 Design a switching network for a room with three doors, one light, and a switch near each door that controls the light. If the position of one switch is changed, the state of the light will change; that is, if the light is on, it will go off, and if it is off, it will go on. Assume that if all switches are closed, the light is on.

Solution: We are given that if all switches are closed, the light is on; that is, current will flow through the network. We shall call the switches A, B, and C, and describe the situation with a truth table. (Fig. 4.27). (Remember that T means the switch is closed, and F means it is open.)

A	B	C	Light
T	T	T	T (on)
T	T	F	F (off)
T	F	T	F (off)
T	F	F	T (on)
F	T	T	F (off)
F	T	F	T (on)
F	F	T	T (on)
F	F	F	F (off)

Figure 4.27

Referring to Figure 4.27, we see on line 1 that when all the switches are closed, the light is on (current will flow). From that starting point, each time we change the position of exactly one switch (lines 2, 3, and 5), the light is off. Each time we change the position of two of the switches (lines 4, 6, and 7), the light is on again. When we change all three switches, the light is off.

Using the truth table in Figure 4.27, we can construct the DNF of a formula that represents the network we are looking for. The DNF is

$$\varphi : ABC \vee A\bar{B}\bar{C} \vee \bar{A}B\bar{C} \vee \bar{A}\bar{B}C.$$

The corresponding network is given in Figure 4.28:

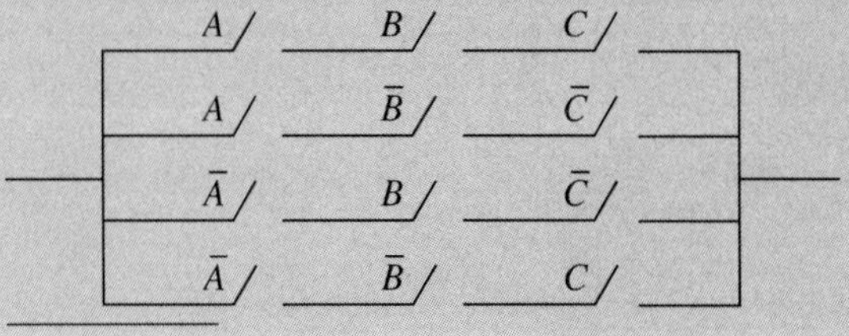

Figure 4.28

A K-map will not help to simplify this formula. We can make a small simplification using distributive rules and show that φ is tautologically equivalent to

$$\psi : A(BC \vee \bar{B}\bar{C}) \vee \bar{A}(B\bar{C} \vee \bar{B}C).$$

We draw the network corresponding to ψ in Figure 4.29:

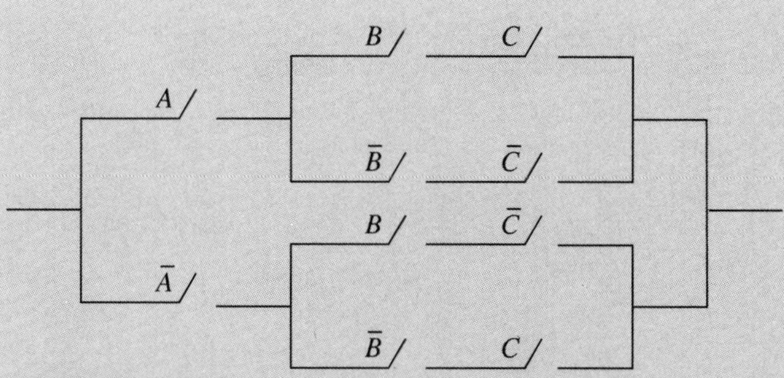

Figure 4.29

Either Figure 4.28 or Figure 4.29 would be a correct solution to the problem.

(We can obtain the same result by reasoning informally. Either A is closed or open. If it is closed, the light will be on iff both B and C are closed, or both B and C are open. If A is open, the light will be on iff B is closed and C is open, or B is open and C is closed.)

4.5 LOGIC NETWORKS

A logic network is an electric circuit that has "gates" instead of the on-off switches of the switching networks. Gates behave like functions. One or more values enter the gate, in the form of electrical impulses, and only one comes out. Each quantity that enters a gate has two possible physical states (say, high or low voltage levels), and the output has one of these states. We shall call the two states 1 and 0.

Several types of gates are generally used. We give their standard symbols in Figure 4.30:

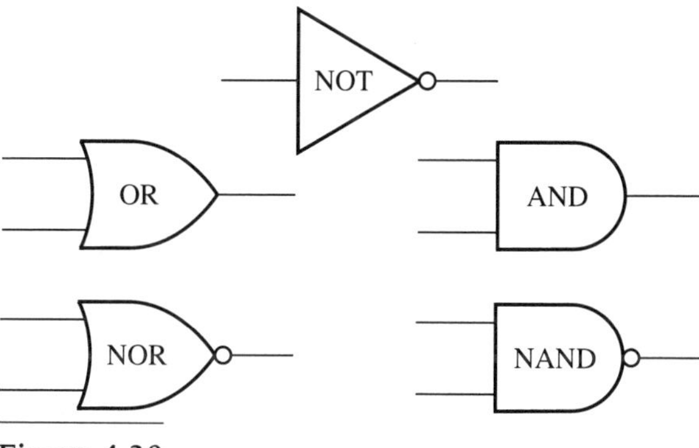

Figure 4.30

We use the convention that the current flows from left to right. The lines on the left, in Figure 4.30, are input lines, and those on the right are output lines. The OR, AND, NOR, and NAND gates are also available with more than two input lines. The table in Figure 4.31 describes the output of these gates:

		$\overline{P}$	$P \vee Q$	PQ	$\overline{P \vee Q}$	$\overline{PQ}$
P	Q	NOT	OR	AND	NOR	NAND
1	1	0	1	1	0	0
1	0	0	1	0	0	1
0	1	1	1	0	0	1
0	0	1	0	0	1	1

Figure 4.31

If we make the convention that the truth value T corresponds to 1, and F to 0, it is clear from the table in Figure 4.31 that we can think of the inputs to these gates as sentential variables and the outputs as formulas of the Sentential Calculus. The convention is that the output is 1 iff the corresponding formula is assigned the truth value T.

The NOT-gate is sometimes called an *inverter*, and it is frequently indicated by a little circle, ○, instead of —NOT▷○— . For example, the output of the network given in Figure 4.32 is $\overline{P\overline{Q}}$ (or $\neg(P \wedge \neg Q)$).

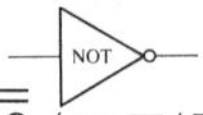

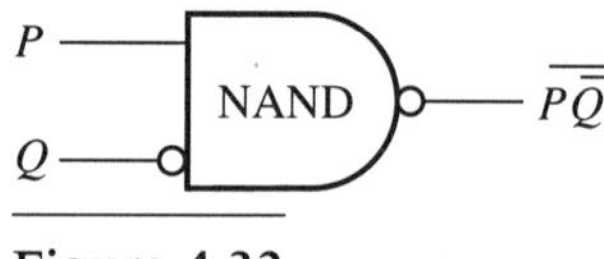

Figure 4.32

Example 4.8

Design logic networks producing the following formulas.

(a) $\varphi_1 : [P(Q \vee R)] \vee (\overline{P}\overline{R})$

(b) $\varphi_2 : (P \vee Q) \wedge \overline{PQ}$

(c) $\varphi_3 : (P \to Q) \leftrightarrow \overline{Q}\overline{R}$

Solution:

(a) The main connective is $\vee$, so our network will be of the form

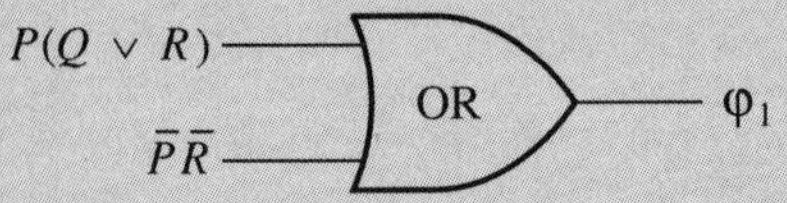

$P(Q \vee R)$ — OR — φ_1
$\overline{P}\overline{R}$

Figure 4.33

The network for $P(Q \vee R)$ is shown in Figure 4.34, and that for $\overline{P}\,\overline{R}$ in Figure 4.35:

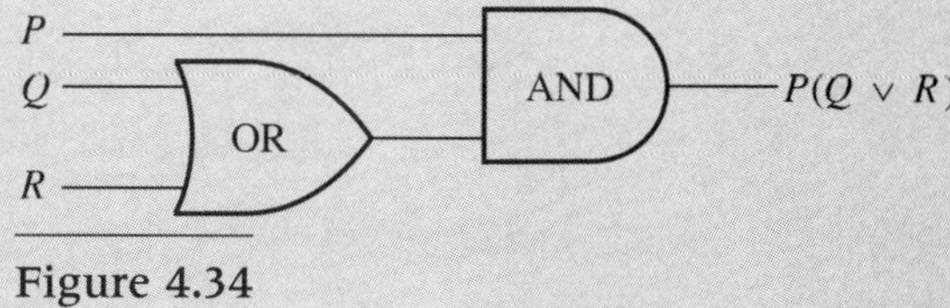

Figure 4.34

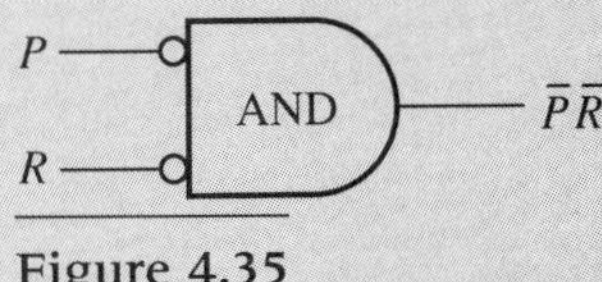

Figure 4.35

Combining Figures 4.33, 4.34, and 4.35, we obtain the network for φ_1:

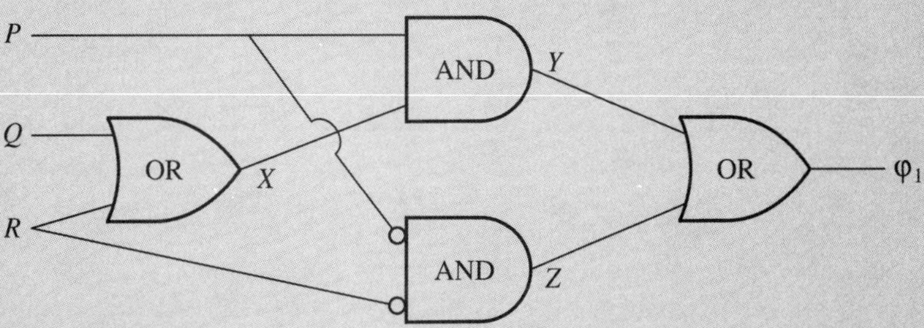

Figure 4.36

In Figure 4.36, the outputs at X, Y, and Z are as follows:

$$X : Q \vee R$$
$$Y : P(Q \vee R)$$
$$Z : \overline{P}\,\overline{R}$$

The disjunction of Y and Z is φ_1.

Because $\overline{P}\,\overline{R} \Leftrightarrow \overline{P \vee R}$, we could replace

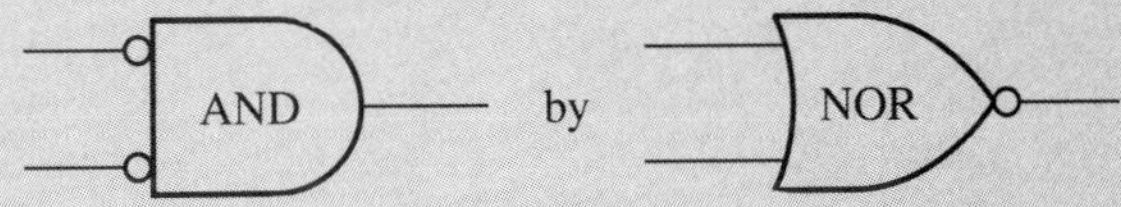

in Figure 4.36.

(b) Figure 4.37 is a network to produce φ_2:

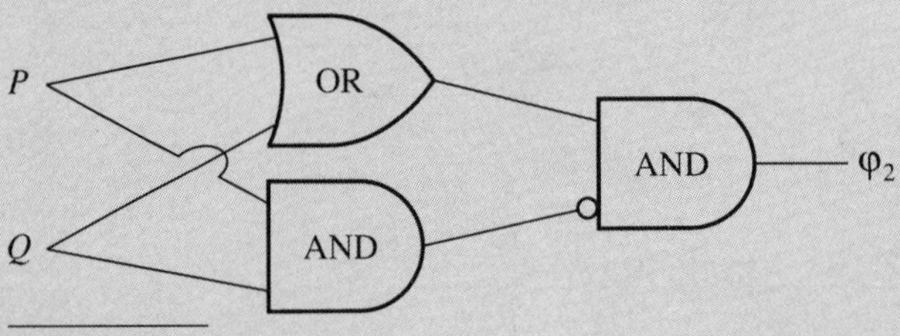

Figure 4.37

We sometimes use the symbol "$P \oplus Q$" as an abbreviation for "$(P \vee Q) \wedge \overline{PQ}$." It is called the *exclusive or*. There is also a gate that produces the exclusive OR. It is called the XOR-gate. Its symbol is given below in Figure 4.38 along with its truth table:

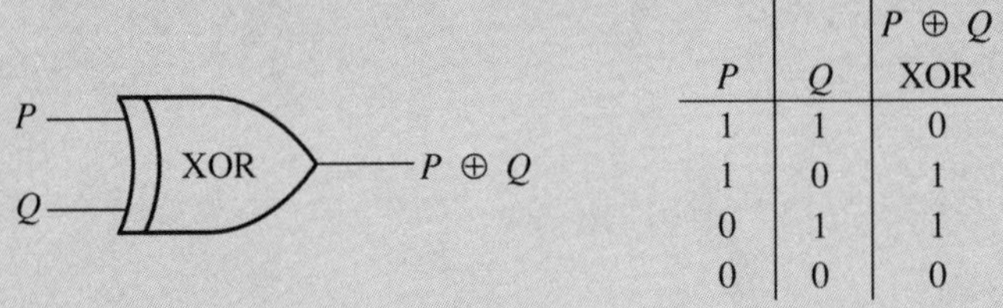

P	Q	$P \oplus Q$ XOR
1	1	0
1	0	1
0	1	1
0	0	0

Figure 4.38

(c) In order to construct a logic network that produces φ_3, we first must find a tautologically equivalent formula that does not contain the symbols $\rightarrow$ and $\longleftrightarrow$. We shall do this by finding its DNF. The following is the truth table for φ_3:

P	Q	R	$P \rightarrow Q$	$\overline{Q}\,\overline{R}$	φ_3
T	T	T	T	F	F
T	T	F	T	F	F
T	F	T	F	F	T
T	F	F	F	T	F
F	T	T	T	F	F
F	T	F	T	F	F
F	F	T	T	F	F
F	F	F	T	T	T

The truth table of φ_3 only has a T on lines 3 and 8, so

$$\varphi_3 \Leftrightarrow P\overline{Q}R \vee \overline{P}\,\overline{Q}\,\overline{R}.$$

Its network is constructed in Figure 4.39:

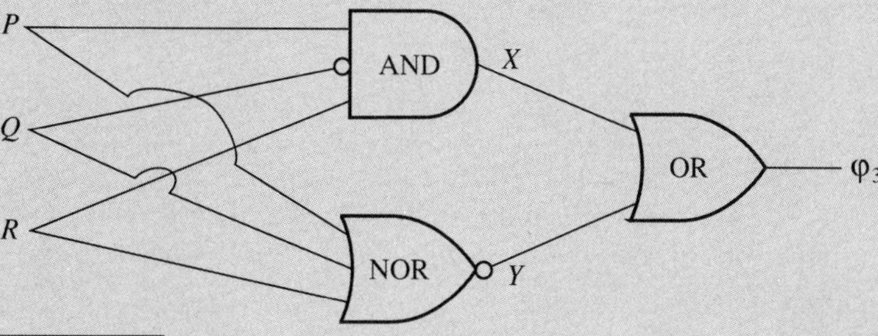

Figure 4.39

At X in Figure 4.39, the output is $P\overline{Q}R$, and at Y, $\overline{P \vee Q \vee R}$, which is tautologically equivalent to $\overline{P}\,\overline{Q}\,\overline{R}$. Sometimes it is difficult to understand a network if there are too many crossing wires. To avoid crossing wires, for example, the network in Figure 4.39 could be redrawn as in Figure 4.40:

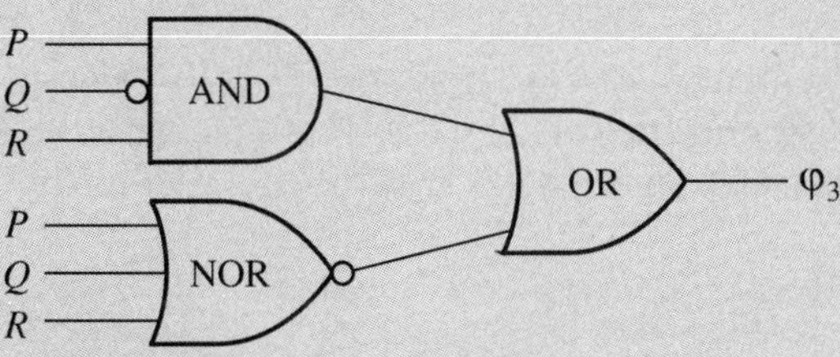

Figure 4.40

Of course, in Figures 4.39 and 4.40, the NOR-gate could be replaced by the AND-gate:

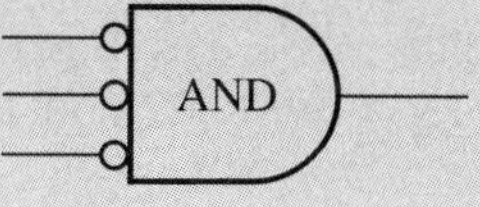

Because the output of each gate in a logic network represents a formula, we can consider logic networks with multiple outputs. In Figure 4.40, for example, we can have an output from both the AND-gate and the NOR-gate. This is drawn in Figure 4.41:

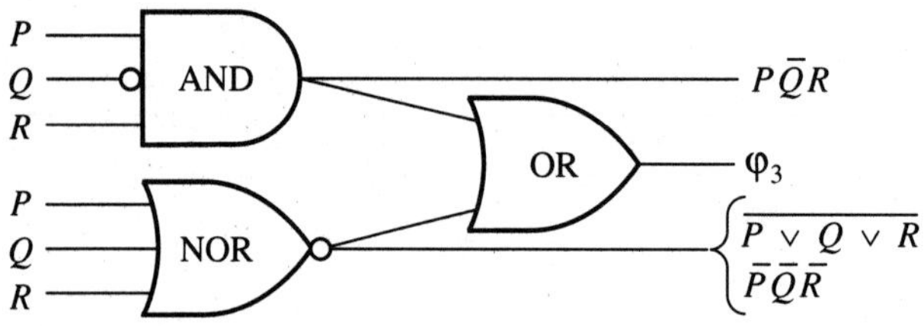

Figure 4.41

Logic networks that can do binary arithmetic have many applications. The sum and carry of two one-digit numbers in binary arithmetic is given in Figure 4.42:

p	q	sum	carry
1	1	0	1
1	0	1	0
0	1	1	0
0	0	0	0

Figure 4.42

The sum 1 + 1 in binary arithmetic is 10, so we indicate this by saying the sum is 0 and the carry is 1. (See line 1 of Figure 4.42.) For the other sums, the carry is 0.

If P is the sentence "$p = 1$" and Q is "$q = 1$," the sum is represented by the formula $P \oplus Q$ and the carry by PQ. The logic network that corresponds to Figure 4.42 is given in Figure 4.43. It is called a *half-adder*:

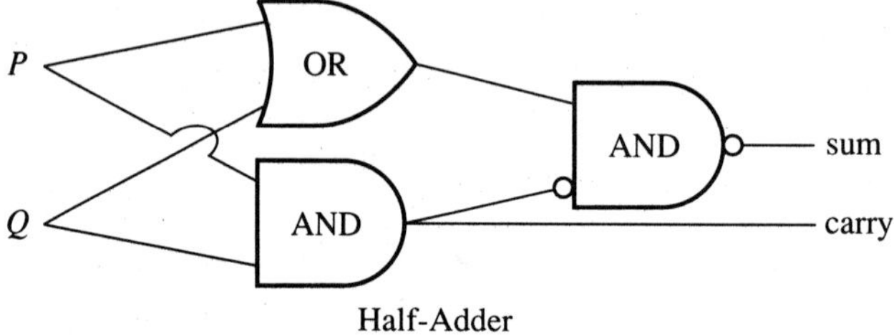

Figure 4.43

In Figure 4.44, we illustrate how to add 29 and 12 in binary arithmetic. First, in order to write 29 and 12 to base 2, we express each as a sum of powers of 2:

$$29 = 2^4 + 2^3 + 2^2 + 2^0 = (11101)_2$$

$$12 = 2^3 + 2^2 = (1100)_2$$

```
carries ------> 1 1 1 0 0
                  1 1 1 0 1   (= 29)
                    1 1 0 0   (= 12)
sum     ------> 1 0 1 0 0 1   (= 41)
```

Figure 4.44

Therefore, to carry out binary addition, we have to be able to add three single-digit binary numbers (the sum of two digits plus the carry from the preceding addition). The table for adding three binary single digit numbers with carry is given in Figure 4.45:

p	q	r	sum	carry
1	1	1	1	1
1	1	0	0	1
1	0	1	0	1
1	0	0	1	0
0	1	1	0	1
0	1	0	1	0
0	0	1	1	0
0	0	0	0	0

Figure 4.45

Again, using the convention that X is the sentence "$x = 1$," we see from Figure 4.45 that

$$\begin{aligned}\text{sum} &\Leftrightarrow PQR \vee P\overline{Q}\,\overline{R} \vee \overline{P}Q\overline{R} \vee \overline{P}\,\overline{Q}R \\ &\Leftrightarrow (P \oplus Q) \oplus R \\ \text{carry} &\Leftrightarrow PQR \vee PQ\overline{R} \vee P\overline{Q}R \vee \overline{P}QR \\ &\Leftrightarrow PQ \vee (P \oplus Q)R\end{aligned}$$

The network corresponding to Figure 4.45 is called a *full-adder* and is pictured in Figure 4.46. In Figure 4.46, we use the symbol "HA" as an abbreviation for the half-adder given in Figure 4.43.

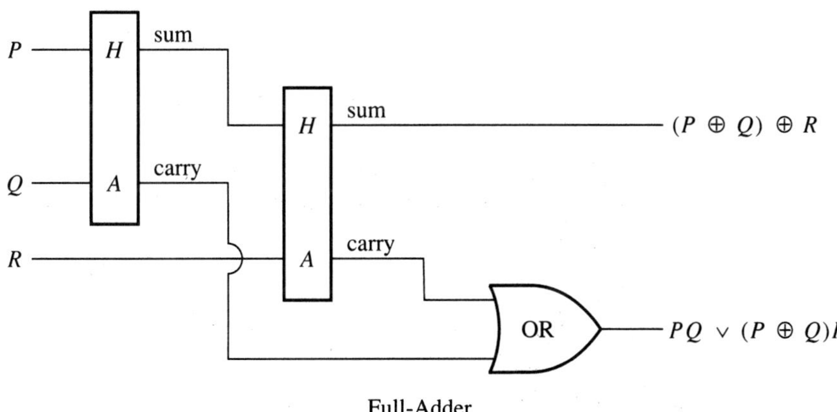

Figure 4.46

Example 4.9 Construct a logic network for adding two three-digit binary numbers $p_2p_1p_0$ and $q_2q_1q_0$.

Solution: Again, P_i is the sentence "$p_i = 1$" and Q_i is "$q_i = 1$." First we add p_0 and q_0. We can use the half-adder to do that. Suppose the sum is s_0 and the carry is c_0. Then we use the full-adder to add c_0, p_1, and q_1. Suppose the sum is s_1 and the carry is c_1. The last step is to use the full-adder again to add c_1, p_2, and q_2. If the sum is s_2 and the carry is c_2, then the sum of the two numbers is

$$c_2s_2s_1s_0$$

in binary notation. The network is given in Figure 4.47.

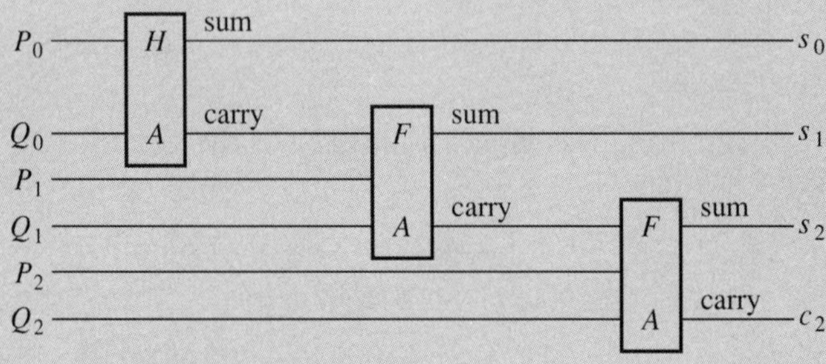

Figure 4.47

We use the following notation:

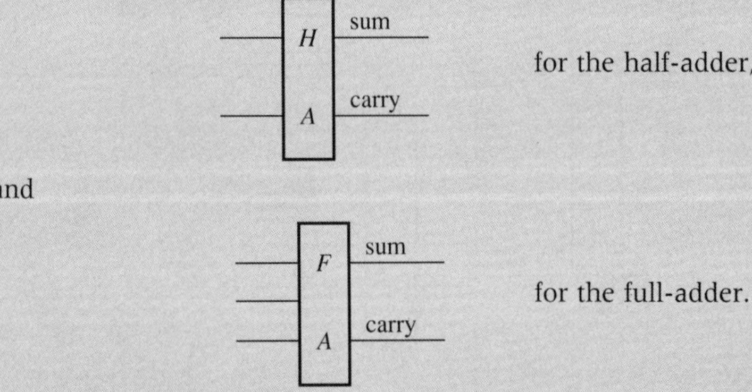

See Figures 4.43 and 4.46.

Example 4.10

A vending machine is to be designed that will dispense one product that costs 15 cents. The machine will have a sliding bar with three coin positions on it for one quarter (25 cents), one dime (10 cents), and one nickel (5 cents). When the bar is pushed in, a logic network inside the machine will make decisions to dispense the product, alert the change-making device to make change, and/or inform the user that he/she did not insert enough money. (Assume that the change machine inside the vending machine computes and dispenses the correct amount of change.)

Design a network that will do the following after the bar is pushed in:

(a) If there is no money, turn on an insufficient funds light (*I*).

(b) If there is exactly the right amount of money, dispense the product (*P*).

(c) If there is too much money, dispense the product (*P*) and signal the change machine (*C*).

(d) If there is some money, but not enough, turn on the insufficient funds light (*I*) and signal the change machine (*C*).

Solution: The form of the solution will be as follows:

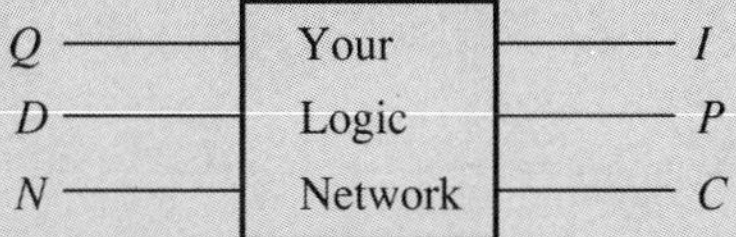

There are three sentential variables:

Q: A quarter is inserted in the slot.

D: A dime is inserted in the slot.

N: A nickel is inserted in the slot.

The possible outcomes are *I*, *P*, *PC*, or *IC*. The table in Figure 4.48 gives the relationship between the input and the output.

Q	D	N	Output
T	T	T	PC
T	T	F	PC
T	F	T	PC
T	F	F	PC
F	T	T	P
F	T	F	IC
F	F	T	IC
F	F	F	I

Figure 4.48

Therefore, using Figure 4.48, we see that

$$
\begin{aligned}
P &\Leftrightarrow QDN \vee QD\overline{N} \vee Q\overline{D}N \vee Q\overline{D}\,\overline{N} \vee \overline{Q}DN \\
&\Leftrightarrow Q \vee DN \\
I &\Leftrightarrow \overline{Q}D\overline{N} \vee \overline{Q}\,\overline{D}N \vee \overline{Q}\,\overline{D}\,\overline{N} \\
&\Leftrightarrow \overline{Q}\,\overline{D} \vee \overline{Q}\,\overline{N} \\
&\Leftrightarrow \overline{Q}(\overline{D} \vee \overline{N}) \\
&\Leftrightarrow \overline{Q}\,\overline{DN} \\
C &\Leftrightarrow QDN \vee QD\overline{N} \vee Q\overline{D}N \vee Q\overline{D}\,\overline{N} \vee \overline{Q}D\overline{N} \vee \overline{Q}\,\overline{D}N \\
&\Leftrightarrow Q \vee D\overline{N} \vee \overline{D}N
\end{aligned}
$$

The network is given in Figure 4.49:

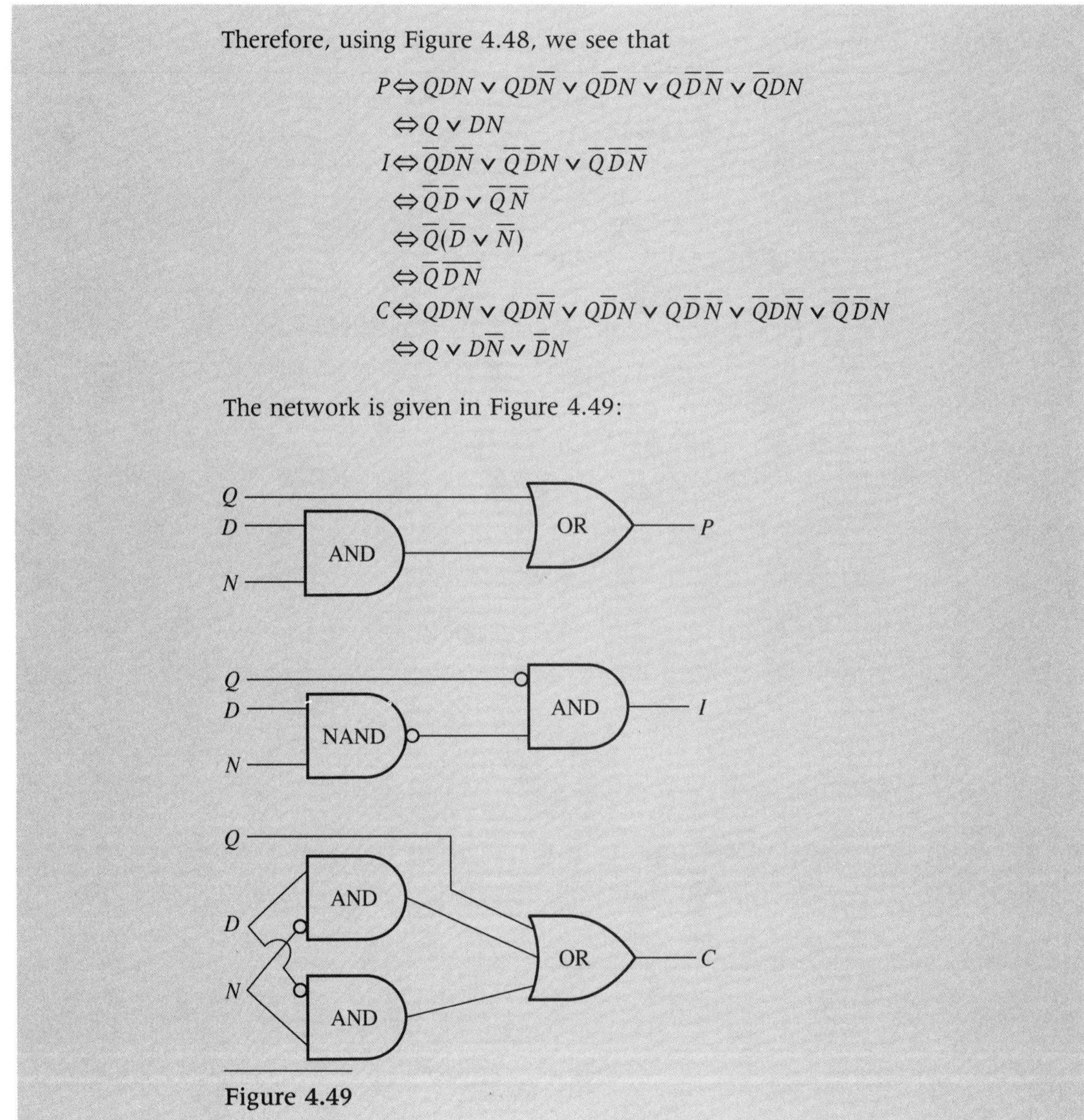

Figure 4.49

(Remark: When constructing networks, it is important to simplify the formula as much as possible. For example, if we did not simplify P, the network that had the output P would have had four additional AND-gates, three additional OR-gates, and many more wires.)

Example 4.11 The following table gives the binary notation for all non-negative integers less than 10:

decimal	0	1	2	3	4	5	6	7	8	9
binary	0000	0001	0010	0011	0100	0101	0110	0111	1000	1001

If $n = p_3p_2p_1p_0$ is a non-negative binary integer less than 10 and P_i is the sentence "$p_i = 1$," design a logic network so that the output is the sentence "n is a perfect square."

Solution: The perfect squares are 0000, 0001, 0100, and 1001. Therefore, the output of the network should be the formula

$$\varphi : \overline{P}_3\overline{P}_2\overline{P}_1\overline{P}_0 \vee \overline{P}_3\overline{P}_2\overline{P}_1P_0 \vee \overline{P}_3P_2\overline{P}_1\overline{P}_0 \vee P_3\overline{P}_2\overline{P}_1P_0$$

Simplifying φ,

$$\begin{aligned}\varphi &\Leftrightarrow \overline{P}_3\overline{P}_1\overline{P}_0 \vee \overline{P}_2\overline{P}_1P_0 \\ &\Leftrightarrow \overline{P}_1(\overline{P}_3\overline{P}_0 \vee \overline{P}_2P_0)\end{aligned}$$

The network is given in Figure 4.50:

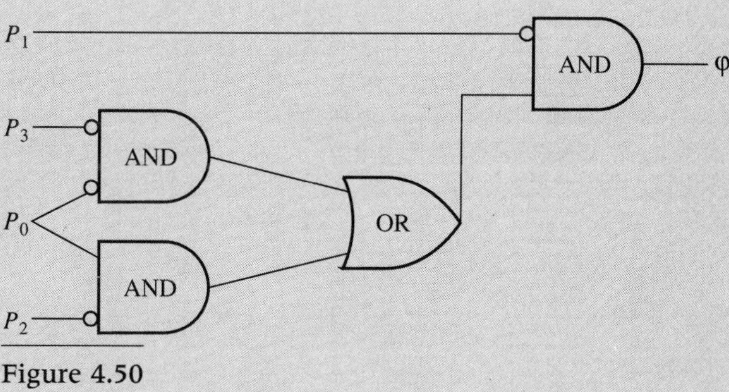

Figure 4.50

(For additional information about switching and logic networks, see *Introduction to Switching Theory and Logical Design* by F. Hill and G. Peterson, John Wiley & Sons, 1968, and *Boolean Algebra and Switching Circuits* by E. Mendelson, McGraw-Hill, Schaum's Outline Series, 1970. Most books on discrete mathematics also discuss these topics. See, for example, *Discrete Mathematics* by K. Ross and C. Wright, Prentice-Hall, 1985, and *Discrete and Combinatorial Mathematics* by A. Hillman, G. Alexanderson, and R. Grassl, Dellen Publishing Co., 1987.)

EXERCISES 4

A. Disjunctive normal form. (Answers to 1(d), 2(a), and 3(b), (h), (l).)

(1) For each of the columns (a) through (e) in the truth table given below, find a formula in DNF that has that truth table:

A	B	C	(a)	(b)	(c)	(d)	(e)
T	T	T	F	T	F	F	T
T	T	F	F	F	F	T	T
T	F	T	T	F	F	T	T
T	F	F	T	F	F	F	T
F	T	T	F	F	T	T	T
F	T	F	F	T	F	F	T
F	F	T	T	F	F	T	T
F	F	F	F	F	F	F	T

(2) For each of the columns (a) through (e) in the truth table given below, find a formula in DNF that has that truth table:

A	B	C	D	(a)	(b)	(c)	(d)	(e)
T	T	T	T	F	T	F	F	T
T	T	T	F	F	F	F	T	T
T	T	F	T	T	T	F	F	T
T	T	F	F	F	F	F	T	T
T	F	T	T	F	F	F	T	T
T	F	T	F	F	T	F	F	T
T	F	F	T	F	F	F	T	T
T	F	F	F	F	F	T	F	T
F	T	T	T	T	F	F	F	F
F	T	T	F	T	F	F	F	F
F	T	F	T	T	T	F	F	F
F	T	F	F	F	F	F	T	F
F	F	T	T	F	F	F	T	F
F	F	T	F	F	T	F	F	F
F	F	F	T	F	F	F	T	F
F	F	F	F	F	F	T	F	F

(3) Find formulas in DNF tautologically equivalent to each of the following formulas:

(a) $A \rightarrow B$
(b) $A \vee \neg B$
(c) $\neg(A \wedge \neg B)$
(d) $A \vee B \leftrightarrow \neg A$
(e) $A \vee B \rightarrow C$
(f) $A \rightarrow (\neg B \rightarrow C)$

(g) $\neg(A \vee B) \leftrightarrow \neg A \wedge C$
(h) $[(\neg A \leftrightarrow B) \vee C] \wedge (\neg B \wedge C)$
(i) $\neg(A \to \neg B \wedge C) \wedge (A \vee B) \wedge \neg C$
(j) $A \vee \neg B \to A \wedge \neg C \wedge D$
(k) $(A \vee B \vee \neg C \vee \neg D) \wedge (\neg A \vee B \vee C \vee \neg D)$
(l) $(A \wedge B) \leftrightarrow (\neg A \vee (C \wedge \neg \mathrm{D}))$
(m) $(A \to B \wedge \neg C) \to (A \vee D \to \neg B \wedge C)$
(n) $\neg[(A \vee B) \wedge (C \to D)] \to A \wedge \neg B \wedge \neg D$

B. Simplify each of the following formulas using

(i) the tautological equivalences given in Figure 4.1, and
(ii) K-maps. (It may be necessary to find the DNF before using a K-map.)

(Answers to 1, 4, 8, 11, 13, 15, 17, 21, and 25.)
(Note that we use "$\bar{P}$" for "$\neg P$," and "PQ" for "$P \wedge Q$.")

(1) $(Q \to P) \to P$
(2) $(P \leftrightarrow \bar{Q}) \vee Q$
(3) $P \to (Q \to P)$
(4) $P(PQ \vee [(\bar{P}\bar{Q}) \vee Q])$
(5) $[\bar{P}(Q \leftrightarrow \bar{P})Q] \to \overline{P \vee Q}$
(6) $\bar{P}Q(\bar{P} \leftrightarrow \overline{P \vee \mathrm{Q}})$
(7) $PQR \vee P\bar{Q}R \vee \bar{P}\bar{Q}R$
(8) $P\bar{Q}R \vee \bar{P}\bar{Q}R \vee QR$
(9) $(P \vee \bar{Q})R \vee P\bar{R} \vee (\bar{Q} \vee R)\bar{R}$
(10) $PQR \vee \bar{P}QR \vee P\bar{Q}R \vee PQ\bar{R}$
(11) $\bar{P}\bar{Q}R \vee \bar{P}Q\bar{R} \vee P\bar{Q}\bar{R} \vee PQ\bar{R}$
(12) $PQR \vee \bar{P}QR \vee PQ\bar{R} \vee \bar{P}Q\bar{R}$
(13) $PQR \vee P\bar{Q}R \vee \bar{P}\bar{Q}R \vee \bar{P}\bar{Q}\bar{R} \vee \bar{P}QR$
(14) $\overline{P \vee Q} \vee R \vee P(QR \vee \bar{Q}\bar{R})$
(15) $(P \vee Q)(\bar{P} \vee R)(Q \vee R)$
(16) $[((P \vee Q)\overline{PQ}) \vee R]\overline{(P \vee Q)\overline{PQ}R}$
(17) $[P \vee (Q \vee R)\overline{QR}]\overline{P \vee (Q \vee R)\overline{QR}}$
(18) $\bar{P}QR\bar{S} \vee \bar{P}Q\bar{R}\bar{S}$
(19) $PQRS \vee PQR\bar{S} \vee \bar{P}QRS \vee \bar{P}QR\bar{S}$
(20) $PQRS \vee P\bar{Q}RS \vee PQR\bar{S} \vee P\bar{Q}R\bar{S} \vee \bar{P}Q\bar{R}\bar{S} \vee \bar{P}Q\bar{R}S$
(21) $\bar{P}\bar{Q} \vee P\bar{Q}RS \vee \bar{P}QR\bar{S} \vee P\bar{Q}\bar{R}\bar{S} \vee \bar{P}Q\bar{R}S$
(22) $PRS \vee \bar{Q}RS \vee PR\bar{S} \vee P\bar{Q}\bar{R}\bar{S}$
(23) $\bar{P}Q \vee RS \vee PR\bar{S} \vee \bar{P}\bar{Q}\bar{R} \vee P\bar{Q}\bar{R}\bar{S}$
(24) $P\bar{Q}RS \vee PQ\bar{R}S \vee \overline{Q \vee R \vee S} \vee \overline{P \vee Q \vee R} \vee P\,(\overline{Q \vee R})$
(25) $P\bar{Q}RS \vee \overline{PQ}R\bar{S} \vee (P \vee \bar{Q})(\overline{R \vee S}) \vee P\bar{Q}\bar{R}S$
(26) $\overline{P\bar{R} \vee Q \vee (P\bar{R} \vee Q)\bar{S}}$
(27) $PQRS \vee \overline{(R \vee S)(Q \vee R \vee \bar{S})(P \vee R \vee \bar{S})}$

C. Simplify each of the following switching networks. Draw the simplified network. (Answers to 2, 5, and 8.)

(1)

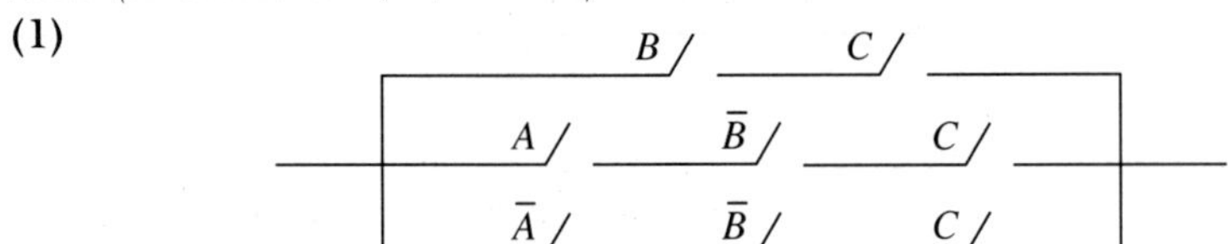

(2)

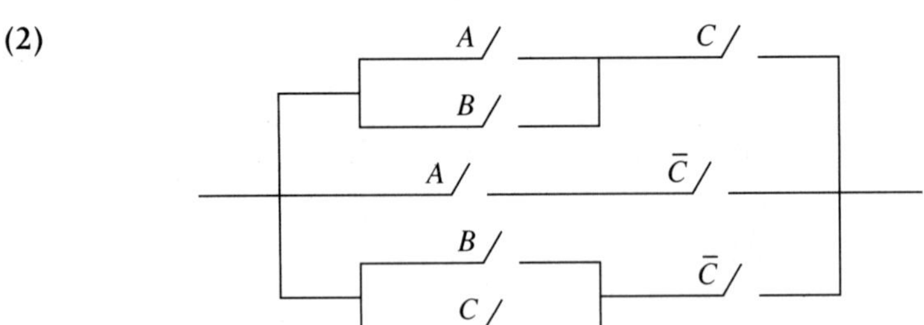

(3)

A B
$\bar{A}$ C
$\bar{A}$ $\bar{C}$

(4)

A B C
A $\bar{B}$ C
$\bar{A}$ B $\bar{C}$
$\bar{A}$ $\bar{B}$ $\bar{C}$

(5)

A $\bar{B}$ C
$\bar{B}$ C
$\bar{B}$ $\bar{C}$

(6)

A C D
A C $\bar{D}$
$\bar{A}$ $\bar{B}$ $\bar{D}$
A $\bar{B}$ $\bar{C}$

(7)

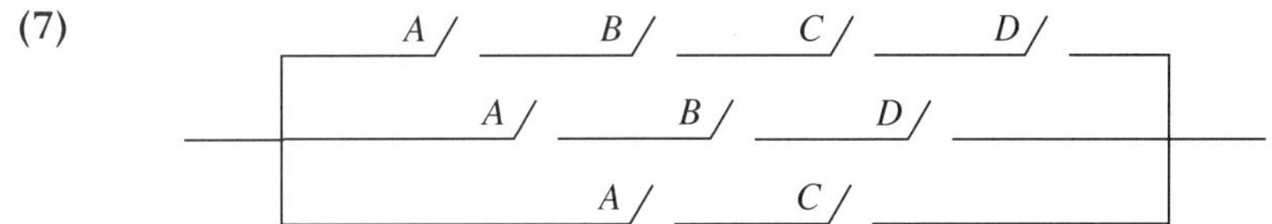

(8)

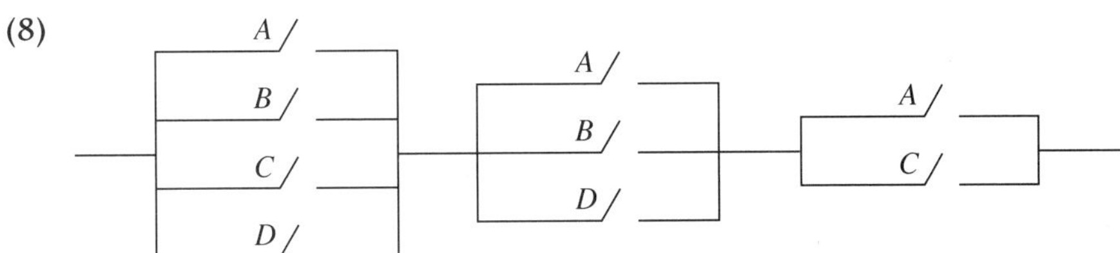

(9)

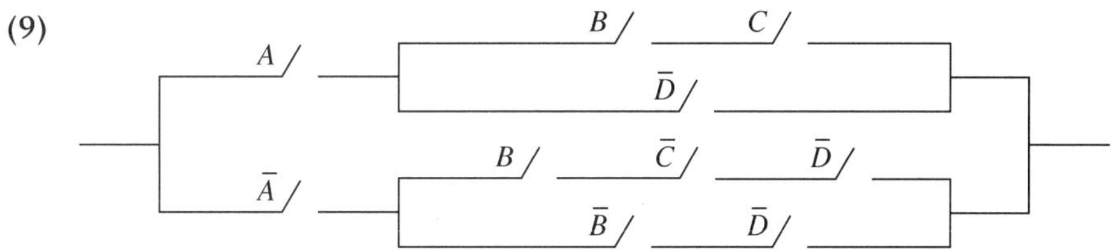

(10)

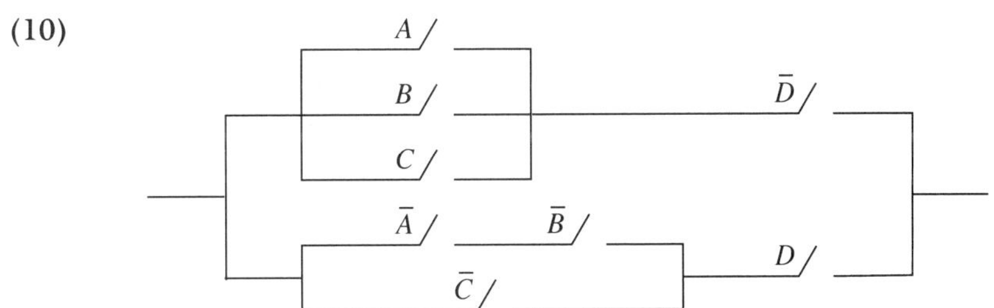

D. Express the output of each of the following logic networks as a formula of the Sentential Calculus. (Answers to 1, 4, 7, 11, and 13.)

(1)

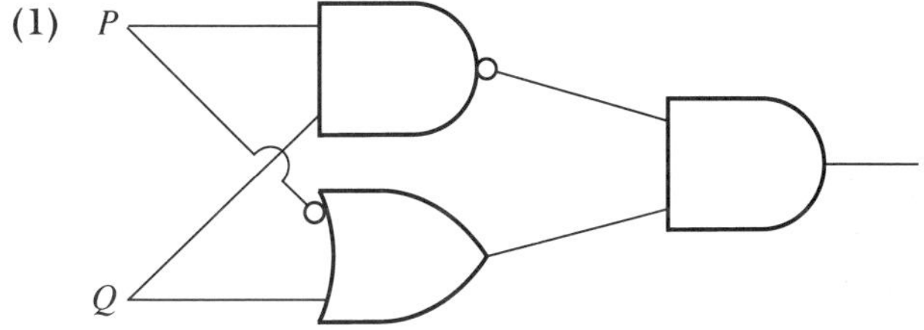

(2)

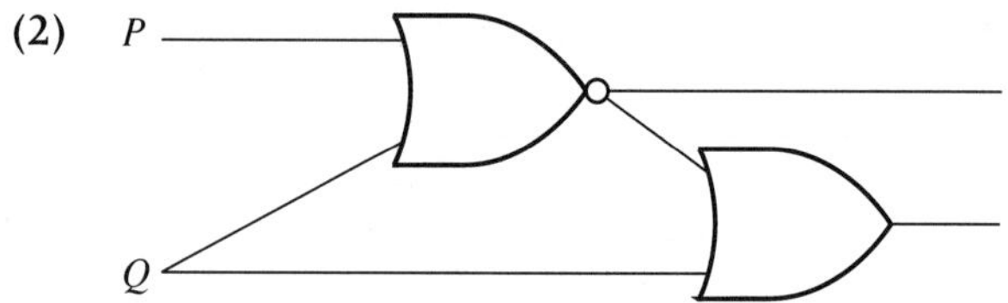

(3)

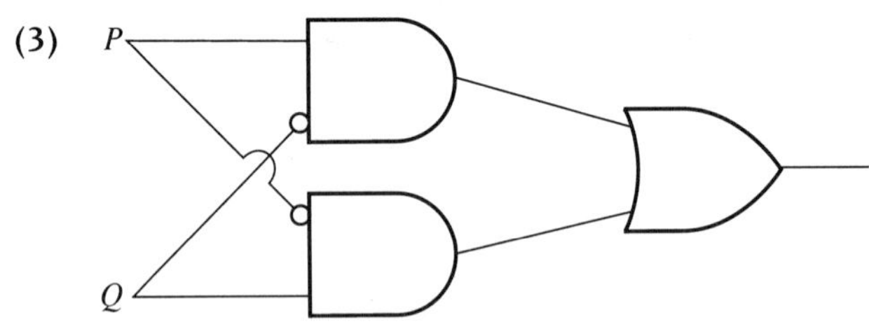

(4)

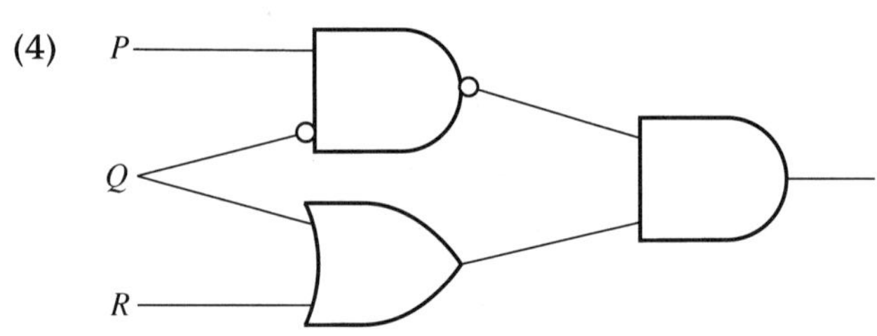

(5)

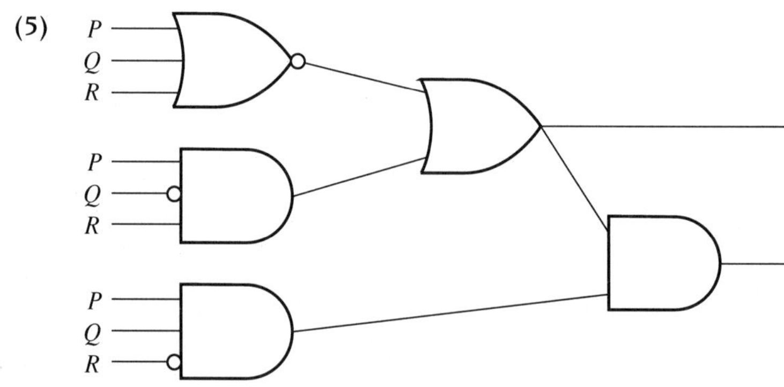

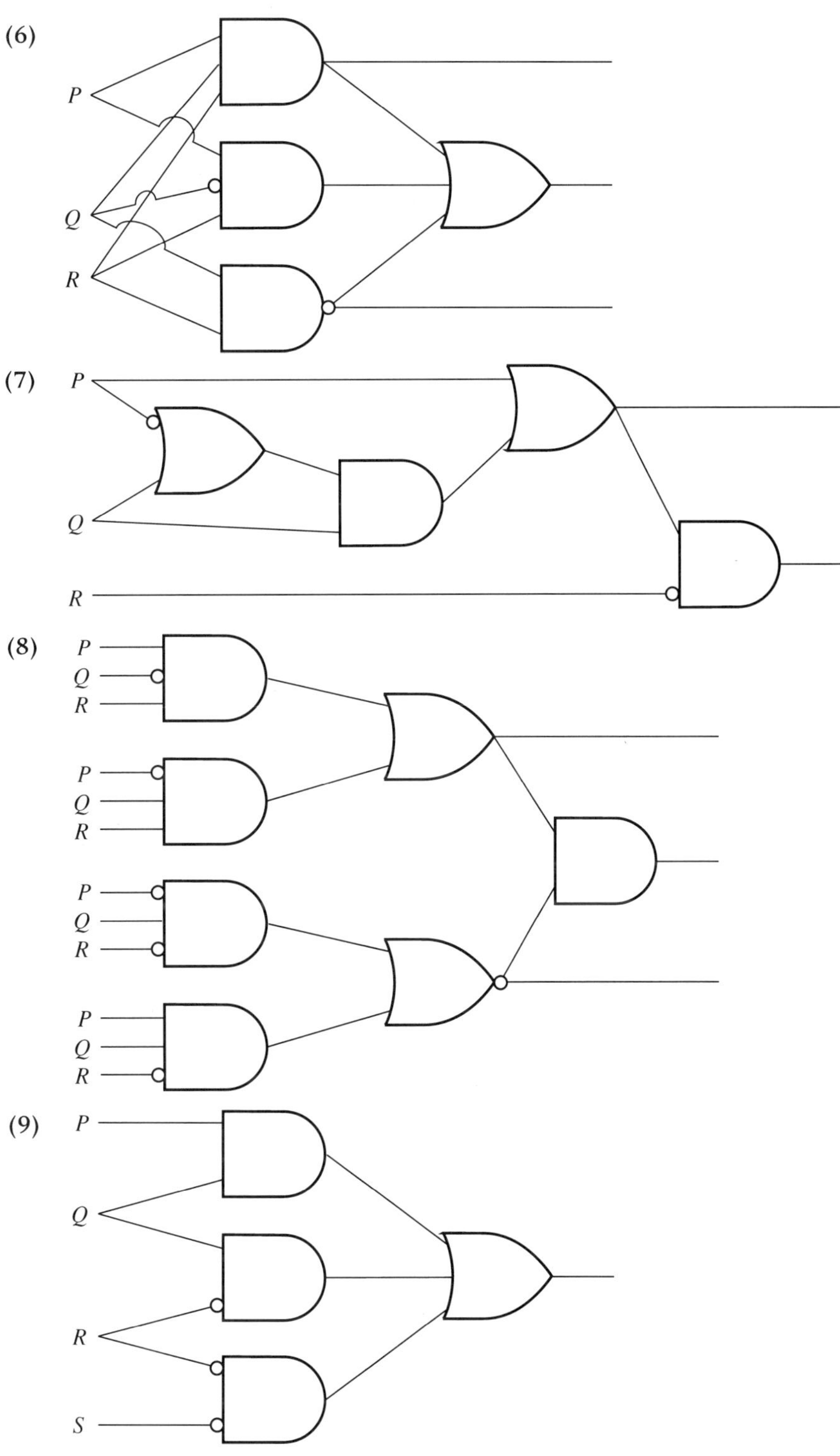
(6)
P
Q
R
(7)
P
Q
R
(8)
P
Q
R
P
Q
R
P
Q
R
P
Q
R
(9)
P
Q
R
S

(10)

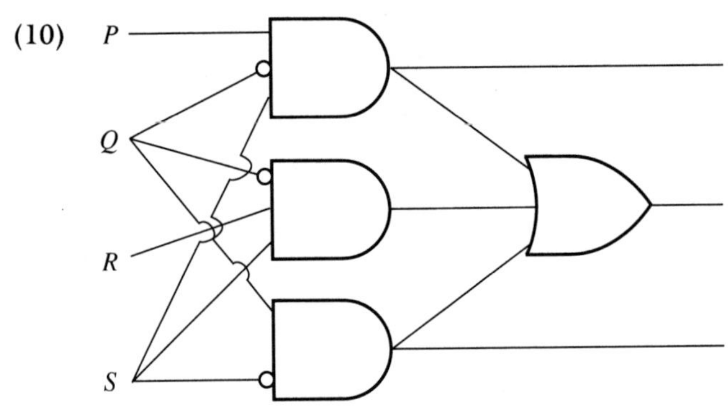

(11)

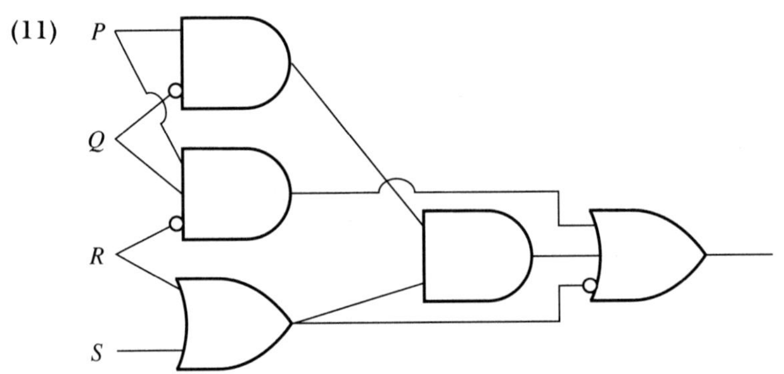

(12)

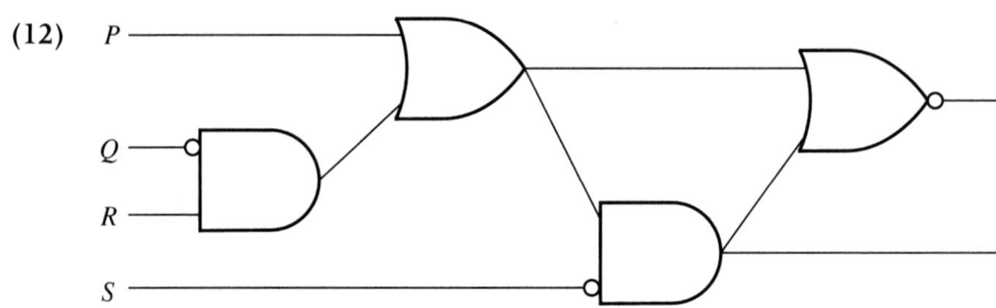

(13)

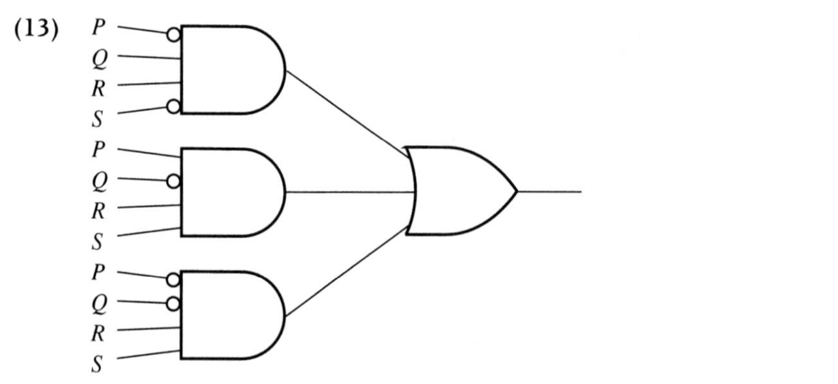

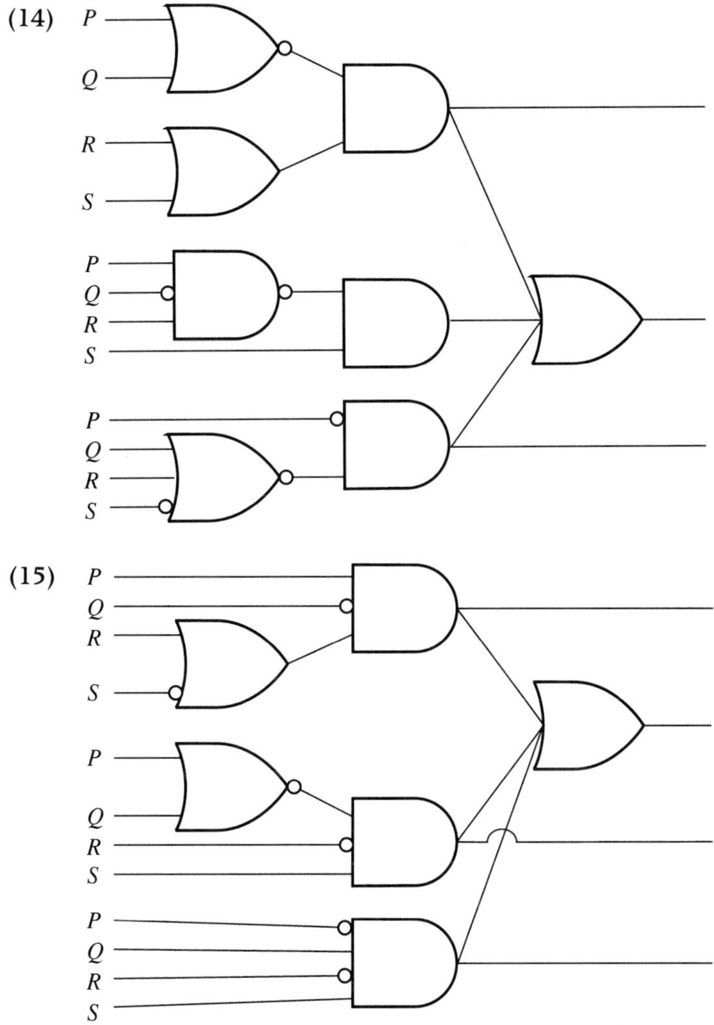

E. Design a switching network for each of the following problems. Simplify your network as much as possible. (Answers to 1, 3, and 6.)

(1) In a small prison, there are four cells for prisoners. On each cell door, there is a switch. Current is to flow through the network that causes an alarm to ring in the main office if any one or more of the cell doors are opened.

(2) In a small minimum-security prison, there are four rooms for prisoners. The jailers want to be sure that, at most, one prisoner is out of his/her room at one time. At each room door, there is a switch. Current is to flow through the network that causes an alarm to ring in the main office if more than one room door is open.

(3) Assume that each of three members of a committee can vote yes by pressing a button. Current is to flow through the network that causes a light to go on if a majority votes yes.
(4) Assume that each of four members of a committee can vote yes by pressing a button. Current is to flow through the network that causes a light to go on if either a majority votes yes, or the chairperson and one other person votes yes.
(5) Assume in Example 4.7 that there are four switches instead of three.
(6) Assume in Example 4.7 that if switch A is closed, then the light will be off if exactly one of B and C is closed. On the other hand, if switch A is open, then switch B alone controls the light.

F. Design a logic network for each of the following problems. Simplify your network as much as possible. (Answers to 1, 3, 4(c), 7(b), 8(a), 9(c), 10(b), and 12(a).) In problems 1 through 6, we use the convention that each sentential variable X represents the sentence "$x = 1$" and lower-case letters represent binary digits, 0 or 1.
(1) The output is the formula that says $p = q$ and $p \neq r$.
(2) The output is the formula that says $p = r$ and $p < q$.
(3) The output is the formula that says exactly two of p, q, and r are equal.
(4) Suppose that $n = p_3p_2p_1p_0$ is a non-negative binary integer less than 10. (See Example 4.11.) The output of the network is a formula that says
(a) n is odd.
(b) n is even.
(c) n is divisible by 3.
(d) n is divisible by 4.
(e) n is a prime. (A *prime* is a positive integer greater than 1 whose only divisors are 1 and itself.)
(5) Suppose $n = p_2p_1p_0$ is a non-negative binary integer. The logic network is to have four outputs, O_1, O_2, O_3, and O_4, where O_1 is the formula that says n is even; O_2 is the formula that says that n, in decimal notation, is 4 or 5; O_3 is the formula that says that n, in decimal notation, is 2, 3, 4, or 6; O_4 is the formula that says that n, in decimal notation, is 4 or 6.
(6) Add two binary integers. (See Example 4.9. You may use half-adders and/or full-adders in your network, as in Figure 4.47.)
(a) $p_1p_0 + q_1q_0$
(b) $p_1p_0 + q_2q_1q_0$
(c) $p_2p_1p_0 + q_3q_2q_1q_0$
(d) $p_3p_2p_1p_0 + q_3q_2q_1q_0$
(7) Solve the vending-machine problem (Example 4.10) under the conditions that the sliding bar has two coin positions on it for a quarter and a dime and that the product costs
(a) 15 cents
(b) 30 cents

(8) Solve the vending-machine problem (Example 4.10) under the conditions that the sliding bar has three positions on it for a quarter, a dime, and a nickel, and that the product costs
   (a) 5 cents
   (b) 30 cents

(9) Solve the vending-machine problem (Example 4.10) under the conditions that the sliding bar has four positions on it for a quarter, two dimes, and a nickel and that the product costs
   (a) 45 cents
   (b) 30 cents
   (c) 20 cents
   (d) 25 cents

(10) A vault is protected by four locks, L_1, L_2, L_3, and L_4. The keys have been distributed to four vice presidents, A, B, C, and D, in the following way.
   (a) A has keys to L_1 and L_2.
   B has keys to L_2 and L_4.
   C has keys to L_3 and L_4.
   D has keys to L_1 and L_3.
   (b) A has a key to L_1.
   B has keys to L_1 and L_2.
   C has keys to L_3 and L_4.
   D has a key to L_2.

   Your network has four inputs, A, B, C, and D, where each is the sentence that the vice president is present. The output is true iff the vault can be opened.

(11) Your network is to allow current to flow so that a car can start iff the driver's seat belt is fastened, the break is off, and either the automatic transmission is in park or neutral (not both).

(12) A mail-order firm sells four products, A, B, C, and D. Under certain conditions, the firm includes one or two bonus gifts with the order.
   (a) The customer receives two bonus gifts, G_1 and G_2, if he/she buys A and at least two of B, C, and D. The customer receives G_1 if he/she just buys A, and G_2 if he or she buys B, C, and D.
   (b) The customer receives two bonus gifts G_1 and G_2 iff he/she buys all products. He/she receives G_1 for buying A and one of the other three, and receives G_2 for buying B, C, and D.

   Your network has four inputs, A, B, C, and D, and two outputs, G_1 and G_2. Current should flow to G_i iff the customer receives the gift G_i. (Note that $i = 1, 2$.)

G. Extend Theorem 4.4 to a block of eight fundamental conjunctions, and prove the extended theorem.

Chapter 5

Language of the Predicate Calculus

We want to extend the language of the Sentential Calculus to a language that is rich enough for mathematics. Mathematics contains expressions of the form: $x < y$; $x^2 + 3x = 1$; $x + y = y + x$, for all integers x and y; there exists an integer between 1 and 3; each integer is either even or odd; if a is a zero of f then $f(a) = 0$, and so forth. Thus, we not only need the sentence letters and connectives of the Sentential Calculus, but we also need symbols for terms ($0, 1, 3, x, y, a, f(a), x^2 + 3x$); predicates and relations (integer, even, odd, between, $<$, $=$); and quantifiers (for all, there exists). The extended system is called the *Predicate Calculus*, and in this chapter, we study its language.

5.1 TERMS AND PREDICATES

Basically, an atomic sentence consists of a subject and a predicate. For example,

(1) Oscar goes to college.
(Subject: Oscar; Predicate: goes to college)

(2) Logic is difficult.
(Subject: Logic; Predicate: is difficult)

(3) The smallest ball in John's store is red.
(Subject: the smallest ball in John's store; Predicate: is red)

In English and mathematics, we also use expressions that have the form of a sentence, but, in our terminology, are not sentences because they are neither true nor false. For example,

(4) He is a professor.

(5) x is a prime.

(6) The derivative of the function f is defined.

If we do not know who "he" is in (4), or what "x" and "f" are in (5) and (6), we cannot determine whether these expressions are true or false. Such expressions will be called *sentence forms*, because if the pronouns are replaced by nouns and the variables replaced by constants, they become sentences. Notice that variables play the same role in mathematics as pronouns do in English.

Grammatically, in (4), "he" is the subject and "is a professor" the predicate; in (5), "x" is the subject and "is a prime" the predicate; and in (6), "the derivative of the function f" is the subject and "is defined" is the predicate.

The subjects of each of the preceding six sentences and sentence forms are examples of what we shall call *terms*. Nouns, pronouns, and variables are terms. Also, certain types of descriptions, as in (3) and (6) above, are terms. "The smallest ball in John's store" names a particular object, and "the derivative of the function f" names a particular object if f is replaced by a particular object. Intuitively, anything that can be the subject of a sentence or sentence form is a term. A little more precisely, a *term* is an expression that is a name, a constant, or a description, or that results in a name, a constant, or a description when the variables or pronouns are replaced by names, constants, or descriptions. Additional examples of terms are the following: Chicago, cat, she, mathematics, y, 5, $\pi^2/3$, the tallest person on x's basketball team, $x + 2$, $\sqrt{x^2 + y^3 - 7}$, f', $\int_a^b f(x)\,dx$.

We shall use the word "predicate" to mean a *property* or *relation*. In the six sentences and sentence forms given above, each of the predicates is a *1-place* predicate; that is, it only applies to one term. In sentence (1) above, for example, Oscar has the property of *going to college*; in (2), logic has the property of *being difficult*; in sentence (3), the smallest ball in John's store has the property of *being red*, and so on. Other predicates may apply to two or more terms. The predicates "older than," "mother of," "equals," "less than," and "admires" are examples of 2-place predicates; and "between," "$x < y$ and $x < z$," and "$x + y = z$" are examples of 3-place predicates.

Formally, in mathematics, predicates and functions are defined as follows:

Definition 5.1 An *n-place predicate* or *relation* is a set of ordered n-tuples.

For example, the relation of less than on the set of real numbers is the set of all ordered pairs of real numbers such that the first coordinate is less than the second coordinate.

If $n > 1$, the *domain* of an n-place predicate, R, is the set of all $n - 1$ tuples $\langle u_1, u_2, \ldots, u_{n-1}\rangle$ such that for some v, $\langle u_1, u_2, \ldots, u_{n-1}, v\rangle \in R$.

Definition 5.2 *f is a function of n variables* if f is an $n+1$-place relation such that for all $\langle u_1, u_2, \ldots, u_n\rangle$ in the domain of f, if $\langle u_1, u_2, \ldots, u_n, v_1\rangle$ and $\langle u_1, u_2, \ldots, u_n, v_2\rangle$ are both elements of f then $v_1 = v_2$.

Thus, a relation is a function if it is single-valued. Normally, we write "$f(u_1, u_2, \ldots, u_n) = v$" instead of "$\langle u_1, u_2, \ldots, u_n, v\rangle \in f$."

5.2 TRANSLATIONS INTO THE PREDICATE CALCULUS

In the formal language of the Predicate Calculus, we shall use lower-case letters (sometimes with subscripts) to represent terms. However, we shall reserve the use of the letters "f," "g," and "h" (sometimes with subscripts or superscripts) for function symbols. Capital letters (sometimes with subscripts or superscripts) will be used for predicate symbols. A 0-place predicate can be considered as a sentence, making this notation consistent with that in the Sentential Calculus, in which sentences were represented by capital letters. The Predicate Calculus should be thought of as an extension of the Sentential Calculus.

To translate a sentence or sentence form into the Predicate Calculus, we first have to identify the terms and predicates. For practice, we shall translate sentences (1) through (3) and sentence forms (4) through (6) above into the language of the Predicate Calculus.

(1) o: Oscar. (term)
Cx: x goes to college. (predicate)
Translation: Co.

Notice that the variable given in the definition of a predicate is just a place holder. If "Cx" means "x goes to college," then "Cy" is translated as "y goes to college." If "m" is the symbol used for "Miriam," then "Cm" is translated as "Miriam goes to college."

(2) l: logic. (term)
Dx: x is difficult. (predicate)
Translation: Dl

(3) b: the smallest ball in John's store. (term)
Rx: x is red. (predicate)
Translation: Rb.

Alternatively,

$f(x)$: the smallest ball in x. (term)
j: John's store. (term)
Rx: x is red. (predicate)
Translation: $Rf(j)$.

(4) x: he. (term)
Px: x is a professor. (predicate)
Translation: Px.

(5) x: x. (term)
Px: x is a prime. (predicate)
Translation: Px.

(6) f: f. (term)
$g(x)$: the derivative of x. (term)
Dx: x is defined. (predicate)
Translation: $Dg(f)$.

Some additional examples are given below. We shall translate the sentences and sentence forms using the constants, predicate symbols, and function symbols listed inside parentheses. Remember that the variables given are just place holders. Different variables or constants may be used in the translation. Compound sentences are translated using sentential connectives. Moreover, we are reserving the letters "f," "g," and "h" for function symbols, and we shall reserve the lower-case letters near the end of the alphabet (u, v, w, x, y, and z) for variables.

(7) Oscar and Virginia go to college. (o, i, Cx)
Solution:
o: Oscar. (term)
i: Virginia. (term)
Cx: x goes to college. (predicate)
Translation: $Co \wedge Ci$.

(8) Oscar will flunk the logic course unless he studies and does his homework. (o, Fx, Sx, Hx)
Solution:
o: Oscar. (term)
Fx: x flunks the logic course. (predicate)
Sx: x studies. (predicate)
Hx: x does his homework. (predicate)
Translation: $\neg(So \wedge Ho) \rightarrow Fo$, or $Fo \vee (So \wedge Ho)$.

(9) If Virginia is older than Miriam, then Miriam is not older than George. (i, m, e, Oxy)
Solution:
i: Virginia. (term)
m: Miriam. (term)
e: George. (term)
Oxy: x is older than y. (predicate)
Translation: $Oim \rightarrow \neg Ome$.

Normally, we shall use the standard mathematical symbols for mathematical expressions. For example, we shall use "2" as the symbol for "two";

"$x < y$" instead of "Lxy" for "x is less than y"; and "$x + y = z$" instead of "$Axyz$" or "$Ef(x, y)z$" for "x plus y equals z."

(10) The smallest prime is less than three.
Solution:
(a) s: the smallest prime.
3: three.
$x < y$: x is less than y.

Translation: $s < 3$.

(b) Since we know that two is the smallest prime, we could also have translated it as $2 < 3$.
(c) $f(x)$: the smallest element in the set x.
p: the set of all primes.
3: three.

Translation: $f(p) < 3$.

(11) Two is between x and y.
Solution:
(a) $Bxyz$: y is between x and z.
2: two.

Translation: $Bx2y$.

(b) 2: two.
$x < y$: x is less than y.

Translation: $(x < 2 \wedge 2 < y) \vee (y < 2 \wedge 2 < x)$.

5.3 QUANTIFIERS

One way to change a sentence form into a sentence is by replacing the variables by names, constants, or descriptions. For example, the sentence form

(1) x lives in the old mansion.

becomes

(2) Virginia lives in the old mansion.

if "x" is replaced by "Virginia." Sentence form (1) becomes

(3) The tallest professor at Harvard lives in the old mansion.

if "x" is replaced by "The tallest professor at Harvard." It becomes

(4) My cat, Whiskers, lives in the old mansion.

if "x" is replaced by "My cat, Whiskers." But consider the following sentence:

(5) Someone lives in the old mansion.

If we restrict the variable x to range over the set of people, sentence (5) means the same as

(6) There is an x such that x lives in the old mansion.

If there is no restriction on the range of the variable x, then sentence (5) means the same as

(7) There is an x such that x is a person and x lives in the old mansion.

The expression "there is an x" is called an *existential quantifier*. If the predicate Lx represents "x lives in the old mansion," we shall translate sentence (5) into our language as $(\exists x)Lx$, as long as the variable x is restricted to range over the set of all people. If this is not the case, we would introduce another predicate, Px, to stand for "x is a person." Then sentence (5) would be translated as $(\exists x)(Px \wedge Lx)$.

Examples of other quantifiers that are used in English and mathematics are "there exist finitely many x's," "there exist infinitely many x's," and "for every x." The most important one is the last one, "for every x." It is called a *universal quantifier*, and it is the one we shall use in our language. Consider the following sentence:

(8) Every ghost in the area lives in the old mansion.

Using quantifiers, sentence (8) can be translated as

(9) For every x, if x is a ghost in the area, then x lives in the old mansion.

If Gx is the predicate "x is a ghost in the area," then sentence (8) (and sentence (9)) can be translated into our language as $(\forall x)(Gx \rightarrow Lx)$. Notice that (8) is *not* translated as $(\forall x)(Gx \wedge Lx)$ because this latter expression would be interpreted as

For every x, x is a ghost in the area, and x lives in the old mansion.

or

Everything is both a ghost in the area and lives in the old mansion.

This interpretation is different from the intended meaning of sentence (8). The following two sentences illustrate the effect of negation on quantifiers:

(10) No ghost in the area lives in the old mansion.

(11) Not all ghosts in the area live in the old mansion.

Sentence (10) may be translated in either of the following two ways:

(i) For any x, if x is a ghost in the area, then x does not live in the old mansion.

$$(\forall x)(Gx \rightarrow \neg Lx).$$

(ii) There does not exist an x such that x is a ghost in the area and x lives in the old mansion.

$$\neg(\exists x)(Gx \wedge Lx).$$

Sentences (i) and (ii) are equivalent; thus, either translation is correct. In the next two chapters, however, we shall usually use the first translation—that given in sentence (i).

Similarly, sentence (11) has two translations:

(i) There is an x such that x is a ghost in the area and x does not live in the old mansion.

$$(\exists x)(Gx \wedge \neg Lx).$$

(ii) It is not the case that for all x, if x is a ghost in the area, then x lives in the old mansion.

$$\neg(\forall x)(Gx \rightarrow Lx).$$

Again, sentences (i) and (ii) are equivalent in English, but, for now, we shall normally use the first translation.

We'll give some more examples below to illustrate the use of the existential and universal quantifiers.

Example 5.1

Translate each of the following sentences into the Predicate Calculus using the letters indicated in parentheses for terms and predicate symbols.

(a) All students who like mathematics also like physics. (Sx, Mx, Px)

(b) There are some students who like Oscar but do not like George. (o, e, Sx, Lxy)

(c) No student who likes mathematics also likes Oscar. (o, Sx, Mx, Lxy)

(d) Each student who attended the meeting was accompanied by at least one of his/her parents. (Sx, Mx, Axy, Pxy)

(e) Students are the only people who attended the meeting accompanied by one of their parents. (Sx, Mx, Axy, Pxy)

(f) Some student who attended the meeting was accompanied by neither Oscar nor by a parent. (Sx, Mx, Axy, Pxy, o) (Assume it is the student's own parent.)

(g) Oscar flies a plane to Chicago with an instructor. (o, c, Ix, $Fxyz$)

(h) If Oscar flies a plane to Chicago with an instructor, then he attends the meeting accompanied by one of his parents. (o, c, Ix, $Fxyz$, Mx, Axy, Pxy)

(i) Not all student pilots who fly a plane to Chicago with an instructor attend a meeting accompanied by the instructor. (Sx, Ix, Mx, Axy, $Fxyz$, c)

Solution: We shall first translate each of the preceding sentences into English using quantifiers and then translate them into the Predicate Calculus. (There may be more than one correct translation.)

(a) For every x, if x is a student and x likes mathematics, then x likes physics.

Sx: x is a student.

Mx: x likes mathematics.

Px: x likes physics.

Translation: $(\forall x)(Sx \wedge Mx \rightarrow Px)$.

Alternatively, sentence (a) could be translated as

For every x, if x is a student, then if x likes mathematics, x likes physics.

Thus, a correct translation into the Predicate Calculus would be

$$(\forall x)(Sx \rightarrow (Mx \rightarrow Px)).$$

Notice that $Sx \wedge Mx \rightarrow Px$ is tautologically equivalent to $Sx \rightarrow (Mx \rightarrow Px)$, so it is not surprising that the two translations are equivalent. In the future, we shall normally not mention equivalent translations of this type.

Also, notice that (a) is **NOT** translated as

(i) $(\forall x)(Sx \wedge Mx \wedge Px)$.

Sentence (i), translated into English, would be

For every x, x is a student, and x likes mathematics, and x likes physics.

Translating into idiomatic English, sentence (i) becomes

(ii) Everyone (or everything) is a student who likes mathematics and likes physics.

This is clearly not the intended meaning of sentence (a) because sentence (ii) is trivially false. Usually, a sentence that begins with a universal quantifier is followed by a formula in which the main connective is the conditional rather than a conjunction.

(b) There is an x such that x is a student and x likes Oscar, but x does not like George. (Note that for us "some" means "one or more," whereas in English, the phrase "some students" usually implies more than one.)

o: Oscar.

e: George.

Sx: x is a student.

Lxy: x likes y.

Translation: $(\exists x)(Sx \wedge Lxo \wedge \neg Lxe)$.

It is important to note that sentence (b) is **NOT** translated as

(iii) $(\exists x)(Sx \rightarrow Lxo \wedge \neg Lxe)$,

and it is **NOT** translated as

(iv) $(\exists x)(Sx \wedge Lxo \rightarrow \neg Lxe)$.

In English, sentence (iii) would be

(v) There is an x such that if x is a student, then x likes Oscar but does not like George.

Sentence (iv) becomes

(vi) There is an x such that if x is a student and x likes Oscar, then x does not like George.

Neither sentence (v) nor (vi) conveys the intended meaning of sentence (b) because sentences (v) and (vi) are trivially true. There does exist something that is not a student. A conditional statement is true if the antecedent is false. An existential quantifier is usually followed by a conjunction, not a conditional.

(c) For every x, if x is a student and x likes mathematics, then x does not like Oscar.

o: Oscar.

Sx: x is a student.

Mx: x likes mathematics.

Lxy: x likes y.

Translation: $(\forall x)(Sx \wedge Mx \rightarrow \neg Lxo)$.

Equivalently,

There does not exist an x such that x is a student and x likes mathematics and x likes Oscar.

Translation: $\neg(\exists x)(Sx \wedge Mx \wedge Lxo)$.

(d) For every x, if x is a student and x attended the meeting, then there is a y such that x is accompanied by y and y is a parent of x.

(In this sentence, the existential quantifier is implied. We translate "at least one" by "there exists.")

Sx: x is a student.

Mx: x attended the meeting.

Axy: x is accompanied by y.

Pxy: x is a parent of y.

Translation: $(\forall x)(Sx \wedge Mx \rightarrow (\exists y)(Axy \wedge Pyx))$.

(Note that if Axy were translated as "x accompanies y," then in the preceding translation, "Axy" would be replaced by "Ayx.")
An equivalent translation for sentence (d) is

> For every x, there is a y such that if x is a student and x attended the meeting, then x is accompanied by y and y is a parent of x.

Translation: $(\forall x)(\exists y)(Sx \wedge Mx \rightarrow Axy \wedge Pyx)$.

(e) For this sentence, we have to change the order of expressions in order to abstract its meaning:

> For every x, if x attended the meeting and there is a y such that x is accompanied by y and y is a parent of x, then x is a student.

That is, if someone attends the meeting and is accompanied by a parent, that person is a student. The predicate symbols are the same as in part (d).

Translation: $(\forall x)[Mx \wedge (\exists y)(Axy \wedge Pyx) \rightarrow Sx]$.

(f) There is an x such that x is a student and x attended the meeting, but neither was x accompanied by Oscar nor does there exist a y so that x is accompanied by y and y is a parent of x.

Translation: $(\exists x)(Sx \wedge Mx \wedge \neg Axo \wedge \neg(\exists y)(Axy \wedge Pyx))$.

In the preceding translation, we could have replaced "nor does there exist a y . . ." by "for all y, if x is accompanied by y then y is not a parent of x." Then the translation becomes

Translation: $(\exists x)(Sx \wedge Mx \wedge \neg Axo \wedge (\forall y)(Axy \rightarrow \neg Pyx))$.

(g) There is an x such that x is an instructor and Oscar flies a plane to Chicago with x.

o: Oscar.

c: Chicago.

Ix: x is an instructor.

$Fxyz$: x flies a plane to y with z.

Translation: $(\exists x)(Ix \wedge Focx)$.

Again, in this problem, the existential quantifier is implied. We translate "an instructor" as "there is an instructor."

(h) If there is an x such that x is an instructor and Oscar flies a plane to Chicago with x, then there is a y such that Oscar attends the meeting and Oscar is accompanied by y and y is a parent of Oscar.

The predicate symbols are translated as in parts (d) and (g) above.

Translation: $(\exists x)(Ix \wedge Focx) \rightarrow (\exists y)(Mo \wedge Aoy \wedge Pyo)$.

Notice the difference in the form of the translation in sentence (h) than in the preceding examples. In sentence (h), the sentential connective, "if · · · then · · ·," is the main connective, whereas in the preceding examples, a quantifier is the main connective.

(i) There is an x such that x is a student pilot, and there is a y so that y is an instructor and x flies to Chicago with y, but it is not the case that x attends a meeting and x is accompanied by y.

c: Chicago.

Sx: x is a student pilot.

Ix: x is an instructor.

Mx: x attends a meeting.

Axy: x is accompanied by y.

$Fxyz$: x flies to y with z.

Translation: $(\exists x)[Sx \wedge (\exists y)[(Iy \wedge Fxcy) \wedge \neg(Mx \wedge Axy)]]$.

Alternatively, sentence (i) may be translated as

> It is not the case that for all x, if x is a student pilot, then for all y if y is an instructor and x flies to Chicago with y, then x attends a meeting and x is accompanied by y.

Translation: $\neg(\forall x)[Sx \rightarrow (\forall y)[(Iy \wedge Fxcy) \rightarrow (Mx \wedge Axy)]]$.

As we mentioned previously, we shall usually use the first translation in both sentences (c) and (i).

Normally, the universal quantifier, $(\forall x)$, is translated as "for every x," "for all x," "for each x," or "for any x," and the structure of the expression is $(\forall x)(\cdots \rightarrow \cdots)$. Key words in an English sentence that usually indicate a universal quantifier are "all," "every," "each," and "any." Also, the word "no," as in Example 5.1(c), and the word "only," as in Example 5.1(e), indicate a universal quantifier. The existential quantifier, $(\exists x)$, is translated as "there exists an x," "there is an x," or "for some x." The structure of an existential statement is usually $(\exists x)(\cdots \wedge \cdots)$. Key words that indicate an existential quantifier are "some," "there is," "there are," and "there exists." It may also be used when the words "not all" occur, as in Example 5.1(i), and "at least one" as in Example 5.1(d).

However, there are no hard and fast rules for translating English into the Predicate Calculus. The articles "a" or "an," for example, in some cases imply a universal quantifier and in others, an existential quantifier. In Example 5.1(g), we interpreted "an instructor" as "there exists an instructor." On the other hand, the sentence

> A student will do well at college if he/she studies.

would be interpreted as

For every x, if x is a student and x studies, then x will do well at college.

Translation: $(\forall x)(Sx \land Dx \rightarrow Wx)$.

Another ambiguity that occurs in English is that plural nouns not preceded by "all" or "some" are sometimes translated using a universal quantifier. At other times, these nouns are translated using an existential quantifier. For example, the sentence

Viruses cause colds.

would normally be interpreted to mean

Some viruses cause colds.

On the other hand, the sentence

Cats are animals.

would be interpreted as

All cats are animals.

The translation depends upon the meaning, and, in some cases, the meaning may not be clear. The problem is that English is ambiguous, whereas the Predicate Calculus is not.

Example 5.2

Translate each of the following formulas of the Predicate Calculus into idiomatic English, using the given interpretations for the constants, function symbols, and predicate symbols:

(a) $(\forall x)(Cx \rightarrow Mx)$; Cx: x is a child, Mx: x has mother.

(b) $(\exists x)(Dx \land (\forall y)(Cy \rightarrow Bxy))$; Dx: x is a dog, Cx: x is a cat, Bxy: x bites y.

(c) $(\exists x)(Tx \land Mx \land (\exists y)(Fyx \land Axy \land Axo))$; o: Oscar, Tx: x is a teacher, Mx: x attends the meeting, Fxy: x is a friend of y, Axy: x is accompanied by y.

(d) $(\forall x)(Sx \rightarrow (\exists y)(\exists z)(Cy \land Iz \land Fxyz))$; Sx: x is a student pilot, Cx: x is a city, Ix: x is an instructor, $Fxyz$: x flies to y with z.

(e) $(\forall x)(Sx \land (\forall y)(Py \rightarrow Vxye_1) \rightarrow Vxh(e_2)e_2)$; e_1: exam 1, e_2: exam 2; Sx: x is a student, Px: x is a problem, $Vxyz$: x solves y on z, $h(x)$: the hardest problem on x.

Solution:

(a) The formal translation is "For every x, if x is a child, then x has a mother." In idiomatic English, it becomes "Every child has a mother."

(b) Formally: There is an x such that x is a dog and for every y, if y is a cat, then x bites y.

Idiomatically: Some dog bites every cat.

(c) Formally: There is an x such that x is a teacher and x attends the meeting, and there is a y such that y is a friend of x and x is accompanied by y and x is accompanied by Oscar.

Idiomatically: Some teacher who attends the meeting is accompanied both by a friend and by Oscar.

(d) Formally: For every x, if x is a student pilot, then there is a y and a z such that y is a city and z is an instructor and x flies to y with z.

Idiomatically: Every student pilot flies to some city with some instructor.

(e) Formally: For every x, if x is a student, and for every y, if y is a problem, then x solves y on exam 1, then x solves the hardest problem on exam 2.

Idiomatically: Every student who solves every problem on exam 1 also solves the hardest problem on exam 2.

Example 5.3

Translate each of the following mathematical statements into the Predicate Calculus, using standard mathematical symbols. Assume that each variable ranges over the set of integers.

(a) There is no largest integer.

(b) If each of two integers has a factor of two, then their sum also has a factor of two.

(c) Two successive positive integers have no common factors different from one.

(d) The associative law for addition of integers.

Solution:

(a) For every integer x, there is an integer y such that y is larger than x:

(1) $$(\forall x)(\exists y)(x<y).$$

Notice that in a formula of the form "$(\forall x)(\exists y)(\ldots)$," the variable y is a function of x. Formula (1), could be written as

(2) $$(\forall x)(x<f(x)),$$

for some function symbol f. Formula (2) illustrates more forcefully that the "y" in formula (1) is dependent upon "x."

On the other hand, if we reverse the quantifiers in formula (1) and obtain

(3) $$(\exists y)(\forall x)(x<y),$$

the interpretation of formula (1) is changed completely. The translation of formula (3) is "There is a y such that for every x, y is larger than x." Equivalently, "There is an integer that is larger than every integer." The order of quantifiers in a formula is important.

Another way to translate sentence (a) is "There does not exist a largest integer." Equivalently, "It is not the case that there is an integer y such that for every integer x, x is less than or equal to y." In the Predicate Calculus, this becomes

$$\neg(\exists y)(\forall x)(x \leq y).$$

(b) $(\forall x)(\forall y)[(\exists u)(\exists v)(x = 2 \cdot u \wedge y = 2 \cdot v) \rightarrow (\exists w)(x + y = 2 \cdot w)]$.

(c) $(\forall x)(\forall y)(x > 0 \wedge y > 0 \wedge y = x + 1 \wedge$
$(\forall z)[(\exists u)(\exists v)(x = z \cdot u \wedge y = z \cdot v) \rightarrow z = 1])$.

(d) $(\forall x)(\forall y)(\forall z)(x + (y + z) = (x + y) + z)$.

5.4 FORMAL LANGUAGE OF THE PREDICATE CALCULUS

The language consists of the following:

Variables: u, v, w, x, y, z, and these letters with subscripts.
(We shall use lower-case letters near the end of the alphabet for variables.)

Constants: $a, b, \ldots, e, i, j, \ldots t$, and these letters with subscripts.

Function symbols: f, g, h, and these letters with subscripts.

Predicate symbols: $A, B, C, \ldots$ and these letters with subscripts or superscripts.
(A 0-place predicate is a sentence.)

(With each function and predicate symbol, we associate the number of arguments of that symbol. Formally, we should use a superscript or a subscript to indicate the number of arguments of the symbol, but these cause the notation to be cumbersome, so we shall omit them. The number of arguments each symbol has will be clear from the context.)

Connectives: $\neg, \wedge, \vee, \rightarrow, \leftrightarrow$.

Quantifiers: $\forall, \exists$.

We use parentheses for grouping.

In the next definition, we define *terms*. Notice that terms are syntactical expressions, certain strings of symbols, unlike predicates and functions, that are mathematical objects.

Definition 5.3 *Terms* are defined as follows:

1. Constants and variables are terms.
2. If f is a function symbol of n variables and $t_1, t_2, \ldots, t_n$ are terms, then $f(t_1, t_2, \ldots, t_n)$ is a term.
3. No other expressions are terms.

(Inductive definitions, like Definition 5.3, will be discussed in more detail in Chapter 15. At this point, it will be sufficient if you can distinguish a term from other strings of symbols.)

In the Sentential Calculus, we described how formulas are constructed from sentential variables using the sentential connectives. Definition 5.3 describes how terms are constructed in the Predicate Calculus from constants and variables using function symbols. For each term, we can construct a tree starting with constants and variables. For example, a tree for the term $\sqrt{x^2 + y^3 - 7}$ is given below.

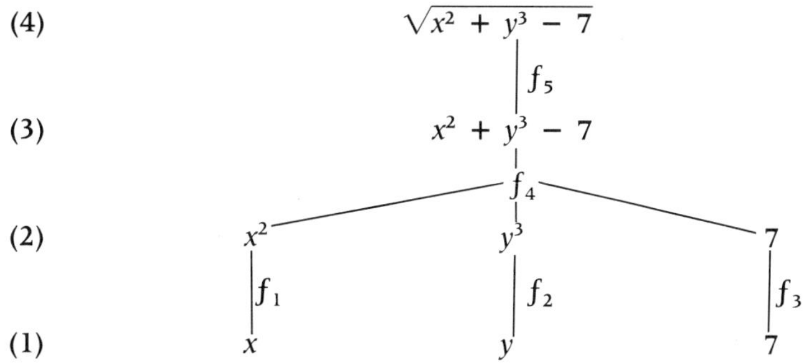

(4) $\sqrt{x^2 + y^3 - 7}$

f_5

(3) $x^2 + y^3 - 7$

f_4

(2) x^2 $\quad$ y^3 $\quad$ 7

f_1 $\quad$ f_2 $\quad$ f_3

(1) x $\quad$ y $\quad$ 7

The functions $f_1, f_2, \ldots, f_5$ are defined as follows: $f_1(u) = u^2$, $f_2(u) = u^3$, $f_3(u) = u$, $f_4(u, v, w) = u + v - w$ and $f_5(u) = \sqrt{u}$. The first level of the tree consists of constants and/or variables. Each subsequent level is constructed from the preceding level by applying some function. A tree for a term need not be unique. For example, a tree for $\sqrt{x^2 + y^3 - 7}$ might just contain two levels. The first level would consist of x, y, and 7, and the second, $\sqrt{x^2 + y^3 - 7}$, which is obtained from the first by applying the function symbol g, where $g(u, v, w) = \sqrt{u^2 + v^3 - w}$.

Next, we shall define what we mean by a *formula* of the Predicate Calculus.

Definition 5.4 If P is an n-place predicate symbol, and $t_1, t_2, \ldots, t_n$ are n terms, then $Pt_1t_2 \ldots t_n$ is called an *atomic formula*.

Each of the following is an example of an atomic formula: San Francisco is north of Los Angeles (*Nsl*); The smallest grape on x's vine is red ($Rf(x)$); Oscar flies to Chicago with x (*Focx*). In mathematics, using mathematical notation, all of the following are examples of atomic formulas: $x^2 + 2y < 3$ (the predicate symbol is $<$, the terms are $x^2 + 2y$ and 3); $\sqrt{x^2 + 1} = x^2 - 3x^3$ (the predicate symbol is $=$, the terms are $\sqrt{x^2 + 1}$ and $x^2 - 3x^3$); $\int_1^\infty f(x)\,dx \geq \sum_{n=2}^\infty f(n)$ (the predicate symbol is $\geq$, the terms are $\int_1^\infty f(x)\,dx$ and $\sum_{n=2}^\infty f(n)$).

Definition 5.5 Formula.

1. An atomic formula is a formula.
2. If φ and ψ are formulas, so are $(\neg\varphi)$, $(\varphi \wedge \psi)$, $(\varphi \vee \psi)$, $(\varphi \rightarrow \psi)$, and $(\varphi \leftrightarrow \psi)$.
3. If v is a variable and φ is a formula, then $((\forall v)\varphi)$ and $((\exists v)\varphi)$ are formulas.
4. No other expressions are formulas.

(This is another inductive definition. See the comment following Definition 5.3.)

Thus, it follows from Definition 5.5 that each formula of the Predicate Calculus has a tree. The first level consists of atomic formulas and each higher level is obtained from preceding levels by applying part 2 or 3 of Definition 5.5. For example, a tree for the formula

$$((\forall x)(Sx \rightarrow ((\exists y)(Cy \wedge Fxya))))$$

is given below.

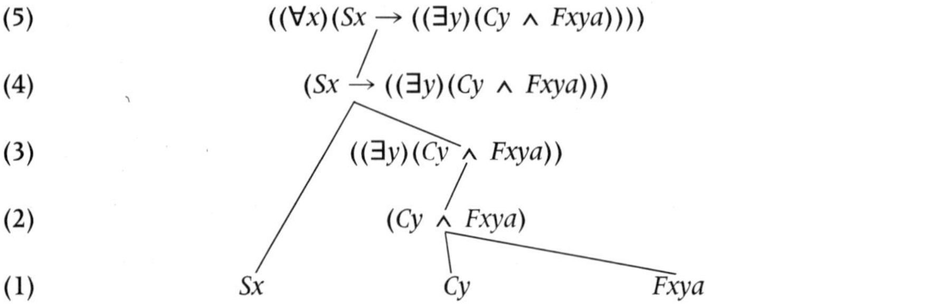

(5) $((\forall x)(Sx \rightarrow ((\exists y)(Cy \wedge Fxya))))$

(4) $(Sx \rightarrow ((\exists y)(Cy \wedge Fxya)))$

(3) $((\exists y)(Cy \wedge Fxya))$

(2) $(Cy \wedge Fxya)$

(1) Sx Cy $Fxya$

All the atomic formulas are on level (1). Level (2) is obtained by applying (b); level (3), by (c); level (4), by (b); and level (5), by (c).

On the other hand, the expression

$$\neg Sx(\forall y) \rightarrow \wedge Ty(\exists x)$$

is a meaningless string of symbols and is not a formula.

5.5 CONVENTIONS FOR PARENTHESES

We shall extend the conventions in the Sentential Calculus for eliminating unnecessary parentheses to the Predicate Calculus. Remember that additional parentheses may always be added to prevent any misunderstanding of the meaning of a formula.

1. The outermost parentheses around a formula will be omitted.
2. Quantifiers bind more closely than any sentential connective.
3. Negation binds more closely than the other sentential connectives.
4. Conjunction and disjunction bind more closely than the conditional and biconditional.

Without the Convention	With the Convention
$(((\forall x)Rx) \rightarrow ((\exists y)(Iy \wedge Bxy)))$	$(\forall x)Rx \rightarrow (\exists y)(Iy \wedge Bxy)$
$((\forall x)((\exists y)Rxy))$	$(\forall x)(\exists y)Rxy$
$((\forall x)((\forall y)((Sx \wedge Ty) \rightarrow Rxy)))$	$(\forall x)(\forall y)(Sx \wedge Ty \rightarrow Rxy)$
$(\neg((\forall x)(\neg(Rx \wedge (\neg Sx)))))$	$\neg(\forall x)\neg(Rx \wedge \neg Sx)$

Note, in particular, the difference between

(1) $$(\forall x)Rx \rightarrow (\exists y)(Iy \wedge Bxy)$$

and

(2) $$(\forall x)(Rx \rightarrow (\exists y)(Iy \wedge Bxy)).$$

In formula (1), the quantifier $(\forall x)$ only affects the variable x, which occurs in Rx, whereas in (2), the quantifier $(\forall x)$ affects the occurrence of x both in Rx, and in Bxy. Moreover, in formula (1), the conditional is the main connective. Formula (1) is interpreted as

> *If* (for each x, Rx) *then* (there is a y such that Iy and Bxy).

On the other hand, formula (2) is interpreted as

> *For every x*, (if Rx then there is a y such that Iy and Bxy).

5.6 SENTENCES AND SENTENCE FORMS

Now that we have a clearer idea of the notion of a formula, we will be able to differentiate between formulas that are *sentences* and those that are not. The latter formulas will be called *sentence forms*. First, we shall define the notion of the *scope* of a quantifier.

Definition 5.6 The *scope of a quantifier* occurring in a formula is the quantifier itself together with the shortest formula immediately following the quantifier.

The parentheses in a formula always make clear what is the shortest formula following a quantifier. In each of the following formulas, we indicate the scope of the quantifier $(\forall x)$.

(1) $$\underbrace{(\forall x)Rx}_{\text{scope}} \rightarrow (\exists y)(Iy \wedge Bxy)$$

(2) $$\underbrace{(\forall x)(Rx \rightarrow (\exists y)(Iy \wedge Bxy))}_{\text{scope}}$$

(3) $$(\exists y)\underbrace{(\forall x)(Rx \wedge Sx \rightarrow Bxy)}_{\text{scope}}$$

(4) $$\underbrace{(\forall x)(\exists y)(x + y = 0)}_{\text{scope}} \rightarrow (\exists x)(\forall y)(y + x = y)$$

(5) $$(\exists y)(y > 0 \wedge \underbrace{(\forall x)(\exists z)(y \cdot x = z))}_{\text{scope}}$$

Definition 5.7 An occurrence of a variable v in a formula is said to be *bound* if it is within the scope of a quantifier $(\forall v)$ or $(\exists v)$. Otherwise the occurrence of v is said to be *free*.

In formula (1) above, for example, the two occurrences of x in $(\forall x)Rx$ are both bound, but the occurrence of x in Bxy is free. All occurrences of y in formula (1) are bound. All occurrences of x and y are bound in formulas (2) through (5).

We say a variable is *free in a formula* if it has at least one free occurrence, and a variable is *bound in a formula* if it has at least one bound occurrence. In formula (1) above, the variable x is both free and bound, whereas y is bound, but not free.

Definition 5.8 A formula is called a *sentence* if it has no free variables. Otherwise, we call it a *sentence form*.

In Section 5.3, we mentioned how sentence forms may be converted to sentences. We shall see later that when an interpretation is given to all the predicate symbols in a sentence, it is either true or false. Thus, the word "sentence" is used in the Predicate Calculus similarly to the way it is used in the Sentential Calculus.

Example 5.4 Classify each of the following expressions as

(i) terms,

(ii) atomic formulas,

(iii) formulas,

(iv) sentences,

(v) none of these.

(Some expressions may be in more than one classification.)

(a) Oscar and Miriam.

(b) x is older than y.

(c) The shortest person in x's class.

(d) Oscar dates every girl in his class.

(e) $\sqrt{x^2 - (y^3 + 7)}$

(f) $3 > 1 + 8$

(g) $(x \wedge y) > 0$

(h) $x < (y < z)$

(i) $(\forall x)(x - x = 0)$

Solution:

(a) "Oscar" and "Miriam" are terms. We cannot connect terms with sentential connectives in our language. Therefore, (a) is neither a term nor a formula, so the answer is (v). (In order to translate into the Predicate Calculus the sentence "Oscar and Miriam are students," we would rewrite it as "Oscar is a student and Miriam is a student.")

(b) This is an atomic formula. It would be translated as Oxy. Because every atomic formula is a formula, (ii) and (iii) is the correct answer.

(c) This expression is a term, so (i) is correct.

(d) This is a sentence. We could translate it into the Predicate Calculus as $(\forall x)(Gx \rightarrow Dox)$. Because every sentence is a formula, (iii) and (iv) is the answer.

(e) This is a term—(i).

(f) The expression "$3 > 1 + 8$" is an atomic formula. The predicate is $>$, and the terms are 3 and $1 + 8$. It is also a sentence, because there are no free variables. (It happens to be false, but that is irrelevant for our purposes here.) Thus, (ii), (iii), and (iv) is the answer.

(g) This expression is meaningless in our language. A sentential connective cannot connect two terms; therefore, as written, the answer is (v). When students write "$(x \wedge y) > 0$," however, they usually mean "$(x > 0) \wedge (y > 0)$." The latter expression is the way it has to be written in our language.

(h) As written, expression (h) is meaningless. The predicate $<$ is a 2-place predicate. The expression "$y < z$" is a formula, not a term, and x is a term. It is meaningless to say that a term is less than a formula. A term can only be less than another term. Thus, (v) is the correct answer. Frequently, in mathematics, we write "$x < y < z$" as an abbreviation for "$x < y$ and $y < z$." The latter expression is a formula.

(i) This is a formula with no free variables. It is not atomic because it contains a quantifier; thus, it is a sentence. The answer is (iii) and (iv).

EXERCISES 5

A. Translate each of the following sentences into the language of the Predicate Calculus, using the indicated letters for constants, function symbols, and predicate symbols. (Answers to 2, 4, 7, 10, 13, 15, 19, 21, 23, 25, 28, 30, 34, 37, 39, 41, 45, and 47.)

(1) All birds are animals. (Bx, Ax)

(2) Some cats do not have tails. (Cx, Tx)

(3) Not all plants need sunlight. (Px, Sx)

(4) No student in the class likes physics. (Sx, Cx, Px)

(5) Some students like physics, and some students like mathematics. (Sx, Px, Mx)

(6) Some students who like physics also like mathematics. (Sx, Px, Mx)

(7) Oscar and Miriam are students who like mathematics. (o, m, Sx, Mx)

(8) The shortest student in the class likes mathematics but does not like physics. (s, Mx, Px)

(9) Only students who like physics like mathematics. (Sx, Px, Mx)

(10) Among the students in the mathematics class, only those who do their homework will receive an *A* in the course. (Sx, Mx, Hx, Ax)

(11) Oscar and George both like Miriam. (o, e, m, Lxy)

(12) If Miriam is older than Virginia and Virginia is older than Theresa, then Miriam is older than Theresa. (m, i, t, Oxy)

(13) Every student who is older than Miriam is also older than Virginia. (m, i, Sx, Oxy)

(14) Every professor is older than every student. (Px, Sx, Oxy)

(15) Every professor who is older than Oscar is also older than some student. (o, Px, Sx, Oxy)

(16) Not every professor who is older than Oscar is older than every student. (o, Px, Sx, Oxy)

(17) No one is older than himself or herself. (Oxy)

(18) If someone is older than Oscar, then all students are older than Oscar. (o, Sx, Oxy)

(19) Any student who likes every professor also likes himself or herself. (Sx, Px, Lxy)

In Exercises (20) through (26), use the following predicate symbols and constants:

Fx: *x* is a faculty member.

Sx: *x* is a student.

Mx: *x* attended the meeting.

Axy: *x* is accompanied by *y*.

Pxy: *x* is a parent of *y*.

e: George.

m: Miriam.

(20) All students and faculty members attended the meeting.
(21) No student who attended the meeting is accompanied by a faculty member.
(22) Some parent of a student did not attend the meeting.
(23) If all faculty members attended the meeting, then some student is accompanied by a faculty member.
(24) George and Miriam attended the meeting, but neither of them were accompanied by a student.
(25) No faculty member who is accompanied by George attended the meeting.
(26) If neither George nor Miriam is accompanied by a student, then no student attended the meeting.

In Exercises (27) through (38), use the following predicate symbols and constants:

Sx: *x* is a student pilot.

Ix: *x* is an instructor.

Cx: *x* is a city.

Fxyz: *x* flies a plane to *y* accompanied by (with) *z*.

c: Chicago.

(27) Some student pilot flies a plane to Chicago accompanied by an instructor.
(28) Some student pilot flies a plane to some city accompanied by an instructor.
(29) Every instructor has some student pilot who flies a plane with him/her to some city.
(30) Anyone who flies a plane to any city accompanied by an instructor is a student pilot. (Note the two different uses of the word "any" in this sentence.)
(31) Not all instructors who fly a plane to some city with an instructor are student pilots.

(32) Student pilots who fly a plane to Chicago accompanied by a student pilot are instructors.
(33) No one but an instructor flies a plane to Chicago accompanied by a student pilot.
(34) There is an instructor such that every student pilot flies a plane to some city with him/her.
(35) No instructor flies a plane to every city accompanied by a student pilot.
(36) Some student pilot who flies a plane to every city by himself or herself is an instructor.
(37) No student pilot who does not fly a plane to any city accompanied by an instructor is an instructor.
(38) Any student pilot who does not fly a plane to all cities by himself or herself is not an instructor.

In Exercises (39) through (50) use the following predicate symbols, function symbols, and constants:

Dx: x is a student.

Ex: x is an exam.

Pxy: x is a problem on exam y.

Sxy: x solves y.

$f(x, y)$: the xth problem on exam y.

$g(x)$: the easiest problem on exam x.

$h(x)$: the hardest problem on exam x.

a: Alex.

n: Nancy.

1: one.

2: two.

3: three.

(39) No student solves problem one on exam three.
(40) Nancy solves the hardest problem on exam one, but does not solve the easiest problem on exam two.
(41) If any student solves the hardest problem on exam two, then both Nancy and Alex solve it.
(42) Every student who solves problem one on exam one also solves the easiest problem on exam one.
(43) If some student solves every problem on every exam, then some student solves the hardest problem on every exam.
(44) Some student who solves every problem on exam one did not solve problem one on exam two.

(Be careful of the order of the quantifiers in Exercises (45) through (50).)

(45) Every student solves some problem on every exam.
(46) Only students solve every problem on some exam.

(47) There is no student who solves every problem on some exam.
(48) Students who solve every problem on some exam also solve every problem on all exams.
(49) Anyone who solves a problem on an exam is a student.
(50) If some student solves some problem on each exam, then some student solves the easiest problem on all exams.

B. Translate each of the following sentences into the language of the Predicate Calculus, using only predicate symbols, constants, and function symbols from the list given below. Assume that each variable ranges over the set of integers. Use the standard symbols for mathematical expressions. For example, "x is a multiple of y" or "y is a factor of x" means "there is a z, so that $x = y \cdot z$"; "x is even" means "x is a multiple of two," "x and y have the same parity" means "x and y are both even or x and y are both odd"; "x is positive" means "x is greater than zero"; and so on (Answers to 1, 3, 6, 9, 11, 15, 18, 21, and 23.)

$x > y$: x is greater than y.

$x \geq y$: x is greater than or equal to y.

$x = y$: x equals y.

(You may also use the convention that "$x \not> y$" is an abbreviation for "$\neg(x > y)$" and similarly for the other predicates.)

$x + y$: the sum of x and y.

$x \cdot y$: the product of x and y.

Also use the standard symbols for integers: "-1" for "negative one"; "0" for "zero"; "1" for "one"; and so on.

(1) The Commutative Law for Addition.
(2) The Distributive Law for Multiplication over Addition.
(3) Some integers are multiples of three.
(4) Not all integers are multiples of five.
(5) Ten is a positive multiple of some integer.
(6) Every positive multiple of seven is greater than five.
(7) No positive multiple of five is less than seven. (Use "$y > x$" for "$x < y$.")
(8) Every positive even integer is greater than or equal to two. (Two is the smallest positive even integer.)
(9) There is a smallest positive odd integer.
(10) There is at least one integer between five and two. (You may assume five is greater than two.)
(11) Every integer has an additive inverse.
(12) Not all integers different from zero have a multiplicative inverse.
(13) The sum of two odd integers is an even integer.
(14) The product of any integer with an even integer is an even integer.
(15) Some odd integer is a factor of every even integer.

(16) Every even integer is a multiple of some odd integer.
(17) There are two integers that have the same parity, and each is a multiple of seven.
(18) Not every pair of integers that has the same parity is a multiple of three.
(19) Every pair of distinct integers that has the same parity has the property that their squares also have the same parity. (Two quantities are *distinct* if and only if they are not equal.)
(20) Each pair of distinct integers that has the same parity has an integer between them.

In Exercises (21) through (24), assume that a linear equation has the form $u \cdot x + v = 0$, where u and v are integers. (It is assumed in this exercise that all variables range over the set of integers.)
(21) Some linear equations have integers as solutions.
(22) A linear equation has an integer solution if and only if the leading coefficient is a factor of the constant term.
(23) No linear equation has two distinct integer solutions.
(24) For any linear equation, if the coefficients are both odd, then the solution is odd unless there is no integer solution.

C. Translate each of the following formulas of the Predicate Calculus into idiomatic English, using the given interpretations for the constants, function symbols and predicate symbols. (Answers to 1, 4, 7, 10, 13, 18, 20, 22, 25, 27, and 29.)

For Exercises (1) through (10), interpret the symbols as follows:

Cx: x is a cat.

Dx: x is a dog.

Tx: x has a tail.

Bxy: x bites y.

Lxy: x likes y.

(1) $(\exists x)(Cx \land \neg Tx)$
(2) $(\forall x)(Dx \rightarrow Tx)$
(3) $(\exists x)(Cx \land (\forall y)(Dy \rightarrow Bxy))$
(4) $(\forall x)(Cx \rightarrow (\forall y)(Dy \rightarrow \neg Lxy))$
(5) $(\exists x)(\exists y)(Cx \land Dy \land Lxy)$
(6) $(\forall x)(Cx \land Tx \rightarrow (\exists y)(Dy \land Bxy))$
(7) $(\forall x)[(\forall y)(Dy \rightarrow Bxy) \rightarrow Cx]$
(8) $(\forall x)(Tx \land Bxx \rightarrow Dx)$
(9) $(\forall x)[Tx \land (\exists y)(Dy \land Bxy) \rightarrow Cx]$
(10) $(\exists x)[Cx \land (\forall y)(Dy \land \neg Ty \rightarrow Bxy \lor \neg Lxy)]$

For Exercises (11) through (20), use the following interpretations:

a: Alex.

n: Nancy.

Pxy: x is a problem on exam y.

Ex: x is an exam.

Mx: x is a man.

Wx: x is a woman.

Sxy: x solves y.

(11) $(\exists x)(Ex \wedge (\forall y)(Pyx \to Say))$
(12) $(\forall x)(Ex \to (\exists y)(Pyx \wedge Sny))$
(13) $(\exists x)(\exists y)(Ex \wedge Pyx \wedge Sny) \to (\exists u)(\exists v)(Eu \wedge Pvu \wedge Sav)$
(14) $(\exists x)[Wx \wedge (\forall y)(\forall z)(Ey \wedge Pzy \to Sxz)]$
(15) $(\forall x)[Mx \to (\exists y)(\exists z)(Ey \wedge Pzy \wedge \neg Sxz)]$
(16) $(\forall x)[(\exists y)(Ey \wedge (\forall z)(Pzy \to Sxz)) \to Mx]$
(17) $(\forall x)[Wx \to (\exists y)(Ey \wedge (\forall z)(Pzy \to Sxz)] \wedge (\exists u)(Eu \wedge (\forall v)(Pvu \to Sav))$
(18) $(\forall x)(Ex \to (\exists y)(Pyx \wedge Sny)) \to (\forall u)[Wu \to (\forall v)(Ev \to (\exists w)(Pwv \wedge Suw))]$
(19) $\neg(\exists x)[Wx \wedge (\exists y)(Ey \wedge (\forall z)(Pzy \wedge Sxz))]$
(20) $(\exists x)[Mx \wedge (\exists y)(\forall z)(Ez \wedge Pyz \to Sxy)]$

For Exercises (21) through (30), use the following interpretations:

Ix: x is an integer.

$x < y$: x is less than y.

$x \leq y$: x is less than or equal to y.

$x = y$: x equals y.

$x + y$: the sum of x and y.

$x - y$: the difference of x and y.

$x \cdot y$: the product of x and y.

Use the standard symbols for real numbers, and assume the variables range over the set of real numbers. Also, "$x \neq y$" is an abbreviation for "$\neg(x = y)$."

(21) $(\forall x)(\forall y)(\forall z)(x \cdot (y \cdot z) = (x \cdot y) \cdot z)$
(22) $(\forall x)(\forall y)[x \neq y \to (\exists z)((x < z \wedge z < y) \vee (y < z \wedge z < x))]$
(23) $(\exists x)(\forall y)(y \leq x)$
(24) $(\forall x)(\exists y)(x < y)$
(25) $(\forall x)[x \neq 0 \to (\exists y)(x \cdot y = 1 \wedge (\forall z)(x \cdot z = 1 \to z = y))]$
(26) $(\forall x)(\forall y)[Ix \wedge Iy \wedge y \neq 0 \to (\exists u)(\exists v)(Iu \wedge Iv \wedge 0 \leq v \wedge v < y \wedge x = u \cdot y + v)]$
(27) $(\exists x)[Ix \wedge 1 < x \wedge (\forall y)(\forall z)(Iy \wedge Iz \wedge x = y \cdot z \to (y = x \wedge z = 1) \vee (y = 1 \wedge z = x))]$
(28) $(\forall x)[(0 \leq x) \to (\exists y)(y \cdot y = x)]$
(29) $(\forall x)[Ix \wedge (\exists y)(Iy \wedge (\exists z)(Iz \wedge y = 2 \cdot z + 1 \wedge x = 5 \cdot y)) \to (\forall u)\ (Iu \to x \neq 2 \cdot u)]$
(30) $(\exists x)[Ix \wedge (\exists y)(Iy \wedge x = 5 \cdot y) \wedge (\forall z)(Iz \to x \neq 7 \cdot z)]$

D. Choose appropriate letters for predicates, constants, and functions, and translate each of the following paragraphs into the language of the Predicate Calculus. (Answers to 3 and 7.)

(1) All instructors who do research like to teach. Smith is an instructor who does not like to teach. There are some instructors who do not do research.

(2) Blacky and Tabby are cats that live in a barn. Blacky chases all mice, but Tabby does not chase any mice. All mice are afraid of Blacky, but some mice are not afraid of Tabby. No mice live in the barn.

(3) Every problem is solved by some student. Oscar solves some problems. Some student solves every problem. No students do not solve any problems.

(4) Every student studies some subject with some instructor. No student studies every subject with every instructor. There is a subject that every student studies with some instructor. Any student who studies every subject with himself or herself is an instructor.

(5) Someone is hit by a car every day. Oscar is hit by a car on Monday. George and Agnes are hit by a car on Tuesday, but no one is hit by a car on Friday.

(6) Some integers are even, and some are odd. No integer is both even and odd. If an integer is even, its immediate successor is odd; and if an integer is odd, its immediate successor is even.

(7) Every integer is a multiple of some integer. There is an integer that is not a multiple of any integer except itself. Every integer is a multiple of itself.

(8) Every integer has at least one additive inverse. No two distinct integers have the same additive inverse. If an integer has x as its additive inverse and another integer has $-x$ as its additive inverse, then the sum of the two integers is zero, for every integer x. (Use the definition that *y is the additive inverse of x* if and only if $x + y = 0$.)

E. In each of the following formulas, circle all free occurrences of the variable x.

(1) $Px \wedge Qxy \rightarrow (\forall x)(Rx \rightarrow Px)$

(2) $(\forall x)(Px \rightarrow (\exists y)(Qy \wedge Rxy))$

(3) $(\exists x)Px \rightarrow (\forall y)(Qy \wedge Rxy)$

(4) $\neg(\exists x)(Px \wedge \neg Rx) \vee Qx$

(5) $(\forall x)Qx \wedge (Rx \vee Pxy)$

(6) $Ray \rightarrow (\forall x)(Qax \rightarrow Px \wedge Rax)$

(7) $(\forall x)Qxa \rightarrow Rxa$

(8) $(\exists y)(Pxy \wedge Qxa \wedge Ray)$

(9) $(\forall x)(\forall y)Rxy \rightarrow Qx$

(10) $(\forall y)[(\exists x)Rxy \rightarrow Qxy]$

(11) $(\forall x)[(\exists y)Pxy \rightarrow (\forall z)(Rz \rightarrow Qxyz)]$

(12) $(\forall y)[Py \rightarrow (\exists z)(Qxz \wedge (\exists x)Ryx \rightarrow Sxyz)]$

(13) $(\exists z)[(\forall y)(\exists x)Rxyz \wedge ((\forall x)(\exists y)Pxyz \to Qxyz)]$
(14) $(\forall x)[(\exists y)(\exists z)Rxyz \to Sxy \wedge Txz]$
(15) $(\forall x)(Px \to (\exists y)(Qxy \wedge Rx)) \wedge (\forall z)Sxz$

F. Classify each of the following expressions as:

(i) terms,
(ii) atomic formulas,
(iii) formulas that are not atomic,
(iv) sentences,
(v) none of these.

(Some expressions may be in more than one classification.) (Answers to 2, 8, 13, 15, 20, 24, and 30.)

(1) Every student in the class will graduate.
(2) The tallest student in the class.
(3) Oscar is older than the student who sits in the first seat in the first row in his class.
(4) Marianne or Peter.
(5) The student with the highest grade point average on x's football team.
(6) x is between 2 and y.
(7) Flowers.
(8) Pink flowers.
(9) Pink.
(10) Flowers imply a cold winter.
(11) The prettiest flower in x's garden is red.
(12) All the flowers in Peter's garden are either pink or red.
(13) Nancy types rapidly and accurately.
(14) Types rapidly.
(15) Accurately.
(16) $x - 2y^2 + z^3$
(17) $(\forall x)(7 < Px)$
(18) $(\exists x)(\forall y)(x + y = 0)$
(19) $x \cdot y = 7x^2 + 2y - 10$
(20) $1 \leq \sqrt{3}$
(21) $(x \vee y) < z$
(22) $Px = Qxy$
(23) $(\forall x)(x + y = 1) \vee (x + y > 1)$
(24) $(x + y = -(x + y))^2$
(25) $(\forall x)[(\forall y)(x + y = 0) \to x = 0]$
(26) $(\forall x)(\forall y)(\exists z)(x + z = y \leftrightarrow z = y - x)$
(27) $(\forall x) \wedge (\forall y)(x + y = y + x)$
(28) $(\forall y)[(\exists x)(x + y = x + y^2) \to y = 0 \vee 1]$
(29) $(\forall x)[(x < Px) \vee (Px < x)]$
(30) $(\forall x)[(\exists y)(x + y = x) \to y = 0]$

G. Construct a tree starting with atomic formulas for each of the following formulas:

(1) $(\forall x)(Px \wedge Qa \rightarrow Rx)$
(2) $(\exists x)((Px \wedge Qx) \wedge \neg Rx)$
(3) $(\forall x)(Px \rightarrow (\exists y)(Qx \wedge Rxy))$
(4) $(\exists x)(Px \wedge (\forall y)(Qxy \rightarrow \neg Rx))$
(5) $(\exists x)Px \rightarrow (\forall y)(Px \wedge Qy \rightarrow Rxy)$
(6) $(\forall x)(\exists y)Pxy \wedge (\exists z)(Pxy \wedge (Qz \vee Rxyz))$
(7) $(\forall x)[(\exists y)(Pxy \wedge Qxyx) \rightarrow Pax \wedge Rabx]$
(8) $(\forall x)(Pxf(a) \rightarrow Qf(x)a) \vee (\exists y)(Pyf(y) \wedge Qaf(a))$
(9) $(\forall x)(Px \wedge Qx) \rightarrow (\exists y)(Py \wedge Ryf(y))$
(10) $(\exists x)[(\forall y)(Px \wedge Py \rightarrow Qxy) \wedge (Px \vee Qxx)]$

Chapter 6

Interpretation Structures

To show an argument in the Sentential Calculus is not valid, we assign truth values to the atomic sentences in such a way that the premises are true and the conclusion is false. Similarly, to show that a set of premises is consistent, we assign truth values to the atomic sentences so that the premises are all true simultaneously. We would like to do something similar in the Predicate Calculus, but how can we make truth assignments in the Predicate Calculus? How can we determine if a formula of the Predicate Calculus is true or false?

6.1 INTERPRETATION STRUCTURES

Consider the existential formula

(1) $(\exists x)(Px \wedge Qxc)$.

Is this formula true or false? We cannot answer this question until we interpret its symbols. First we must decide the range of the variables, that is, what is the universe? What values may the variables assume? We would expect an existential formula to mean "something exists." By specifying the universe, we specify what these "things" could be—people, numbers, functions, lawyers, plants, and so on.

Next, we must specify what the predicate symbols, constants, and function symbols (if any) mean in the universe. Thus, an interpretation structure for formula (1) is a structure

$$\mathbf{S} = \langle \mathbf{U}, \mathbf{P}, \mathbf{Q}, \mathbf{c} \rangle,$$

where **U** is the universe, a nonempty set, and **P**, **Q**, and **c** are interpretations of *P*, *Q*, and *c*, respectively.

For example, let **U** = $\mathbb{R}$, the set of all real numbers; **Px**: **x** is rational; **Qxy** : **x > y**; and **c** : **0**. Then the interpretation of formula (1) in the structure **S** is

(2) Some real number is both rational and positive.

Sentence (2) is a true statement in **S** because, for example, **3** is a real number that is both rational and positive. So, there does exist such a number in $\mathbb{R}$.

On the other hand, we could take another interpretation structure

$$\mathbf{S}' = <\mathbf{U}', \mathbf{P}', \mathbf{Q}', \mathbf{c}'>$$

for formula (1), where **U′**, **Q′**, and **c′** are the same as **U, Q**, and **c**, respectively, but

$$\mathbf{P}'\mathbf{x} : \mathbf{x} \text{ is negative.}$$

Then in **S′**, formula (1) is interpreted as

(3) Some real number is both negative and positive.

Sentence (3) is, of course, false in **S′**. There does not exist any real number that is both negative and positive.

An interpretation structure for a formula in the Predicate Calculus is a generalization of an assignment of truth values to the sentential variables in a formula in the Sentential Calculus.

Let us next look at an interpretation structure for the following universal formula:

(4) $(\forall x)(Px \wedge Qx \rightarrow Rxf(c))$.

In formula (4), f is a function symbol and c is a constant. An interpretation structure, **S**, for formula (4) has the following form: $\mathbf{S} = <\mathbf{U}, \mathbf{P}, \mathbf{Q}, \mathbf{R}, \mathbf{f}, \mathbf{c}>$. Suppose we make the following interpretations:

$$\mathbf{U} = \mathbb{Z}, \text{the set of all integers.}$$

$$\mathbf{Px} : \mathbf{x} \text{ is positive.}$$

$$\mathbf{Qx} : \mathbf{x} \text{ is even.}$$

$$\mathbf{Rxy} : \mathbf{x} > \mathbf{y}.$$

$$\mathbf{c} = \mathbf{0}$$

$$\mathbf{f(x)} = \mathbf{x} + \mathbf{1}.$$

Then in **S**, formula (4) is interpreted as

(5) Every positive even integer is greater than one.

Sentence (5) is true in this interpretation because the positive even integers are **2**, **4**, **6**, . . . , and each of them is greater than **1**.

On the other hand, suppose we change the structure **S** to a different structure **S′** for formula (4) by defining

$$\mathbf{c}' = \mathbf{4}$$

but we do not change any of the other quantities. Then we would interpret formula (4) in $\mathbf{S}'$ as

(6) Every positive even integer is greater than five.

However, sentence (6) is false in $\mathbf{S}'$ because we can find a positive even integer, namely two or four, that is not greater than five. Therefore, it is not true that every positive even integer is greater than five.

How should we interpret a formula that has a free variable? For example, "x" is free in

(7) $Pxa \rightarrow (\exists y)Qf(x, y)b.$

An interpretation structure, **S**, for formula (7) has the following form: $\mathbf{S} = < \mathbf{U}, \mathbf{P}, \mathbf{Q}, \mathbf{f}, \mathbf{a}, \mathbf{b} >$, where **P** and **Q** are 2-place predicates on **U**, **f** is a function of two variables from **U** to **U** (in symbols, $\mathbf{f} : \mathbf{U} \times \mathbf{U} \rightarrow \mathbf{U}$), and **a** and **b** are elements of **U**. To give a specific example, suppose

$$\mathbf{U} = \mathbb{R}, \text{the set of real numbers.}$$

$$\mathbf{Pxy} : \mathbf{x} < \mathbf{y}.$$

$$\mathbf{Qxy} : \mathbf{x} = \mathbf{y}.$$

$$\mathbf{f(x, y)} = \mathbf{x} \cdot \mathbf{y}.$$

$$\mathbf{a} = \mathbf{2}.$$

$$\mathbf{b} = \mathbf{1}.$$

If we interpret formula (7) without giving any interpretation to "x," it becomes the following sentence form:

(8) If $\mathbf{x} < \mathbf{2}$, then there is a real number **y** so that $\mathbf{x} \cdot \mathbf{y} = \mathbf{1}$.

That is,

(9) If $\mathbf{x} < \mathbf{2}$, then **x** has a multiplicative inverse.

The sentence form (8) (and equivalently (9)) is sometimes true and sometimes false. For example, if **x** is replaced by $-\mathbf{4}$, sentence form (8) is true because there is a real number **y**—namely, $\mathbf{y} = -\mathbf{1/4}$—so that $\mathbf{x} \cdot \mathbf{y} = \mathbf{1}$. It is also true if **x** is replaced by **3**, because then we have a conditional statement in which the antecedent is false. However, it is false if **x** is replaced by **0**, because in this case the antecedent of the conditional statement is true but the conclusion is false; there is no real number **y** so that $\mathbf{0} \cdot \mathbf{y} = \mathbf{1}$.

Let **S** be an interpretation structure for φ and suppose φ has n free variables $x_1, x_2, \ldots, x_n$. Let $\mathbf{a} = < \mathbf{a}_1, \mathbf{a}_2, \ldots, \mathbf{a}_n >$ be an n-tuple of elements from **U**, the universe of **S**. We say **a** *satisfies* φ if, whenever φ is interpreted in **S** with each free x_i interpreted as $\mathbf{a}_i$, the resulting sentence is true in **S**. We shall give a more precise definition of truth and satisfiability a little later.

Each of the following would be an interpretation of formula (7) in **S**:

(10) If $\mathbf{-4} < \mathbf{2}$, then there is a real number $\mathbf{y}$ so that $\mathbf{-4} \cdot \mathbf{y} = \mathbf{1}$.

(11) If $\mathbf{3} < \mathbf{2}$, then there is a real number $\mathbf{y}$ so that $\mathbf{3} \cdot \mathbf{y} = \mathbf{1}$.

(12) If $\mathbf{0} < \mathbf{2}$, then there is a real number $\mathbf{y}$ so that $\mathbf{0} \cdot \mathbf{y} = \mathbf{1}$.

As we mentioned above, (10) and (11) are true, but (12) is false. Thus, we say $< \mathbf{-4} >$ (or just $\mathbf{-4}$) satisfies formula (7) in **S** and so does **3**, but **0** does not.

Notice that in formula (7), the x in Pxa means the same as the x in $(\exists y)Qf(x, y)b$. Thus, in each interpretation, we must replace each free occurrence of x by the same quantity. The same remark is true for predicate symbols, function symbols, and constants. Thus, for example, the following is **NOT** an interpretation of formula (7):

If $\mathbf{1} < \mathbf{2}$, then there is a real number $\mathbf{y}$ so that $\mathbf{6} \cdot \mathbf{y} = \mathbf{1}$.

Similarly, we **CANNOT** interpret the formula

(13) $(\exists x)(Px \wedge (\forall y)(Py \rightarrow Qxy))$

as

Some prime is smaller than every integer.

That is, we translated "Px" as "x is a prime," but we translated "Py" as "y is an integer." This is **WRONG**! Wherever the symbol "P" occurs, it must be interpreted the same way. Thus, the following sentence is a *correct* interpretation of formula (13):

(14) Some prime is smaller than every prime.

In this case, P is interpreted as the property of "being a prime." Therefore, Px is "x is a prime" and Py is "y is a prime."

We shall give more examples of interpretations a little later in this chapter. First, we shall give a mathematical definition of an interpretation structure.

Suppose that we want an interpretation structure for a set of formulas Γ. First we shall list all the predicate symbols, function symbols, and constants that occur in some formula in Γ[1]:

Predicate symbols: $P_1, P_2, \ldots, P_k$

Function symbols: $f_1, f_2, \ldots, f_\ell$

Constants: $c_1, c_2, \ldots, c_m$.

Then $\mathbf{S} = < \mathbf{U}, \mathbf{P}_1, \mathbf{P}_2, \ldots, \mathbf{P_k}, \mathbf{f}_1, \mathbf{f}_2, \ldots, \mathbf{f}_\ell, \mathbf{c}_1, \mathbf{c}_2, \ldots \mathbf{c_m} >$ is an *interpretation structure for* Γ if the following conditions are satisfied:

(i) **U** is a nonempty set.

---

[1]If Γ is infinite, some of the lists of symbols may also be infinite. In this case **S** would be an infinite sequence.

(ii) If P_i is an n-place predicate symbol then $\mathbf{P}_i$ is a set of ordered n-tuples of elements of $\mathbf{U}$; that is,

$$\mathbf{P}_i \subseteq \mathbf{U}^n = \mathbf{U} \times \mathbf{U} \times \cdots \times \mathbf{U}, n \text{ times.}$$

(This is the same as saying that $\mathbf{P}_i$ is an n-place predicate or relation on $\mathbf{U}$. See Definition 5.1.) In particular, if P_i is a 1-place predicate symbol, then $\mathbf{P}_i \subseteq \mathbf{U}$.

(iii) If f_i is a function symbol that has n variables, then $\mathbf{f}_i$ is a function from $\mathbf{U}^n$ to $\mathbf{U}$,

$$\mathbf{f}_i : \mathbf{U}^n \to \mathbf{U}.$$

(See Definition 5.2.)

(iv) If c_i is a constant, then $\mathbf{c}_i \in \mathbf{U}$.

For example, suppose Γ is the following set of formulas:

(a) $(\forall x)[Px \to Pf(x) \wedge (\exists y)(Qxy \wedge \neg Rxyf(x))]$, (b) Pa, (c) $(\forall x)(Qax \to Rax\, f(x))$.

In this set, we have one 1-place predicate symbol, P; one 2-place predicate symbol, Q; one 3-place predicate symbol, R; one 1-place function symbol, f; and one constant, a. An interpretation structure for Γ has the following form: $\mathbf{S} = \langle \mathbf{U}, \mathbf{P}, \mathbf{Q}, \mathbf{R}, \mathbf{f}, \mathbf{a} \rangle$, where $\mathbf{P} \subseteq \mathbf{U}$, $\mathbf{Q} \subseteq \mathbf{U} \times \mathbf{U}$, $\mathbf{R} \subseteq \mathbf{U} \times \mathbf{U} \times \mathbf{U}$, $\mathbf{f} : \mathbf{U} \to \mathbf{U}$ and $\mathbf{a} \in \mathbf{U}$. For example, let

$$\mathbf{U} = \text{the set of real numbers,}$$
$$\mathbf{P} = \text{the set of all rational numbers,}$$

(That is, "**Px**" is interpreted as "**x** is rational.")

$$\mathbf{Q} = \{\langle \mathbf{r}, \mathbf{s} \rangle : \mathbf{r}, \mathbf{s} \in \mathbf{U} \text{ and } \mathbf{r} < \mathbf{s}\}.$$

(Therefore, "**Qxy**" is interpreted as "**x** < **y**.")

$$\mathbf{R} = \{\langle \mathbf{q}, \mathbf{r}, \mathbf{s} \rangle : \mathbf{q}, \mathbf{r}, \mathbf{s} \in \mathbf{U} \text{ and } \mathbf{r} \text{ is between } \mathbf{q} \text{ and } \mathbf{s}\}.$$

(**Rxyz** is interpreted as "**y** is between **x** and **z**.")

$$\text{For each } \mathbf{r} \in \mathbf{U}, \qquad \mathbf{f(r)} = \mathbf{r} + \mathbf{1}, \text{ and } \mathbf{a} = \mathbf{1}.$$

With this interpretation structure, the formulas in Γ are interpreted as follows:

(a) For each rational number **x**, **x** + **1** is rational and there is a real number **y** that is greater than **x** (**x** < **y**) such that **y** is not between **x** and **x** + **1**.

(b) One is a rational number.

(c) Each real number, **x**, that is greater than **1** is between **1** and **x** + **1**.

It is important to remember that each occurrence of a predicate symbol, function symbol, and constant in Γ is interpreted the same way. Thus, for example, Px is interpreted as "**x** is rational," $Pf(x)$ is interpreted as "**x** + **1** is rational," and

Pa is interpreted as "**1** is rational." Similarly for the others, *Q* is always interpreted as **<**; *R* as "between"; *a* as "**1**"; and *f*(*x*) as "**x + 1**."

All our interpretations, up to this time, have been numerical; that is, the domain is some set of numbers. However, interpretation structures do not have to be numerical. We can construct them so that they have more similarity to the truth tables of the Sentential Calculus. Suppose we wanted to find an interpretation structure in which the formulas

(15) $(\exists x)(Px \wedge Qx)$

(16) $(\exists x)(Qx \wedge Rx)$

are true, but

(17) $(\exists x)(Px \wedge Rx)$

is false.

If (17) is to be false in an interpretation structure **S** $=<$ **U, P, Q, R** $>$, then there can be no element **a** in the universe **U** of **S** such that **a** $\in$ **P** and **a** $\in$ **R** (that is, no **a** such that **a** has both the property **P** and the property **R**). However, if (15) and (16) are to be true, neither **P**, **Q**, nor **R** can be empty. (There have to be elements in **U** with these properties.) It follows that **U** has to have at least two elements. Let **U** = {**a, b**}, **Q** = {**a, b**}, **P** = {**a**}, and **R** = {**b**}. (Remember that *P*, *Q*, and *R* are 1-place predicate symbols, so that their interpretations, **P**, **Q**, and **R**, respectively, are subsets of **U**.) This situation can be described by a truth table as follows:

U = {a, b}

	P	Q	R
a	*T*	*T*	*F*
b	*F*	*T*	*T*

Figure 6.1

Now we can use the truth table in Figure 6.1 to show that the interpretation structure has the required property—(15) and (16) are true, but (17) is false. Formula (15) is true in **S** because **Pa** and **Qa** are both true. (**Px** is true if and only if **x** $\in$ **P** and similarly for the other predicates.) There is some **x**—namely, **x** = **a**—so that **Px** and **Qx** are true. Similarly, formula (16) is true because **Qb** and **Rb** are both true. However, (17) is false because we see from the truth table in Figure 6.1 that there is no **x** $\in$ **U**, so that both **Px** and **Rx** are true.

For our purposes, the types of interpretation structures mentioned above (numerical and those with finite domains) are sufficient. However, because we used other types of interpretations in preceding chapters, we shall give a few more examples.

Example 6.1

For each of the formulas listed below, give an interpretation structure in which it is true and one in which it is false. (We shall give interpretations that are not numerical and that do not use truth tables.)

(a) $(\forall x)(Ax \wedge Bxa \rightarrow Bxb)$

(b) $(\forall x)[(\exists y)(\exists z)(Cy \wedge Dz \wedge Exyz) \rightarrow Fx]$

(c) $(\forall x)(\forall y)[(\exists u)(\exists v)(Gxu \wedge Hyv) \rightarrow Ix \wedge Jy]$

(d) $(\exists y)(\exists z)(Kx \wedge Ly \wedge Mz \wedge Nxy \wedge Oxz)$

Solution:

(a) An interpretation structure for formula (a) has the form

(18) $\mathbf{S} = <\mathbf{U}, \mathbf{A}, \mathbf{B}, \mathbf{a}, \mathbf{b}>$,

where **U** is a nonempty set, $\mathbf{A} \subseteq \mathbf{U}$, $\mathbf{B} \subseteq \mathbf{U} \times \mathbf{U}$, and $\mathbf{a}, \mathbf{b} \in \mathbf{U}$.

First, we shall give an interpretation in which formula (a) is true. Let **U** be the set of all cities in the world; let $\mathbf{A} = \mathbf{U}$; $<\mathbf{x}, \mathbf{y}> \in \mathbf{B}$ iff **x** is north of **y**; let **a** be New York; and **b** be Miami. Then the interpretation of (a) is "Any city that is north of New York is north of Miami." Of course, this is a true sentence.

On the other hand, if we take another interpretation with the same structure as (18) and **U, A, B** the same as above, but we take **a** to be Miami and **b** to be New York, then the translation of (a) is "Any city that is north of Miami is north of New York." Now, this is false because, for example, Charlotte, North Carolina, is a city that is north of Miami, but not north of New York.

(b) A structure for (b) has the following form: $\mathbf{S} = <\mathbf{U}, \mathbf{C}, \mathbf{D}, \mathbf{E}, \mathbf{F}>$, where **U** is a nonempty set; each of **C, D,** and **F** are subsets of **U**; and $\mathbf{E} \subseteq \mathbf{U} \times \mathbf{U} \times \mathbf{U}$.

To construct an interpretation in which formula (b) is true, let **U** be the set of all things and people in the world. In addition, let **Cx** : **x** is a course; **Dx** : **x** is a university; **Exyz** : **x** takes **y** at **z**; and **Fx** : **x** is a student. Then (b) may be interpreted as "Only students take courses at universities."

On the other hand, if we change the interpretation of **Fx** to "**x** is a professor" and do not change anything else, then (b) becomes the false statement "Only professors take courses at universities."

(c) The structure $\mathbf{S} = <\mathbf{U}, \mathbf{G}, \mathbf{H}, \mathbf{I}, \mathbf{J}>$ is an interpretation structure for formula (c) if **G** and **H** are 2-place predicates, and **I** and **J** are 1-place predicates.

For formula (c) to be true, let **U** be the set of all people in the world; **Gxy** : **x** is the natural father of **y**; **Hxy** : **x** is the natural mother of **y**; **Ix** : **x** is a male; and **Jx** : **x** is a female. Then we can translate (c) as "Natural fathers are males, and natural mothers are females."

If we take the interpretation of **J** to be the same as that for **I**—that is, **Jx** : **x** is a male—and leave all other interpretations the same, then the interpretation of (c) is "Natural fathers are males, and natural mothers are males." Thus, this interpretation of formula (c) is false.

(d) The structure for this formula has the form: **S** =< **U, K, L, M, N, O** >, where **U** is a nonempty set; **K, L**, and **M** are 1-place predicates; and **N** and **O**, 2-place predicates.

We shall take as our interpretation structure the following: **U** is the set of all animals in the world; **Kx** : **x** is a tiger; **Lx** : **x** is an antelope; **Mx** : **x** is a hyena; **Nxy** : **x** is fiercer than **y**; and **Oxy** : **x** is prettier than **y**. Then the sentence form (d) would be translated as follows:

(19) **x has the property that x is a tiger, x** is fiercer than some antelope, and **x is prettier than some hyena.**

Then (19) will be true if we take **x** to be one of the tigers that are elements of **U**—say, the fiercest tiger in the San Diego Zoo. On the other hand, if we take **x** to be the tallest giraffe in the San Diego Zoo, (19) will be false.

6.2 TRUTH AND FALSITY IN INTERPRETATION STRUCTURES

If φ is a formula in the Predicate Calculus, what does it mean for φ to be true or false in an interpretation structure **S** for φ? We have indicated above informally with examples how we decide truth or falsity and, if the formula has free variables, how we define satisfaction. We will discuss the problem in more detail in this section. We shall give a definition of satisfaction and also define truth and falsity.

In Chapter 5, we defined a formula of the Predicate Calculus. Each formula is built up from atomic formulas by using sentential connectives and quantifiers in an appropriate manner, and each atomic formula is built from a predicate letter and terms. (See Definitions 5.3 and 5.5.) Therefore, it is natural to decide first how to interpret a term in an interpretation structure. (Recall that we use boldface type for interpretations in interpretation structures.)

Let φ be a formula with, at most, n free variables, $x_1, x_2, \ldots, x_n$, let **S** be an interpretation structure for φ, and let $\mathbf{a} = <\mathbf{a}_1, \mathbf{a}_2, \ldots, \mathbf{a}_n>$ be an n-tuple of elements in **U**, the universe of **S**.

(I) Suppose t is a term in φ.

Case 1: t is x_i or a_i then **t** is $\mathbf{a}_i$.

Case 2: t is $f(t_1, t_2, \ldots, t_m)$ then **t** is $\mathbf{f}(\mathbf{t}_1, \mathbf{t}_2, \ldots, \mathbf{t}_m)$.

(II) Suppose φ is an atomic formula. In this case, φ has the form $Rt_1t_2 \ldots t_m$, where $t_1, t_2, \ldots, t_m$ are m terms. Let **R** be the interpretation of R in **S** and $\mathbf{t}_1, \mathbf{t}_2, \ldots, \mathbf{t}_m$ be the interpretations of $t_1, t_2, \ldots, t_m$, respectively. Then, the n-tuple **a** *satisfies* φ in **S** iff $< \mathbf{t}_1, \mathbf{t}_2, \ldots, \mathbf{t}_m > \in \mathbf{R}$.

In the case that φ has no free variables or, equivalently, φ is a sentence, the interpretation of φ is independent of the sequence **a**. Thus, either every sequence satisfies φ or no sequence does.

For example, suppose there are three atomic formulas, *Px*, *Pf*(*a*), and *Qf*(*y*)*b*, and the interpretation structure **S**, is defined as

$$\mathbf{S} = <\mathbf{U}, \mathbf{P}, \mathbf{Q}, \mathbf{f}, \mathbf{a}, \mathbf{b}, \mathbf{c}>,$$

where **U** = {**a, b, c**}, **P** = {**a, c**}, **Q** = {**<a, b>**, **<b, b>**, **<b, c>**}, and **f(a)** = **c, f(b)** = **b, f(c)** = **a**. (We assume that the interpretation of *a* is **a**, *b* is **b**, and *c* is **c**.) The truth table for this structure is given in Figure 6.2. (The truth table for **Q** is similar to a multiplication table.)

Thus, for example, **<a, b>** satisfies both *Px* and *Qf*(*y*)*b* (**Pa** and **Qf**(**b**)**b** are true, **f**(**b**) = **b**). (Notice that **a** replaces *x* and **b** replaces *y*.) On the other hand, **<b, a>** satisfies neither of them (**Pb** and **Qf**(**a**)**b** are false, **f**(**a**) = **c**). (Here, **b** replaces *x* and **a** replaces *y*.) However, because **f**(**a**) = **c** and *Pc* is true in this interpretation structure, every sequence satisfies *Pf*(*a*).

U = {a, b, c}

	P
a	*T*
b	*F*
c	*T*

Q	a	b	c
a	*F*	*T*	*F*
b	*F*	*T*	*T*
c	*F*	*F*	*F*

	f
a	c
b	b
c	a

Figure 6.2

(III) Suppose the formula φ is constructed from formulas ψ and χ using the sentential connectives. We have five cases to consider, corresponding to the five sentential connectives, ¬, ∨, ∧, →, and ↔. We proceed just as we did when we constructed truth tables for the sentential connectives.

Case 1: φ is ¬ψ
a *satisfies* φ in **S** iff **a** does not satisfy ψ in **S**.

Case 2: φ is ψ ∨ χ
a *satisfies* φ in **S** iff **a** satisfies ψ in **S** or **a** *satisfies* χ in **S**.

Case 3: φ is ψ ∧ χ
a *satisfies* φ in **S** iff **a** satisfies both ψ and χ in **S**.

Case 4: φ is ψ → χ
a *satisfies* φ in **S** iff **a** does not satisfy ψ in **S** or **a** does satisfy χ in **S**.

Case 5: φ is ψ ↔ χ
a *satisfies* φ in **S** iff **a** satisfies both ψ and χ in **S** or **a** satisfies neither ψ nor χ in **S**.

For example, in the structure defined in Figure 6.2, $< \mathbf{a}, \mathbf{c} >$ satisfies $\neg Qxy$; $< \mathbf{b}, \mathbf{c} >$ satisfies $Px \vee Qxy$; $< \mathbf{b}, \mathbf{a}, \mathbf{c} >$ satisfies $Px \rightarrow Qyz$; and $< \mathbf{a}, \mathbf{c}, \mathbf{b} >$ satisfies $Px \leftrightarrow Py \vee Qxz$.

(IV) The last case to consider is that the formula φ either has the form $(\forall v)\psi$ or $(\exists v)\psi$. Let $\mathbf{a}' = < \mathbf{a}_1, \mathbf{a}_2, \ldots, \mathbf{a}_n, \mathbf{b} >$. (The formula ψ may have v as a free variable. **b** is substituted for v in the interpretations below.)

Case 1: φ is $(\forall v)\psi$
a *satisfies* φ in **S** iff **a**′ satisfies ψ for every $\mathbf{b} \in \mathbf{U}$.

Case 2: φ is $(\exists v)\psi$
a *satisfies* φ in **S** iff **a**′ satisfies ψ for some $\mathbf{b} \in \mathbf{U}$.

Definition 6.1 Suppose φ is a formula with at most n free variables, **S** is an interpretation structure for φ, and **U** is the universe of **S**.

1. φ is *true* in **S** iff every n-tuple of elements from **U** satisfies φ.
2. φ is *false* in **S** iff no n-tuple of elements from **U** satisfies φ.

It follows from Definition 6.1 that there may be formulas that are neither true nor false in some interpretation structure. For example, we have shown above that there is an interpretation structure for formula (7) in which it is neither true nor false. It also follows that in each interpretation structure, a sentence is either true or false, but not both.

Definition 6.2 A formula φ is *valid* or *universally valid* if φ is true in every interpretation structure for φ.

The valid formulas of the Predicate Calculus correspond to the tautologies of the Sentential Calculus. The following are examples of valid formulas: $(\forall x)(Px \vee \neg Px)$, $(\exists x)(\forall y)Rxy \rightarrow (\forall y)(\exists x)Rxy$, and $(\forall x)(Px \rightarrow Qx) \rightarrow ((\exists x)Px \rightarrow (\exists x)Qx)$. (See Section 10.2.)

As additional examples, let us decide on the truth of each of the following sentences in the interpretation structure given in Figure 6.2:

(1) $(\exists x)(Px \wedge Qxb)$

Formula (1) is true in **S** iff we can find an $\mathbf{x} \in \mathbf{U}$ so that both **Px** and **Qxb** are true. Since **Pa** and **Qab** are both true, formula (1) is true in **S**. (The "**x**" that makes **Px** ∧ **Qxb** true is "**a**.")

(2) $(\exists x)(Px \rightarrow Qxa)$

Even though **Qxa** is false in **S** for every choice of $\mathbf{x} \in \mathbf{U}$, formula (2) is true in **S** because there is an $\mathbf{x} \in \mathbf{U}$ so that **Px** is false—namely, $\mathbf{x} = \mathbf{b}$. Therefore, **Pb** → **Qba** is true in **S**, so it follows that (2) is true in **S**. (The "**x**" that makes **Px** → **Qxa** true is "**b**.")

(3) $(\exists x)(Px \wedge Qxa)$

This formula is false in **S** because there is no $\mathbf{x} \in \mathbf{U}$ so that both **Px** and **Qxa** are true in **S**. From the truth table for **P**, **x** must be either **a** or **c** to make **Px** true, but neither **Qaa** nor **Qca** is true.

(4) $(\exists x)(\exists y)(Px \wedge Py \wedge Qxy)$

Formula (4) is false in **S** because **Qxy** is only true if

(a) $\mathbf{x} = \mathbf{a}$ and $\mathbf{y} = \mathbf{b}$, in which case **Pb** is false so the conjunction is false.

(b) $\mathbf{x} = \mathbf{b}$ and $\mathbf{y} = \mathbf{b}$; again, **Pb** is false.

(c) $\mathbf{x} = \mathbf{b}$ and $\mathbf{y} = \mathbf{c}$; again, in this case, **Pb** is false.

Therefore, there are no **x** and **y** in **U** such that $\mathbf{Px} \wedge \mathbf{Py} \wedge \mathbf{Qxy}$ is true.

(5) $(\forall x)(Px \rightarrow \neg Qxf(c))$

A universal formula is true in **S** iff it is true for *all* elements **x** in **U**. We have three cases to consider—**x** is **a**, **b**, or **c**.

Case 1: $\mathbf{x} = \mathbf{a}$. Since **Qaa** is false, ($\mathbf{f}(\mathbf{c}) = \mathbf{a}$) $\neg$**Qac** is true, so when $\mathbf{x} = \mathbf{a}$, the conditional

(*) $$\mathbf{Px} \rightarrow \neg \mathbf{Qxf}(\mathbf{c})$$

is true.

Case 2: $\mathbf{x} = \mathbf{b}$. In this case, (*) is again true because its antecedent, **Pb**, is false.

Case 3: $\mathbf{x} = \mathbf{c}$. This case is similar to case 1 because **Qca** is false, so (*) is true.

Therefore, (5) is true in **S**.

(6) $(\forall x)(Pf(x) \wedge \neg Qxa)$

Formula (6) is false in **S** because, for example, if $\mathbf{x} = \mathbf{b}$, the conjunction $\mathbf{Pf}(\mathbf{x}) \wedge \neg\mathbf{Qxa}$ is false ($\mathbf{f}(\mathbf{b}) = \mathbf{b}$ and **Pb** is false in **S**). A universal formula $(\forall x)\ \psi(x)$ is false in an interpretation structure if there is *one* element $\mathbf{x} \in \mathbf{U}$ such that $\psi(\mathbf{x})$ is false.

(7) $(\forall x)[Px \rightarrow (\forall y)(Py \rightarrow Qxy)]$

Again, this is a universal formula. Thus, it is false if

(*) $$\mathbf{Px} \rightarrow (\forall \mathbf{y})(\mathbf{Py} \rightarrow \mathbf{Qxy})$$

is false for some $\mathbf{x} \in \mathbf{U}$. If $\mathbf{x} = \mathbf{b}$, (*) is true because its antecedent is false. Suppose $\mathbf{x} = \mathbf{a}$. Then the antecedent of (*) is true so that (*) itself is true iff

(**) $$\mathbf{Py} \rightarrow \mathbf{Qay}$$

is true for each $\mathbf{y} \in \mathbf{U}$. However, if $\mathbf{y} = \mathbf{a}$, (**) is false. Thus, formula (7) is false in **S**.

(8) $(\forall x)[Px \rightarrow (\exists y)(Py \wedge \neg Qxy)]$

In this case, we shall show that

(*) $$\mathbf{Px} \rightarrow (\exists \mathbf{y})(\mathbf{Py} \wedge \neg \mathbf{Qxy})$$

is true for all $\mathbf{x} \in \mathbf{U}$, from which it follows that (8) is true in **S**.

Case 1: $\mathbf{x} = \mathbf{a}$. Then the antecedent of (*) is true and, since $\mathbf{Pa} \wedge \neg \mathbf{Qaa}$ is true ($\mathbf{y} = \mathbf{a}$), the consequent of (*) is true. Thus, (*) is true.

Case 2: $\mathbf{x} = \mathbf{b}$. Then the antecedent of (*) is false, so (*) is true.

Case 3: $\mathbf{x} = \mathbf{c}$. In this case, the antecedent of (*) is true. Also, $\mathbf{Pa} \wedge \neg \mathbf{Qca}$ is true ($\mathbf{y} = \mathbf{a}$). Therefore, the consequent of (*) is true.

It follows that (8) is true in **S**.

(9) $(\exists x)[Px \wedge (\forall y)(Py \rightarrow \neg Qxy)]$

Formula (9) is true in **S** if there is some $\mathbf{x} \in \mathbf{U}$ such that

(*) $$\mathbf{Px} \wedge (\forall \mathbf{y})(\mathbf{Py} \rightarrow \neg \mathbf{Qxy})$$

is true in **S**. If $\mathbf{x} = \mathbf{b}$, then (*) is false because the first conjunct is false. If $\mathbf{x} = \mathbf{a}$ or $\mathbf{x} = \mathbf{c}$, the first conjunct is true. If $\mathbf{x} = \mathbf{c}$, then **Qxy** is false for all $\mathbf{y} \in \mathbf{U}$. Thus,

(**) $$\mathbf{Py} \rightarrow \neg \mathbf{Qcy}$$

is true for all $\mathbf{y} \in \mathbf{U}$. Consequently, (*) is true if $\mathbf{x} = \mathbf{c}$, from which it follows that (9) is true in **S**.

As another example, we shall take a numerical interpretation structure **S** and determine the truth or falsity of each of the formulas (1) through (9) in **S**. Let $\mathbf{S} = \langle \mathbf{U}, \mathbf{P}, \mathbf{Q}, \mathbf{f}, \mathbf{a}, \mathbf{b}, \mathbf{c} \rangle$, where **U** = the set of all integers; **Px** : **x** is a prime; **Qxy** : $\mathbf{x} > \mathbf{y}$; $\mathbf{a} = \mathbf{5}$; $\mathbf{b} = \mathbf{0}$; $\mathbf{c} = \mathbf{5}$; and $\mathbf{f}(\mathbf{x}) = \mathbf{x} + \mathbf{1}$. (Two different symbols may have the same interpretation, but the same symbol may *not* have two different interpretations in a given structure.)

The interpretation of formula (1) in **S** is

Some prime is positive.

This is true in **S** because, for example, **2** is a positive prime.

Formula (2) has the interpretation

There is an integer such that if it is a prime, then it is greater than five.

Formula (2) is also true in **S** because there is an integer in **U** that is not a prime. For example, if we take the integer to be four, the following statement is true because the antecedent is false:

If four is a prime, then four is greater than five.

Another way to show that formula (2) is true in **S** is to choose an integer in **U** that is a prime and is greater than five. For example, take the integer to be seven. Then the following sentence is true:

If seven is a prime, then seven is greater than five.

Thus, (2) is true in **S**.

Formula (3) is also true in **S** because there is a prime that is greater than five. For example, seven is a prime and **7 > 5**.

Formula (4) is interpreted in **S** as

There are two integers that are primes, and the first is larger than the second.

This sentence is true in **S** because, for example, **7** and **5** are primes in **U** and **7 > 5**.

Formula (5) is interpreted as

No prime is greater than six.

This is, of course, false because **7 ∈ U, 7** is a prime and **7 > 6**.

The interpretation of (6) is

Every integer x has the property that $x + 1$ is a prime and x is not greater than five.

Again, this is false because we can find an integer in **U**, say **3**, such that **3 + 1** is not a prime. This is sufficient to make (6) false in **S**. (We can also find an integer **x** such that **x + 1** is a prime and **x** is greater than **5**—say, **x** = **6**; we can also find an integer **x** so that **x + 1** is not a prime and **x** is greater than **5**—say, **x** = **7**. Either of these are also counter-examples to (6).)

Formula (7) is also false in **S**. Its interpretation is

Every pair of primes has the property that the first is larger than the second.

If the first prime (x) is taken to be **5** and the second prime (y) is taken to be **7**, it is not the case that **5 > 7**.

The interpretation of (8) is true:

For each prime, there is a prime that is not larger than the first one.

This sentence is true because whatever prime **x ∈ U** that we choose, we can always find a prime—namely, **x** itself—that is not larger than **x**.

As was mentioned earlier (see Example 5.3 (a)), another way of expressing formula (8) is as follows:

(*) $$(\forall x)[Px \rightarrow Pg(x) \wedge \neg Qxg(x)].$$

In an interpretation for (*), interpret P and Q as in **S** above, and interpret g to be the function **g**: **U** → **U** so that **g(x)** = **x** for all **x ∈ U**. Then, (*) is true.

Formula (9) is true in **S**. Its interpretation is

There is a prime that has the property that no prime is smaller than it.

In other words,

There is a smallest prime.

This is true in **S**. Two is the smallest prime.

The results involving quantifiers can be summarized as follows:

A universal formula, $(\forall v)\varphi$, is *true* in an interpretation **S** for φ iff φ is true in **S** for *every* interpretation of v in **S**; $(\forall v)\varphi$ is *false* in **S** iff φ is false in **S** for *some* interpretation of v in **S**.

An existential formula, $(\exists v)\varphi$, is *true* in an interpretation structure **S** for φ iff φ is true in **S** for *some* interpretation of v in **S**; $(\exists v)\varphi$ is *false* in **S** iff φ is false for *every* interpretation of v in **S**.

Although we do have a precise definition of truth, we may not be able to tell whether a sentence is true or false in an interpretation structure. Consider the sentence "There are an infinite number of twin primes." (Twin primes are consecutive odd primes like 3 and 5, 5 and 7, 11 and 13, and so forth.) We can translate this sentence into the language of the Predicate Calculus as φ: $(\forall x)(\exists y)(\exists z)(x < y \wedge y < z \wedge Tyz)$. An interpretation structure for φ is $\mathbf{S} = \langle \mathbf{U}, \mathbf{T}, < \rangle$, where **U** is the set of positive integers, **Txy**: **x** and **y** are twin primes, and **<** is the less-than relation on **U**. It is not known whether φ is true or false in **S**.

An interpretation structure is an extension of a truth assignment in the Sentential Calculus, but unlike the Sentential Calculus, there is no algorithm for determining whether a sentence φ of the Predicate Calculus is true or false in an interpretation structure for φ.

EXERCISES 6

A. For each of the following formulas, find an interpretation structure in which it is true and one in which it is false. (Answers to 1, 4, 7, 9, 12, and 15.)

(1) $(\forall x)(Px \wedge Qx \rightarrow Rx)$
(2) $(\exists x)(Px \wedge \neg Qx \wedge \neg Rx)$
(3) $(\forall x)(Px \rightarrow Qx) \vee (\exists x)(Px \wedge \neg Rx)$
(4) $Pa \wedge (\exists x)(Px \wedge \neg Qxb)$
(5) $(\exists x)(Px \wedge Qxa) \rightarrow (\forall y)(Py \wedge Ry \rightarrow Qby)$
(6) $(\forall x)[Px \wedge Qx \rightarrow (\exists y)(Ry \wedge \neg Sxy)]$
(7) $(\exists x)[Px \wedge \neg Qx \wedge (\forall y)(Rxy \rightarrow \neg Sxy)]$
(8) $(\exists x)Px \rightarrow (\forall y)(Qy \rightarrow Rxy)$
(9) $(\forall x)(Px \rightarrow Qxa) \rightarrow (\exists y)(Py \wedge Ry \wedge \neg Qxy)$
(10) $(\forall x)(\forall y)[Px \wedge Qy \rightarrow (\exists z)(Rxy \wedge Sxyz)]$
(11) $(\exists x)(\forall y)[Px \wedge Ry \wedge Sxy \wedge (\exists z)(Pz \wedge Rx \wedge \neg Qxaz)]$
(12) $(\forall x)[Px \wedge Qxf(x) \rightarrow (\exists y)(\exists z)(Py \wedge \neg Qyz \wedge Rf(x)yz)]$
(13) $(\forall x)Rxf(x) \wedge (\forall x)(\forall y)(Rxy \rightarrow Ryx) \wedge$
$(\forall x)(\forall y)(\forall z)(Rxy \wedge Ryz \rightarrow Rxz)$

(14) $(\forall x)Rxx \vee (\forall x)(\forall y)(Rxy \rightarrow Ryx) \vee$
$(\forall x)(\forall y)(\forall z)(Rxy \wedge Ryz \rightarrow Rxz)$

(15) $(\forall x)(\forall y)(Rxy \vee Ryx) \wedge (\forall x)(\exists y)Rxy \wedge (\forall y)(\exists x)\neg Rxy$

B. For each of the following sets of formulas, find an interpretation structure in which the last formula is false, but the other formulas are true. (Answers to 1, 4, 8, 12, and 15.)

(1) $(\forall x)(Ax \rightarrow Bx)$, $(\exists x)(Ax \wedge Bx)$

(2) $(\exists x)(Px \wedge \neg Qx)$, $(\forall x)(Px \rightarrow \neg Qx)$

(3) $(\forall x)(Ax \rightarrow Bx)$, $(\forall x)(Cx \rightarrow \neg Ax)$,
$(\exists x)(Cx \wedge \neg Bx)$, $(\forall x)(Cx \rightarrow \neg Bx)$

(4) $(\forall x)(Px \rightarrow Qx)$, $\neg Pa$, $\neg Qa$

(5) $(\exists x)Px \rightarrow (\exists x)Qx$, Qa, $(\forall x)(Px \rightarrow Qx)$

(6) $(\exists x)Dx \rightarrow Eab$, $Fa \rightarrow (\exists x)Dx$, $(\exists x)Fx$, $(\exists x)Exb$

(7) $(\forall x)(Px \rightarrow (\exists y)(Qy \wedge Rxy))$, $(\exists x)(Px \wedge (\forall y)(Qy \rightarrow Rxy))$

(8) $(\forall x)(Lxa \rightarrow \neg Lxb)$, $(\forall x)(\exists y)Lxy$, $(\exists x)\neg Lxb$

(9) $(\forall x)[Px \rightarrow (\forall y)(Qy \rightarrow Rxy)]$, $(\forall x)[Px \rightarrow (\exists y)(\neg Rxy \wedge Sy)]$,
$(\exists x)(Sx \wedge \neg Qx)$

(10) $(\exists x)[Ex \wedge (\forall y)(Fy \rightarrow Gxy)]$, $(\forall x)(\forall y)[Ex \rightarrow (Gxy \leftrightarrow Hy)]$,
$(\forall x)(Hx \leftrightarrow Fx)$

(11) $(\forall x)[(\forall y)(Rxy \rightarrow Sy) \rightarrow Sx]$, $Rab \wedge Sa$, Sb

(12) $(\forall x)(\forall y)(\forall z)(Rxy \wedge Ryz \rightarrow Rxz)$, $(\forall x)Rxx$,
$(\forall x)(\exists y)Rxy$, $(\forall x)(\forall y)(Rxy \rightarrow \neg Ryx)$

(13) $(\forall x)Rxf(x)$, $(\exists x)(\forall y)Rxy$

(14) $(\forall x)(\forall y)(Px \wedge Py \rightarrow Rxy)$, $(\exists x)Px$, $(\exists x)Pf(x)$, $(\exists x)Rf(x)x$

(15) $(\forall x)(\forall y)(Px \wedge Py \rightarrow Pf(x, y))$, $(\exists x)(Px \wedge \neg Qx)$,
$(\forall x)(Px \rightarrow (\exists y)(Py \wedge \neg Qf(x, y))$

(16) $(\forall x)[Ax \wedge Bx \rightarrow (\forall y)(By \wedge Cy \leftrightarrow Dxy)]$,
$(\exists x)[Ax \wedge Bx \wedge (\forall y)(By \wedge Ey \leftrightarrow \neg Dxy)]$,
$(\exists x)(Ax \wedge Bx \wedge \neg Ex)$

Chapter 7

Proofs Involving Universal Quantifiers

It follows from results of Alonzo Church in 1936,[1] based on results of Kurt Gödel, that there is no algorithm that is analogous to the truth tables of the Sentential Calculus to tell us whether an argument in the Predicate Calculus is valid. Thus, proofs are more important in the Predicate Calculus than in the Sentential Calculus. A proof for the Sentential Calculus is defined in Definition 2.1. For this definition to be useful for the Predicate Calculus, we must add some more rules of inference to those rules we already have, Rules *P, T, CP*, and *IP*.

In the beginning, proofs in the Predicate Calculus will have the following form:

1. Drop quantifiers.
2. Give a proof in the Sentential Calculus.
3. Add quantifiers.

In this chapter, we discuss the rules for adding and dropping universal quantifiers. The rationale for these rules and the rules given for existential quantifiers in Chapter 8 is based on the interpretation of quantifiers given in Chapter 6.

---

[1]Journal of Symbolic Logic, vol. 37, pp. 1–18; Correction, *ibid*, pp. 101–102.

7.1 UNIVERSAL SPECIFICATION (DROPPING A UNIVERSAL QUANTIFIER)

Suppose we take as a premise,

All giraffes have long necks.

Translating this sentence into the language of the Predicate Calculus,

(1) $(\forall x)(Gx \rightarrow Lx)$,

where Gx is "x is a giraffe" and Lx is "x has a long neck." We should be able to derive from (1), each of the following:

If Jerry is a giraffe, then Jerry has a long neck.

(2) $Gj \rightarrow Lj$

If x is a giraffe, then x has a long neck.

(3) $Gx \rightarrow Lx$

If the animal that belongs to y is a giraffe, then the animal that belongs to y has a long neck.

(4) $Gf(y) \rightarrow Lf(y)$

In general, for any term t,

If t is a giraffe, then t has a long neck.

(5) $Gt \rightarrow Lt$

Therefore, if everything has a certain property, we should like to be able to deduce that any particular thing has that property; that is, if $\mathcal{P}$ is any property, from $(\forall x)\mathcal{P}(x)$, deduce $\mathcal{P}(t)$, where t is a term.

Let us apply this rule to another example. Consider the following sentence:

For each integer, there is a larger integer.

(6) $(\forall x)[Ix \rightarrow (\exists y)(Iy \wedge y > x)]$

Applying the above rule, we should be able to deduce each of the following from (6). Substitute 1 for x in (6).

(7) $I1 \rightarrow (\exists y)(Iy \wedge y > 1)$

(If one is an integer, then some integer is greater than one.)

Substitute $2x^2 - 3x + 8$ for x in (6).

(8) $I2x^2 - 3x + 8 \rightarrow (\exists y)(Iy \wedge y > 2x^2 - 3x + 8)$

(If $2x^2 - 3x + 8$ is an integer, then some integer is greater than $2x^2 - 3x + 8$.)

Substitute u for x in (6).

(9) $Iu \rightarrow (\exists y)(Iy \wedge y > u)$

(If u is an integer, then some integer is greater than u.)
Substitute y for x.

(10) $Iy \rightarrow (\exists y)(Iy \wedge y > y)$

(If y is an integer, then some integer is greater than itself.)

Something is wrong with (10)! One of the fundamental properties of a proof is that if the premises are true, then everything that is deduced from the premises must be true. *Proofs preserve truth*. In our example, however, (6) is true but (10) is false.

In sentence (6), the property $\mathcal{P}(x)$ is

$\mathcal{P}(x)$: $Ix \rightarrow (\exists y)(Iy \wedge y > x)$.

In "$\mathcal{P}(x)$," there is a free occurrence of x within

(11) $(\exists y)(Iy \wedge y > x)$.

Formula (11) is translated as

(12) Some integer is greater than x.

The variable x is free in (11), but it is within the scope of the quantifier $(\exists y)$. Therefore, if we substitute any term for x that contains a free occurrence of y, it will no longer be free after the substitution. Notice how such a substitution changes the meaning of (11) in the following examples:

Substitute y for x.

(13) $(\exists y)(Iy \wedge y > y)$

(Some integer is greater than itself.)
Substitute $y - y^2$ for x.

(14) $(\exists y)(Iy \wedge y > y - y^2)$

(There is an integer whose square is positive.)
Substitute $y + 1$ for x.

(15) $(\exists y)(Iy \wedge y > y + 1)$

(Some integer is greater than its own immediate successor.)
Substitute $y - 1$ for x.

(16) $(\exists y)(Iy \wedge y > y - 1)$

(Some integer is greater than its own immediate predecessor.)

However, if t is any term that contains no free occurrence of y and we substitute t for x in (11), we obtain

(17) $(\exists y)(Iy \wedge y > t)$,

which is translated as

(18) Some integer is greater than t.

When deducing formulas from a universal formula like (6), we must avoid substitutions like those in (13) through (16) because such substitutions change the meaning of the expression. We shall call them *invalid* substitutions. We must restrict ourselves to substitutions similar to those given in (7), (8), (9), and (17). We shall call them *valid* substitutions.

Definition 7.1 If φ is a formula, t is a term, and v is a variable, then $\varphi[t|v]$ is the formula obtained from φ by replacing each free occurrence of v by t. The substitution of t for v in φ is called a *valid substitution* as long as no occurrence of a variable in t is bound in $\varphi[t|v]$. (Every occurrence of a variable in a term is free because terms contain no quantifiers.)

Keep in mind that there are two important restrictions involved in the definition of a valid substitution in Definition 7.1: First, *no occurrence of a variable in t becomes bound after the substitution*. Second, *each free occurrence of v is replaced by t*.

Now we can state the rule for dropping a universal quantifier, *Universal Specification*, or Rule *US*.

Rule *US* (Rule of Universal Specification) *If φ is a formula, t is a term and v is a variable, then $\varphi[t|v]$ is derivable from $(\forall v)\varphi$ as long as the substitution of t for v in φ is valid. Equivalently, $(\forall v)\varphi \vdash \varphi[t|v]$, if the substitution of t for v in φ is valid, or, if Γ is a set of formulas, then if $\Gamma \vdash (\forall v)\varphi$ then $\Gamma \vdash \varphi[t|v]$, if the substitution of t for v in φ is valid.*

For example, we would document the proof of formula (4) from (1) as follows:

{1}	1.	$(\forall x)(Gx \to Lx)$	P
{1}	2.	$Gf(y) \to Lf(y)$	1 *US* $[f(y)|x]$

When we use the Rule *US*, we shall indicate in parentheses the term that was substituted for the variable.

The formal proof of (7) from (6) is given below:

{1}	1.	$(\forall x)[Ix \to (\exists y)(Iy \wedge y > x)]$	P
{1}	2.	$I1 \to (\exists y)(Iy \wedge y > 1)$	1 *US* $[1|x]$

Each of the following illustrates an **incorrect** use of Rule *US*:

{1}	1.	$(\forall x)[Ix \to (\exists y)(Iy \wedge y > x)]$	P
{1}	2.	$I1 \to (\exists y)(Iy \wedge y > x)$	1 *US* $(1|x)$

(This is **wrong** because we did not substitute 1 for x in **every** free occurrence of x.)

{1}	1.	$(\forall x)[Ix \to (\exists y)(Iy \wedge y > x)]$	P
{1}	2.	$I\sqrt{|y|} \to (\exists y)(Iy \wedge y > \sqrt{|y|})$	1 *US* $(\sqrt{|y|}|x)$

(Here, the **error** is that we made an **invalid substitution**.)

{1}	1.	$(\forall x)Ix \rightarrow (\exists y)(Iy \wedge y > x)$	P
{1}	2.	$Ix \rightarrow (\exists y)(Iy \wedge y > x)$	1 US $(x\|x)$

(In this case, the scope of $(\forall x)$ is **NOT** the whole formula, so that Rule *US* does **NOT** apply. If the formula on line 1 was replaced by $(\forall x)[Ix \rightarrow (\exists y)(Iy \wedge y > x)]$, the proof would have been correct.)

7.2 UNIVERSAL GENERALIZATION (ADDING A UNIVERSAL QUANTIFIER)

If φ is a formula and v a variable, then we shall say the formula $(\forall v)\varphi$ is obtained from φ by *generalizing on the variable v*. In this section, we shall decide when generalization on a variable is valid. To be valid, it must preserve truth; that is, if φ is true, then $(\forall v)\varphi$ must also be true.

First of all, *we cannot generalize on a constant*. It does not necessarily follow from "John went to the play" that "Everyone went to the play."

However, we cannot always generalize on a variable either. Consider the following example. Suppose we are trying to prove that every prime larger than two is odd. (A prime is an integer greater than one whose only factors are one and itself.) We might argue as follows:

Suppose x is a prime that is larger than two. Then, using an indirect proof, suppose x is not odd. Then x must be even. Each even number has two as a factor. Since x is larger than two, this implies x has a factor that is different from one and itself. Therefore, x is not a prime. Thus, by the Rule of Indirect Proof, x is odd.

We deduced "x is odd." Assuming the variables range over the set of integers, is it valid to deduce "Every integer is odd"? Certainly not! Even though x is a variable and we deduced that x is odd, x was not a *generic* variable. We had stated originally in a premise that x had some special properties—namely, x is a prime and x is larger than two. It was just for these x's that we deduced that x is odd.

If the informal proof given above were made formal, it would start with a premise,

{1}	1.	$Px \wedge (x > 2)$	P

Then, the line "x is odd"—say, line n—is deduced using premise 1 and perhaps other premises also.

$\{1, k_1, k_2, \ldots, k_m\}$	n.	Ox	(*Reasons*)

The x on line n represents the same integer as the x on line 1. It is not a generic variable. It has the property that it is a prime greater than 2. Thus, we *cannot* deduce from line n that $(\forall x)Ox$. However, we can deduce from lines 1 and n using the Rule of Conditional Proof,

$\{k_1, k_2, \ldots, k_m\}$	$n + 1.$	$Px \wedge (x > 2) \rightarrow Ox$	$1, n$ *CP*.

Line $n + 1$ no longer depends on premise 1. Line $n + 1$ would be translated as "If x is a prime larger than two, then x is odd." This statement is true for every x. Therefore, from line $n + 1$, we may deduce

$\{k_1, \ldots, k_m\}$	$n + 2.$	$(\forall x)(Px \wedge (x > 2) \rightarrow Ox)$	$n + 1$ *UG*.

On line $n + 1$, we can generalize on x because here it is a generic variable. There are no conditions on it.

We see from the preceding example that if a variable is free in a premise, on that line it is not a generic variable. Some restrictions are placed on it. We are going to call attention to such variables in a proof by "flagging" them. A flagged variable is *not* a generic variable. We cannot generalize on it.

Definition 7.2 Flagged Variables.
If φ is a formula in a derivation, a variable v is said to be *flagged* in φ if either

1. φ is a premise and v is free in φ, or
2. there is a premise ψ in which v is free, v is free in φ, and ψ is used to derive φ.

The Rule of Universal Generalization or Rule *UG*, states that we can generalize on a variable *as long as it is not flagged.*

Rule *UG* (Rule of Universal Generalization) *If φ is a formula and v a variable, then $(\forall v)\varphi$ is derivable from φ, as long as v is not flagged in φ. Equivalently, if Γ is a set of formulas and if $\Gamma \vdash \varphi$, then $\Gamma \vdash (\forall v)\varphi$, if v is not free in Γ.*

The most important thing to remember about Rule *UG* is that generalization is only permitted on *variables that are not flagged.* We shall designate flagged variables in formal proofs by listing them to the right of the formula. Thus, in the preceding example, lines 1, n, $n + 1$, and $n + 2$ would be documented as follows:

$\{1\}$	1.	$Px \wedge (x > 2)$	x *P* (for *CP*)
$\{1, k_1, \ldots, k_m\}$	$n.$	Ox	x Reason
$\{k_1 \ldots, k_m\}$	$n + 1.$	$Px \wedge (x > 2) \rightarrow Ox$	$1, n$ *CP*
$\{k_1, \ldots, k_m\}$	$n + 2.$	$(\forall x)(Px \wedge (x > 2) \rightarrow Ox)$	$n + 1$ *UG*

Thus, on line $n + 1$, x is not flagged, so we can apply Rule *UG* and add the quantifier. Also note that on line $n + 2$, the whole formula on line $n + 1$ is within the scope of the quantifier $(\forall x)$. To indicate this, put parentheses around the formula on line $n + 1$. If the parentheses are omitted, line $n + 2$ would have been

(*) $$(\forall x)Px \wedge (x > 2) \rightarrow Ox.$$

In this case, the only formula within the scope of the quantifier is Px. Formula (*) has a different meaning than the formula on line $n + 2$. The latter formula is translated as

Every prime larger than two is odd.

On the other hand, formula (*) is translated as

If every number is a prime and x is larger than two, then x is odd.

Deriving (*) from line $n + 1$ is an **incorrect** use of the Rule *UG*. The universal quantifier $(\forall x)$ must really be "in front" of the whole formula, as in line $n + 1$ above, not just appear to be "in front," as in (*) above.

7.3 PROOFS INVOLVING UNIVERSAL QUANTIFIERS

In all the examples and exercises in this chapter, all premises will be written using only universal quantifiers. For example, formulas of the form "No *A* is *B*" will be translated as "$(\forall x)$ (if x is an *A*, then x is not a *B*)." Similarly, "only *A*'s are *B*'s" is translated as "$(\forall x)$ (if x is a *B*, then x is an *A*)."

Example 7.1 Translate each of the following valid arguments into the language of the Predicate Calculus, using the indicated constants and predicate symbols, and give a formal proof. The only quantifiers required are universal quantifiers.

(a) All cats have whiskers. Tuffy does not have whiskers. Therefore, Tuffy is not a cat. (Cx, Wx, t)

(b) All ghosts who live in the old mansion like Oscar. Oscar does not like anyone who likes him. Therefore, any ghost who Oscar likes does not live in the old mansion. (Gx, Mx, Lxy, o)

(c) No student likes any instructor. Each student likes all students. Therefore, no instructor is a student. (Sx, Ix, Lxy)

Solution:

(a) Cx: x is a cat; Wx: x has whiskers; t: Tuffy.

Premises: $(\forall x)(Cx \rightarrow Wx)$; $\neg Wt$

Conclusion: $\neg Ct$

Proof:

{1}	1.	$(\forall x)(Cx \rightarrow Wx)$	P
{2}	2.	$\neg Wt$	P
{1}	3.	$Ct \rightarrow Wt$	1 US $[t\|x]$
{1, 2}	4.	$\neg Ct$	2, 3 T

(b) Gx: x is a ghost; Mx: x lives in the old mansion; Lxy: x likes y; o: Oscar.

Premises: $(\forall x)(Gx \wedge Mx \rightarrow Lxo)$; $(\forall y)(Lyo \rightarrow \neg Loy)$

Conclusion: $(\forall z)(Gz \wedge Loz \rightarrow \neg Mz)$

Proof:

{1}	1.	$(\forall x)(Gx \wedge Mx \rightarrow Lxo)$	P
{2}	2.	$(\forall y)(Lyo \rightarrow \neg Loy)$	P

We want to prove the conditional formula $Gz \wedge Loz \rightarrow \neg Mz$ for each z. We will use a conditional proof—take $Gz \wedge Loz$ as a premise, and derive $\neg Mz$.

{3}	3.	$Gz \wedge Loz$	z P (for CP)
{2}	4.	$Lzo \rightarrow \neg Loz$	2 US $[z\|y]$
{2, 3}	5.	$\neg Lzo$	z 3, 4 T
{1}	6.	$Gz \wedge Mz \rightarrow Lzo$	1 US $[z\|x]$
{1, 2, 3}	7.	$\neg(Gz \wedge Mz)$	z 5, 6 T
{1, 2, 3}	8.	$\neg Gz \vee \neg Mz$	z 7 T
{1, 2, 3}	9.	$\neg Mz$	z 3, 8 T
{1, 2}	10.	$Gz \wedge Loz \rightarrow \neg Mz$	3, 9 CP
{1, 2}	11.	$(\forall z)(Gz \wedge Loz \rightarrow \neg Mz)$	10 UG

Notice that, in line 3, z is flagged, and in each line that depends on line 3 in which z is free (lines 5, 7, 8, and 9), z is flagged. Wherever z is flagged, it tells us that z is not a generic variable. (In this example, on the lines where z is flagged, z is "a ghost who Oscar likes.") Thus, we cannot generalize on the variable z on any of the lines where it is flagged. On line 10, however, z is no longer flagged, because after using the Rule CP, line 10 no longer depends on premise 3. Thus, on line 10, z is a generic variable, so we can apply Rule UG.

Also, notice that it did not matter what symbols we used for the bound variables in the premises or in the conclusion. For example, it would be equivalent if we wrote the second premise as "$(\forall x)(Lxo \rightarrow \neg Lox)$" instead of "$(\forall y)(Lyo \rightarrow \neg Loy)$," because each formula is deducible from the other using Rules US and UG.

(c) Sx: x is a student; Ix: x is an instructor; Lxy: x likes y.

Premises: $(\forall x)(Sx \rightarrow (\forall y)(Iy \rightarrow \neg Lxy))$; $(\forall x)(Sx \rightarrow (\forall y)(Sy \rightarrow Lxy))$
Conclusion: $(\forall x)(Ix \rightarrow \neg Sx)$

In this case, we shall give an indirect proof. We shall take as a premise $\neg(Ix \rightarrow \neg Sx)$ or, what is tautologically equivalent, $Ix \wedge Sx$, and then derive a contradiction.

{1}	1.	$(\forall x)(Sx \rightarrow (\forall y)(Iy \rightarrow \neg Lxy))$	P
{2}	2.	$(\forall x)(Sx \rightarrow (\forall y)(Sy \rightarrow Lxy))$	P
{3}	3.	$Ix \wedge Sx$	x P (for IP)
{1}	4.	$Sx \rightarrow (\forall y)(Iy \rightarrow \neg Lxy)$	1 US $[x\|x]$
{2}	5.	$Sx \rightarrow (\forall y)(Sy \rightarrow Lxy)$	2 US $[x\|x]$
{1, 3}	6.	$(\forall y)(Iy \rightarrow \neg Lxy)$	x 3, 4 T
{2, 3}	7.	$(\forall y)(Sy \rightarrow Lxy)$	x 3, 5 T
{1, 3}	8.	$Ix \rightarrow \neg Lxx$	x 6 US $[x\|y]$
{2, 3}	9.	$Sx \rightarrow Lxx$	x 7 US $[x\|y]$
{1, 2, 3}	10.	$Lxx \wedge \neg Lxx$	x 3, 8, 9 T
{1, 2}	11.	$\neg(Ix \wedge Sx)$	3, 10 IP
{1, 2}	12.	$Ix \rightarrow \neg Sx$	11 T
{1, 2}	13.	$(\forall x)(Ix \rightarrow \neg Sx)$	12 UG

Here, the premise for Rule *IP* is given on line 3, and the contradiction is derived on line 10. Thus, line 11 no longer depends on premise 3, so x is not flagged on line 11.

Note that it is **not valid** to deduce

$$Sy \rightarrow (\forall y)(Iy \rightarrow \neg Lyy)$$

from line 1 because it is an **incorrect** use of Rule *US*. The substitution of y for x is **not** a valid substitution.

7.4 ORDERING RELATIONS

In the next example and in the exercises, we shall use some terminology that is defined below.

Definition 7.3 Suppose that R is a binary (2-place) relation (predicate). Let Dx represent "x is in the domain of R," and assume the domain of R is contained in its range. (The *domain* of R is $\{u : (\exists v)(< u, v >\in R)\}$ and the *range* of R is $\{v : (\exists u)(< u, v >\in R\})$.)

1. R is *reflexive on D* iff $(\forall x)(Dx \rightarrow Rxx)$.
2. R is *irreflexive on D* iff $(\forall x)(Dx \rightarrow \neg Rxx)$.

3. R is *symmetric on D* iff $(\forall x)(\forall y)(Dx \wedge Dy \rightarrow (Rxy \rightarrow Ryx))$.
4. R is *asymmetric on D* iff $(\forall x)(\forall y)(Dx \wedge Dy \rightarrow (Rxy \rightarrow \neg Ryx))$.
5. R is *antisymmetric on D* iff $(\forall x)(\forall y)(Dx \wedge Dy \rightarrow (Rxy \wedge Ryx \rightarrow x = y))$.
6. R is *transitive on D* iff $(\forall x)(\forall y)(\forall z)(Dx \wedge Dy \wedge Dz \rightarrow (Rxy \wedge Ryz \rightarrow Rxz))$.
7. R is an *equivalence relation on D* iff R is reflexive, symmetric, and transitive on D.
8. R is a *partial ordering on D* iff R is reflexive, antisymmetric, and transitive on D.
9. R is *connected on D* iff $(\forall x)(\forall y)(Dx \wedge Dy \rightarrow (Rxy \vee Ryx \vee x = y))$.
10. R is a *linear ordering on D* iff R is a connected partial ordering on D.

For example, the relation $<$ on the set of integers is irreflexive, asymmetric, and transitive. The relation $\leq$ is a linear ordering on the set of integers. Suppose $\mathbb{Q}$ is the set of fractions p/q, where p and q are integers and $q \neq 0$. Define a relation E on $\mathbb{Q}$ so that $Ep/q, r/s$ iff $p \cdot s = q \cdot r$, then E is an equivalence relation on $\mathbb{Q}$. (Two fractions are equivalent if they represent the same number.)

Example 7.2

Give a formal proof of each of the following. (Assume that the domain of R is contained in its range.)

(a) If there is some element a in the domain of R (that is, Da is a premise), then R cannot be both reflexive and asymmetric.

(b) If a and b are in the domain of R, Rab, R is symmetric and transitive then Raa.

Solution:

(a) Part (a) is equivalent to showing that the following set of premises is inconsistent:

$$Da;\ (\forall x)(Dx \rightarrow Rxx);\ (\forall x)(\forall y)(Dx \wedge Dy \rightarrow (Rxy \rightarrow \neg Ryx)).$$

We shall give a formal proof of a contradiction:

{1}	1.	Da	P
{2}	2.	$(\forall x)(Dx \rightarrow Rxx)$	P
{3}	3.	$(\forall x)(\forall y)(Dx \wedge Dy \rightarrow (Rxy \rightarrow \neg Ryx))$	P
{2}	4.	$Da \rightarrow Raa$	2 US $[a\|x]$
{1, 2}	5.	Raa	1, 4 T
{3}	6.	$(\forall y)(Da \wedge Dy \rightarrow (Ray \rightarrow \neg Rya))$	3 US $[a\|x]$
{3}	7.	$Da \wedge Da \rightarrow (Raa \rightarrow \neg Raa)$	6 US $[a\|y]$
{1, 3}	8.	$Raa \rightarrow \neg Raa$	1, 7 T
{1, 3}	9.	$\neg Raa$	8 T
{1, 2, 3}	10.	$Raa \wedge \neg Raa$	5, 9 T

(b) **Premises:** $Da \land Db \land Rab$; $(\forall x)(\forall y)(Dx \land Dy \to (Rxy \to Ryx))$; $(\forall x)(\forall y)(\forall z)(Dx \land Dy \land Dz \to (Rxy \land Ryz \to Rxz))$

Conclusion: Raa

Proof:

{1}	1.	$Da \land Db \land Rab$	P
{2}	2.	$(\forall x)(\forall y)(Dx \land Dy \to (Rxy \to Ryx))$	P
{3}	3.	$(\forall x)(\forall y)(\forall z)(Dx \land Dy \land Dz \to (Rxy \land Ryz \to Rxz))$	P
{2}	4.	$(\forall y)(Da \land Dy \to (Ray \to Rya))$	2 US $[a\|x]$
{2}	5.	$Da \land Db \to (Rab \to Rba)$	4 US $[b\|y]$
{1, 2}	6.	$Rab \to Rba$	1, 5 T
{1}	7.	Rab	1 T
{1, 2}	8.	Rba	6, 7 T
{3}	9.	$(\forall y)(\forall z)(Da \land Dy \land Dz \to (Ray \land Ryz \to Raz))$	3 US $[a\|x]$
{3}	10.	$(\forall z)(Da \land Db \land Dz \to (Rab \land Rbz \to Raz))$	9 US $[b\|y]$
{3}	11.	$Da \land Db \land Da \to (Rab \land Rba \to Raa)$	10 US $[a\|z]$
{1, 3}	12.	$Rab \land Rba \to Raa$	1, 11 T
{1, 2, 3}	13.	Raa	7, 8, 12 T

After the Rule *US* is completely understood, students may omit line 4 of the proof and replace the documentation for line 5 by

2 ***US*** **$[a|x][b|y]$.**

This means that the Rule *US* was used twice: the first time, *a* replaces *x*, and the second time, *b* replaces *y*. Lines 9 and 10 may also be omitted if line 11 is documented as follows:

3 ***US*** **$[a|x][b|y][a|z]$.**

EXERCISES 7

A. Give a formal proof of each of the following arguments. (Answers to 1, 5, 9, and 12.)

(1) **Premises:** $(\forall x)(Ax \to Bx)$, $(\forall x)(Cx \to \neg Bx)$

Conclusion: $(\forall x)(Ax \to \neg Cx)$

(2) **Premises:** $(\forall x)(Ax \to Bx \lor Cx)$, $(\forall x)(Bx \to Cx)$

Conclusion: $(\forall x)(Ax \to Cx)$

(3) **Premises:** $(\forall x)(Ax \to \neg Bx)$, $(\forall x)(Cx \lor \neg Dx \to Bx)$, Aa
Conclusion: $\neg Ca \land Da$

(4) **Premises:** $(\forall x)(Ax \lor Bx \to \neg Cx)$, $(\forall x)(Bx \to Dx)$, $(\forall x)(Dx \to Ax)$, Ca
Conclusion: $\neg Da$

(5) **Premises:** $(\forall x)(Ax \to Bx \lor \neg Cx)$, $(\forall x)(Ax \land Bx \to Dx)$, $(\forall x)(Ax \land \neg Cx \to \neg Ex)$
Conclusion: $(\forall x)(Ax \land Ex \to Dx)$

(6) **Premises:** $(\forall x)(Ax \to (\forall y)(By \to Cxy))$, $(\forall x)(Dx \to Bx)$
Conclusion: $(\forall x)(Ax \to (\forall y)(Dy \to Cxy))$

(7) **Premises:** $(\forall x)(Ax \to (\forall y)(By \to Cxy))$, $(\forall x)(Ax \to Bx)$, $Ba \land \neg Cba$
Conclusion: $\neg Ab$

(8) **Premises:** $(\forall x)(Ax \to Bx)$, $(\forall y)(By \to (\forall x)Ax)$
Conclusion: $(\forall x)Bx \leftrightarrow Ba$

(9) **Premises:** $(\forall x)(\forall y)(\forall z)(Rxy \land Ryz \to Rxz)$, $(\forall x)(\forall y)(Rxy \to Ryx)$
Conclusion: $(\forall x)(\forall y)(Rxy \to Rxx)$

(10) **Premises:** $(\forall x)(\forall y)(\forall z)(Rxy \land Ryz \to Rxz)$, $(\forall x)(\forall y)(Rxy \to Ryx)$
Conclusion: $(\forall x)Rxa \to (\forall x)Rxx$

(11) **Premises:** $(\forall x)(\forall y)(Ax \land By \to \neg Rxy)$, Aa, $(\forall y)(Cy \to Ray)$
Conclusion: $(\forall z)(Cz \to \neg Bz)$

(12) **Premises:** $(\forall x)(Ax \to \neg Rxa)$, $Ab \land Ba$, $(\forall x)(Cx \to Bx)$, $(\forall x)(Ax \to (\forall y)(By \land Cy \to Rxy))$
Conclusion: $\neg Ca$

(13) **Premises:** $(\forall x)(\forall y)(Rxy \land Ryx \to Sxy)$
Conclusion: $(\forall x)(\forall y)[(\forall z)(Rxz \land Ryz) \to Sxy]$

(14) **Premises:** $(\forall x)(\forall y)(Rxy \to Rf(y)f(x))$, $(\forall x)Rxf(x)$
Conclusion: $(\forall x)Rf(f(x))f(x)$

(15) **Premises:** $(\forall x)(\forall y)(\forall z)(Rxz \land Ryz \to Rxy)$, $(\forall x)Rxf(x)$
Conclusion: $Rab \land Rbc \to Rac$
(*Hint*: First show R is reflexive and symmetric.)

B. Translate each of the following valid arguments into the language of the Predicate Calculus, using the given predicate symbols and constants, and give a formal proof. Use only universal quantifiers for the translations. (Answers to 1, 7, 9, and 11.)

(1) All birds have wings. No fish have wings. Therefore, no fish are birds. (Bx, Wx, Fx)

(2) All physicists are mathematicians. George is not a mathematician. Therefore, George is not a physicist. (Px, Mx, e)

(3) Doctors are college graduates. Professors are college graduates. Therefore, either doctors or professors are college graduates. (Dx, Cx, Px)

(4) No Democrat or Republican voted. Miriam voted. Therefore, Miriam is not a Republican. (Dx, Rx, Vx, m)

(5) The only animals that have feathers are birds. All birds sing. Therefore, no animal that does not sing has feathers. (Ax, Fx, Bx, Sx)

(6) Every house has either a washing machine or a dishwasher. Every house with a washing machine also has a clothes dryer. Every house with a dishwasher also has a radio. Therefore, every house that does not have a radio has a clothes dryer. (Hx, Wx, Dx, Cx, Rx)

(7) Only ghosts live in the haunted house. No happy ghost is intelligent. No ghost is happy. Therefore, anyone who lives in the haunted house is either intelligent or unhappy. (Gx, Lx, Hx, Ix)

(8) Every mathematician likes every chemist. No mathematician likes any philosopher. Carl Friedrich Gauss was a mathematician. Therefore, no philosopher is a chemist. (Mx, Cx, Px, Lxy, c)

(9) Every professor is older than every student. Oscar is not older than Theresa, who is a student. Therefore, Oscar is not a professor. (Px, Sx, Oxy, o, t)

(10) Men who are athletes are both baseball and football fans. Athletes who are baseball fans are men who are football fans. Anyone who is not an athlete is a man who is a baseball fan. Therefore, all men are baseball fans and conversely. (Mx, Ax, Bx, Fx)

(11) Fathers and mothers like children. Norman is a father who doesn't like Peter. No mother likes Sarah. Rita is a mother. Therefore, neither Peter nor Sarah are children. (Fx, Mx, Cx, Lxy, n, p, r, s)

(12) Anyone who likes all lawyers is a criminal. Only liars like all criminals. Lawyers like lawyers, and criminals like criminals. Therefore, all lawyers are liars. (Wx, Cx, Ax, Lxy)

(13) Any woman who accompanies a child to a meeting is a parent. Leslie accompanies Alex to every meeting. Therefore, if Leslie is a woman, Alex is a child and m is a meeting, then Leslie is a parent. (Wx, Cx, Mx, Px, $Axyz$, ℓ, a, m)

(14) Geniuses solve every problem on every exam. Anyone who solves a problem on an exam is a student. Arlene solves every problem on every exam. Therefore, if Arlene is a genius, then she is a student, as long as there is a problem, p, and an exam, e. (Gx, Px, Ex, Dx, $Sxyz$, a, p, e)

(15) No student borrows from his or her enemy. Every student borrows from his or her roommate. Therefore, no student has a roommate who is his or her enemy. (Sx, Exy, Rxy, Bxy)

C. In these exercises, all relations are binary relations and the terminology is defined in Definition 7.3. Give a formal proof of each of the following statements. (Answers to 1 and 5.)

(1) If R is asymmetric on D, then R is irreflexive on D.

(2) If R is transitive and irreflexive on D, then it is asymmetric on D.

(3) If R is a linear ordering on D and x and y are any two distinct elements of D, then either Rxy or Ryx but not both.

In Exercises (4) through (7), assume all variables range over D.

(4) If R is reflexive on D and S is any binary relation on D such that

$$(\forall x)(\forall y)(\forall z)(Sxy \rightarrow (Rxz \rightarrow Ryz)),$$

then

$$(\forall x)(\forall y)(Sxy \rightarrow Ryx).$$

(5) If E is an equivalence relation on D, then

$$(\forall x)(\forall y)(\forall z)(Exy \wedge Exz \rightarrow Eyz).$$

(6) If R is connected on D and E is an equivalence relation on D such that

$$(\forall x)(\forall y)(Exy \wedge Rxy \rightarrow Ryx),$$

then

$$(\forall x)(\forall y)(\forall z)(Exy \wedge Eyz \wedge Rxy \wedge Ryz \wedge x \neq z \rightarrow Rxz).$$

(7) If R is reflexive on D, S is a binary relation on D, and f is a function from D to D such that

$$(\forall x)(\forall y)(\forall z)[Sxy \rightarrow (Rf(x)z \leftrightarrow Rf(y)z)],$$

then

$$(\forall x)(\forall y)(Sxy \rightarrow Rf(x)f(y)).$$

D. Show that each of the following sets of premises is inconsistent by giving a formal proof of a contradiction. (Answers to 1, 5, 10, and 16.)

(1) $(\forall x)(Px \rightarrow Qx)$, $(\forall x)(Px \rightarrow \neg Qx)$, Pa

(2) $(\forall x)(\forall y)(Px \wedge Qy \rightarrow Rxy)$, $(\forall x)(Qx \rightarrow \neg Rxa)$, $Pa \wedge Qa$

(3) $(\forall x)(Qx \rightarrow Px)$, $(\forall x)(\neg Qx \rightarrow Rx)$, $(\forall x)\neg Rx$, $\neg(\forall x)Px$

(4) $(\forall x)(\forall y)Rxy$, $\neg(\forall y)(\forall x)Rxy$

(5) $(\forall x)(Px \rightarrow Qx \wedge Rx)$, $(\forall x)(Sx \rightarrow \neg Rx)$, $\neg(\forall x)(Px \rightarrow \neg Sx)$

(6) $(\forall x)(Px \leftrightarrow Qx \vee Rx)$, $(\forall x)(Qx \leftrightarrow \neg Sx)$, $(\forall x)(Rx \leftrightarrow \neg Tx)$, $(\forall x)(Px \leftrightarrow Sx \wedge Tx)$

(7) $(\forall x)[(\forall y)(Qy \wedge Rxy) \rightarrow Px]$, $\neg Pa$, $(\forall x)(Qx \wedge Rax)$

(8) $(\forall x)(Px \rightarrow (\forall y)(Qy \rightarrow Rxy))$, Pb, Qa, $\neg Rba$

(9) $(\forall x)(\forall y)(Px \wedge Qy \rightarrow Rxy)$, $(\forall x)(Sx \rightarrow \neg Rax)$, Pa, $\neg(\forall x)(Sx \rightarrow \neg Qx)$

(10) $(\forall x)(\forall y)(\forall z)(Rxy \wedge Ryz \rightarrow Rxz)$, $(\forall x)(\forall y)(Rxy \rightarrow Ryx)$, Rab, $\neg Rbb$

(11) $Pa \wedge Sa$, $Qf(a) \wedge Raf(a)$, $(\forall x)[Sx \rightarrow (\forall y)(Px \wedge Qy \rightarrow \neg Rxy)]$

(12) $(\forall x)[Px \rightarrow (\forall y)(Qy \rightarrow Rxy)]$, $(\forall x)(Tx \rightarrow Qx)$, $(\forall x)(\forall y)(\forall z)\ (Rxy \wedge Ryz \rightarrow Rxz)$, $Pa \wedge Sb \wedge Tc \wedge \neg Rab$, $(\forall x)(Sx \rightarrow Rcx)$

(13) $(\forall x)(\forall y)(Px \wedge Qxy \rightarrow Rxy)$, $(\forall x)(Sx \rightarrow Px)$, $(\forall x)[Sx \rightarrow (\forall y)\ (Qxy \rightarrow Txy)]$, $(\forall x)(Rxb \rightarrow \neg Txb)$, $Sa \wedge Qab$

(14) $(\forall x)[(\forall y)(Py \rightarrow Rxy) \rightarrow Qa \wedge \neg Sxa]$, $(\forall x)(\forall y)(Py \rightarrow Rxy)$, $(\forall x)(\forall y)(Qy \rightarrow Sxy)$

(15) $(\forall x)(\forall y)[Py \wedge Qxy \rightarrow (\forall z)(Sz \wedge \neg Rxz)]$, Pa, Qba, Rba

(16) Tab, $(\forall x)(\forall y)(Txy \rightarrow Qx \wedge Ry)$, $(\forall x)(\forall y)(\forall z)(Sxyz \rightarrow Px)$, $(\forall x)(\forall y)(Qx \wedge Ry \rightarrow Scyx)$, $\neg Pc$

(17) $(\forall x)(\forall y)(Qx \vee Txy \rightarrow Py)$, $(\forall x)[Rx \rightarrow (\forall y)(Py \rightarrow Qa \wedge Sxya)]$, $Rb \wedge Tbc$, $(\forall x)(Pc \rightarrow \neg Sbcx)$

E. Each of the following proofs is correct. Fill in the missing documentation, including the flagged variables. (Answers to 1 and 3.)

(1)

{1}	1.	$(\forall x)(Ax \rightarrow Bx)$	P
{2}	2.	$(\forall y)(Cy \rightarrow \neg By)$	P

Deduce $(\forall z)(Az \rightarrow \neg Cz)$.

3.	Az
4.	$Az \rightarrow Bz$
5.	Bz
6.	$Cz \rightarrow \neg Bz$
7.	$\neg Cz$
8.	$Az \rightarrow \neg Cz$
9.	$(\forall z)(Az \rightarrow \neg Cz)$

(2)

{1}	1.	$(\forall x)(Ax \wedge Bx \rightarrow Cx)$	P
{2}	2.	$(\forall x)(Bx \rightarrow \neg Cx \vee Dx)$	P
{3}	3.	$(\forall x)(Ax \rightarrow Bx)$	P

Deduce $(\forall x)(Ax \rightarrow Dx)$.

4.	$Ax \wedge \neg Dx$
5.	$Ax \rightarrow Bx$
6.	Bx
7.	$Ax \wedge Bx$
8.	$Ax \wedge Bx \rightarrow Cx$
9.	Cx
10.	$Cx \wedge \neg Dx$
11.	$\neg(\neg Cx \vee Dx)$
12.	$Bx \rightarrow \neg Cx \vee Dx$
13.	$\neg Bx$
14.	$Bx \wedge \neg Bx$
15.	$\neg(Ax \wedge \neg Dx)$
16.	$Ax \rightarrow Dx$
17.	$(\forall x)(Ax \rightarrow Dx)$

(3)

{1}	1.	$(\forall x)(\forall y)(Axy \wedge Ayx \rightarrow Bxy)$	P

Deduce $(\forall x)(\forall y)[(\forall z)(Axz \wedge Ayz) \rightarrow Bxy]$.

2.	$(\forall z)(Axz \wedge Ayz)$
3.	$Axy \wedge Ayy$
4.	$Axx \wedge Ayx$
5.	$Axy \wedge Ayx$
6.	$Axy \wedge Ayx \rightarrow Bxy$

		7.	Bxy	
		8.	$(\forall z)(Axz \wedge Ayz) \rightarrow Bxy$	
		9.	$(\forall y)[(\forall z)(Axz \wedge Ayx) \rightarrow Bxy]$	
		10.	$(\forall x)(\forall y)[(\forall z)(Axz \wedge Ayz) \rightarrow Bxy]$	

(4)	{1}	1.	$(\forall x)[Ax \rightarrow (\forall y)(Bxy \rightarrow Cy]$	P
	{2}	2.	$(\forall x)[Ax \rightarrow (\forall y)(Dxy \rightarrow Bxy)]$	P

Deduce $(\forall x)(\forall y)(Ax \wedge Dxy \rightarrow Cy)$.

3.	$Ax \wedge Dxy$
4.	$Ax \rightarrow (\forall y)(Bxy \rightarrow Cy)$
5.	$Ax \rightarrow (\forall y)(Dxy \rightarrow Bxy)$
6.	$(\forall y)(Bxy \rightarrow Cy)$
7.	$(\forall y)(Dxy \rightarrow Bxy)$
8.	$Bxy \rightarrow Cy$
9.	$Dxy \rightarrow Bxy$
10.	Bxy
11.	Cy
12.	$Ax \wedge Dxy \rightarrow Cy$
13.	$(\forall y)(Ax \wedge Dxy \rightarrow Cy)$
14.	$(\forall x)(\forall y)(Ax \wedge Dxy \rightarrow Cy)$

(5)	{1}	1.	Pa	P
	{2}	2.	$(\forall x)(Qx \rightarrow Rax)$	P
	{3}	3.	$(\forall x)(\forall y)(Px \wedge Rxy \rightarrow Sy)$	P

Deduce $(\forall x)(Qx \rightarrow Sx)$.

4.	Qx
5.	$Qx \rightarrow Rax$
6.	Rax
7.	$Pa \wedge Rax$
8.	$(\forall y)(Pa \wedge Ray \rightarrow Sy)$
9.	$Pa \wedge Rax \rightarrow Sx$
10.	Sx
11.	$Qx \rightarrow Sx$
12.	$(\forall x)(Qx \rightarrow Sx)$

F. Each of the following is an *incorrect* "proof." First flag the variables that should be flagged. Then state explicitly where the error or errors occur and explain which rule or rules are used incorrectly. (Answers to 1, 5, and 9.)

(1)	{1}	1.	$(\forall x)(Px \rightarrow Qx \wedge Rx)$	P
	{2}	2.	$(\forall x)(Qx \rightarrow Sa)$	P
	{3}	3.	$(\forall x)(Rx \rightarrow Ta)$	P
	{1}	4.	$Px \rightarrow Qx \wedge Rx$	1 US $[x\|x]$
	{2}	5.	$Qx \rightarrow Sa$	2 US $[x\|x]$
	{3}	6.	$Rx \rightarrow Ta$	3 US $[x\|x]$
	{2, 3}	7.	$Qx \wedge Rx \rightarrow Sa \wedge Ta$	5, 6 T

	{1, 2, 3}	8.	$Px \rightarrow Sa \wedge Ta$	4, 7 *T*
	{1, 2, 3}	9.	$(\forall x)Px \rightarrow Sa \wedge Ta$	8 *UG*

(2)	{1}	1.	$(\forall x)(Sx \rightarrow Qx \vee Rx)$	*P*
	{2}	2.	$(\forall x)(Qx \rightarrow Sx)$	*P*
	{3}	3.	Sx	*P* (for *CP*)
	{2}	4.	$Qx \rightarrow Sx$	2 *US* $[x\|x]$
	{2, 3}	5.	$\neg Qx$	3, 4 *T*
	{1}	6.	$Sx \rightarrow Qx \vee Rx$	1 *US* $[x\|x]$
	{1, 3}	7.	$Qx \vee Rx$	3, 6 *T*
	{1, 2, 3}	8.	Rx	5, 7 *T*
	{1, 2}	9.	$Sx \rightarrow Rx$	3, 8 *CP*
	{1, 2}	10.	$(\forall x)(Sx \rightarrow Rx)$	9 *UG*

(3)	{1}	1.	$(\forall x)(\forall y)(Px \wedge Qy \rightarrow Rxy)$	*P*
	{2}	2.	Pa	*P*
	{3}	3.	Qy	*P* (for *CP*)
	{1}	4.	$(\forall y)(Pa \wedge Qy \rightarrow Rxy)$	1 *US* $[a\|x]$
	{1}	5.	$Pa \wedge Qy \rightarrow Rxy$	4 *US* $[y\|y]$
	{1, 2, 3}	6.	Rxy	2, 3, 5 *T*
	{1, 2}	7.	$Qy \rightarrow Rxy$	3, 6 *CP*
	{1, 2}	8.	$(\forall y)(Qy \rightarrow Rxy)$	7 *UG*
	{1, 2}	9.	$(\forall x)(\forall y)(Qy \rightarrow Rxy)$	8 *UG*

(4)	{1}	1.	$(\forall x)(Px \rightarrow Qx)$	*P*
	{2}	2.	Px	*P* (for *CP*)
	{1}	3.	$Py \rightarrow Qy$	1 *US* $[y\|x]$
	{2}	4.	$(\forall x)Px$	2 *UG*
	{2}	5.	Py	4 *US* $[y\|x]$
	{1, 2}	6.	Qy	3, 5 *T*
	{1, 2}	7.	$(\forall y)Qy$	6 *UG*
	{1}	8.	$Px \rightarrow (\forall y)Qy$	2, 7 *CP*
	{1}	9.	$(\forall x)Px \rightarrow (\forall y)Qy$	8 *UG*

(5)	{1}	1.	$(\forall x)Px \rightarrow (\forall x)Qx$	*P*
	{2}	2.	$(\forall x)(Qx \rightarrow Rx)$	*P*
	{3}	3.	Px	*P* (for *CP*)
	{1}	4.	$Px \rightarrow (\forall x)Qx$	1 *US* $[x\|x]$
	{1, 3}	5.	$(\forall x)Qx$	3, 4 *T*
	{1, 3}	6.	Qx	5 *US* $[x\|x]$
	{2}	7.	$Qx \rightarrow Rx$	2 *US* $[x\|x]$
	{1, 2, 3}	8.	Rx	6, 7 *T*
	{1, 2}	9.	$Px \rightarrow Rx$	3, 8 *CP*
	{1, 2}	10.	$(\forall x)(Px \rightarrow Rx)$	9 *UG*

(6)	{1}	1.	Pa	*P*
	{2}	2.	$(\forall x)[(\forall y)(Qy \wedge Rxy) \rightarrow Px]$	*P*

{2}	3.	$(\forall y)(Qy \wedge Ray) \rightarrow Pa$	2 *US* $[a/x]$
{1, 2}	4.	$(\forall y)(Qy \wedge Ray)$	1, 3 *T*
{1, 2}	5.	$Qx \wedge Rax$	4 *US* $[x/y]$
{1, 2}	6.	Rax	5 *T*
{1, 2}	7.	$(\forall x)Rax$	6 *UG*

(7) {1}	1.	$(\forall y)(Px \wedge Qy \rightarrow Rxy)$	*P*
{2}	2.	$(\forall x)[(\forall y)(Px \wedge Rxy) \rightarrow Sx]$	*P*
{3}	3.	$Px \wedge Qy$	*P* (for *CP*)
{1}	4.	$Px \wedge Qy \rightarrow Rxy$	1 *US* $[y/y]$
{1, 3}	5.	Rxy	3, 4 *T*
{2}	6.	$(\forall y)(Px \wedge Rxy) \rightarrow Sx$	2 *US* $[x/x]$
{2}	7.	$Px \wedge Rxy \rightarrow Sx$	6 *US* $[y/y]$
{1, 2, 3}	8.	Sx	3, 5, 7 *T*
{1, 2}	9.	$Px \wedge Qy \rightarrow Sx$	3, 8 *CP*
{1, 2}	10.	$(\forall y)(Px \wedge Qy \rightarrow Sx)$	9 *UG*
{1, 2}	11.	$(\forall x)(\forall y)(Px \wedge Qy \rightarrow Sx)$	10 *UG*

(8) {1}	1.	$(\forall x)(\forall y)(Px \rightarrow (Qy \rightarrow Rxy))$	*P*
{2}	2.	$(\forall x)(\forall y)(Px \rightarrow (Rxy \rightarrow Sy))$	*P*
{1}	3.	$(\forall y)(Px \rightarrow (Qy \rightarrow Rxy))$	1 *US* $[x/x]$
{1}	4.	$Px \rightarrow (Qy \rightarrow Rxy)$	3 *US* $[y/y]$
{2}	5.	$(\forall y)(Px \rightarrow (Rxy \rightarrow Sy))$	2 *US* $[x/x]$
{2}	6.	$Px \rightarrow (Rxy \rightarrow Sy)$	5 *US* $[y/y]$
{1, 2}	7.	$Qy \rightarrow Sy$	4, 6 *T*
{1, 2}	8.	$(\forall y)(Qy \rightarrow Sy)$	7 *UG*

(9) {1}	1.	$(\forall x)[Px \rightarrow (\forall y)(Qy \rightarrow Rxy)]$	*P*
{2}	2.	$(\forall x)(Sx \rightarrow \neg Px)$	*P*
{3}	3.	Sa	*P*
{2}	4.	$Sa \rightarrow \neg Pa$	2 *US* $[a/x]$
{2, 3}	5.	$\neg Pa$	3, 4 *T*
{1}	6.	$Pa \rightarrow (\forall y)(Qy \rightarrow Rxy)$	1 *US* $[a/x]$
{1, 2, 3}	7.	$(\forall y)(Qy \rightarrow Rxy)$	5, 6 *T*
{1, 2, 3}	8.	$(\forall x)(\forall y)(Qy \rightarrow Rxy)$	7 *UG*

(10) {1}	1.	$(\forall x)(Px \wedge Qy \rightarrow Rxy)$	*P*
{2}	2.	$(\forall x)(Px \wedge Sy \rightarrow Rxy)$	*P*
{3}	3.	Pa	*P*
{1}	4.	$Pa \wedge Qy \rightarrow Ray$	1 *US* $[a/x]$
{1, 3}	5.	$Qy \rightarrow Ray$	3, 4 *T*
{2}	6.	$Pa \wedge Sy \rightarrow Ray$	2 *US* $[a/x]$
{2, 3}	7.	$Sy \rightarrow Ray$	3, 6 *T*
{1, 2, 3}	8.	$Qy \vee Sy \rightarrow Ray$	5, 7 *T*
{1, 2, 3}	9.	$(\forall y)(Qy \vee Sy \rightarrow Ray)$	8 *UG*

Chapter 8

General Proofs

In this chapter, we introduce rules for adding and dropping the existential quantifier and illustrate proofs using both the rules for the universal quantifier and the rules for the existential quantifier.

8.1 EXISTENTIAL GENERALIZATION (ADDING AN EXISTENTIAL QUANTIFIER)

Stated informally, the rule for adding an existential quantifier is as follows: If a property $\mathcal{P}$ holds for some particular term t, then there exists a v for which $\mathcal{P}$ holds. For example, let Ix be "x is an integer," then the formula

(1) $I2 \wedge (\exists y)(Iy \wedge y \neq 2)$

is translated as

(2) Two is an integer, and some integer is different from two.

We should be able to deduce from (1) that

(3) $(\exists x)[Ix \wedge (\exists y)(Iy \wedge y \neq x)]$

and

(4) $(\exists u)[Iu \wedge (\exists y)(Iy \wedge y \neq u)]$.

Both (3) and (4) are translated as follows: "There is an integer such that some integer is different from the first one." In other words, if two has the property described by (1), then there exists something that has that property. However, consider the following formula:

(5) $(\exists y)[Iy \wedge (\exists y)(Iy \wedge y \neq y)]$.

Formula (5) is obtained from (1) by substituting y for 2, and it is always false. There is no integer that is different from itself; however, (1) is true. Because the rules of inference must preserve truth, we cannot deduce (5) from (1).

Let us see what went wrong. Formula (1) is interpreted to mean that two has a certain property—namely, the property described in (2). The formula obtained from (5) by omitting the existential quantifier,

(6) $Iy \wedge (\exists y)(Iy \wedge y \neq y)$,

is interpreted to mean "y is an integer, and there is an integer that is different from itself." The variable y in (6) does *not* have the same property as 2 in (1). That is where the problem lies. We want to say that if a property $\mathcal{P}$ holds for some term t, then there exists a v that has the *same* property $\mathcal{P}$. Thus, we must insure that if $(\exists v)\varphi$ is derived from ψ, then ψ is $\varphi[t|v]$ and the substitution of t for v in φ is valid. Informally, this means that φ says the same thing about v as $\varphi[t|v]$ does about t. The rule for adding an existential quantifier, which we call Existential Generalization, or Rule *EG*, is stated as follows:

Rule *EG* (Rule for Existential Generalization) *If φ is a formula, t a term, and v a variable, then $(\exists v)\varphi$ is derivable from $\varphi[t|v]$, if the substitution of t for v in φ is valid. Equivalently, $\varphi[t|v] \vdash (\exists v)\varphi$, if the substitution of t for v in φ is valid. Or, if Γ is a set of formulas, then if $\Gamma \vdash \varphi[t|v]$, then $\Gamma \vdash (\exists v)\varphi$, if the substitution of t for v in φ is valid.*

We shall show how this rule eliminates the difficulty described above. (In Chapter 16, we shall show that this rule and all the other rules we use here are valid; that is, they always preserve truth.)

To show that we can deduce (3) from (1) using Rule *EG*, let the variable v in Rule *EG* be x, and let the formula φ in Rule *EG* be

(7) $Ix \wedge (\exists y)(Iy \wedge y \neq x)$.

The term t is 2, so the substitution of 2 for x in (7) is valid. The formula $\varphi[2|x]$ is formula (1). Thus, Rule *EG* allows us to deduce (3) from (1). Similarly, we can show that (4) is derivable from (1).

On the other hand, if we take the term t to be y or, in fact, if t is any term in which y occurs, then the substitution of t for x in (7) is not valid, so that Rule *EG* is not applicable.

In order to be sure that Rule *EG* is applicable in deducing $(\exists v)\varphi$ from some formula ψ, check that

1. replacing every free occurrence of v in φ by t is a valid substitution, and
2. if every free occurrence of v in φ is replaced by t, then φ becomes ψ.

Example 8.1

Give a formal proof of each of the following valid arguments:

(a) **Premises:** $(\forall x)(Gx \rightarrow Sx)$; Ga

Conclusion: $(\exists x)Sx$

(b) **Premises:** $(\forall x)[Px \rightarrow (\forall y)(Qy \rightarrow \neg Rxy)]$; $(\forall x)[Px \rightarrow (\forall y)(\neg Sx \rightarrow Qy)]$; Pb

Conclusion: $(\exists x)[Px \wedge (\forall y)(Rxy \rightarrow Sx)]$

Solution: The normal procedure to use to prove an existential statement, $(\exists x)\varphi(x)$, is to find a term t so that $\varphi[t|x]$ holds and then use Rule *EG*.

(a) To prove $(\exists x)Sx$, we shall derive Sa and then use Rule *EG*:

{1}	1.	$(\forall x)(Gx \rightarrow Sx)$	P
{2}	2.	Ga	P
{1}	3.	$Ga \rightarrow Sa$	1 US $[a\|x]$
{1, 2}	4.	Sa	2, 3 T
{1, 2}	5.	$(\exists x)Sx$	4 EG $[a\|x]$

Notice, to check if Rule *EG* is applicable, we check whether the substitution of a for x in Sx is valid. Thus, we use the documentation $[a|x]$ on line 5.

(b) Again, in this case, we shall follow the normal procedure to prove an existential formula and find an appropriate term.

{1}	1.	$(\forall x)[Px \rightarrow (\forall y)(Qy \rightarrow \neg Rxy)]$	P
{2}	2.	$(\forall x)[Px \rightarrow (\forall y)(\neg Sx \rightarrow Qy)]$	P
{3}	3.	Pb	P
{1}	4.	$Pb \rightarrow (\forall y)(Qy \rightarrow \neg Rby)$	1 US $[b\|x]$
{1, 3}	5.	$(\forall y)(Qy \rightarrow \neg Rby)$	3, 4 T
{2}	6.	$Pb \rightarrow (\forall y)(\neg Sb \rightarrow Qy)$	2 US $[b\|x]$
{2, 3}	7.	$(\forall y)(\neg Sb \rightarrow Qy)$	3, 6 T
{1, 3}	8.	$Qy \rightarrow \neg Rby$	5 US $[y\|y]$
{2, 3}	9.	$\neg Sb \rightarrow Qy$	7 US $[y\|y]$
{1, 3}	10.	$Rby \rightarrow \neg Qy$	8 T
{2, 3}	11.	$\neg Qy \rightarrow Sb$	9 T
{1, 2, 3}	12.	$Rby \rightarrow Sb$	10, 11 T
{1, 2, 3}	13.	$(\forall y)(Rby \rightarrow Sb)$	12 UG
{1, 2, 3}	14.	$Pb \wedge (\forall y)(Rby \rightarrow Sb)$	3, 13 T
{1, 2, 3}	15.	$(\exists x)[Px \wedge (\forall y)(Rxy \rightarrow Sx)]$	14 EG $[b\|x]$

In Example 8.1 (b), the appropriate term to use is b, because one of the premises, premise 3, involves b. Note that we had to derive the universal statement, $(\forall y)(Rby \rightarrow Sb)$, before using Rule *EG* because, in the conclusion, the formula $(\forall y)(Rxy \rightarrow Sx)$ is within the scope of the quantifier $(\exists x)$. Also, when Rule *EG* was applied to obtain line 15, *every* occurrence of b on line 14 was replaced by x, and the scope of $(\exists x)$ on line 15 is the *whole formula*. We indicate the scope of $(\exists x)$ by an appropriate positioning of parentheses.

8.2 EXISTENTIAL PROOF (DROPPING AN EXISTENTIAL QUANTIFIER)

In this section, we describe how to give a proof if one or more of the premises are existential formulas. For example, consider the following argument:

All cats are animals. Some cats have tails. Therefore, some animals have tails.

Translating it to the Predicate Calculus:

Premises: $(\forall x)(Cx \rightarrow Ax)$; $(\exists x)(Cx \wedge Tx)$

Conclusion: $(\exists x)(Ax \wedge Tx)$

We might argue informally as follows: Because there exists a cat with a tail, let us call it something—say, y. Assume $Cy \wedge Ty$. Because all cats are animals, we can deduce from premise 1 that $Cy \rightarrow Ay$. Thus, from the assumption $Cy \wedge Ty$ and the implication $Cy \rightarrow Ay$, we can deduce Ay. We now have Ay and, from the assumption, we can deduce Ty. So we deduce $Ay \wedge Ty$. Using Rule *EG*, we obtain $(\exists x)(Ax \wedge Tx)$.

However, we did *not* prove the conclusion from the original premises. We replaced premise 2, $(\exists x)(Cx \wedge Tx)$, by $Cy \wedge Ty$. We derived the conclusion using premise 1 and $Cy \wedge Ty$. This is a very natural way to argue, however; if something exists, give it a name, at least temporarily. The rule we shall use, which we call the Rule of Existential Proof, or Rule *EP*, will validate this type of argument. It implies that proving the conclusion from the temporary premise is sufficient. The preliminary version of the Rule *EP* has the following form:

If Γ is a set of formulas, φ and ψ are formulas, then

(*) If $\Gamma, \psi[v|u] \vdash \varphi$, then $\Gamma, (\exists u)\psi \vdash \varphi$.

The variable v is called the *EP variable* and u the *existential variable*. (Remember that this is the preliminary version of Rule *EP*. Later we shall add some conditions to this rule.)

We illustrate (*) by giving a formal proof of the argument given above:

{1}	1.	$(\forall x)(Cx \rightarrow Ax)$	P
{2}	2.	$(\exists x)(Cx \wedge Tx)$	P
{3}	3.	$Cy \wedge Ty$	y P (for *EP*)
{1}	4.	$Cy \rightarrow Ay$	1 *US* $[y\|x]$
{3}	5.	Cy	y 3 T
{1, 3}	6.	Ay	y 4, 5 T
{3}	7.	Ty	y 3 T
{1, 3}	8.	$Ay \wedge Ty$	y 6, 7 T
{1, 3}	9.	$(\exists x)(Ax \wedge Tx)$	8 *EG* $[y\|x]$
{1, 2}	10.	$(\exists x)(Ax \wedge Tx)$	2, 3, 9 *EP* $[y\|x]$

On line 3, we have the premise for *EP*. We are assuming that y is a cat with a tail. The variable y is the *EP* variable. Notice that y is flagged on line 3. It is also flagged on each line that depends on premise 3 and on which y is free (lines 5, 6, 7, and 8). Also notice that even though y is flagged on line 8 and we may not use universal generalization on y, we may use existential generalization on y. Line 9 depends on premise 3, but y is not free on line 9, so it is not flagged. On line 10, we used Rule *EP* to replace premise 3 by premise 2. The documentation on line 10 indicates where the existential premise is (line 2), where the temporary premise for *EP* is (line 3), and where the conclusion is (line 9). Also, $[y|x]$ on line 10 indicates that y is the *EP* variable, x is the existential variable, and the substitution of y for x is a valid substitution.

At this point, stop a minute to be sure that you understand Rule *EP* as described in (*) above. How was (*) used? What are Γ, φ, and ψ in this proof?

The Rule *EP* is a rule like Rule *IP*, which allows us to replace one premise by another. In the proof above, we replaced premise 3 by premise 2. Therefore, if the formula φ is derived using Rule *EP*, the form of the proof is as follows:

α	$i.$	$(\exists v)\psi$	(Reason)
$\{j\}$	$j.$	$\psi[u\|v]$	u P (for *EP*)
.	.	.	.
$\beta \cup \{j\}$	$k.$	φ	(Reason)
$\beta \cup \alpha$	$k + 1.$	φ	i, j, k *EP* $[u\|x]$

We assume α and β are sets of positive numbers, $j \notin \alpha$ and $j \notin \beta$. The existential formula on line i is frequently a premise, but it does not have to be. If it is a premise, then $\alpha = \{i\}$. On line j, we have the *EP* premise, then φ was derived on line k. The Rule *EP* allowed us to replace the premise on line j by the premises that were used to derive the existential formula $(\exists v)\psi$ on line i.

Also notice that the substitution of the *EP* variable v for u in ψ in rule (*) must be a valid substitution. If not, we could obtain a false conclusion from true premises. Suppose, for example, that the existential premise is the true statement "There is an integer such that every positive integer is greater than it,"

$$(\exists x)[Ix \wedge (\forall y)(Py \wedge Iy \rightarrow y > x)].$$

If y is taken to be the *EP* variable, the premise for *EP* would be

$$Iy \wedge (\forall y)(Py \wedge Iy \rightarrow y > y).$$

From this, we can deduce the false statement

$$(\forall y)(Iy \wedge Py \rightarrow y > y),$$

which is translated as "Every positive integer is greater than itself." To prevent this type of invalid argument, the substitution of the *EP* variable for the existential variable must be valid.

In choosing the *EP* variable, we have to be sure that it is not a variable that has previously been chosen to have some other properties. For example, suppose our premises are Cx : x is a cat and $(\exists y)\neg Cy$: Something is not a cat. If we take the *EP* variable to be x, then the *EP* premise would be $\neg Cx$. Then we can deduce the contradiction $Cx \wedge \neg Cx$. However, the original premises were not contradictory, so the argument is not valid.

Also, if the conclusion has a free variable, we can construct an invalid argument by choosing the *EP* variable to be that free variable. A simple example of such an invalid argument is taking the premise to be $(\exists x)Cx$: There is a cat, and deduce Cy : y is a cat. It certainly does not follow from the premise that y is a cat, but if we took y to be the *EP* variable, we could prove Cy. It follows that the *EP* variable cannot be free in any of the previous premises, except the *EP* premise, and cannot be free in the conclusion. It turns out, as we shall prove later, these are the only restrictions on the Rule *EP*. Thus, the formal statement of the rule is as follows:

Rule EP (Rule of Existential Proof) *If Γ is a set of premises, u and v are variables, and φ and ψ are formulas, then φ is derivable from Γ and $(\exists u)\psi$ if φ is derivable from Γ and $\psi[v|u]$, as long as the substitution of v for u in ψ is valid, v is not free in φ and v is not free in any premise in Γ. In symbols: If $\Gamma, \psi[v|u] \vdash \varphi$, then $\Gamma, (\exists u)\psi \vdash \varphi$, as long as the substitution of v for u in ψ is valid, v is not free in φ and v is not free in any premise in Γ.*

The variable v is called the *EP variable*, u is called the *existential variable*, and $\psi[v|u]$ is called the *EP premise*.

Remember that the idea behind the Rule *EP* is that we are told something exists (an existential premise), so in the proof we give it a name (the *EP* variable). However, we cannot give it a name that was used for something else. Thus, the *EP* variable cannot be free in the conclusion or in any of the premises used to derive the conclusion. Also remember that the *EP* variable has to be flagged. It is free in a premise, so it is not a generic variable. We cannot generalize on an *EP* variable.

To avoid any possible misuse of the Rule *EP*, in the examples that follow we shall choose a new variable for the *EP* variable—a variable that has not occurred anywhere else in the proof, bound or unbound. Students should do the same.

8.3 GENERAL PROOFS

In the following examples, we illustrate the use of the four rules, *US*, *UG*, *EG*, and *EP*. The exercises provide additional examples.

Example 8.2 Give a formal proof of each of the following valid arguments:

(a) **Premises:** $(\forall x)(Ax \wedge Bx \rightarrow Dx)$; $(\exists x)(Ax \wedge Bx \wedge Cx)$

Conclusion: $(\exists x)(Ax \wedge Cx \wedge Dx)$

(b) **Premises:** $(\exists x)[Px \wedge (\forall y)(Qy \to \neg Rxy)]$; $(\forall x)[Px \to (\forall y)(Sy \to Rxy)]$
Conclusion: $(\forall x)(Qx \to \neg Sx)$

(c) **Premises:** $(\forall x)(Px \to Qx)$
Conclusion: $(\forall x)[(\exists y)(Py \wedge Rxy) \to (\exists z)(Qz \wedge Rxz)]$

(d) **Premises:** $(\exists x)[Px \wedge (\forall y)(Py \wedge Qya \to Rxy)]$; $(\exists x)(Px \wedge Qxa)$
Conclusion: $(\exists x)(\exists y)(Px \wedge Py \wedge Rxy)$

(e) **Premises:** $(\forall x)[Px \to (\forall y)(\neg Qy \to Rxy)]$; $(\forall x)[Px \to (\exists y)(Sy \wedge \neg Rxy)]$
Conclusion: $(\exists x)Px \to (\exists x)(Sx \wedge Qx)$

Solution: For each argument, we shall first give an informal proof.

(a) Informal proof: Take as a temporary premise (for *EP*), (i) $Ay \wedge By \wedge Cy$. Then from premise 1, deduce (using Rule *US*), (ii) $Ay \wedge By \to Dy$. From (i), deduce (iii) $Ay \wedge By$. From (ii) and (iii), deduce (iv) Dy. From (i), deduce (v) $Ay \wedge Cy$, and from (iv) and (v), deduce (vi) $Ay \wedge Cy \wedge Dy$. Finally, use Rule *EG* to obtain the conclusion, and then use Rule *EP* so that the conclusion follows from the original premises. The formal proof follows:

{1}	1.	$(\forall x)(Ax \wedge Bx \to Dx)$	P
{2}	2.	$(\exists x)(Ax \wedge Bx \wedge Cx)$	P
{3}	3.	$Ay \wedge By \wedge Cy$	y P (for EP)
{1}	4.	$Ay \wedge By \to Dy$	1 US $[y\|x]$
{3}	5.	$Ay \wedge By$	y 3 T
{1, 3}	6.	Dy	y 4, 5 T
{3}	7.	$Ay \wedge Cy$	y 3 T
{1, 3}	8.	$Ay \wedge Cy \wedge Dy$	y 6, 7 T
{1, 3}	9.	$(\exists x)(Ax \wedge Cx \wedge Dx)$	8 EG $[y\|x]$
{1, 2}	10.	$(\exists x)(Ax \wedge Cx \wedge Dx)$	2, 3, 9 EP $[y\|x]$

To ensure that the Rule *EP* was used correctly on line 10, check the following:

1. Line 2 is an existential formula. The scope of the existential quantifier must be the *whole formula*.
2. After dropping the existential quantifier on line 2, the substitution of y for x is *valid*, and the resulting formula, the *EP* premise, is on line 3.
3. The *EP* variable y is not free on line 9 and also is not free in any of the premises, except the *EP* premise, on which line 9 depends. Line 9 depends on premises 1 and 3. Premise 3 is the *EP* premise, and y is not free in premise 1.

If proofs are documented, as we shall document them in all our examples, it will not be difficult to check if the rules are used correctly.

(b) Informal proof: Assume (i) Qx (as a premise for Rule *CP*) and assume (ii) $Pz \wedge (\forall y)(Qy \to \neg Rzy)$ (as a premise for *EP*). From premise 2, deduce (iii)

$Pz \to (\forall y)(Sy \to Rzy)$. Then from (ii), we can deduce both (iv) Pz and (v) $(\forall y)(Qy \to \neg Rzy)$. From (iii) and (iv), deduce (vi) $(\forall y)(Sy \to Rzy)$. From (v), deduce (vii) $Qx \to \neg Rzx$ and from (i) and (vii), deduce (viii) $\neg Rzx$. From (vi), deduce (ix) $Sx \to Rzx$, then from (viii) and (ix), deduce $\neg Sx$. By applying the appropriate logical rules, deduce the conclusion from premises 1 and 2. The formal proof follows:

{1}	1.	$(\exists x)[Px \wedge (\forall y)(Qy \to \neg Rxy)]$	P
{2}	2.	$(\forall x)[Px \to (\forall y)(Sy \to Rxy)]$	P
{3}	3.	Qx	x P (for CP)
{4}	4.	$Pz \wedge (\forall y)(Qy \to \neg Rzy)$	z P (for EP)
{2}	5.	$Pz \to (\forall y)(Sy \to Rzy)$	2 US $[z\|x]$
{4}	6.	Pz	z 4 T
{4}	7.	$(\forall y)(Qy \to \neg Rzy)$	z 4 T
{2, 4}	8.	$(\forall y)(Sy \to Rzy)$	z 5, 6 T
{4}	9.	$Qx \to \neg Rzx$	z 7 US $[x\|y]$
{3, 4}	10.	$\neg Rzx$	x, z 3, 9 T
{2, 4}	11.	$Sx \to Rzx$	z 8 US $[x\|y]$
{2, 3, 4}	12.	$\neg Sx$	x 10, 11 T
{2, 4}	13.	$Qx \to \neg Sx$	3, 12 CP
{2, 4}	14.	$(\forall x)(Qx \to \neg Sx)$	13 UG
{1, 2}	15.	$(\forall x)(Qx \to \neg Sx)$	1, 4, 14 EP $[z\|x]$

Because z is the EP variable, not x, in this proof, it would have been correct to apply Rule EP to either line 12 or line 13. The last three lines of the proof could have been either of the following:

{1, 2, 3}	13.	$\neg Sx$	x 1, 4, 12 EP $[z\|x]$
{1, 2}	14.	$Qx \to \neg Sx$	3, 13 CP
{1, 2}	15.	$(\forall x)(Qx \to \neg Sx)$	14 UG

or

{2, 4}	13.	$Qx \to \neg Sx$	3, 12 CP
{1, 2}	14.	$Qx \to \neg Sx$	1, 4, 13 EP $[z\|x]$
{1, 2}	15.	$(\forall x)(Qx \to \neg Sx)$	14 UG

The informal proof describes lines 1 through 12 of the formal proof. The remaining three lines were applications of the rules of logic. Normally, in an informal proof, we omit mentioning the rules of logic, because the rules of logic are usually natural and self-evident.

In the remaining proofs, we shall sometimes use several tautologies at once. For example, we may omit line 6 in the formal proof of part (b). Then line 8 would be documented as "z 4, 5 T." However, we shall *not* omit line 7, because we use Rule US on line 7. The Rule US requires that the scope of the quantifier is the *whole formula*. Rule US is *not* a rule of unrestricted cancellation of quantifiers. It does *not* imply that a quantifier in the middle of a formula can be dropped.

Also, remember that whenever a variable is free in a premise, it is flagged. Thus, x is flagged on line 3 of the preceding proof, and z is flagged on line 4. Also, x is flagged on every line that depends on premise 3 and on which x is free (lines 10 and 12), and z is flagged on every line that depends on premise 4 and on which z is free (lines 6 through 11). Even though lines 12 through 14 depend on premise 4, z is not free on these lines, so it is not flagged. Even though x is free on line 13, line 13 does not depend on premise 3, so x is not flagged on line 13. Flagging a variable indicates that it is *not* a generic variable, a variable that has some conditions placed on it. We can not apply Rule *UG* on a flagged variable, but we can apply Rule *EG*.

(c) Informal proof: Assume (i) $Pu \wedge Rxu$ for some u. Then deduce from premise 1 (ii) $Pu \to Qu$. From (i) and (ii), deduce (iii) $Qu \wedge Rxu$. It follows from (iii) that $(\exists z)(Qz \wedge Rxz)$. The formal proof follows:

{1}	1.	$(\forall x)(Px \to Qx)$	*P*
{2}	2.	$(\exists y)(Py \wedge Rxy)$	x *P* (for *CP*)
{3}	3.	$Pu \wedge Rxu$	x, u *P* (for *EP*(u))
{1}	4.	$Pu \to Qu$	1 *US* $[u \vert x]$
{1, 3}	5.	Qu	u 3, 4 *T*
{1, 3}	6.	$Qu \wedge Rxu$	x, u 3, 5 *T*
{1, 3}	7.	$(\exists z)(Qz \wedge Rxz)$	x 6 *EG* $[u \vert z]$
{1, 2}	8.	$(\exists z)(Qz \wedge Rxz)$	x 2, 3, 7 *EP* $[u \vert y]$
{1}	9.	$(\exists y)(Py \wedge Rxy) \to (\exists z)(Qz \wedge Rxz)$	2, 8 *CP*
{1}	10.	$(\forall x)[(\exists y)(Py \wedge Rxy) \to (\exists z)(Qz \wedge Rxz)]$	9 *UG*

On line 3, there are two free variables, but u is the *EP* variable. We indicate this by writing "(for *EP*(u))." Thus, it is correct to apply Rule *EP* to line 7 even though x is free, because x is not the *EP* variable. Also, notice that in this proof we must apply Rule *EP* before Rule *CP* because we are using Rule *CP* to eliminate premise 2, *not* premise 3; that is, we are trying to prove

$$(\exists y)(Py \wedge Rxy) \to (\exists z)(Qz \wedge Rxz),$$

not

$$(Pu \wedge Rxu) \to (\exists z)(Qz \wedge Rxz).$$

(d) Informal proof: Take, as premises, (i) $Pu \wedge (\forall y)(Py \wedge Qya \to Ruy)$ and (ii) $Pv \wedge Qva$. From (i), deduce (iii) $(\forall y)(Py \wedge Qya \to Ruy)$, and from (iii), deduce (iv) $Pv \wedge Qva \to Ruv$. Then from (ii) and (iv), deduce (v) Ruv. Therefore, from (i), (ii), and (v), deduce (vi) $Pu \wedge Pv \wedge Ruv$, from which we may deduce the desired conclusion. The formal proof is as follows:

{1}	1.	$(\exists x)[Px \wedge (\forall y)(Py \wedge Qya \rightarrow Rxy)]$	P
{2}	2.	$(\exists x)(Px \wedge Qxa)$	P
{3}	3.	$Pu \wedge (\forall y)(Py \wedge Qya \rightarrow Ruy)$	u P (for EP)
{4}	4.	$Pv \wedge Qva$	v P (for EP)
{3}	5.	$(\forall y)(Py \wedge Qya \rightarrow Ruy)$	u 3 T
{3}	6.	$Pv \wedge Qva \rightarrow Ruv$	u 5 US $[v\|y]$
{3, 4}	7.	Ruv	u, v 4, 6 T
{3, 4}	8.	$Pu \wedge Pv \wedge Ruv$	u, v 3, 4, 7 T
{3, 4}	9.	$(\exists y)(Pu \wedge Py \wedge Ruy)$	u 8 EG $[v\|y]$
{3, 4}	10.	$(\exists x)(\exists y)(Px \wedge Py \wedge Rxy)$	9 EG $[u\|x]$
{2, 3}	11.	$(\exists x)(\exists y)(Px \wedge Py \wedge Rxy)$	2, 4 10 EP $[v\|x]$
{1, 2}	12.	$(\exists x)(\exists y)(Px \wedge Py \wedge Rxy)$	1, 3, 11 EP $[u\|x]$

In this proof, because we have two existential premises, we have to use Rule *EP* twice. If we used the same variable (u), for example, as the *EP* variable on both lines 3 and 4, then u is free in both premises 3 and 4 so that it would have been an incorrect use of Rule *EP* on line 11. (Why?)

Remember that a bound variable is a "dummy" variable; that is,

(*) $$(\exists x)(Px \wedge Qxa)$$

and

(**) $$(\exists z)(Pz \wedge Qza)$$

mean exactly the same thing—there is something with a certain property. If, for example, a: 0; Px: x is an integer; and $Qxy : x > y$, then both (*) and (**) would be translated as "There exists a positive integer." Thus, even though we used the same existential variable in both premises 1 and 2, it does *not* mean that the object that exists is the same in both cases, so we should *not* use the same letter to represent both these object.

In this proof, we applied Rule *EP* first for the variable v and then u, but it also would have been correct to do it in the reverse order, because u is not free in premise 4. If v were free in premise 3, then we would have to apply Rule *EP* in the reverse order. (Why?) Note, also, that the "a" that occurs in the premises in (d) is a constant, so it is not flagged.

(e) The conclusion in this example, is a conditional statement. We shall take the antecedent, $(\exists x)Px$, as a premise and deduce the consequent, $(\exists x)(Sx \wedge Qx)$. We start by taking Pz as a premise. Then deduce (i) $Pz \rightarrow (\forall y)(\neg Qy \rightarrow Rzy)$ from premise 1, and (ii) $Pz \rightarrow (\exists y)(Sy \wedge \neg Rzy)$ from premise 2. From Pz, (i), and (ii), deduce (iii) $(\forall y)(\neg Qy \rightarrow Rzy)$ and (iv) $(\exists y)(Sy \wedge \neg Rzy)$, respectively. Now, we take as a premise (v) $Su \wedge \neg Rzu$. From (iii), deduce (vi) $\neg Qu \rightarrow Rzu$. Thus, from (v) and (vi), deduce (vii) $Su \wedge Qu$, from which we obtain (viii) $(\exists x)(Sx \wedge Qx)$. The formal proof follows:

{1}	1.	$(\forall x)[Px \to (\forall y)(\neg Qy \to Rxy)]$	P
{2}	2.	$(\forall x)[Px \to (\exists y)(Sy \wedge \neg Rxy)]$	P
{3}	3.	$(\exists x)Px$	P (for CP)
{4}	4.	Pz	z P (for EP)
{1}	5.	$Pz \to (\forall y)(\neg Qy \to Rzy)$	1 US $[z\|x]$
{2}	6.	$Pz \to (\exists y)(Sy \wedge \neg Rzy)$	2 US $[z\|x]$
{1, 4}	7.	$(\forall y)(\neg Qy \to Rzy)$	z 4, 5 T
{2, 4}	8.	$(\exists y)(Sy \wedge \neg Rzy)$	z 4, 6 T
{9}	9.	$Su \wedge \neg Rzu$	u, z P (for $EP(u)$)
{1, 4}	10.	$\neg Qu \to Rzu$	z 7 US $[u\|y]$
{1, 4, 9}	11.	$Su \wedge Qu$	u 9, 10 T
{1, 4, 9}	12.	$(\exists x)(Sx \wedge Qx)$	11 EG $[u\|x]$
{1, 2, 4}	13.	$(\exists x)(Sx \wedge Qx)$	8, 9, 12 EP $[u\|y]$
{1, 2, 3}	14.	$(\exists x)(Sx \wedge Qx)$	3, 4, 13 EP $[z\|x]$
{1, 2}	15.	$(\exists x)Px \to (\exists x)(Sx \wedge Qx)$	3, 14 CP

The premise for Rule *CP* (line 3) is existential, so we continue by taking the *EP* premise (line 4) and call the object that exists, "z." Then, on line 8, we deduce another existential formula. To drop the existential quantifier, we take another *EP* premise and, in this case, call the object that exists "u." On line 12, we deduce the consequent of the conditional statement, but it does not depend on premise 3. Premise 3 is the antecedent of the conditional statement. Thus, before applying the Rule *CP*, we must apply Rule *EP* to replace premise 4 by premise 3. However, z is free both in premise 4 (where z is the *EP* variable) and premise 9 (where u is the *EP* variable). Therefore, we must use $EP[u|y]$ before $EP[z|x]$. (Why?) Note that $EP[u|y]$ allowed us to replace premise 9 by premises 2 and 4, and $EP[z|x]$ allowed us to replace premise 4 by premise 3.

8.4 GUIDELINES (CONTINUED)

We presented some guidelines for constructing proofs in the Sentential Calculus (see Section 2.6). Here we shall give some additional guidelines for the Predicate Calculus. All these guidelines will be listed again in Chapter 12.

7. To prove an existential formula, $(\exists x)\varphi$, prove $\varphi[t|x]$ for some term t such that the substitution of t for x in φ is valid, then use Rule *EG*.
8. To prove a universal formula, $(\forall x)\varphi$, try one of the following:
   (a) If φ is a conditional, $\psi \to \chi$, take ψ as a premise, and deduce χ. Then use the Rule *CP* to derive $\psi \to \chi$. Then use Rule *UG*.
   (b) If x is not free in a premise, take $\neg\varphi$ as a premise and derive a contradiction. Then use the Rule *IP* to derive φ. Then use Rule *UG*.

9. Drop existential quantifiers before dropping universal quantifiers; that is, if any of the premises are existential formulas, choose *EP* premises and *EP* variables before using Rule *US* to drop the universal quantifiers.

We shall explain how we used these guidelines for each of the arguments in Example 8.2.

(a) The conclusion is an existential statement, so, using guideline 7, we try to find a term t such that $At \wedge Ct \wedge Dt$. We also have an existential premise, so our first step (after stating the given premises) is to choose the *EP* premise (line 3) and the *EP* variable y (guideline 9). We deduce $Ay \wedge Cy \wedge Dy$ by using Rule *US* and tautologies. Thus, y is the term we are looking for. Finally, we apply Rules *EG* and *EP*.

(b) The conclusion is a universal statement, $(\forall x)(Qx \rightarrow \neg Sx)$. Here, using guideline 8(a), our first step is to take Qx as a premise for Rule *CP* (line 3) and try to deduce $\neg Sx$. Next, because we have an existential premise, using guideline 9, we take a premise for *EP* (line 4). Then, using Rule *US* and tautologies, we deduce $\neg Sx$. At the end, we apply Rules *CP*, *UG*, and *EP*.

(c) The conclusion is a universal conditional statement, so, similarly to (b), we take the hypothesis as a premise for Rule *CP*. Next, because the premise is an existential formula, we take the *EP* premise. Then, using Rule *US* and tautologies, we deduce the conclusion of the conditional statement. Finally, we finish with Rules *EP*, *CP*, and *UG*.

(d) In this argument, the conclusion is existential, but here we have to find two terms, t_1 and t_2, such that $Rt_1 \wedge Pt_2 \wedge Rt_1t_2$, (guideline 7). We have two existential premises, so we start the proof by taking two *EP* premises (guideline 9). Then, using Rule *US* and tautologies, we show that the required terms, t_1 and t_2, are the two *EP* variables u and v. We complete the proof by using Rule *EG* twice and Rule *EP* twice.

(e) In this case, the conclusion is a conditional formula, so we start by using guideline 2 for the Sentential Calculus and take the antecedent as a premise. Now we have to prove an existential formula, so by guideline 7, we have to find a term t and deduce $St \wedge Qt$. To do this, using guideline 9, we first drop the existential quantifier in premise 3 and take an appropriate *EP* premise (line 4). In the course of using Rules *US* and *T*, we obtain another existential formula (line 8). We replace it by the appropriate *EP* premise (line 9). Then, again using Rule *US*, we deduce $Su \wedge Qu$. To finish the proof, we use Rule *EG*, Rule *EP* twice, and, finally, Rule *CP*.

To illustrate guideline 8(b), we shall give an indirect proof of the argument in Example 8.2(b):

{1}	1.	$(\exists x)[Px \wedge (\forall y)(Qy \rightarrow \neg Rxy)]$	*P*
{2}	2.	$(\forall x)[Px \rightarrow (\forall y)(Sy \rightarrow Rxy)]$	*P*
{3}	3.	$Qx \wedge Sx$	x *P* (for *IP*)
{4}	4.	$Pz \wedge (\forall y)(Qy \rightarrow \neg Rzy)$	z *P* (for *EP*)

{4}	5.	$(\forall y)(Qy \rightarrow \neg Rzy)$	z 4 T
{4}	6.	$Qx \rightarrow \neg Rzx$	z 5 US $[x\|y]$
{3, 4}	7.	$\neg Rzx$	x, z 3, 6 T
{2}	8.	$Pz \rightarrow (\forall y)(Sy \rightarrow Rzy)$	2 US $[z\|x]$
{2, 4}	9.	$(\forall y)(Sy \rightarrow Rzy)$	z 4, 8 T
{2, 4}	10.	$Sx \rightarrow Rzx$	z 9 US $[x\|y]$
{2, 3, 4}	11.	Rzx	x, z 3, 10 T
{2, 3, 4}	12.	$Rzx \wedge \neg Rzx$	x, z 7, 11 T

Line 3, our premise for Rule *IP*, is tautologically equivalent to $\neg(Qx \rightarrow \neg Sx)$. We have derived a contradiction using the *EP* premise (line 4) instead of the existential premise (line 1). At this point, however, we cannot use Rule *EP* to replace the *EP* premise by the existential premise because z, the *EP* variable, is free on line 12. To avoid this difficulty, we use the tautology (called the Rule of Absurdity) $P \wedge \neg P \rightarrow Q$, a contradiction implies every formula. Thus, from line 12, we can deduce a contradiction in which z is not a free variable. Let A be a sentence (no free variables):

{2, 3, 4}	13.	$A \wedge \neg A$	12 T
{1, 2, 3}	14.	$A \wedge \neg A$	1, 4, 13 EP $[z\|x]$
{1, 2}	15.	$\neg(Qx \wedge Sx)$	3, 14 IP
{1, 2}	16.	$Qx \rightarrow \neg Sx$	15 T
{1, 2}	17.	$(\forall x)(Qx \rightarrow \neg Sx)$	16 UG

Because z is not free on line 13 and is not free in premises 2 or 3, we may use Rule *EP* at this point to replace the *EP* premise by the existential premise. Next, we use Rule *IP* to derive line 15, which is tautologically equivalent to line 16. The variable x is not flagged on line 16, so we complete the proof by using Rule *UG*.

In this proof, we could have avoided the difficulty with Rule *EP* that occurred on line 12 by using Rule *IP* before Rule *EP*. Then the last lines of the proof would be as follows:

{2, 4}	13.	$\neg(Qx \wedge Sx)$	3, 12 IP
{2, 4}	14.	$Qx \rightarrow \neg Sx$	13 T
{2, 4}	15.	$(\forall x)(Qx \rightarrow \neg Sx)$	14 UG
{1, 2}	16.	$(\forall x)(Qx \rightarrow \neg Sx)$	1, 4, 15 EP $[z\|x]$

However, there are examples of indirect proofs in which it is very difficult to avoid this "trick" using the Rule of Absurdity so that Rule *EP* is applicable.

In general, the guidelines recommend that you first look at the formula you are trying to prove. If it's existential, use guideline 7. If it's universal, use guideline 8. In the process, drop existential quantifiers before universal quantifiers. If the conclusion is neither existential nor universal (that is, the quantifier is not the main connective), then use guidelines 1 through 6, which were presented in Section 2.6.

8.5 VALID ARGUMENTS VERSUS PROOFS

There is a dichotomy involved in the discussion of the structure of our language and its semantics. On the one hand, we have the notion of a (*formal*) *proof* (Definition 2.1, along with Rules *P, T, CP, IP, US, UG, EG,* and *EP*); on the other hand, we have the notion of a *valid argument* (an argument is valid iff whenever the premises are true, the conclusion is also true). Little distinction was made between these two concepts in the Sentential Calculus and rightly so because they are equivalent concepts. In the Sentential Calculus, an argument with premises $\varphi_1, \varphi_2, \ldots, \varphi_n$ and conclusion ψ is valid iff

$$\varphi_1 \wedge \varphi_2 \wedge \cdots \wedge \varphi_n \Rightarrow \psi.$$

(That is, $\varphi_1 \wedge \varphi_2 \wedge \cdots \wedge \varphi_n \rightarrow \psi$ is a tautology.) Thus, by the Rule of Tautology, there is a formal proof of ψ from $\varphi_1, \varphi_2, \ldots, \varphi_n$. Conversely, in the Sentential Calculus, we proved that all the rules of inference preserve truth. Thus, if there is a formal proof of an argument, it is valid.

Note that the notion of "truth" in the Predicate Calculus is a generalization of that notion in the Sentential Calculus. An interpretation structure is a generalization of a truth assignment. An argument in the Predicate Calculus with premises Γ and conclusion φ is *valid* if, for every interpretation structure $\mathbf{S}$ for $\Gamma \cup \{\varphi\}$, whenever all the formulas in Γ are true in $\mathbf{S}$, then φ is true in $\mathbf{S}$. (However, as we remarked at the end of Chapter 6 and in the introduction to Chapter 7, unlike the Sentential Calculus, there is no algorithm to determine whether a formula of the Predicate Calculus is true in an interpretation structure.)

In the Predicate Calculus, although the notions of provability and validity are also equivalent, it is much more difficult to prove. (See the Soundness Theorem and the Completeness Theorem in Chapters 9 and 16.) At this point, we shall accept these facts without proof:

1. A formal proof preserves truth. (the Soundness Theorem)

and

2. If an argument is valid, then it has a formal proof. (the Completeness Theorem)

We shall, however, distinguish the two notions of provability and validity in our discussion. The notion of provability deals with the structure or syntax of the language, whereas that of validity deals with the semantics.

EXERCISES 8

A. Give a formal proof of each of the following valid arguments. (Answers to 1, 4, 8, 11, 14, 18, and 23.)

(1) **Premises:** $(\forall x)(Ax \wedge Bx \rightarrow Cx)$, $(\exists x)(Bx \wedge \neg Cx)$
Conclusion: $(\exists x)(\neg Ax)$

(2) **Premises:** $(\forall x)(Px \rightarrow \neg Qx \vee \neg Rx)$, $(\exists x)(Qx \wedge Rx)$
Conclusion: $(\exists x)(\neg Px)$

(3) **Premises:** $(\forall x)(Ax \rightarrow Cx)$, $(\forall x)(Bx \rightarrow Dx)$, $(\exists x)(Ax \wedge Bx)$
Conclusion: $(\exists x)(Cx \wedge Dx)$

(4) **Premises:** $(\forall x)(Px \rightarrow Qx)$, $(\exists x)(Rx \wedge \neg Qx)$, $(\forall x)(Rx \rightarrow Px \vee Sx)$
Conclusion: $(\exists x)(Rx \wedge Sx)$

(5) **Premises:** $(\forall x)(\neg Ax \rightarrow Bx)$, $(\forall x)(\neg Bx \vee Cx)$, $(\exists x)(\neg Cx) \wedge (\exists x)Cx$
Conclusion: $Ca \vee (\exists x)Ax$

(6) **Premises:** $(\forall x)(Px \vee Qx \vee Rx)$, $(\forall x)(Px \rightarrow \neg Sx)$, $\neg Ra \wedge Sa$
Conclusion: $(\exists x)(\neg Px \wedge Qx)$

(7) **Premises:** $(\forall x)(\forall y)(Ax \wedge By \rightarrow Cxy)$, $(\exists x)(Ax \wedge Cax)$, $(\exists x)(Bx \wedge \neg Cax)$
Conclusion: $(\exists x)(\neg Ax)$

(8) **Premise:** $(\exists y)(\forall x)Rxy$
Conclusion: $(\forall x)(\exists y)Rxy$

(9) **Premises:** $(\exists x)Ax$, $(\exists x)Bx$, $(\forall x)[Ax \rightarrow (\forall y)(By \rightarrow Cxy)]$
Conclusion: $(\exists x)(\exists y)Cxy$

(10) **Premises:** $(\exists x)[Px \wedge (\forall y)(Qy \rightarrow Rxy)]$, $(\forall x)(\forall y)(Px \wedge Rxy \rightarrow Sy)$, $(\forall x)(Sx \rightarrow Qx)$
Conclusion: $(\forall x)(Qx \leftrightarrow Sx)$

(11) **Premises:** $(\forall x)(\forall y)(Dx \wedge Ey \rightarrow Fxy)$, $(\forall x)(\forall y)(Dx \wedge Fxy \rightarrow Gy)$
Conclusion: $(\exists x)Dx \rightarrow (\forall x)(Ex \rightarrow Gx)$

(12) **Premises:** $(\forall x)(\forall y)(Jx \wedge \neg Ky \rightarrow Lxy)$, $(\exists x)[Jx \wedge (\forall y)(My \rightarrow \neg Lxy)]$
Conclusion: $(\forall x)(Mx \rightarrow Kx)$

(13) **Premises:** $(\exists x)[Qx \wedge (\forall y)(Py \rightarrow Rxy)]$, $(\forall x)[Px \rightarrow (\forall y)(Qy \wedge Sy \rightarrow \neg Ryx)]$
Conclusion: $(\exists x)Px \rightarrow (\exists x)(\neg Sx)$

(14) **Premises:** $(\exists x)[Px \wedge (\forall y)(Py \wedge Rxy \rightarrow Qya)]$, $(\exists x)(Px \wedge \neg Qxa)$, $(\exists x)(\neg Px \wedge Qxa)$
Conclusion: $(\exists x)(\exists y)(Px \wedge Py \wedge \neg Rxy)$

(15) **Premises:** $(\exists x)(\forall y)(Fx \rightarrow Gy)$
Conclusion: $(\forall x)Fx \rightarrow (\forall x)Gx$

(16) **Premises:** $(\forall x)[(\exists y)(\neg Ay \wedge Bxy) \rightarrow Cx]$, $(\exists x)Bxa$
Conclusion: $Aa \vee (\exists x)(\exists y)(Cx \wedge Bxy)$

(17) **Premises:** $(\forall x)(Px \rightarrow (\exists y)Qxy)$, $(\exists x)[Px \wedge (\forall y)(Qxy \rightarrow Rxy)]$
Conclusion: $(\exists x)(\exists y)Rxy$

(18) **Premises:** $(\forall x)(\forall y)(\forall z)(Rxy \wedge Ryz \rightarrow Rxz)$, $(\forall x)(\forall y)(Rxy \rightarrow Ryx)$
Conclusion: $(\forall x)(\exists y)Rxy \rightarrow (\forall x)Rxx$

(19) **Premises:** $(\forall x)(\forall y)(\forall z)(Rxz \wedge Ryz \rightarrow Rxy)$, $(\forall x)(\exists y)Rxy$
Conclusion: $Rab \wedge Rbc \rightarrow Rac$.
(*Hint*: Show R is reflexive and symmetric.)

(20) **Premises:** $(\forall x)(\forall y)[Px \wedge Qy \rightarrow (\exists z)(Rz \wedge Sxyz)]$, $(\forall x)(Qx \vee Rx)$, $(\exists x)[Px \wedge (\forall y)(Ry \rightarrow (\exists z)(Qz \wedge Sxyz))]$
Conclusion: $(\exists x)(\forall y)(\exists z)Sxyz$

(21) **Premises:** $(\exists x)(\exists y)Txy$, $(\forall x)(\forall y)(Txy \rightarrow Qx \wedge Ry)$, $(\forall x)(\forall y)(\forall z)(Sxyz \rightarrow Px)$
Conclusion: $(\forall x)[(\forall y)(\forall z)(Qy \wedge Rz \rightarrow Sxyz) \rightarrow Px]$

(22) **Premises:** $(\forall x)(Px \rightarrow Qx)$, $(\forall x)(\forall y)(Qx \vee Txy \rightarrow Py)$, $(\forall x)[Rx \rightarrow (\forall y)(Py \rightarrow (\exists z)(Qz \wedge Sxyz))]$
Conclusion: $(\forall x)(\forall y)[Rx \wedge Txy \rightarrow (\exists z)(Py \wedge Sxyz)]$

(23) **Premises:** $(\forall x)(\forall y)(Px \leftrightarrow Sxy)$, $(\forall x)(\forall y)(Rxy \rightarrow Qy)$, $(\forall x)(Qx \rightarrow Tx)$, $(\exists x)Tx \rightarrow (\forall y)(\forall z)Ryz$
Conclusion: $(\forall x)[(\exists y)(Px \wedge Rxy) \leftrightarrow (\exists z)(Qz \wedge Sxz)]$

B. Translate each of the following valid arguments into the language of the Predicate Calculus, using the given predicate symbols, constants, and function symbols, and give a formal proof of validity. (Translate sentences of the form "No P's are Q's" as "$(\forall x)(Px \rightarrow \neg Qx)$" and "Not all P's are Q's" as "$(\exists x(Px \wedge \neg Qx)$".) (Answers to 1, 5, 8, 11, and 16.)

(1) All cats are mammals. Some cats do not have tails. Therefore, some mammals do not have tails. (Cx, Mx, Tx)

(2) Some animals do not have wings. Only animals with wings can fly. Therefore, some animals cannot fly. (Ax, Wx, Fx)

(3) All rational numbers are real numbers. Not all rational numbers are integers. Therefore, not all real numbers are integers. (Qx, Rx, Ix)

(4) Everyone who is older than Alex is older than Betsy. Only senior citizens are older than Betsy. Therefore, only senior citizens are older than Alex. (Sx, Oxy, a, b)

(5) Deer and sheep are grazing animals. Grazing animals are prey animals. No prey animal eats meat. Therefore, no deer eats meat. (Dx, Sx, Gx, Px, Ex)

(6) Politicians and attorneys are liars. Some politician lives in town, and some attorney lives in town. Either all liars are friendly or no liar is friendly. Therefore, either some politician is friendly or some attorney is not friendly. (Px, Ax, Lx, Tx, Fx)

(7) All functions that are differentiable are continuous. Not all integrable functions are continuous. Therefore, not all integrable functions are differentiable. (Fx, Cx, Dx, Ix)

(8) All students admire all scholars. There is some student who admires some teacher. Some professor is both a scholar and a teacher. Therefore, there is a student who admires some professor. (Sx, Cx, Tx, Px, Axy)

(9) There is a doctor whom no administrator likes. Anyone who dislikes a doctor is a hypochondriac. Therefore, all administrators are hypochondriacs. (Dx, Ax, Hx, Lxy)

(10) All fathers are men. Therefore, anyone who kills a father, kills a man. (Fx, Mx, Kxy)

(11) Every picture in the museum is a masterpiece. Any artist who paints a masterpiece is a genius. Some unknown artist painted a picture in the museum. Therefore, some unknown artist is a genius. (Px, Mx, Ax, Gx, Ux, Txy : x paints y)

(12) Every private who is hardworking has a sergeant who likes him/her. There is a private who is a college graduate who is only liked by college graduates. All college graduates are hardworking. Therefore, there is a sergeant who is a college graduate. (Px, Hx, Sx, Cx, Lxy)

(13) Any man who loves a woman has some woman who loves him. Any man who has some woman who loves him, loves himself. No man loves himself. Therefore, no man loves a woman. (Mx, Wx, Lxy) (*Hint*: Translate the conclusion as $(\forall x)(Mx \rightarrow \neg(\exists y)(Wy \wedge Lxy))$ and use an indirect proof.)

(14) There is a professor such that some student studies every subject with him/her. Some subject has the property that no student studies it with a professor. Therefore, some student studies some subject with a professor. (Px, Dx : x is a student; Sx : x is a subject; $Txyz$: x studies y with z)

(15) A symmetric matrix is positive definite if and only if all its eigenvalues are positive. Some matrices are not symmetric. Some matrices have eigenvalues that are not all positive. Every symmetric positive definite matrix has a positive determinant. Therefore, every symmetric matrix whose eigenvalues are all positive has a positive determinant. (Sx, Mx, Dx : x is positive definite; Exy : x is an eigenvalue of y, Px : x is positive; $f(x)$: the determinant of x)

(16) A set is open if and only if it contains a neighborhood of each of its points. A set is closed if and only if its complement is open. The complement of a set is a set. Therefore, any set whose complement contains a neighborhood of each of its points is closed. (Sx, Ox, Txy : x contains

y; $Nxy : x$ is a neighborhood of y; $Pxy : x$ is a point of y; $Cx : x$ is closed; $f(x)$: the complement of x)

(17) Two formulas have the same truth table if and only if they are tautologically equivalent. There are different formulas that have the same truth table. Therefore, some tautologically equivalent formulas are different. (Fx, $Sxy : x$ is the same as y; $Txy : x$ is tautologically equivalent to y; $f(x)$: the truth table of x)

C. Show that each of the following arguments is valid; that is, choose appropriate predicate symbols and constants, translate each into the language of the Predicate Calculus, and give a formal proof of validity. (See the parenthetic remark about translations in Exercise B.) (Answers to 1, 5, and 9.)

(1) Every student who studies, passes the course. Some student did not pass the course. Therefore, some student did not study.

(2) Only citizens can vote. Not all residents are citizens. Therefore, some residents cannot vote.

(3) All doctors and lawyers are college graduates. Not all intelligent people are college graduates. Therefore, there is someone who is neither a doctor nor a lawyer but is intelligent.

(4) All triangles are polygons. Therefore, anyone who draws a triangle draws a polygon.

(5) Some odd integer divides all even integers. Therefore, every even integer is divisible by some odd integer.

(6) Teenagers who play tennis are in good health. Teenagers who play football have team spirit. Anyone who has team spirit is in good health. All teenagers who go to college either play tennis or football. Therefore, if some teenager goes to college, then some teenager is in good health.

(7) Some cats like all dogs. Every dog likes some person. Spot is a dog. Any cat who likes Spot likes every person whom Spot likes. Therefore, some cat likes some person.

(8) Each boy at the party liked each girl at the party with whom he danced. Some boys at the party do not like any girl at the party who is drinking alcohol. No boy at the party likes any girl at the party who is smoking cigarettes. There are girls at the party who are either drinking alcohol or smoking cigarettes. Therefore, some boys at the party do not dance with some girls at the party.

(9) Parallel lines that are not identical do not intersect. If two lines are parallel, either they have no points in common or all points in common. Two lines that have all points in common intersect. Therefore, two parallel lines that are not identical have no points in common.

(10) Functions have a smooth graph if and only if there are no breaks and no sharp points on their graphs. If the graph of a function does not have a break, then the function is continuous. Therefore, functions that have smooth graphs are continuous.

D. Each of the following proofs is correct. Fill in the missing documentation, including the flagged variables. (Answers to 1 and 4.)

(1)	{1}	1.	$(\forall x)(Ax \wedge Bx \rightarrow Cx)$	P
	{2}	2.	$(\exists x)(Ax \wedge \neg Cx)$	P

Deduce $(\exists x)(Ax \wedge \neg Bx)$.

	3.	$Ay \wedge \neg Cy$
	4.	$Ay \wedge By \rightarrow Cy$
	5.	$\neg Cy$
	6.	$\neg(Ay \wedge By)$
	7.	$\neg Ay \vee \neg By$
	8.	Ay
	9.	$\neg By$
	10.	$Ay \wedge \neg By$
	11.	$(\exists x)(Ax \wedge \neg Bx)$
{1, 2}	12.	$(\exists x)(Ax \wedge \neg Bx)$

(2)	{1}	1.	$(\forall x)[(\exists y)Rxy \rightarrow (\forall z)Rzx]$	P

Deduce $(\forall x)(\forall y)(Rxy \rightarrow Ryx)$.

	2.	Rxy
	3.	$(\exists y)Rxy$
	4.	$(\exists y)Rxy \rightarrow (\forall z)Rzx$
	5.	$(\forall z)Rzx$
	6.	Ryx
	7.	$Rxy \rightarrow Ryx$
	8.	$(\forall y)(Rxy \rightarrow Ryx)$
{1}	9.	$(\forall x)(\forall y)(Rxy \rightarrow Ryx)$

(3)	{1}	1.	$(\exists x)[Px \wedge (\forall y)(Qy \rightarrow Rxy)]$	P
	{2}	2.	$(\forall z)(Sz \rightarrow Qz)$	P

Deduce $(\exists x)[Px \wedge (\forall y)(Sy \rightarrow Rxy)]$.

	3.	$Pu \wedge (\forall y)(Qy \rightarrow Ruy)$
	4.	Pu
	5.	$(\forall y)(Qy \rightarrow Ruy)$
	6.	$Sy \rightarrow Qy$
	7.	$Qy \rightarrow Ruy$
	8.	$Sy \rightarrow Ruy$
	9.	$(\forall y)(Sy \rightarrow Ruy)$
	10.	$Pu \wedge (\forall y)(Sy \rightarrow Ruy)$
	11.	$(\exists x)[Px \wedge (\forall y)(Sy \rightarrow Rxy)]$
{1, 2}	12.	$(\exists x)[Px \wedge (\forall y)(Sy \rightarrow Rxy)]$

(4)	{1}	1.	$(\forall x)[Px \rightarrow (\exists y)(Qy \wedge Rxy)]$	P
	{2}	2.	$(\forall x)[Px \rightarrow (\forall y)(Rxy \rightarrow Sx)]$	P

Deduce $(\forall x)(Px \to Sx)$.

		3.	Px	
		4.	$Px \to (\exists y)(Qy \land Rxy)$	
		5.	$Px \to (\forall y)(Rxy \to Sx)$	
		6.	$(\exists y)(Qy \land Rxy)$	
		7.	$(\forall y)(Rxy \to Sx)$	
		8.	$Qz \land Rxz$	
		9.	$Rxz \to Sx$	
		10.	Sx	
		11.	Sx	
		12.	$Px \to Sx$	
	{1, 2}	13.	$(\forall x)(Px \to Sx)$	

(5)	{1}	1.	$(\exists x)(Px \land Qx)$	P
	{2}	2.	$(\exists z)[Pz \land (\forall y)(Qy \to \neg Rzy)]$	P

Deduce $(\exists x)[Px \land (\exists y)(Py \land \neg Ryx)]$.

		3.	$Pu \land Qu$	
		4.	$Pv \land (\forall y)(Qy \to \neg Rvy)$	
		5.	$(\forall y)(Qy \to \neg Rvy)$	
		6.	$Qu \to \neg Rvu$	
		7.	$\neg Rvu$	
		8.	$Pv \land \neg Rvu$	
		9.	$(\exists y)(Py \land \neg Ryu)$	
		10.	$Pu \land (\exists y)(Py \land \neg Ryu)$	
		11.	$(\exists x)[Px \land (\exists y)(Py \land \neg Ryx)]$	
		12.	$(\exists x)[Px \land (\exists y)(Py \land \neg Ryx)]$	
	{1, 2}	13.	$(\exists x)[Px \land (\exists y)(Py \land \neg Ryx)]$	

(6)	{1}	1.	$(\exists x)Ax$	P
	{2}	2.	$(\forall x)(Ax \to \neg Bxa)$	P
	{3}	3.	Ca	P
	{4}	4.	$(\forall x)[Ax \to (\forall y)(Cy \land Dy \to Bxy)]$	P

Deduce $\neg Da$.

		5.	Da	
		6.	Az	
		7.	$Az \to (\forall y)(Cy \land Dy \to Bzy)$	
		8.	$(\forall y)(Cy \land Dy \to Bzy)$	
		9.	$Ca \land Da \to Bza$	
		10.	Bza	
		11.	$Az \to \neg Bza$	
		12.	$Bza \land \neg Bza$	
		13.	$P \land \neg P$	
		14.	$P \land \neg P$	
	{1, 2, 3, 4}	15.	$\neg Da$	

(7)

{1}	1.	$(\forall x)[Px \to (\forall y)(Qy \to \neg Rxy)]$	P
{2}	2.	$(\forall x)[Px \to (\forall y)(Sy \to Rxy)$	P

Deduce $(\exists x)(Px \wedge Qx) \to (\exists y)\neg Sy$.

	3.	$(\exists x)(Px \wedge Qx)$
	4.	$Pz \wedge Qz$
	5.	$Pz \to (\forall y)(Qy \to \neg Rzy)$
	6.	$(\forall y)(Qy \to \neg Rzy)$
	7.	$Qz \to \neg Rzz$
	8.	$\neg Rzz$
	9.	$Pz \to (\forall y)(Sy \to Rzy)$
	10.	$(\forall y)(Sy \to Rzy)$
	11.	$Sz \to Rzz$
	12.	$\neg Sz$
	13.	$(\exists y)\neg Sy$
	14.	$(\exists y)\neg Sy$
{1, 2}	15.	$(\exists x)(Px \wedge Qx) \to (\exists y)\neg Sy$

(8)

{1}	1.	$(\exists x)[Px \wedge (\forall y)(Sy \to Rxy)]$	P
{2}	2.	$(\forall x)(Qx \to Sx)$	P
{3}	3.	$(\forall x)(Rxx \wedge Sx \to Qx)$	P
{4}	4.	$(\forall x)[Qx \to (\forall y)(Qy \wedge Rxy)]$	P

Deduce $(\forall x)(Px \to Qx) \leftrightarrow (\exists x)(Rxx \wedge Sx)$.

	5.	$(\forall x)(Px \to Qx)$
	6.	$Pz \wedge (\forall y)(Sy \to Rzy)$
	7.	$Pz \to Qz$
	8.	$Qz \to Sz$
	9.	Sz
	10.	$(\forall y)(Sy \to Rzy)$
	11.	$Sz \to Rzz$
	12.	Rzz
	13.	$Rzz \wedge Sz$
	14.	$(\exists x)(Rxx \wedge Sx)$
	15.	$(\forall x)(Px \to Qx) \to (\exists x)(Rxx \wedge Sx)$
{1, 2}	16.	$(\forall x)(Px \to Qx) \to (\exists x)(Rxx \wedge Sx)$
	17.	$(\exists x)(Rxx \wedge Sx)$
	18.	$Ruu \wedge Su$
	19.	$Ruu \wedge Su \to Qu$
	20.	Qu
	21.	$Qu \to (\forall y)(Qy \wedge Ruy)$
	22.	$(\forall y)(Qy \wedge Ruy)$
	23.	$Qx \wedge Rux$
	24.	Qx
	25.	$Px \to Qx$
	26.	$(\forall x)(Px \to Qx)$

	27.	$(\forall x)(Px \rightarrow Qx)$	
$\{3, 4\}$	28.	$(\exists x)(Rxx \wedge Sx) \rightarrow (\forall x)(Px \rightarrow Qx)$	
$\{1, 2, 3, 4\}$	29.	$(\forall x)(Px \rightarrow Qx) \leftrightarrow (\exists x)(Rxx \wedge Sx)$	

E. Each of the following is an *incorrect* "proof." First flag the variables that should be flagged. Then state explicitly where the error or errors occur and explain which rule or rules are used incorrectly. (Remember that any formula may be taken as a premise; therefore, on any line where there is a formula and the documentation is *P*, there cannot be an error. Below we omit the reasons for using Rule *P*.) (Answers to 1, 4, and 10.)

(1)

$\{1\}$	1.	$(\forall x)(Px \rightarrow Qx)$	*P*

Deduce $(\exists x)Px \rightarrow (\forall y)Qy$.

$\{2\}$	2.	$(\exists x)Px$	*P*
$\{3\}$	3.	Pz	*P*
$\{3\}$	4.	Py	3 *US* $[y\|z]$
$\{1\}$	5.	$Py \rightarrow Qy$	1 *US* $[y\|x]$
$\{1, 3\}$	6.	Qy	4, 5 *T*
$\{1, 3\}$	7.	$(\forall y)Qy$	6 *UG*
$\{1, 2\}$	8.	$(\forall y)Qy$	2, 3, 7 *EP* $[z\|x]$
$\{1\}$	9.	$(\exists x)Px \rightarrow (\forall y)Qy$	2, 8 *CP*

(2)

$\{1\}$	1.	$(\forall x)(Px \wedge Qx \rightarrow Rx)$	*P*
$\{2\}$	2.	$(\exists x)Px \wedge (\exists x)Qx$	*P*

Deduce $(\exists x)Rx$.

$\{3\}$	3.	$Py \wedge Qy$	*P*
$\{1\}$	4.	$Py \wedge Qy \rightarrow Ry$	1 *US* $[y\|x]$
$\{1, 3\}$	5.	Ry	3, 4 *T*
$\{1, 3\}$	6.	$(\exists x)Rx$	5 *EG* $[y\|x]$
$\{1, 2\}$	7.	$(\exists x)Rx$	2, 3, 6 *EP* $[y\|x]$

(3)

$\{1\}$	1.	$(\forall x)Pxa$	*P*
$\{2\}$	2.	$(\forall x)[(\forall y)Pxy \rightarrow Qx]$	*P*

Deduce $(\exists x)Qx$.

$\{1\}$	3.	Paa	1 *US* $[a\|x]$
$\{2\}$	4.	$(\forall y)Pay \rightarrow Qa$	2 *US* $[a\|x]$
$\{2\}$	5.	$Paa \rightarrow Qa$	4 *US* $[a\|y]$
$\{1, 2\}$	6.	Qa	3, 5 *T*
$\{1, 2\}$	7.	$(\exists x)Qx$	6 *EG* $[a\|x]$

(4)

$\{1\}$	1.	$(\forall x)(Px \rightarrow Qx)$	*P*
$\{2\}$	2.	$(\forall x)(\exists y)(\neg Qx \rightarrow Rxy \wedge Sy)$	*P*

Deduce $(\exists x)Px \rightarrow (\exists x)\neg(Rxx \wedge Sx)$.

{3}	3.	$(\exists x)Px$	P
{4}	4.	Pz	P
{1}	5.	$Pz \rightarrow Qz$	1 US $[z\|x]$
{1, 4}	6.	Qz	4, 5 T
{2}	7.	$(\exists y)(\neg Qz \rightarrow Rzy \wedge Sy)$	2 US $[z\|x]$
{8}	8.	$\neg Qz \rightarrow Rzz \wedge Sz$	P
{1, 4, 8}	9.	$\neg(Rzz \wedge Sz)$	6, 8 T
{1, 4, 8}	10.	$(\exists x)\neg(Rxx \wedge Sx)$	9 EG $[z\|x]$
{1, 4, 8}	11.	$(\exists x)Px \rightarrow (\exists x)\neg(Rxx \wedge Sx)$	10 T
{1, 2, 4}	12.	$(\exists x)Px \rightarrow (\exists x)\neg(Rxx \wedge Sx)$	7, 8, 11 EP $[z\|x]$
{1, 2, 3}	13.	$(\exists x)Px \rightarrow (\exists x)\neg(Rxx \wedge Sx)$	3, 4, 12 EP $[z\|x]$

(5) {1} 1. $(\forall x)(\exists y)(Px \rightarrow Qy)$ P

Deduce $(\exists y)(\forall x)(Px \rightarrow Qy)$.

{1}	2.	$(\exists y)(Px \rightarrow Qy)$	1 US $[x\|x]$
{3}	3.	$Px \rightarrow Qz$	P
{3}	4.	$(\forall x)(Px \rightarrow Qz)$	3 UG
{3}	5.	$(\exists y)(\forall x)(Px \rightarrow Qy)$	4 EG $[z\|y]$
{1}	6.	$(\exists y)(\forall x)(Px \rightarrow Qy)$	2, 3, 5 EP $[z\|y]$

(6) {1} 1. $(\forall x)(\exists y)(Px \leftrightarrow Qy)$ P

Deduce $(\exists y)(\forall x)(Px \leftrightarrow Qy)$.

{1}	2.	$(\exists y)(Py \leftrightarrow Qy)$	1 US $[y\|x]$
{3}	3.	$Px \leftrightarrow Qx$	P
{3}	4.	$(\forall x)(Px \leftrightarrow Qx)$	3 UG
{3}	5.	$(\exists y)(\forall x)(Px \leftrightarrow Qy)$	4 EG $[x\|y]$
{1}	6.	$(\exists y)(\forall x)(Px \leftrightarrow Qy)$	2, 3, 5 EP $[x\|y]$

(7) {1} 1. $(\exists x)(Px \wedge (\forall y)(Qy \rightarrow Rxy)$ P
{2} 2. $(\exists x)(Sx \wedge (\forall y)(Py \rightarrow \neg Rxy)$ P

Deduce $(\exists x)(Sx \wedge \neg Qx)$.

{3}	3.	$Pz \wedge (\forall y)(Qy \rightarrow Rzy)$	P
{3}	4.	$(\forall y)(Qy \rightarrow Rzy)$	3 T
{5}	5.	$Sz \wedge (\forall y)(Py \rightarrow \neg Rzy)$	P
{5}	6.	$(\forall y)(Py \rightarrow \neg Rzy)$	5 T
{3}	7.	$Qz \rightarrow Rzz$	4 US $[z\|y]$
{5}	8.	$Pz \rightarrow \neg Rzz$	6 US $[z\|y]$
{3, 5}	9.	$\neg Rzz$	3, 8 T
{3, 5}	10.	$\neg Qz$	7, 9 T
{3, 5}	11.	$Sz \wedge \neg Qz$	5, 10 T
{3, 5}	12.	$(\exists x)(Sx \wedge \neg Qx)$	11 EG $[z\|x]$
{2, 3}	13.	$(\exists x)(Sx \wedge \neg Qx)$	2, 5, 12 EP $[z\|x]$
{1, 2}	14.	$(\exists x)(Sx \wedge \neg Qx)$	1, 3, 13 EP $[z\|x]$

(8)	{1}	1.	$(\exists x)Px \wedge (\forall y)(Qy \to Rxy)$	P
	{2}	2.	$(\forall x)(Px \to Qx)$	P

Deduce $(\forall z)Rzz$.

{3}	3.	$Pz \wedge (\forall y)(Qy \to Rzy)$	P
{2}	4.	$Pz \to Qz$	2 US $[z\mid x]$
{2, 3}	5.	Qz	3, 4 T
{3}	6.	$(\forall y)(Qy \to Rzy)$	3 T
{3}	7.	$Qz \to Rzz$	6 US $[z\mid y]$
{2, 3}	8.	Rzz	5, 7 T
{2, 3}	9.	$(\forall z)Rzz$	8 UG
{1, 2}	10.	$(\forall z)Rzz$	1, 3, 9 EP $[z\mid x]$

(9)	{1}	1.	$(\forall x)[Px \to (\forall y)(Qy \to Rxy)]$	P
	{2}	2.	$(\exists x)(Px \wedge Qx)$	P

Deduce $(\exists x)Rxx$.

{3}	3.	$Py \wedge Qy$	P
{1}	4.	$Py \to (\forall y)(Qy \to Ryy)$	1 US $[y\mid x]$
{1, 3}	5.	$(\forall y)(Qy \to Ryy)$	3, 4 T
{1, 3}	6.	$Qy \to Ryy$	5 US $[y\mid y]$
{1, 3}	7.	Ryy	3, 6 T
{1, 3}	8.	$(\exists x)Rxx$	7 EG $[y\mid x]$
{1, 2}	9.	$(\exists x)Rxx$	2, 3, 8 EP $[y\mid x]$

(10)	{1}	1.	$(\exists x)[Px \wedge (\forall y)(Qy \to Rxy)]$	P

Deduce $(\exists x)Qx \to (\exists x)Rxx$.

{2}	2.	Qx	P
{3}	3.	$Py \wedge (\forall y)(Qy \to Ryy)$	P
{3}	4.	$(\forall y)(Qy \to Ryy)$	3 T
{3}	5.	$Qx \to Rxx$	4 US $[x\mid y]$
{2, 3}	6.	Rxx	2, 5 T
{2, 3}	7.	$(\exists x)Rxx$	6 EG $[x\mid x]$
{1, 2}	8.	$(\exists x)Rxx$	1, 3, 7 EP $[y\mid x]$
{1}	9.	$Qx \to (\exists x)Rxx$	2, 8 CP
{1}	10.	$(\exists x)Qx \to (\exists x)Rxx$	9 EG $[x\mid x]$

(11)	{1}	1.	$(\forall x)Px \vee (\exists y)(Qy \wedge Rxy)$	P

Deduce $(\exists y)(\forall x)(Qy \wedge Rxy)$.

{1}	2.	$Px \vee (\exists y)(Qy \wedge Rxy)$	1 US $[x\mid x]$
{1}	3.	$(\exists y)(Qy \wedge Rxy)$	2 T
{4}	4.	$Qz \wedge Rxz$	P
{4}	5.	$(\forall x)(Qz \wedge Rxz)$	4 UG
{4}	6.	$(\exists y)(\forall x)(Qy \wedge Rxy)$	5 EG $[z\mid y]$
{1}	7.	$(\exists y)(\forall x)(Qy \wedge Rxy)$	3, 4, 6 EP $[z\mid y]$

(12)

{1}	1.	$(\forall x)(\forall y)(Pxy \to \neg Qy)$	P
{2}	2.	$(\forall x)(\forall y)(\neg Rx \to Pxy)$	P

Deduce $(\forall x)[(\exists y)Qy \to Rx]$.

{3}	3.	$(\exists y)Qy$	P
{4}	4.	Qz	P
{1}	5.	$(\forall y)(Pxy \to \neg Qy)$	1 US $[x\|x]$
{1}	6.	$Pxz \to \neg Qz$	5 US $[z\|y]$
{1, 4}	7.	$\neg Pxz$	4, 6 T
{2}	8.	$(\forall y)(\neg Rx \to Pxy)$	2 US $[x\|x]$
{2}	9.	$\neg Rx \to Pxz$	8 US $[z\|y]$
{1, 2, 4}	10.	Rx	7, 9 T
{1, 2}	11.	$Qz \to Rx$	4, 10 CP
{1, 2}	12.	$(\exists y)Qy \to Rx$	11 EG $[z\|y]$
{1, 2}	13.	$(\forall x)[(\exists y)Qy \to Rx]$	12 UG

(13)

{1}	1.	$(\exists x)(Px \wedge Qx \wedge Rx)$	P
{2}	2.	$(\forall x)(Px \wedge Qx \to Sx)$	P

Deduce $(\forall x)Sx$.

{3}	3.	$Py \wedge Qy \wedge Ry$	P
{3}	4.	$Py \wedge Qy$	3 T
{1}	5.	$Py \wedge Qy$	1, 3, 4 EP $[y\|x]$
{2}	6.	$Py \wedge Qy \to Sx$	2 US $[y\|x]$
{1, 2}	7.	Sx	5, 6 T
{1, 2}	8.	$(\forall x)Sx$	7 UG

(14)

{1}	1.	$(\exists x)(\exists y)(Px \vee Qxy)$	P
{2}	2.	$(\forall x)(Px \to Rx)$	P
{3}	3.	$(\forall x)(Qxx \to \neg Sx)$	P

Deduce $(\exists x)(Rx \wedge \neg Sx)$.

{4}	4.	$Pz \vee Qzz$	P
{2}	5.	$Pz \to Rz$	2 US $[z\|x]$
{3}	6.	$Qzz \to \neg Sz$	3 US $[z\|x]$
{2, 3, 4}	7.	$Rz \wedge \neg Sz$	4, 5, 6 T
{2, 3, 4}	8.	$(\exists x)(Rx \wedge \neg Sx)$	7 EG $[z\|x]$
{1, 2, 3}	9.	$(\exists x)(Rx \wedge \neg Sx)$	1, 4, 8 EP $[z\|x]$

Chapter 9

Applications of Interpretation Structures

In the Sentential Calculus, we prove that an argument is invalid by finding a truth assignment for the atomic sentences so that the premises are true and the conclusion is false. We also prove that a set of premises is consistent by finding a truth assignment for the atomic sentences so that all the premises are true. In the Predicate Calculus, we will do the same thing, except "truth assignment" will be replaced by "interpretation structure."

9.1 INVALID ARGUMENTS

Definition 9.1 An argument with premises Γ and conclusion φ is *valid* iff for every interpretation structure **S** for $\Gamma \cup \{\varphi\}$ and every sequence **a** from **S** if **a** satisfies Γ, then **a** satisfies φ. Otherwise, the argument is said to be *invalid*. (It is implicitly assumed that if the argument has n free variables, then the sequence **a** has at least n coordinates.)

It follows from Definition 9.1 and from the definition of satisfaction that if all the formulas in $\Gamma \cup \{\varphi\}$ are sentences, then the argument with premises Γ and conclusion φ is valid iff for every interpretation structure **S** for $\Gamma \cup \{\varphi\}$, if all the sentences in Γ are true in **S**, then φ must also be true in **S**. In most of the arguments we discuss we will be dealing with sentences, so we shall use the notion of truth rather than satisfaction. It follows, then, that to show such an argument is invalid, we must find an interpretation structure in which all the sentences in Γ are true, but φ is false. This type of interpretation structure is called a *counter-example*.

It may not be easy to show that an argument is invalid. The procedure to follow is either try to construct a counter-example directly or try to construct a proof. If something goes wrong with the proof, it may give you an idea on how to construct a counter-example. We shall give illustrations in the examples. (Even if we are given an interpretation structure, it is not always easy to decide if a given sentence is true or false in that structure. See, for example, the sentence about twin primes and the interpretation structure given at the end of Chapter 6.)

Example 9.1

Show that each of the following arguments is invalid by constructing a counter-example:

(a) Some cats have stripes. Hefalump is a cat. Therefore, Hefalump has stripes.

(b) All lions are mammals. All lions have hair. Therefore, all mammals have hair.

(c) Every even prime divides every even integer. There is an even prime that does not divide any integer that is a multiple of three. Therefore, there is an even prime that is not a multiple of three.

(d) Some student pilot flies to some city with every instructor. Every student pilot who flies to some city with an instructor is happy. Therefore, every instructor is happy.

Solution: In each case, we shall translate the argument into an argument in the Predicate Calculus and then try to find a counter-example or, if that fails, try to give a proof. The reason the proof fails may give us an idea of how to construct a counter-example. (Remember that the validity of an argument depends on its structure. If an argument is given in English, there is no relationship between the validity of the argument and the truth or falsity of the conclusion.)

(a) We will symbolize this argument as follows: Cx : x is a cat; Sx : x has stripes; p : Hefalump.

Premises: $(\exists x)(Cx \wedge Sx)$, Cp

Conclusion: Sp

In this case, we shall try to give a counter-example directly. We would like **Sp** to be false and **Cp** to be true. Thus, to make the first premise true, we must find some other element in the universe (**a**, for example) so that both **Ca** and **Sa** are true. (An existential formula $(\exists x)\varphi$ is *true* in **S** iff φ is true in **S** for *some* interpretation of x in **S**.) In interpretation structure that satisfies these conditions is given in Figure 9.1. The first premise is true because **Ca** and **Sa** are both true, so **a** is the element that "exists." The second premise is true because **Cp** is true, but the conclusion is false because **Sp** is false. Thus, the argument is invalid.

U = {a, p}

	C	S
a	*T*	*T*
p	*T*	*F*

Figure 9.1

(b) This argument is symbolized as follows: Lx : x is a lion; Mx : x is a mammal; Hx : x has hair.

Premises: $(\forall x)(Lx \rightarrow Mx)$, $(\forall x)(Lx \rightarrow Hx)$

Conclusion: $(\forall x)(Mx \rightarrow Hx)$

Again, in this case, it is easy to construct a counter-example directly. To make the conclusion false, there must be some element **a** in the universe so that **Ma** is true and **Ha** is false. (A universal formula $(\forall x)\varphi$ is *false* in **S** iff φ is false in **S** for *some* interpretation of x in **S**.) Both premises will be true if **La** is false. (A universal formula $(\forall x)\varphi$ is *true* in **S** iff φ is true in **S** for *all* interpretations of x in **S**.) The corresponding interpretation structure is given in Figure 9.2:

U = {a}

	M	L	H
a	*T*	*F*	*F*

Figure 9.2

(c) We shall symbolize the predicates and the argument as follows: Px : x is a prime; Ix : x is an integer; Ex : x is even; Mx : x is a multiple of three; Dxy : x divides y.

Premises: $(\forall x)[Px \wedge Ex \rightarrow (\forall y)(Iy \wedge Ey \rightarrow Dxy)]$,
$(\exists x)[Px \wedge Ex \wedge (\forall y)(Iy \wedge My \rightarrow \neg Dxy)]$

Conclusion: $(\exists x)(Px \wedge Ex \wedge \neg Mx)$

In this case, we shall try to give a proof. We shall try to find a term t so that $Pt \wedge Et \wedge \neg Mt$. We have an existential premise, so we start by taking a premise for Rule *EP*.

Suppose (i) $Pz \wedge Ez \wedge (\forall y)(Iy \wedge My \rightarrow \neg Dzy)$. We can deduce $Pz \wedge Ez$, so to get the conclusion, we should try to deduce $\neg Mz$. From (i), we can deduce (ii) $(\forall y)(Iy \wedge My \rightarrow \neg Dzy)$, and from (ii), we can deduce (iii) $Iz \wedge Mz \rightarrow \neg Dzz$.

From premise 1, we can deduce (iv) $Pz \wedge Ez \rightarrow (\forall y)(Iy \wedge Ey \rightarrow Dzy)$. Thus, from (i) and (iv), we can deduce $(\forall y)(Iy \wedge Ey \rightarrow Dzy)$, which implies (v) $Iz \wedge Ez \rightarrow Dzz$.

Notice that if Iz has a false interpretation, then formulas (iii) and (v) have a true interpretation. So there is no reason why Mz must have a false interpretation. Thus, we can again give an interpretation structure in which the universe has one element and the two premises are true but the conclusion is false (see Fig. 9.3).

It doesn't matter whether **Daa** is true or false. In either case, both premises are true and the conclusion is false. The conclusion is false because $\mathbf{Pa} \wedge \mathbf{Ea} \wedge \neg\mathbf{Ma}$ is false. (An existential formula $(\exists x)\varphi$ is *false* in **S** iff φ is false for *every* interpretation of x in **S**.) Premise 1 is true because

$$\mathbf{Pa} \wedge \mathbf{Ea} \rightarrow (\mathbf{Ia} \wedge \mathbf{Ea} \rightarrow \mathbf{Daa})$$

is true, and premise 2 is true because

$$\mathbf{Pa} \wedge \mathbf{Ea} \wedge (\mathbf{Ia} \wedge \mathbf{Ma} \rightarrow \neg\mathbf{Daa})$$

is true.

U = {a}

	P	E	I	M
a	*T*	*T*	*F*	*T*

D	a
a	*T* or *F*

Figure 9.3

(d) We shall use the following predicate symbols: Sx : x is a student pilot; Ix : x is an instructor; Cx : x is a city; $Fxyz$: x flies to y with z; Hx : x is happy. Then the argument becomes

Premises: $(\exists x)(\exists y)[Sx \wedge Cy \wedge (\forall z)(Iz \rightarrow Fxyz)]$,
$(\forall x)[Sx \wedge (\exists y)(\exists z)(Cy \wedge Iz \wedge Fxyz) \rightarrow Hx]$

Conclusion: $(\forall x)(Ix \rightarrow Hx)$

To attempt a proof, assume Ix as a premise for Rule *CP*. The first premise is existential, so assume (i) $Su \wedge Cv \wedge (\forall z)(Iz \rightarrow Fuvz)$ as a premise for *EP*. From (i), we may deduce $(\forall z)(Iz \rightarrow Fuvz)$, which implies (ii) $Ix \rightarrow Fuvx$. Then from (ii) and Ix, we can deduce $Fuvx$. From premise 2, we can deduce (iii) $Sx \wedge (\exists y)(\exists z)(Cy \wedge Iz \wedge Fxyz) \rightarrow Hx$ and (iv) $Su \wedge (\exists y)(\exists z)(Cy \wedge Iz \wedge Fuyz) \rightarrow Hu$. From the assumption Ix, from (i), and from $Fuvz$, we may deduce (v) $Cv \wedge Ix \wedge Fuvx$.

Using Rule *EG* twice on (v), we may deduce (vi) $(\exists y)(\exists z)(Cy \wedge Iz \wedge Fuyz)$. Then from (i), (vi), and (iv), we may deduce (vii) Hu. However, we were trying to deduce Hx.

Remember that each premise and everything that is deduced from a premise has to be true in every interpretation structure that is used to show that the argument is invalid. Therefore, each of the formulas (i) through (vii) must be true, but the interpretation of *Hx* could be false. We'll try an interpretation structure in which the universe has three elements—**U** = {**a, b, c**}. To make the conclusion false, we will make **Ia** true but **Ha** false. Because **Ha** is false, the interpretation of the antecedent of (iii) must be false when *x* is interpreted as **a**. Thus, we shall require that **Sa** be false. All other atomic formulas will have a true interpretation. The interpretation structure is given in Figure 9.4.

U = {a, b, c}

	S	I	C	H
a	*F*	*T*	*T*	*F*
b	*T*	*T*	*T*	*T*
c	*T*	*T*	*T*	*T*

F = U × U × U
(**Fxyz** is true for all **x, y, z** ∈ **U**)

Figure 9.4

Premise 1 is true because **Sb** ∧ **Cc** is true, and (∀**z**)(**Iz** → **Fbcz**) is true because **Fbcz** is true for all **z** ∈ **U**. (Actually, we could have taken a two-element universe, because the two elements that are postulated to exist in premise 1 do not have to be distinct.) To show that premise 2 is true, we have to show that the interpretation of formula (iii) is true for all **x** ∈ **U**. If **x** ≠ **a**, then the consequent of the conditional is true, and if **x** = **a**, because **Sa** is false, the antecedent is false. In either case, (iii) is true, so premise 2 is true. Finally, the conclusion is false because there is an **x** ∈ **U**—namely, **x** = **a**—so that **Ix** is true and **Hx** is false.

Alternatively, we can find the counter-example directly for part (d) by arguing as follows: The conclusion is false if there is an element **a** in **U** so that **Ia** is true and **Ha** is false. Premise 2 is true if **Sa** is false and **Hx** is true for all **x** in **U**, **x** ≠ **a**. Premise 1 is true if **Fxyz** is true for all **x**, **y**, **z** in **U**, and there is some element **b** in **U** so that **Sb** and **Cb** are each true.

Sometimes, in order to show an argument is not valid, it is easier to give a numeric interpretation structure rather than the truth-table structures that we used in Example 9.1. This is frequently true when the formulas contain 2-place or higher place predicate symbols. Also, there are times when an interpretation structure that has a finite universe does not serve the purpose.

Example 9.2

Show that the following argument is invalid.

Premises: $(\forall x)\neg Rxx$, $(\forall x)(\forall y)(\forall z)(Rxy \wedge Ryz \rightarrow Rxz)$, $(\forall x)(\exists y)Rxy$
Conclusion: $(\forall x)(\exists y)Ryz$

Solution: We will show informally that there is no interpretation structure with a finite universe in which the three premises are true. Suppose **U** is the universe and $\mathbf{a} \in \mathbf{U}$. By premise 3, there is some $\mathbf{b} \in \mathbf{U}$ such that **Rab** is true. But $\mathbf{b} \neq \mathbf{a}$ by premise 1. Then, by premise 3, there is some $\mathbf{c} \in \mathbf{U}$ such that **Rbc** is true. The element $\mathbf{c} \neq \mathbf{b}$ by premise 1 and $\mathbf{c} \neq \mathbf{a}$ by premise 2. (If $\mathbf{c} = \mathbf{a}$, then both **Rab** and **Rba** would be true, so by premise 2, we could deduce that **Raa** is true. But if **Raa** is true, we contradict premise 1.)

Now we have three distinct elements **a**, **b**, and **c** in **U**. We continue this argument. By premise 2, there is a $\mathbf{d} \in \mathbf{U}$ such that **Rcd** is true. Then using an argument similar to that used above, we deduce that $\mathbf{d} \neq \mathbf{c}$, $\mathbf{d} \neq \mathbf{a}$, and $\mathbf{d} \neq \mathbf{b}$. Thus, **U** must have at least four distinct elements. This process can be continued as long as we like, so it follows that **U** cannot be finite.

In order for the three premises to be true in an interpretation structure, the universe must contain a sequence

$$\mathbf{a}_1, \mathbf{a}_2, \mathbf{a}_3, \mathbf{a}_4, \ldots$$

such that

$$\mathbf{Ra}_i\mathbf{a}_j \text{ is true iff } i < j.$$

As mentioned above, for each **a** in **U**, **Raa** is false. Therefore, the conclusion $(\forall x)(\exists y)Ryx$ is false if the universe has a "smallest" element.

There is a natural numeric interpretation structure **S** in which the premises are true and the conclusion is false—namely, take $\mathbf{S} = <\mathbf{U}, \mathbf{R}>$, where **U** is the set of positive integers and **R** is $<$. It follows from properties of the positive integers that $<$ is irreflexive $((\forall \mathbf{x})\neg(\mathbf{x} < \mathbf{x}))$ and transitive $((\forall \mathbf{x})(\forall \mathbf{y})(\forall \mathbf{z})(\mathbf{x} < \mathbf{y} \wedge \mathbf{y} < \mathbf{z} \rightarrow \mathbf{x} < \mathbf{z}))$ on **U** and also that for each positive integer, there is a larger one. (If **x** is a positive integer, then $\mathbf{x} < \mathbf{x} + \mathbf{1}$, so take $\mathbf{y} = \mathbf{x} + \mathbf{1}$.) Thus, premise 3 is true in **S**. However, the conclusion is interpreted in **S** as "There is no smallest positive integer." The conclusion, therefore, is false in **S**. (When we use a numeric interpretation structure, we assume all the well-known properties of the integers, rationals, and the reals to be true.)

9.2 CONSISTENT SETS OF FORMULAS

A set of formulas, Γ, is consistent if a contradiction cannot be derived from Γ. Thus, in the Predicate Calculus, as in the Sentential Calculus, we have the following definition.

Definition 9.2 If Γ is a set of formulas, Γ is *consistent* if there is no formula φ such that $\Gamma \vdash \varphi \wedge \neg\varphi$. Otherwise, Γ is said to be *inconsistent.*

Because our rules of inference preserve truth, as we will show in Chapter 16, if $\Gamma \vdash \varphi \wedge \neg\varphi$ (that is, if a contradiction can be derived from Γ using the rules of inference), then there is no interpretation structure for Γ in which a sequence satisfies all the formulas in Γ. (If there were such an interpretation structure **S**, then there would be a sentence that was both true and false in **S**, and that is impossible.) Thus, to show that a set of formulas Γ is consistent, it is sufficient to show that there is an interpretation structure and a sequence in that structure that satisfies all the formulas in Γ. In all our examples, the formulas are sentences, so in this case, to show Γ is consistent, we must find an interpretation structure in which all the formulas in Γ are true.

Because it is sometimes difficult to construct an interpretation structure with a given property, to show a set of formulas is consistent, it is sometimes easier to try to derive a contradiction. (Similarly, to show an argument is invalid, we sometimes try to prove it.) Then, if the proof of a contradiction cannot be completed, the process itself may give you an idea of how to construct an interpretation structure in which all the premises are true. We give some illustrations in Example 9.3.

Example 9.3 Show by the method of interpretation that each of the following sets of sentences is consistent:

(a) $(\exists x)(Ax \wedge Bx)$, $(\exists x)(Ax \wedge \neg Bx)$, $(\exists x)(\neg Ax \wedge Bx)$, $(\exists x)(\neg Ax \wedge \neg Bx)$

(b) $(\forall x)Rxx$, $(\forall x)(\forall y)(Rxy \rightarrow Ryx)$, $(\exists x)(\exists y)(\exists z)(Rxy \wedge Ryz \wedge \neg Rxz)$

(c) $(\forall x)(\forall y)[Py \wedge Qxy \rightarrow (\exists z)(Sz \wedge \neg Rxz)]$, $(\exists x)(Px \wedge Qax \wedge Rax)$

(d) $(\forall x)Rxf(x)$, $(\forall x)(\exists y)\neg Rxy$, $(\forall x)(\exists y)\neg Ryx$

Solution:

(a) We have four existential premises. As long as we choose four *different* *EP* variables, we avoid a contradiction. Thus, the universe of the interpretation structure must have four elements. A truth-table interpretation structure is given in Figure 9.5, in which all the formulas in (a) are true.

U = {a, b, c, d}

	A	B
a	*T*	*T*
b	*T*	*F*
c	*F*	*T*
d	*F*	*F*

Figure 9.5

Each of **Aa** ∧ **Ba**, **Ab** ∧ ¬**Bb**, ¬**Ac** ∧ **Bc**, and ¬**Ad** ∧ ¬**Bd** is true, so each of the existential formulas in (a) is true in this interpretation.

(b) If these three formulas imply a contradiction, then every 2-place relation that is reflexive (premise 1) and symmetric (premise 2), would have to be transitive (the negation of premise 3). (See Definition 6.3.) Thus, we have to find an interpretation structure < **U, R** > where **R** is reflexive and symmetric, but not transitive (see Fig. 9.6).

U = {a, b, c}

R	a	b	c
a	*T*	*T*	*F*
b	*T*	*T*	*T*
c	*F*	*T*	*T*

Figure 9.6

We see from the table in Figure 9.6 that **R** is reflexive because there are *T*'s along the main diagonal (the diagonal that starts in the upper left corner and ends in the lower right corner); **R** is symmetric because the table is symmetric with respect to the main diagonal; but **R** is not transitive because **Rab** and **Rbc** are true, but **Rac** is false. So **a**, **b**, and **c** are the elements that exist in **U** that give premise 3 a true interpretation.

(c) The first step in the attempt to derive a contradiction from the formulas in (c) is to take (i) *Pu* ∧ *Qau* ∧ *Rau* as a premise for *EP*. Then, from premise 1, derive, using *US* $[a|x][u|y]$, (ii) *Pu* ∧ *Qau* → (∃*z*)(*Sz* ∧ ¬*Raz*).

Then from (i) and (ii), deduce (iii) (∃*z*)(*Sz* ∧ ¬*Raz*). Therefore, as long as we do *not* take *u* to be the *EP* variable when we take a premise for *EP* for (iii), we avoid a contradiction. See Figure 9.7 for an interpretation structure in which both formulas in (c) are true.

The first formula in (c) is true because **Sb** and **Rxb** are true for all **x** ∈ **U**, and the second formula is true because **Pa** ∧ **Qaa** ∧ **Raa** is true.

U = {a, b}

R	a	b
a	*T*	*F*
b	*T*	*F*

	P	S
a	*T*	*T*
b	*T*	*T*

Q = U × U
(Qxy is true for all x, y ∈ U)

Figure 9.7

(d) From the three formulas in (d), we can deduce (i) $Rxf(x)$; (ii) $(\exists y)\neg Rxy$; and (iii) $(\exists y)\neg Ryx$. We can avoid a contradiction if we take the *EP* variable for (ii) different from $f(x)$. If we choose the interpretation of R to be an irreflexive relation, then (ii) and (iii) are both true. Thus, to avoid a contradiction, choose **f** so that $\mathbf{f(x)} \neq \mathbf{x}$. Let's take a numeric interpretation, $\mathbf{S} = <\mathbf{U}, \mathbf{R}, \mathbf{f}>$, where **U** is the set of positive integers, **R** is $<$, and $\mathbf{f(x)} = \mathbf{x} + \mathbf{1}$ for all $\mathbf{x} \in \mathbf{U}$. For each positive integer $\mathbf{x}$, $\mathbf{x} < \mathbf{x} + \mathbf{1}$, so premise 1 is true. Premises 2 and 3 are true because, for each positive integer **x**, it is not the case that $\mathbf{x} < \mathbf{x}$ (that is, the **y** that exists is **x** itself).

Because we will frequently have to find an interpretation structure **S** for a set of sentences Γ that has the property that each sentence in Γ is true in **S**, it is convenient to give such a structure a name.

Definition 9.3 If **S** is an interpretation structure for a set of formulas Γ, **S** is called a *model* for Γ (symbolically $\mathbf{S} \models \Gamma$) if all the formulas in Γ are true in **S**. (In the case that there is just one formula φ in Γ, we write $\mathbf{S} \models \varphi$ instead of $\mathbf{S} \models \{\varphi\}$.)

Thus, for example, it follows from Example 9.3(d) that if

$$\Gamma = \{(\forall x)Rxf(x), (\forall x)(\exists y)\neg Rxy, (\forall z)(\exists y)\neg Ryx\},$$

then $\mathbf{S} \models \Gamma$, where $\mathbf{S} = <\mathbf{U}, \mathbf{R}, \mathbf{f}>$; $\mathbf{U}$ = set of positive integers; **R** is $<$; and $\mathbf{f(x)} = \mathbf{x} + \mathbf{1}$ for all $\mathbf{x} \in \mathbf{U}$.

Notice the difference between the two symbols ⊢ and ⊨. The first symbol is used when there is a *proof* of a formula from some set of formulas. Thus, $\Gamma \vdash \varphi$ means that there is a formal proof of φ from the formulas in Γ using the rules of inference. It deals with the structure or syntax of the language. On the other hand, we only use ⊨ when we are talking about truth or satisfaction in an interpretation structure. This symbol deals with the meaning or semantics of the language.

9.3 *THE COMPLETENESS THEOREM (PRELIMINARY DISCUSSION)

The word "complete" is used in several different ways in mathematics. For example, we may say a set of formulas Γ is *complete* if, for any formula φ, it is not the case that $\Gamma \vdash \varphi$ (symbolically, $\Gamma \nvdash \varphi$), then $\Gamma \cup \{\varphi\}$ is inconsistent. So Γ is complete in the sense that any formula that is consistent with Γ is provable from Γ.

The Completeness Theorem for the Predicate Calculus is due to Kurt Gödel.[1] Here the word "complete" is used in the sense that the rules of logic are sufficient to prove all "true" formulas. In order to state Gödel's Completeness Theorem, it is convenient to extend the definition of $\models$, which we defined in the preceding section (Definition 9.3).

Definition 9.4 If Γ is a set of formulas and φ is a formula, Γ *semantically implies* φ(symbolically, $\Gamma \models \varphi$) if the argument with premises Γ and conclusion φ is valid (see Definition 9.1). If $\Gamma = \{\psi\}$, we frequently write $\psi \models \varphi$ instead of $\{\psi\} \models \varphi$.

In the case that all the formulas in $\Gamma \cup \{\varphi\}$ are sentences, $\Gamma \models \varphi$ iff for every interpretation structure $\mathbf{S}$ for $\Gamma \cup \{\varphi\}$, if $\mathbf{S} \models \Gamma$, then $\mathbf{S} \models \varphi$.

We have emphasized the fact that the rules of inference that we use preserve truth. In other words, whenever the premises are true, any formula that is derived from the premises using the rules of inference must also be true. This is called the *Soundness Theorem*. We can state the Soundness Theorem in a compact form using our new notation.

Soundness Theorem

If Γ is a set of sentences and φ is a sentence, then

$$\text{if } \Gamma \vdash \varphi, \text{ then } \Gamma \models \varphi.$$

The *Completeness Theorem* is the converse of the Soundness Theorem.

Completeness Theorem

If Γ is a set of sentences and φ is a sentence, then

$$\text{if } \Gamma \models \varphi, \text{ then } \Gamma \vdash \varphi.$$

---

[1]Gödel gave the proof in his doctoral dissertation at the University of Vienna, 1930.

The Completeness Theorem states that whenever Γ semantically implies φ, there is a formal proof of φ from the formulas in Γ using the rules of logic. We shall give a proof of the Completeness Theorem in Chapter 16. In this section, we give another form of the Completeness Theorem and give one of its important consequences.

Theorem 9.1 The following two statements are equivalent:

(a) For any set of sentences Γ and for every sentence φ, if $\Gamma \models \varphi$, then $\Gamma \vdash \varphi$ (Gödel's Completeness Theorem).

(b) For any set of sentences Δ, if Δ is consistent, then Δ has a model.

Proof: First we shall show that (a) implies (b). We shall give an indirect proof and assume (a) and the negation of (b). So assume Δ is consistent but does not have a model. Then, for any interpretation structure **S** for $\Delta \cup \{\varphi\}$, of which there are in fact none, it is true that

$$\text{if } \mathbf{S} \models \Delta, \text{ then } \mathbf{S} \models \varphi \wedge \neg\varphi,$$

because the hypothesis is false. Thus, by Definition 9.4, $\Delta \models \varphi \wedge \neg\varphi$. Now, if we apply (a), we obtain $\Delta \vdash \varphi \wedge \neg\varphi$. Then, it follows from Definition 9.2, that Δ is not consistent, and this contradicts one of our original assumptions. We have now shown that (a) implies (b).

Conversely, we wish to show that (b) implies (a). We shall again give an indirect proof. Suppose $\Gamma \models \varphi$, but it is not the case that $\Gamma \vdash \varphi$ (in symbols, $\Gamma \nvdash \varphi$). Then, we claim that $\Gamma \cup \{\neg\varphi\} \nvdash \psi \wedge \neg\psi$. Suppose there is a proof of $\psi \wedge \neg\psi$, using Γ and $\neg\varphi$ as premises. Thus, because $\psi \wedge \neg\psi \rightarrow \varphi$ is a tautology, we have

$$\Gamma \cup \{\neg\varphi\} \vdash \varphi.$$

By the rule of conditional proof,

$$\Gamma \vdash \neg\varphi \rightarrow \varphi.$$

But $\neg\varphi \rightarrow \varphi$ is tautologically equivalent to φ, so we deduce $\Gamma \vdash \varphi$, which contradicts our assumption that $\Gamma \nvdash \varphi$. Thus, by the Rule of Indirect Proof, we deduce

$$\Gamma \cup \{\neg\varphi\} \nvdash \psi \wedge \neg\psi.$$

Consequently, it follows from Definition 9.2 that $\Gamma \cup \{\neg\varphi\}$ is consistent.

Now we shall apply (b). (Use Rule *US* on (b), replacing Δ by "$\Gamma \cup \{\neg\varphi\}$.") Then, we obtain that $\Gamma \cup \neg\varphi$ has a model, **S**; that is, $\mathbf{S} \models \Gamma$, but φ is false in **S**. By hypothesis, $\Gamma \models \varphi$, so $\mathbf{S} \models \Gamma$ implies $\mathbf{S} \models \varphi$, which means φ is true in **S**. Thus, we have shown that the sentence φ is both true and false in **S** and that is impossible. This completes the proof that (b) implies (a). ■

The following theorem follows from the Completeness Theorem, Theorem 9.1, and Definition 9.2:

Theorem 9.2 If Γ is a set of sentences, then Γ is consistent if and only if Γ has a model.

The next consequence of the Completeness Theorem is called the *Compactness Theorem.*

Compactness Theorem

If Γ is a set of sentences and φ is a sentence, then if $\Gamma \models \varphi$, there is a finite subset $\Delta \subseteq \Gamma$ such that $\Delta \models \varphi$.

Proof: Suppose $\Gamma \models \varphi$. Then, by the Completeness Theorem, $\Gamma \vdash \varphi$. Because a proof is finite in length (see the definition of a proof in Chapter 12), there is a finite subset $\Delta \subseteq \Gamma$ such that $\Delta \vdash \varphi$. It follows from the Soundness Theorem that $\Delta \models \varphi$.

■

It is important to be able to prove that a set of formulas is consistent. Using the fact that a proof is finite in length, we can prove that if Γ is an infinite set of formulas, then to show Γ is consistent, it is sufficient to show that each *finite* subset of Γ is consistent, giving a syntactic version of the Compactness Theorem.

Theorem 9.3 If Γ is an infinite set of sentences, then Γ is consistent if and only if each finite $\Delta \subseteq \Gamma$ is consistent.

Proof: Suppose Γ is consistent and Δ is a finite subset of Γ. If Δ is not consistent, then by Definition 9.2, Δ proves a contradiction. However, it follows from the definition of a proof, if Δ proves a contradiction, so does Γ. This implies that Γ is not consistent, which is a contradiction.

Conversely, suppose each finite subset of Γ is consistent but Γ is inconsistent. If Γ is inconsistent, then there is a formula φ such that $\Gamma \vdash \varphi \wedge \neg\varphi$. Because a proof is finite in length, there is a finite subset Δ of Γ such that $\Delta \vdash \varphi \wedge \neg\varphi$. This contradicts our assumption that every finite subset of Γ is consistent.

■

The following semantical form of Theorem 9.3 is a consequence of the Compactness Theorem.

Theorem 9.4 If Γ is a set of sentences such that every finite subset of Γ has a model, then Γ has a model.

Proof: Let Γ be a set of sentences such that every finite subset of Γ has a model. Suppose Γ has no model. Then $\Gamma \models \varphi$ for every sentence φ. In particular, $\Gamma \models \psi \land \neg\psi$ for some sentence ψ. The Compactness Theorem implies that there is a finite set $\Delta \subseteq \Gamma$ such that $\Delta \models \psi \land \neg\psi$. By hypothesis, Δ has a model. Because Δ has a model and $\Delta \models \psi \land \neg\psi$, it follows that $\psi \land \neg\psi$ has a model. This is, of course, a contradiction. Consequently, Γ must have a model. ■

Theorem 9.4 is also frequently called the Compactness Theorem.

As a simple application of compactness, we shall prove that there exists an infinite set. Let

$$\Gamma = \{\varphi_2, \varphi_3, \varphi_4, \ldots, \varphi_n, \ldots\},$$

where φ_n is a formula that means "there exist n distinct objects." (Such sentences can be stated in the language of the Predicate Calculus with Identity; see Chapter 11.)

Each finite subset $\Delta \subseteq \Gamma$ has a model, because if $\Delta = \{\varphi_{m_1}, \varphi_{m_2}, \ldots, \varphi_k\}$, where $m_1 < m_2 < \cdots < k$, then the set $\mathbf{U} = \{\mathbf{1, 2, 3, \ldots, k}\}$ is a model for Δ. $\mathbf{U}$ contains k distinct elements. Thus, it follows from the Compactness Theorem that Γ has a model. Any model of Γ is an infinite set.

Using some of the ideas in the preceding example, we can show, for example, that if Δ is a set of sentences that has arbitrarily large finite models, then Δ must have an infinite model (see Exercise E 8).

For another example, let $\mathbb{N} = \{\mathbf{0, 1, 2, \ldots}\}$ be the set of non-negative integers, and let Γ be the set of all sentences that describe the properties of the non-negative integers. Then the set of non-negative integers with the appropriate constants, predicates, and functions is a model for Γ. Let us call this model $\mathbf{M}$. Now we claim that there must be a model of Γ that contains an "infinite" number—that is, a number that is larger than each element of $\mathbb{N}$. Let ω be a new constant symbol in our language, and let Θ be the following infinite set of sentences:

$$0 < \omega, 1 < \omega, 2 < \omega, \ldots.$$

Every finite subset of $\Gamma \cup \Theta$ has a model because if we take a finite subset of Θ that consists of the sentences

$$k_1 < \omega, k_2 < \omega, \ldots, k_n < \omega, \qquad \text{where } k_1 < k_2 < \cdots < k_n,$$

and if we interpret ω to be $k_n + 1$, then $\mathbf{M}$ itself is a model for this system. Thus, the Compactness Theorem implies that $\Gamma \cup \Theta$ has a model. In this model, $\boldsymbol{\omega}$ is an infinite number.

We shall give additional applications of the Completeness Theorem and Compactness Theorem in the Exercises and in later chapters.

EXERCISES 9

A. Translate each of the following arguments into the language of the Predicate Calculus, using the indicated predicate symbols, constants, and function symbols. If the argument is valid, give a formal proof, and if the argument is not valid, give a counter-example (an interpretation structure in which the premises are true and the conclusion false). (Answers to 1, 4, 7, 11, and 15.)

(1) All children like dogs. All dogs are faithful. Therefore, all children are faithful. (*Cx, Dx, Fx, Lxy*)

(2) Some children are happy, and some are not. Those children who are happy are successful. Some children who are not happy are not successful. Therefore, there are children who are not happy but they are successful. (*Cx, Hx, Sx*)

(3) Some people like Arthur, and some like Betty. No one likes Charles. Arthur, Betty, and Charles are people. Therefore,
(a) There is some person who likes no one.
(b) There is some person whom no one likes.
(c) There is some person who likes both Arthur and Betty.
(*Px, Lxy, a, b, c*)

(4) Every happy person likes someone. Dorothy is an unhappy person who likes Ed, who is a happy person. Therefore, Ed likes Dorothy. (*Px, Hx, Lxy, d, e*)

(5) No state is west of California. Nevada is a state. Therefore, there is a state that is not west of California. (*Sx, Wxy, c, n*).

(6) Montana is north of Wyoming. Wyoming is north of Colorado. Colorado is north of New Mexico. Therefore, Montana is north of New Mexico. (*Nxy, m o, c, n*)

(7) Paintings that are painted in oil are valuable unless they are painted by an unknown artist. Not all paintings that are painted in oil are valuable. Therefore, there is an unknown artist. (*Px; Vx; Ox* : *x* is painted in oil; *Ux* : *x* is an unknown artist; *Axy* : *x* is painted by *y*)

(8) Some student pilot flies to some city with an instructor. Every student pilot who flies to some city with an instructor is happy. Therefore,
(a) Every student pilot is happy.
(b) Some student pilot is happy.
(*Sx, Cx, Ix, Hx, Fxyz*)

(9) (a) Some people read some mystery stories. Some people read some love stories. Therefore, there are people who read both mystery stories and love stories.
(b) Some people read some love stories. Anyone who reads some mystery stories also reads some love stories. Therefore, some people read some mystery stories.

(c) Some people read some love stories. Anyone who reads some love stories also reads some mystery stories. Therefore, some people read some mystery stories.
(Px, Mx, Lx, Rxy)

(10) A Cadillac is more expensive than both a Dodge and a Ford. A Ford is not more expensive than a Dodge. Therefore, a Dodge is more expensive than a Ford. (Mxy, c, d, o)

(11) In some states, all automobile passengers wear seat belts or else they are subject to arrest. There are automobile passengers in all states who do not wear seat belts. Therefore,
(a) Some automobile passengers are subject to arrest.
(b) If there is a state, then some automobile passengers are subject to arrest.
(c) If there is no state, then no automobile passenger is subject to arrest.
(Px; Sx; Ixy: x is in y; Bx; Ax)

(12) Someone who met Smith last night is the murderer, unless Smith is the murderer. Jones met Smith last night. Therefore, either Smith or Jones is the murderer. (Mxy : x met y last night; Ux : x is the murderer; s; j)

(13) Everyone who met Smith last night was involved in some crime. Jones did not meet Smith last night. Therefore, Jones was not involved in any crime. (Mxy, Ixy, Cx, s, j)

(14) Anyone who solves the hardest question on an exam also solves the easiest question on that exam. Oscar solved the hardest question on some exam. Therefore, Oscar solved the easiest question on some exam. (Ex; $Sxyz$; $f(x)$: the hardest problem on exam x; $g(x)$: the easiest problem on exam x; o)

(15) Geniuses solve every problem on every exam. Anyone who solves a problem on an exam is a student. Marianne solves every problem on every exam. Therefore, if Marianne is a genius, then she is a student. (Gx, Px, Ex, Dx, $Sxyz$, m)

(16) Only women who are parents accompany a child to a meeting. Therefore, if some woman accompanies a child to a meeting, then some woman is a parent. (Wx, Px, Cx, Mx, $Axyz$)

B. For each of the following, if the argument is valid, give a formal proof; otherwise give a counter-example. (Answers to 1, 7, 10, and 12.)

(1) **Premises:** $(\forall x)(Ax \to Cx)$, $(\forall x)(Bx \to Dx)$, $(\exists x)Ax$, $(\exists x)\neg Dx$
Conclusion: $(\exists x)(Cx \wedge \neg Bx)$

(2) **Premises:** $(\forall x)(Px \to \neg Qx)$, $(\forall x)(Qx \vee \neg Rx)$, $(\exists x)\neg Qx \vee (\exists x)Qx$
Conclusion: $(\exists x)(\neg Px \vee \neg Rx)$

(3) **Premises:** $(\exists x)(Ax \wedge Bx \wedge \neg Cx)$, $(\exists x)(Bx \wedge \neg Cx \wedge Dx)$
Conclusion: $(\exists x)(Ax \wedge Dx)$

(4) **Premises:** $(\forall x)(Px \wedge Qx \rightarrow Rx)$, $(\forall x)(Px \wedge \neg Qx \rightarrow Sx)$
Conclusion: $(\forall x)(Rx \vee Sx)$

(5) **Premises:** $(\forall x)(Px \wedge Qx \rightarrow Rx \vee Sx)$, $(\exists x)(Px \wedge \neg Rx)$
Conclusion: $(\exists x)Sx$

(6) **Premises:** $(\forall x)[Px \rightarrow (\forall y)(Qy \rightarrow Rxy)]$, $(\forall x)[Px \rightarrow (\exists y)(\neg Rxy \wedge Sy)]$
Conclusion: $(\exists x)(Sx \wedge \neg Qx)$

(7) **Premises:** $(\forall x)[Ax \rightarrow (\forall y)(\neg By \rightarrow Cxy)]$, $(\exists x)[Ax \wedge (\forall y)(Dy \rightarrow \neg Cxy)]$
Conclusion: $(\forall x)(Dx \rightarrow Bx)$

(8) **Premises:** $(\exists x)[Ex \wedge (\forall y)(Fy \rightarrow Gxy)]$, $(\forall x)(\forall y)[Ex \rightarrow (Gxy \leftrightarrow Hy)]$
Conclusion: $(\forall x)(Fx \leftrightarrow Hx)$

(9) **Premises:** $(\forall x)[Px \rightarrow (\forall y)(Qy \leftrightarrow \neg Rxy)]$, $(\exists x)[Px \wedge (\forall y)(\neg Sy \rightarrow Rxy)]$, $(\forall x)(\forall y)(Px \wedge Sy \rightarrow \neg Rxy)$
Conclusion: $(\forall x)(Qx \leftrightarrow Sx)$

(10) **Premises:** $(\forall x)[Ax \rightarrow (Bx \rightarrow Cx)]$, $(\exists x)(Ax \wedge Cx)$, $(\forall x)[Bx \rightarrow (Ax \rightarrow \neg Cx)]$
Conclusion: $(\exists x)(Ax \wedge Bx)$

(11) **Premises:** $(\forall x)[(\forall y)(Axy \rightarrow Ry) \rightarrow Rx]$, $Aab \wedge Ra$
Conclusion: Rb

(12) **Premises:** $(\forall x)(\forall y)(\forall z)(Rxy \wedge Ryz \rightarrow Rxz)$, $(\forall x)(\exists y)Rxy$, $(\forall x)Rxx$
Conclusion: $(\forall x)(\forall y)(Rxy \rightarrow Ryx)$

(13) **Premises:** $(\forall x)(\forall y)(Fx \wedge Gy \rightarrow Hxy)$, $(\exists x)(\forall y)[Fx \wedge (Jy \rightarrow \neg Hxy)]$
Conclusion: $(\forall x)(Gx \rightarrow \neg Jx)$

(14) **Premises:** $(\forall x)(\forall y)(\forall z)(Rxy \wedge Ryz \rightarrow Rxz)$, $(\forall x)Rxx$, $(\forall x)(\forall y)(\exists z)(Rxz \wedge Ryz)$
Conclusion: $(\forall x)(\forall y)(Rxy \vee Ryx)$

(15) **Premises:** $(\exists x)(Jx \wedge Kx)$, $(\exists z)[Jz \wedge (\forall y)(Ky \rightarrow \neg Lzy)]$
Conclusion: $(\exists x)[Jx \wedge (\exists z)(Jz \wedge \neg Lzx)]$

C. Choose appropriate predicate symbols and constants, and translate each of the following consistent sets of premises into the Predicate Calculus. Then show each set of premises is consistent by choosing an appropriate numeric or truth-table interpretation structure. (Answers to 3, 8, 11, 15, and 17.)

(1) All unicorns are white. No unicorn is white.
(2) Some deer have horns. There are deer that do not have horns.
(3) All stags have horns. All stags are animals. Some animals do not have horns.
(4) All lions are predators. Some lions do not have mates. Some predators have mates.
(5) Only students can take a course. Alex cannot take a course. Alex is a student.

(6) All birds have feathers. No mammal is a bird. There are mammals with feathers, and there are mammals without feathers.
(7) Everyone is looking at somebody. There is somebody whom nobody is looking at.
(8) Everyone is looking at somebody. Somebody is not being looked at by everybody.
(9) Every student likes some professor. No student likes all professors.
(10) No one who likes Betty also likes Arthur. Everyone likes someone. Everyone likes Arthur.
(11) There is a problem that only mathematicians can solve. No mathematician can solve every problem.
(12) Some problem can be solved by every mathematician. Some problem cannot be solved by any mathematician. Alice is a mathematician who can solve some problem.
(13) Some genius can solve a problem, if anyone can. There is a problem that no genius can solve.
(14) Only geniuses can solve all problems. Some genius cannot solve all problems.
(15) No student was accompanied by his/her father to a meeting. Only their fathers accompanied students to meetings. Some students have fathers, and some students do not have fathers.
(16) Every student who was accompanied by his/her father to a meeting was happy. There are some happy students who were not accompanied by their fathers to a meeting. If there is an unhappy student, then there is a student who was accompanied by his/her father to a meeting.
(17) Every student who went to the meeting was accompanied by his/her father. Alex is a student who went to the meeting and was accompanied by his mother. Betty is a student who was not accompanied by her father. Someone went to the meeting who does not have a father.

D. For each of the following sets of premises, if it is consistent find an interpretation structure to prove it, otherwise give a formal proof of a contradiction. (Answers to 3, 5, 9, 13, and 16.)

(1) $(\forall x)(Ax \to Cx)$, $(\forall x)(Bx \to Cx)$, $(\exists x)(Ax \wedge \neg Bx)$
(2) $(\forall x)(Dx \to \neg Ex)$, $(\forall x)(Fx \to Ex \vee \neg Dx)$, $(\exists x)(Dx \wedge Fx)$
(3) $(\forall x)(Px \vee Qx)$, $(\exists x)(Px \to Rx)$, $(\exists x)(Qx \to Rx)$, $(\forall x)\neg Rx$
(4) $(\exists x)(Fx \wedge Gx)$, $(\exists x)(Fx \wedge \neg Gx)$, $(\forall x)(Fx \vee Gx)$
(5) $(\exists x)Px \to (\forall x)Qx$, $Qa \wedge (\exists x)(Px \wedge \neg Qx)$
(6) $(\forall x)(Px \to \neg Qx)$, $(\forall x)(\neg Rxa \to Qx)$, $Pa \wedge \neg Raa$
(7) $(\exists x)(Px \to Qx)$, $(\exists x)(Px \to Rx)$, $(\forall x)Px$, $(\forall x)(Qx \to \neg Rx)$
(8) $(\forall x)[Ux \wedge Vx \to Wx]$, $(\exists x)(Ux \wedge Wx)$, $(\forall x)[Vx \to (Ux \to \neg Wx)]$, $(\forall x)(Ux \to \neg Vx)$
(9) $(\forall x)[(\forall y)Py \to Qx]$, $(\forall x)Px$, $(\exists y)\neg Qy$
(10) $(\forall x)[Fx \to (\exists y)(Gy \wedge \neg Hxy)]$, $(\forall x)(Jx \to Fx)$, $(\exists x)[Jx \wedge (\forall y)(Gy \to Hxy)]$
(11) $(\forall x)(\forall y)(Rxy \to Ryx)$, $(\forall x)(\forall y)(\forall z)(Rxy \wedge Ryz \to Rxz)$, $(\exists x)\neg Rxx$

(12) $(\exists x)[Px \wedge (\forall y)(Qy \rightarrow Rxy)], (\exists y)[Qy \wedge (\forall x)(Px \rightarrow \neg Rxy)]$
(13) $(\forall x)(\exists y)(Px \wedge Qy \wedge Rxy), (\forall y)(\exists x)(Px \wedge Qy \rightarrow \neg Rxy)$
(14) $(\forall x)(\forall y)(Fx \wedge Fy \rightarrow Gxy), (\exists x)(\exists y)(\neg Fx \wedge \neg Gxy)$,
$(\exists x)[Fx \wedge (\forall y)(Fy \rightarrow Gyx)]$
(15) $(\forall x)(\forall y)(Fx \wedge Gy \rightarrow Hxy), (\exists x)(\exists y)(Fx \wedge \neg Hxy), (\forall x)(Fx \rightarrow Gx)$
(16) $(\forall x)Rxf(x), (\forall x)(\exists y)\neg Rxy$
(17) $(\forall x)(\forall y)(\forall z)(Fxy \wedge Fyz \rightarrow Fxz), (\forall x)(\forall y)(Fxy \rightarrow Fyx)$,
$(\forall x)Fxf(x) \wedge (\exists x)\neg Fxx$
(18) $(\forall x)(\forall y)(Px \wedge Py \rightarrow Pf(x, y)), (\forall x)(Px \rightarrow \neg Qx)$,
$(\exists x)[Px \wedge (\forall y)(Py \rightarrow Qf(y, x))]$
(19) $(\forall x)(\forall y)(Px \wedge Py \rightarrow Rxy), (\exists x)Px, (\exists x)Pf(x), (\forall x)\neg Rf(x)x$
(20) $(\forall x)[Rx \wedge \neg Sx \rightarrow Txf(x) \wedge Uf(x)]$,
$(\exists x)[Vx \wedge Rx \wedge (\forall y)(Txy \rightarrow Vy)], (\forall x)(Vx \rightarrow \neg Sx), (\forall x)(Vx \rightarrow \neg Ux)$

*E. (Answers to 1, 4, and 7(a).)

(1) Suppose that φ and ψ are sentences such that $\varphi \vdash \psi$. Does there exist an interpretation structure in which φ is true and ψ is false? Give reasons for your answer.

(2) Suppose that φ and ψ are sentences such that $\varphi \vdash \psi \wedge \neg\psi$. Does φ have a model? Give reasons for your answer.

(3) Suppose that φ and ψ are sentences such that whenever **S** is an interpretation structure for $\{\varphi, \psi\}$ such that if φ is false in **S**, then ψ is true in **S**. Does $\neg\psi \vdash \varphi$? Give reasons for your answer.

(4) Suppose that Γ is an infinite set of formulas such that x is not free in any formula in Γ. Prove that if $\Gamma \vdash \varphi$, then $\Gamma \vdash (\forall x)\varphi$.

(5) (a) Give an example of two sets of sentences Γ and Δ such that each of Γ and Δ is consistent but $\Gamma \cup \Delta$ is inconsistent.

(b) Prove that if Γ and Δ are sets of sentences such that there is a sentence φ such that $\Gamma \models \varphi$ and $\Delta \models \neg\varphi$, then $\Gamma \cup \Delta$ is inconsistent.

(6) Let $\Gamma = \{(\exists x)\neg Px, Pc_1, Pc_2, Pc_3, \ldots\}$. Is Γ consistent? Give reasons for your answer.

(7) Show that each of the following sets of sentences Γ of the Predicate Calculus is consistent. ($\mathbb{N}$ is the set of positive integers.)

(a) $\Gamma = \{\varphi_n : n \in \mathbb{N}\}$, where φ_n is
$(\forall x_1)(\forall x_2) \ldots (\forall x_n)(Px_1 \wedge Px_2 \wedge \cdots \wedge Px_n \rightarrow (\exists y)\neg Py)$.

(b) $\Gamma = \{(\forall x)(\forall y)(Rxy \rightarrow \neg Ryx), (\forall x)(\forall y)(\forall z)(Rxy \wedge Ryz \rightarrow Rxz)\}$
$\cup \{\psi_n : n \in \mathbb{N}\}$, where ψ_n is
$(\forall x)$ ("there exist n distinct y's") $(Rxy_1 \wedge Rxy_2 \wedge \cdots \wedge Rxy_n)$.

(8) Suppose Λ is a set of sentences of the Predicate Calculus that has arbitrarily large finite models. Prove that Γ has an infinite model.

(9) Let Γ be a set of sentences of the Predicate Calculus so that in each interpretation structure **S** for Γ, at least one sentence in Γ is true. Prove that there are finitely many formulas $\varphi_1, \varphi_2, \ldots \varphi_n \in \Gamma$ such that $\varphi_1 \vee \varphi_2 \vee \cdots \vee \varphi_n$ is true in every interpretation structure for Γ.

Chapter 10

Logical Theorems and Valid Formulas

Up to this chapter, we have been talking about proofs of formulas φ from non-empty sets of premises Γ. If Γ is the empty set (that is, the proof of φ does not depend on any premises), we say φ is a *logical theorem*. If φ is true in every interpretation structure for φ, we say φ is *valid*. It follows from the Completeness Theorem and the Soundness Theorem (see Section 9.3) that φ is a logical theorem iff φ is valid. We discuss these notions in this chapter.

10.1 LOGICAL THEOREMS

Definition 10.1 Suppose φ and ψ are formulas.

1. φ is called a *logical theorem* if there is a proof of φ from the empty set of formulas (in symbols, $\vdash \varphi$).
2. φ *logically implies* ψ if $\varphi \rightarrow \psi$ is a logical theorem (in symbols, $\vdash \varphi \rightarrow \psi$).
3. φ and ψ are *logically equivalent* if $\varphi \leftrightarrow \psi$ is logical theorem (in symbols, $\vdash \varphi \leftrightarrow \psi$).

For example, it follows from the Rule of Tautology that if φ is a tautology, then φ is a logical theorem. (A "tautology" in the Predicate Calculus is a substitution instance of a tautology in the Sentential Calculus. For example, $(\exists x)Px \vee \neg(\exists x)Px$ is a tautology.) However, there are many other formulas of the Predicate Calculus, besides tautologies, that are logical theorems. We give some below, and we also give a list of important logical theorems in Section 10.6.

Earlier, when we were translating English into the Predicate Calculus, we made the comment that sentences like the following two mean the same thing:

(1) No student likes history.

(2) There does not exist a student who likes history.

Translating these two sentences into the language of the Predicate Calculus, (1) becomes

(3) $(\forall x)(Sx \rightarrow \neg Hx)$,

and (2) becomes

(4) $\neg(\exists x)(Sx \wedge Hx)$.

Note that $\neg(Sx \wedge Hx) \Leftrightarrow (Sx \rightarrow \neg Hx)$. Thus, we would expect that formulas of the form $\neg(\exists x)\varphi$ and $(\forall x)\neg\varphi$ are logically equivalent. This is indeed the case.

Theorem 10.1 If φ is any formula, and x is any variable, then

$$\vdash \neg(\exists x)\varphi \leftrightarrow (\forall x)\neg\varphi.$$

Proof: We shall first prove that $\neg(\exists x)\varphi \rightarrow (\forall x)\neg\varphi$ and then the converse. In each case, we shall give an indirect proof.

{1}	1.	$\neg(\exists x)\varphi$	*P* (for *CP*)
{2}	2.	φ	*x P* (for *IP*)
{2}	3.	$(\exists x)\varphi$	2 *EG* $[x\|x]$
{1, 2}	4.	$(\exists x)\varphi \wedge \neg(\exists x)\varphi$	1, 3 *T*
{1}	5.	$\neg\varphi$	2, 4 *IP*
{1}	6.	$(\forall x)\neg\varphi$	5 *UG*
	7.	$\neg(\exists x)\varphi \rightarrow (\forall x)\neg\varphi$	1, 6 *CP*

We are assuming that x is free in φ. If it is not, then x would not be flagged on line 2. Notice that x is not flagged on line 5 because line 5 does not depend on premise 2. Thus, we can use Rule *UG* on line 6. Line 7 does not depend on any premises, so we have shown that $\vdash \neg(\exists x)\varphi \rightarrow (\forall x)\neg\varphi$.

In the next part of the proof, we assume u is a variable that does not occur in φ and ψ is any formula with no free variables—a sentence:

{8}	8.	$(\forall x)\neg\varphi$	*P* (for *CP*)
{9}	9.	$(\exists x)\varphi$	*P* (for *IP*)
{10}	10.	$\varphi[u\|x]$	*u P* (for *EP*)
{8}	11.	$\neg\varphi[u\|x]$	8 *US* $[u\|x]$
{8, 10}	12.	$\varphi[u\|x] \wedge \neg\varphi[u\|x]$	*u* 10, 11 *T*
{8, 10}	13.	$\psi \wedge \neg\psi$	12 *T*
{8, 9}	14.	$\psi \wedge \neg\psi$	9, 10, 13 *EP* $[u\|x]$
{8}	15.	$\neg(\exists x)\varphi$	9, 14 *IP*
	16.	$(\forall x)\neg\varphi \rightarrow \neg(\exists x)\varphi$	8, 15 *CP*
	17.	$\neg(\exists x)\varphi \leftrightarrow (\forall x)\neg\varphi$	7, 16 *T*

On line 12, we derived a contradiction, but u, the *EP* variable, is free. Thus, we use the tautology

$$\varphi[u|x] \wedge \neg\varphi[u|x] \rightarrow \psi \wedge \neg\psi$$

to derive line 13. On line 13, we have a contradiction, but u is not free. (Remember, we cannot apply Rule *EP* if the *EP* variable is free in the conclusion. So we use the tautology 9 in Chapter 2, Rule of Absurdity,

$$P \wedge \neg P \rightarrow Q$$

to remedy the situation.) Notice, that we use the Rule *EP* before Rule *IP* because we want to contradict the existential premise on line 9, not the *EP* premise on line 10. Line 16 does not depend on any premises, so we have shown that $(\forall x)\neg\varphi$ logically implies $\neg(\exists x)\varphi$. Neither line 7 nor line 16 depends upon any premises, so line 17 does not depend on any premises either. Thus, line 17 gives us the result that $\neg(\exists x)\varphi$ is logically equivalent to $(\forall x)\neg\varphi$. (In the next section, we shall give a semantical proof of Theorem 10.1.)

■

10.2 VALID FORMULAS

The notion of a logical theorem depends upon the definition of a formal proof, so it is a grammatical or structural notion. On the other hand, the notion of validity depends upon the notion of truth, so it is a semantical notion. (See Definitions 9.1, 9.3 and 9.4.)

Definition 10.2 Suppose φ and ψ are formulas.

1. φ is *valid* if φ is true in every interpretation structure for φ. In symbols, $\models \varphi$.
2. φ *semantically implies* ψ, $\varphi \models \psi$, if the argument with premise φ and conclusion ψ is valid.
3. φ is *semantically equivalent* to ψ if $\varphi \models \psi$ and $\psi \models \varphi$.

If follows from the Soundness and Completeness Theorems that if φ is a sentence, then

$$(*) \qquad \vdash \varphi \text{ iff } \models \varphi.$$

That is, φ is a logical theorem if and only if φ is valid. (Note that if φ is not a sentence, but φ' is the sentence obtained from φ by generalizing on all the free variables in φ, then φ' is a logical theorem (valid) iff φ is a logical theorem (valid). Thus, $(*)$ also holds if φ is not a sentence.)

It follows from (*) that exactly one of the following alternatives is true:

φ is a logical theorem

or

There is an interpretation structure for φ in which φ is false.

As our first example, we shall give a semantical proof of Theorem 10.1. We must show that if φ is any formula and x is any variable, then

(1) $\models \neg(\exists x)\varphi \leftrightarrow (\forall x)\neg\varphi$,

or, equivalently,

(2) $\neg(\exists x)\varphi \models (\forall x)\neg\varphi$

and

(3) $(\forall x)\neg\varphi \models \neg(\exists x)\varphi$.

To prove (2), suppose a sequence **a** satisfies $\neg(\exists x)\varphi$. Then, **a** does not satisfy $(\exists x)\varphi$. Thus, **a** does not satisfy φ for any interpretation of x. Consequently, **a** satisfies $\neg\varphi$ for every interpretation of x. This implies that **a** satisfies $(\forall x)\neg\varphi$. The proof of (3) is similar. Suppose **a** satisfies $(\forall x)\neg\varphi$. Then **a** satisfies $\neg\varphi$ for every interpretation of x. Therefore, **a** does not satisfy φ for any interpretation of x. This implies that **a** does not satisfy $(\exists x)\varphi$ or, equivalently, **a** does satisfy $\neg(\exists x)\varphi$.

Example 10.1

For each of the following formulas, determine whether it is a logical theorem. If it is a logical theorem, give a formal proof (from the empty set of premises); if not, give a counter-example, an interpretation structure in which the formula is false.

(a) $(\forall x)(Px \to Qx) \to ((\exists x)Px \to (\exists x)Qx)$

(b) $(\forall x)(\forall y)(Px \vee \neg Py)$

(c) $(\exists x)Px \wedge (\exists x)Qx \to (\exists x)(Px \wedge Qx)$

(d) $(\forall x)(Px \vee Qx) \to (\forall x)Px \vee (\forall x)Qx$

Solution: The procedure for each formula is either to find a counter-example directly or to try to give a proof. If the proof breaks down, then try to find a counter-example.

(a) Take (i) $(\forall x)(Px \to Qx)$ and $(\exists x)Px$ as premises for Rule *CP*. Then take (ii) Py as a premise for *EP*. From (i), deduce (iii) $Py \to Qy$, and from (ii) and (iii), deduce Qy. Then, using Rules *EG*, *EP*, and *CP*, we obtain the desired result. Formally,

{1}	1.	$(\forall x)(Px \to Qx)$	*P* (for *CP*)
{2}	2.	$(\exists x)Px$	*P* (for *CP*)
{3}	3.	Py	*y P* (for *EP*)
{1}	4.	$Py \to Qy$	1 *US* $[y\|x]$
{1, 3}	5.	Qy	*y* 3, 4 *T*
{1, 3}	6.	$(\exists x)Qx$	5 *EG* $[y\|x]$
{1, 2}	7.	$(\exists x)Qx$	2, 3, 6 *EP* $[y\|x]$
{1]	8.	$(\exists x)Px \to (\exists x)Qx$	2, 7 *CP*
	9.	$(\forall x)(Px \to Qx) \to ((\exists x)Px \to (\exists x)Qx)$	1, 8 *CP*

(b) Let's try an indirect proof and assume $\neg(Px \vee \neg Py)$ or, what is logically equivalent, (i) $\neg Px \wedge Py$. There does not seem any way to derive a contradiction from (i) because there could be an interpretation structure in which *Px* is false and *Py* is true, so (i) is true. This is the idea for constructing the counter example. See Figure 10.1.

U = {a, b}

	P
a	*F*
b	*T*

Figure 10.1

In this structure, (b) is false because **Pa** $\wedge$ $\neg$**Pb** is false. (That is, (b) is not true for *all x* and *y*.)

(c) Intuitively, we see that (c) is not a theorem because each of $(\exists x)Px$ and $(\exists x)Qx$ could be true in an interpretation structure without there being *one x*, so that both *Px* and *Qx* are true. Thus, if our interpretation structure is as in Figure 10.2, (c) is false because **Pa** and **Qb** are both true, but $(\exists x)(Px \wedge Qx)$ is false.

U = {a, b}

	P	**Q**
a	*T*	*F*
b	*F*	*T*

Figure 10.2

(d) Using reasoning similar to that in (c), we see that formula (d) is also false in the interpretation structure given in Figure 10.2 because the disjunction $Px \vee Qx$ is true for each *x* but neither is *Px* true for every *x* nor is *Qx* true for every *x*. Therefore, (d) is not a logical theorem.

Later we shall show that the converses of both (c) and (d) are logical theorems, but first we shall introduce some additional rules of inference in order to simplify some proofs.

10.3 ADDITIONAL RULES OF INFERENCE

We shall extend the Rule for Tautologies, Rule *T*, to a rule for logical theorems that will permit us to use previously obtained results. For example, if $\neg(\exists x)\varphi$ occurs on a line of a proof, we should be able to deduce $(\forall x)\neg\varphi$, because in Theorem 10.1 we proved that $\vdash \neg(\exists x)\varphi \rightarrow (\forall x)\neg\varphi$.

Rule *LT* (Rule for Logical Theorems) *A formula φ may occur on any line of a proof if there is a constructive proof of φ from the empty set of premises or if there are formulas $\psi_1, \psi_2, \ldots, \psi_n$ preceding φ such that there is a constructive proof of $\psi_1 \wedge \psi_2 \wedge \cdots \wedge \psi_n \rightarrow \varphi$.*

The only "constructive" proofs we shall use here are those that use Rule *T* or those that we have given earlier. Thus, Rule *LT* supersedes Rule *T*. Rule *LT* is valid because of the Soundness Theorem: for any formula χ if $\vdash \chi$, then $\models \chi$. Thus, Rule *LT* preserves truth.

Another rule that will be useful in the Predicate Calculus is the Rule for Logically Equivalent Formulas, or Rule *LE*.

Rule *LE* (Rule for Logically Equivalent Formulas) *If φ_1, φ_2, ψ_1, and ψ_2 are formulas such that φ_2 is obtained from φ_1 by replacing some occurrences of ψ_1 in φ_1 by ψ_2, and if ψ_1 and ψ_2 are logically equivalent ($\vdash \psi_1 \leftrightarrow \psi_2$), then φ_2 is deducible from φ_1 (in fact, $\vdash \varphi_1 \leftrightarrow \varphi_2$).*

A "formula" is defined in Definition 5.5. A "subformula" of a formula is a consecutive sequence of symbols from the formula that, itself, is a formula. The subformulas of a formula φ are all the formulas that appear on some level of the tree for φ (see Section 5.4). In Rule *LE*, ψ_1 is a subformula of φ_1 and ψ_2 is a subformula of φ_2. Rule *TE*, which was used in Chapter 4, is a special case of Rule *LE*.

Rules *LT* and *LE* are derived rules of inference. They are consequences of the other rules of inference. (See Theorem 16.10 and the remarks preceding this theorem.) They are useful because they allow us to use previously proved results to obtain new results and, thus, shorten proofs. We illustrate these rules in the next theorem.

Theorem 10.2 *If φ is any formula and x is any variable, then*

(a) $\vdash \neg(\forall x)\varphi \leftrightarrow (\exists x)\neg\varphi$,

(b) $\vdash \neg(\exists x)\neg\varphi \leftrightarrow (\forall x)\varphi$,

(c) $\vdash \neg(\forall x)\neg\varphi \leftrightarrow (\exists x)\varphi$.

Proof: (a) We shall use Theorem 10.1 in the following proof:

1.	$\neg(\exists x)\neg\varphi \leftrightarrow (\forall x)\neg(\neg\varphi)$	*LT* (Theorem 10.1)
2.	$(\exists x)\neg\varphi \leftrightarrow \neg(\forall x)\neg(\neg\varphi)$	1 *T*
3.	$(\exists x)\neg\varphi \leftrightarrow \neg(\forall x)\varphi$	2 *LE* (*T*)

On line 1, we use Theorem 10.1 with $\neg\varphi$ substituted for φ. We obtained line 2 by negating both sides of line 1. Then to obtain line 3, we substituted φ for $\neg(\neg\varphi)$ on line 2. This is an application of Rule *LE* because

$$\vdash \neg(\neg\varphi) \leftrightarrow \varphi.$$

Every tautology is a logical theorem. The Rules *LT* and *LE* allow us to use known results to obtain new results.

The proofs of (b) and (c) are similar. To obtain (b), start the same way as in (a). Then to obtain (b), substitute φ for $\neg(\neg\varphi)$ on line 1. To obtain (c), take the negation of both sides of Theorem 10.1. The proofs are as follows:

(b)	1.	$\neg(\exists x)\neg\varphi \leftrightarrow (\forall x)\neg(\neg\varphi)$	*LT* (Theorem 10.1)
	2.	$\neg(\exists x)\neg\varphi \leftrightarrow (\forall x)\varphi$	1 *LE* (*T*)
(c)	1.	$\neg(\exists x)\varphi \leftrightarrow (\forall x)\neg\varphi$	*LT* (Theorem 10.1)
	2.	$(\exists x)\varphi \leftrightarrow \neg(\forall x)\neg\varphi$	1 *T*

Additional examples of the use of Rules *LT* and *LE* are given in the next example. (As an exercise, give a semantical proof of Theorem 10.2.)

Example 10.2 Give a formal proof of each of the following:

(a) $\vdash [(A \wedge \neg B \wedge C) \vee (\neg A \wedge \neg B \wedge C) \vee (B \wedge C)] \leftrightarrow C$

(b) $\neg(\exists x)[Px \wedge (\forall y)(Qy) \rightarrow Rxy)] \vdash (\forall x)[Px \rightarrow (\exists y)(Qy \wedge \neg Rxy)]$

Solution: We shall give a formal proof of (a). (See Example 4.3 (a).) Let φ be the formula

$$[(A \wedge \neg B \wedge C) \vee (\neg A \wedge \neg B \wedge C) \vee (B \wedge C)].$$

1.	$\varphi \leftrightarrow [((A \vee \neg A) \wedge (\neg B \wedge C)) \vee (B \wedge C)]$	*LE* (*T*)
2.	$\varphi \leftrightarrow [(\neg B \wedge C) \vee (B \wedge C)]$	*LE* (*T*)
3.	$\varphi \leftrightarrow [(\neg B \vee B) \wedge C]$	*LE* (*T*)
4.	$\varphi \leftrightarrow C$	*LE* (*T*)

The documentation for each line, *LE* (*T*), means that we used the Rule *LE* and the logical theorem we used was a tautology.
The following is a formal proof of (b):

{1}	1.	$\neg(\exists x)[Px \wedge (\forall y)(Qy \rightarrow Rxy)]$	*P*
{1}	2.	$(\forall x)\neg[Px \wedge (\forall y)(Qy \rightarrow Rxy)]$	1 *LT* (Theorem 10.1)
{1}	3.	$(\forall x)[Px \rightarrow \neg(\forall y)(Qy \rightarrow Rxy)]$	2 *LE* (*T*)
{1}	4.	$(\forall x)[Px \rightarrow (\exists y)\neg(Qy \rightarrow Rxy)]$	3 *LE* (Theorem 10.2(*a*))
{1}	5.	$(\forall x)[Px \rightarrow (\exists y)(Qy \wedge \neg Rxy)]$	4 *LE* (*T*)

We should like to add a word of warning on the use of the Rule *LE*. It is absolutely necessary that if ψ_2 is to replace ψ_1, then $\psi_1 \leftrightarrow \psi_2$ must be a logical theorem. The following is an **INCORRECT** use of Rule *LE*:

{1}	1.	$Px \leftrightarrow Qx$	*x P*
{2}	2.	$(\exists x)(Qx \wedge \neg Px)$	*P*
{1, 2}	3.	$(\exists x)(Qx \wedge \neg Qx)$	1, 2 *LE*

This is **WRONG**. The premises on lines 1 and 2 are not inconsistent, but we derived a contradiction on line 3 by an incorrect use of Rule *LE*. On line 1, $Px \leftrightarrow Qx$ is a premise, **NOT** a logical theorem. Thus, the substitution of Qx for Px on line 2 was an **INCORRECT** use of Rule *LE*.

The logical theorems given in Theorems 10.1 and 10.2, along with the Rules *LT* and *LE*, give us rules for negating expressions containing quantifiers. Because

$$\neg(\varphi \rightarrow \psi) \Leftrightarrow (\varphi \wedge \neg\psi)$$

and

$$\neg(\varphi \wedge \psi) \Leftrightarrow (\varphi \rightarrow \neg\psi),$$

we obtain as corollaries of Theorem 10.2(a) and 10.1,

(1) $\vdash \neg(\forall x)(\varphi \rightarrow \psi) \leftrightarrow (\exists x)(\varphi \wedge \neg\psi)$

and

(2) $\vdash \neg(\exists x)(\varphi \wedge \psi) \leftrightarrow (\forall x)(\varphi \rightarrow \neg\psi)$.

For example, the negation of

All cats have tails, $(\forall x)(Cx \rightarrow Tx)$

is

Some cats do not have tails, $(\exists x)(Cx \wedge \neg Tx)$.

The negation of

Some unicorns have horns, $(\exists x)(Ux \wedge Hx)$

is

No unicorns have horns, $(\forall x)(Ux \rightarrow \neg Hx)$.

A little more complicated example[1] from mathematics is to negate the expression

$$(3)\ L = \lim_{n \to \infty} a_n,$$

where L and each a_n are real numbers. Formula (3) is an abbreviation for

L is the limit of a_n as n increases without limit.

If we assume ϵ is a variable that ranges over the real numbers and n and N are variables that range over the positive integers, (3) is defined by

$$(4)\ (\forall \epsilon)[\epsilon > 0 \rightarrow (\exists N)(\forall n)(n > N \rightarrow |L - a_n| < \epsilon)].$$

Translating (4) into English: For each positive real number ϵ, there is a positive integer N such that for each positive integer n larger than N, $|L - a_n| < \epsilon$. (In the tail end of the sequence, the terms are arbitrarily close to L.) Thus, $L \neq \lim_{n \to \infty} a_n$ is equivalent to

$$(5)\ \neg(\forall \epsilon)[\epsilon > 0 \rightarrow (\exists N)(\forall n)(n > N \rightarrow |L - a_n| < \epsilon)].$$

To understand the meaning of (5), we use the rules for negating quantifiers and obtain

$$(6)\ (\exists \epsilon)[\epsilon > 0 \wedge \neg(\exists N)(\forall n)(n > N \rightarrow |L - a_n| < \epsilon)]$$
$$(7)\ (\exists \epsilon)[\epsilon > 0 \wedge (\forall N)\neg(\forall n)(n > N \rightarrow |L - a_n| < \epsilon)]$$
$$(8)\ (\exists \epsilon)[\epsilon > 0 \wedge (\forall N)(\exists n)(n > N \wedge |L - a_n| \not< \epsilon)]$$

Formula (6) is obtained from (5) by using (1) (and Rule *LT*); (7) is obtained from (6) by using Theorem 10.1 (and Rule *LE*); and, finally, (8) is obtained from (7) by using (1) again (and Rule *LE*). Thus, (5) and (8) are logically equivalent. Translating formula (8) into English: There is a positive real number ϵ such that for each positive integer N, there is a larger positive integer n such that $|L - a_n| \geq \epsilon$.

That is:

No matter how far out we go in the sequence, there is a term a_n, further out, that is further than ϵ from L. (Also, see Exercises 10 C (25) and (26).)

On the list of useful logical theorems in Section 10.6, the first six are the theorems dealing with the interchange of negation and quantifiers mentioned in this section.

---

[1]This example is important for mathematics students, but it may be omitted by others.

10.4 ADDITIONAL LOGICAL THEOREMS

The sentence "All cats have tails" could be translated into the Predicate Calculus as

$$(\forall x)(Cx \rightarrow Tx).$$

However, $(\forall y)(Cy \rightarrow Ty)$ or $(\forall z)(Cz \rightarrow Tz)$ would also be translated into English as "All cats have tails." A bound variable is called a "dummy" variable. It does not matter what name is given to it.

Similarly, in mathematics, the variable of summation is a dummy variable. For example,

$$\sum_{i=1}^{5} \frac{2}{i} = \sum_{j=1}^{5} \frac{2}{j} = \sum_{k=1}^{5} \frac{2}{k}$$
$$= \frac{2}{1} + \frac{2}{2} + \frac{2}{3} + \frac{2}{4} + \frac{2}{5} = \frac{137}{30}.$$

However, in changing the name of a bound variable, we have to be careful that we do not make an invalid substitution, because, as mentioned earlier, this can change the meaning of the formula. For example, consider the English sentence

(1) All students like some professor.

Translating (1) into the Predicate Calculus, we get

(2) $(\forall x)[Sx \rightarrow (\exists y)(Py \wedge Lxy)]$.

Formulas (3) and (4) would be translated into English the same way as sentence (1).

(3) $(\forall z)[Sz \rightarrow (\exists y)(Py \wedge Lzy)]$

(4) $(\forall u)[Su \rightarrow (\exists y)(Py \wedge Luy)]$

However, if we replace the x in (2) by y, we obtain

(5) $(\forall y)[Sy \rightarrow (\exists y)(Py \wedge Lyy)]$.

Sentence (5) would be translated as

(6) For each student, there is a professor who likes himself or herself.

Clearly (1) and (6) are not equivalent. Note that the substitution of y for x in

(7) $Sx \rightarrow (\exists y)(Py \wedge Lxy)$

was *not* a valid substitution because there is a free occurrence of x within the scope of the quantifier $(\exists y)$.

Theorem 10.3 *If φ is any formula, and x and y are any variables such that y is not free in φ and the substitution of y for x in φ is valid, then*

(a) $\vdash (\forall x)\varphi \leftrightarrow (\forall y)\varphi[y|x]$,

(b) $\vdash (\exists x)\varphi \leftrightarrow (\exists y)\varphi[y|x]$.

Proof:

(a)

{1}	1.	$(\forall x)\varphi$	P (for CP)
{1}	2.	$\varphi[y|x]$	1 US $[y|x]$
{1}	3.	$(\forall y)\varphi[y|x]$	2 UG
	4.	$(\forall x)\varphi \rightarrow (\forall y)\varphi[y|x]$	1, 3 CP
{5}	5.	$(\forall y)\varphi[y|x]$	P (for CP)
{5}	6.	φ	5 US $[x|y]$
{5}	7.	$(\forall x)\varphi$	6 UG
	8.	$(\forall y)\varphi[y|x] \rightarrow (\forall x)\varphi$	5, 8 CP
	9.	$(\forall x)\varphi \leftrightarrow (\forall y)\varphi[y|x]$	4, 8 T

Notice that if the substitution of y for x in φ is not valid, then we could not use Rule US to obtain line 2, and if y were free in φ, then we could not use Rule UG to obtain line 3. Moreover, if the substitution of y for x in φ is valid and y is not free in φ, then if all free occurrences of y in $\varphi[y|x]$ are replaced by x, we get back to the formula we started with, φ, on line 6.

(b)

{1}	1.	$(\exists x)\varphi$	P (for CP)
{2}	2.	$\varphi[y|x]$	y P (for EP)
{2}	3.	$(\exists y)\varphi[y|x]$	2 EG $[y|y]$
{1}	4.	$(\exists y)\varphi[y|x]$	1, 2, 3 EP $[y|x]$
	5.	$(\exists x)\varphi \rightarrow (\exists y)\varphi[y|x]$	1, 4 CP
{6}	6.	$(\exists y)\varphi[y|x]$	P (for CP)
{7}	7.	φ	x P (for EP)
{7}	8.	$(\exists x)\varphi$	7 EG $[x|x]$
{6}	9.	$(\exists x)\varphi$	6, 7, 8 EP $[x|y]$
	10.	$(\exists y)\varphi[y|x] \rightarrow (\exists x)\varphi$	6, 9 CP
	11.	$(\exists x)\varphi \leftrightarrow (\exists y)\varphi[y|x]$	5, 10 T

On line 2, we took y to be the EP variable. Because y is not free on line 3, this is a valid use of Rule EP on line 4. Similarly, on line 7, we took x to be the EP variable. (Line 7 would not be correct if y is free in φ.) The variable x is not free on line 8, so the use of Rule EP on line 9 was again correct. If we would have, for example, used z as the EP variable on line 2, the proof would have proceeded as follows:

{2}	2.	$\varphi[z|x]$	z P (for EP)
{2}	3.	$(\exists y)\varphi[y|x]$	2 EG $[z|y]$

and so forth.

■

If the predicate symbols Sx, Px, and Lxy are translated as in (1) and (2) above, the sentence

(8) All students like all professors.

would be translated as

(9) $(\forall x)(\forall y)(Sx \land Py \rightarrow Lxy)$.

Formally, (9) is translated as

(10) For every x and every y, if x is a student and y a professor, then x likes y.

But (10) means exactly the same as

(11) For every y and every x, if x is a student and y a professor, then x likes y.

Sentence (11) is translated into the Predicate Calculus as

(12) $(\forall y)(\forall x)(Sx \land Py \rightarrow Lxy)$.

Similarly, the sentence

(13) Some students like some professors.

can be translated either as

(14) $(\exists x)(\exists y)(Sx \land Py \land Lxy)$

or

(15) $(\exists y)(\exists x)(Sx \land Py \land Lxy)$.

Formulas (9) and (12) should be logically equivalent, and formulas (14) and (15) should be logically equivalent, and indeed they are, as is shown in the next theorem.

Theorem 10.4 *If φ is any formula, and x and y are any variables, then*

(a) $\vdash (\forall x)(\forall y)\varphi \leftrightarrow (\forall y)(\forall x)\varphi$,

(b) $\vdash (\exists x)(\exists y)\varphi \leftrightarrow (\exists y)(\exists x)\varphi$.

Proof: To prove (a), drop the quantifiers using Rule *US* and add them in the reverse order using Rule *UG*. Similarly, for (b), drop the quantifiers by taking premises for *EP*, then add them in the reverse order using Rule *EG*. We leave the details to the reader. ■

We shall next consider the relationship between the following two formulas:

(16) $(\exists x)(\forall y)\varphi$,

(17) $(\forall y)(\exists x)\varphi$.

Theorem 10.5 *If x and y are variables, then*

(a) *For every formula* φ, $\vdash (\exists x)(\forall y)\varphi \to (\forall y)(\exists x)\varphi$.

(b) *There is a formula* φ *for which* $(\forall y)(\exists x)\varphi \to (\exists x)(\forall y)\varphi$ *is not a logical theorem.*

Proof: The proof of (a) is similar to the proof in Exercise 8 A(8), and to prove (b), take φ to be Rxy and use the interpretation structure given in Example 9.2. ∎

Thus, interchanging a universal and existential quantifier may not lead to an equivalent formula.

The following logical theorems are useful when it is necessary to move quantifiers around in a formula.

Theorem 10.6 *If* φ *and* ψ *are any formulas and x is any variable that is not free in* φ, *then*

(a) $\vdash (\forall x)(\varphi \wedge \psi) \leftrightarrow [\varphi \wedge (\forall x)\psi]$

(b) $\vdash (\exists x)(\varphi \wedge \psi) \leftrightarrow [\varphi \wedge (\exists x)\psi]$

(c) $\vdash (\forall x)(\varphi \vee \psi) \leftrightarrow [\varphi \vee (\forall x)\psi]$

(d) $\vdash (\exists x)(\varphi \vee \psi) \leftrightarrow [\varphi \vee (\exists x)\psi]$

(e) $\vdash (\forall x)(\varphi \to \psi) \leftrightarrow [\varphi \to (\forall x)\psi]$

(f) $\vdash (\exists x)(\varphi \to \psi) \leftrightarrow [\varphi \to (\exists x)\psi]$

(g) $\vdash (\forall x)(\psi \to \varphi) \leftrightarrow [(\exists x)\psi \to \varphi]$

(h) $\vdash (\exists x)(\psi \to \varphi) \leftrightarrow [(\forall x)\psi \to \varphi]$

Before proving Theorem 10.6, we shall give some examples. Each of the following pairs of sentences have equivalent meanings:

An example of Theorem 10.6 (b):

(i) Some student likes Smith, and Smith is a professor. $[(\exists x)(Sx \wedge Lxs) \wedge Ps]$

(ii) Some student likes Smith, who is a professor. $[(\exists x)(Sx \wedge Lxs \wedge Ps)]$

(We are using the fact that $\vdash (\varphi \wedge (\exists x)\psi) \leftrightarrow ((\exists x)\psi \wedge \varphi)$.)

An example of (c):

(i) Either Smith is promoted or everyone resigns. $[Ps \vee (\forall x)Rx]$

(ii) For every x, either Smith is promoted or x resigns. $[(\forall x)(Ps \vee Rx)]$

An example of (f):

(i) Every professor likes some student. $[(\forall x)(Px \to (\exists y)(Sy \wedge Lxy))]$

(ii) For every professor, there is some student whom the professor likes. $[(\forall x)(\exists y)(Px \to Sy \wedge Lxy)]$

An example of (g):

(i) If some element belongs to A, then A is not empty. $[(\exists x)(x \in A) \to A \neq 0]$

(ii) For every x, if x belongs to A, then A is not empty.
$[(\forall x)(x \in A \rightarrow A \neq 0)]$

Proof of Theorem 10.6: We shall prove parts (b), (c), (f), and (g) and leave the others as exercises.

(b) Let z be a variable that does not appear in either φ or in ψ.

{1}	1.	$(\exists x)(\varphi \wedge \psi)$	P (for CP)
{2}	2.	$\varphi \wedge \psi[z\|x]$	z P (for EP)
{2}	3.	$\psi[z\|x]$	z 2 T
{2}	4.	$(\exists x)\psi$	3 EG $[z\|x]$
{2}	5.	$\varphi \wedge (\exists x)\psi$	2, 4 T
{1}	6.	$\varphi \wedge (\exists x)\psi$	1, 2, 5 EP $[z\|x]$
	7.	$(\exists x)(\varphi \wedge \psi) \rightarrow [\varphi \wedge (\exists x)\psi]$	1, 6 CP

On line 2, we have the premise for EP. Because x is not free in φ, φ does not change if z is substituted for each free occurrence of x. Then, on line 5, since z, the EP variable, is not free, we may apply Rule EP to change premise 2 to premise 1.

{8}	8.	$\varphi \wedge (\exists x)\psi$	P (for CP)
{8}	9.	$(\exists x)\psi$	8 T
{10}	10.	$\psi[z\|x]$	z P (for EP)
{8, 10}	11.	$\varphi \wedge \psi[z\|x]$	z, 8, 10 T
{8, 10}	12.	$(\exists x)(\varphi \wedge \psi)$	11 EG $[z\|x]$
{8}	13.	$(\exists x)(\varphi \wedge \psi)$	8, 10, 12 EP $[z\|x]$
	14.	$[\varphi \wedge (\exists x)\psi] \rightarrow (\exists x)(\varphi \wedge \psi)$	8, 13 CP
	15.	$(\exists x)(\varphi \wedge \psi) \leftrightarrow [\varphi \wedge (\exists x)\psi]$	7, 14 T

Notice that if x were free in φ, the Rule EG would not have been used correctly on line 12. (Why?)

(c) {1}	1.	$(\forall x)(\varphi \vee \psi)$	P (for CP)
{2}	2.	$\neg\varphi$	P (for CP)
{1}	3.	$\varphi \vee \psi$	1 US $[x\|x]$
{1, 2}	4.	ψ	2, 3 T
{1, 2}	5.	$(\forall x)\psi$	4 UG
{1}	6.	$\neg\varphi \rightarrow (\forall x)\psi$	2, 5 CP
{1}	7.	$\varphi \vee (\forall x)\psi$	6 T
	8.	$(\forall x)(\varphi \vee \psi) \rightarrow [\varphi \vee (\forall x)\psi]$	1, 7 CP
{9}	9.	$\varphi \vee (\forall x)\psi$	P (for CP)
{10}	10.	$\neg(\varphi \vee \psi)$	x P (for IP)
{10}	11.	$\neg\varphi \wedge \neg\psi$	x 10 T
{9, 10}	12.	$(\forall x)\psi$	9, 11 T
{9, 10}	13.	ψ	x 12 US $[x\|x]$
{9, 10}	14.	$\psi \wedge \neg\psi$	x 11, 13 T
{9}	15.	$\varphi \vee \psi$	10, 14 IP

{9}	16.	$(\forall x)(\varphi \vee \psi)$	15 *UG*
	17.	$[\varphi \vee (\forall x)\psi] \rightarrow (\forall x)(\varphi \vee \psi)$	9, 16 *CP*
	18.	$(\forall x)(\varphi \vee \psi) \leftrightarrow [\varphi \vee (\forall x)\psi]$	8, 17 *T*

In the first part of the proof, we are proving a disjunction, $\varphi \vee (\forall x)\psi$. Because $[\varphi \vee (\forall x)\psi] \Leftrightarrow [\neg\varphi \rightarrow (\forall x)\psi]$, we took $\neg\varphi$ as a premise and then derived $(\forall x)\psi$. In the second part, we use an indirect proof. (Where in this proof did we use the fact that x is not free in φ?)

(f) {1}	1.	$(\exists x)(\varphi \rightarrow \psi)$	*P* (for *CP*)
{2}	2.	φ	*P* (for *CP*)
{3}	3.	$\varphi \rightarrow \psi[z\|x]$	z *P* (for *EP*)
{2, 3}	4.	$\psi[z\|x]$	z 2, 3 *T*
{2, 3}	5.	$(\exists x)\psi$	4 *EG* $[z\|x]$
{3}	7.	$\varphi \rightarrow (\exists x)\psi$	2, 5 *CP*
{1}	8.	$\varphi \rightarrow (\exists x)\psi$	1, 3, 7 *EP* $[z\|x]$
	9.	$(\exists x)(\varphi \rightarrow \psi) \rightarrow [\varphi \rightarrow (\exists x)\psi]$	1, 8 *CP*
{10}	10.	$\varphi \rightarrow (\exists x)\psi$	*P* (for *CP*)
{11}	11.	$\neg(\exists x)(\varphi \rightarrow \psi)$	*P* (for *IP*)
{11}	12.	$(\forall x)(\varphi \wedge \neg\psi)$	11 *LT*
{11}	13.	$\varphi \wedge \neg\psi[z\|x]$	12 *US* $[z\|x]$
{10, 11}	14.	$(\exists x)\psi$	10, 13 *T*
{15}	15.	$\psi[z\|x]$	z *P* (for *EP*)
{11, 15}	16.	$\psi[z\|x] \wedge \neg\psi[z\|x]$	z 13, 15 *T*
{11, 15}	17.	$P \wedge \neg P$	16 *T*
{10, 11}	18.	$P \wedge \neg P$	14, 15, 17 *EP* $[z\|x]$
{10}	19.	$(\exists x)(\varphi \rightarrow \psi)$	11, 18 *IP*
	20.	$[\varphi \rightarrow (\exists x)\psi] \rightarrow (\exists x)(\varphi \rightarrow \psi)$	10, 19 *CP*
	21.	$(\exists x)(\varphi \rightarrow \psi) \leftrightarrow [\varphi \rightarrow (\exists x)\psi]$	9, 20 *T*

(Answer the following questions about this proof: (i) Where did we use the fact that x is not free in φ? (ii) What logical theorem was used on line 12? (iii) What is the purpose of line 17?).

(g) {1}	1.	$(\forall x)(\psi \rightarrow \varphi)$	*P* (for *CP*)
{2}	2.	$(\exists x)\psi$	*P* (for *CP*)
{3}	3.	$\psi[z\|x]$	z *P* (for *EP*)
{1}	4.	$\psi[z\|x] \rightarrow \varphi$	1 *US* $[z\|x]$
{1, 3}	5.	φ	3, 4 *T*
{1, 2}	6.	φ	2, 3, 5 *EP* $[z\|x]$
{1}	7.	$(\exists x)\psi \rightarrow \varphi$	2, 6 *CP*
	8.	$(\forall x)(\psi \rightarrow \varphi) \rightarrow [(\exists x)\psi \rightarrow \varphi]$	1, 7 *CP*
{9}	9.	$(\exists x)\psi \rightarrow \varphi$	*P* (for *CP*)
{10}	10.	$\psi \wedge \neg\varphi$	*P* (for *IP*)
{10}	11.	$\neg\varphi$	10 *T*
{9, 10}	12.	$\neg(\exists x)\psi$	9, 11 *T*
{9, 10}	13.	$(\forall x)\neg\psi$	12 *LT*

{9, 10}	14.	$\neg\psi$	13 *US* $[x\vert x]$
{9, 10}	15.	$\psi \wedge \neg\psi$	10, 14 *T*
{9}	16.	$\neg(\psi \wedge \neg\varphi)$	10, 15 *IP*
{9}	17.	$\psi \to \varphi$	16 *T*
{9}	18.	$(\forall x)(\psi \to \varphi)$	17 *UG*
	19.	$[(\exists x)\psi \to \varphi] \to (\forall x)(\psi \to \varphi)$	9, 18 *CP*
	20.	$(\forall x)(\psi \to \varphi) \leftrightarrow [(\exists x)\psi \to \varphi]$	8, 19 *T*

(Answer the following questions about this proof: (i) On line 6, could we have used Rule *CP* instead of Rule *EP* and then use Rule *EP* on line 7? That is, can we interchange lines 6 and 7? (ii) Why did we choose this particular premise on line 10? (iii) Where in the proof did we use the fact that x is not free in φ?) ■

Example 10.3

Give a formal proof of each of the following:

(a) $(\exists x)(Px \wedge (\forall y)Rxy) \vdash (\forall y)(\exists x)(Px \wedge Rxy)$

(b) $\vdash [(\forall x)Qx \to \neg(\exists y)(\exists z)Ryz] \leftrightarrow (\exists x)(\forall y)(\forall z)(Qz \to \neg Ryz)$

Solution:

(a)

{1}	1.	$(\exists x)(Px \wedge (\forall y)Rxy)$	*P*
{2}	2.	$Pz \wedge (\forall y)Rzy$	z *P* (for *EP*)
{2}	3.	$(\forall y)(Pz \wedge Rzy)$	z 2 *LT*
{2}	4.	$Pz \wedge Rzy$	z 3 *US* $[y\vert y]$
{2}	5.	$(\exists x)(Px \wedge Rxy)$	4 *EG* $[z\vert x]$
{2}	6.	$(\forall y)(\exists x)(Px \wedge Rxy)$	5 *UG*
{1}	7.	$(\forall y)(\exists x)(Px \wedge Rxy)$	1, 2, 6 *EP* $[z\vert x]$

(b)

1.	$[(\forall x)Qx \to \neg(\exists y)(\exists z)Ryz] \leftrightarrow$ $[(\forall x)(Qx \to (\forall y)\neg(\exists z)Ryz]$	*LE* (1)
2.	$[(\forall x)Qx \to \neg(\exists y)(\exists z)Ryz] \leftrightarrow$ $[(\forall x)Qx \to (\forall y)(\forall z)\neg Ryz]$	1 *LE* (1)
3.	$[(\forall x)(Qx \to \neg(\exists y)(\exists z)Ryz] \leftrightarrow$ $(\exists x)(Qx \to (\forall y)(\forall z)\neg Ryz)$	2 *LE* (19)
4.	$[(\forall x)(Qx \to \neg(\exists y)(\exists z)Ryz] \leftrightarrow$ $(\exists x)(\forall y)(Qx \to (\forall z)\neg Ryz)$	3 *LE* (16)
5.	$[(\forall x)Qx \to \neg(\exists y)(\exists z)(Ryz] \leftrightarrow$ $(\exists x)(\forall y)(\forall z)(Qx \to \neg Ryz)$	4 *LE* (16)

The number in parentheses at the end of each line of the proof of (b) is the number of the logical theorem in the list in Section 10.6 that is used.

10.5 *PRENEX NORMAL FORM

It is sometimes convenient to have a standard or normal form for formulas in the Predicate Calculus. One such form has all the quantifiers in front of the formula and is called *prenex normal form*. For example, each of the following three formulas are in prenex normal form:

(1) $(\forall x)(\exists y)(Px \wedge Qy \rightarrow \neg Rxy)$

(2) $(\forall x)(\forall y)(\exists z)(Px \rightarrow (\neg Qy \rightarrow Ruz \vee Suz))$

(3) $(\exists x)(\forall y)(\exists z)(\exists u)\neg(Pxu \wedge Qyz \wedge Rxyz)$

However, the following three formulas are *not* in prenex normal form:

(4) $(\forall x)[(\exists y)Py \rightarrow Qx]$

(5) $(\exists x)(\forall y)(Py \rightarrow (\exists z)Qxz)$

(6) $(\exists z)[(\forall y)Py \rightarrow (Qx \rightarrow Rxy)]$

A formula is in prenex normal form if it has the following structure:

(7) $(Q_1x_1)(Q_2x_2)\ \ldots\ (Q_nx_n)\varphi,$

where each Q_i is either $\exists$ or $\forall$ and the formula φ contains no quantifiers. The scope of each of the quantifiers Q_i includes all of the formula φ. (Notice that in formula (4) the scope of $(\exists y)$ does not include Qx; in (5), the scope of $(\exists z)$ does not include Py; and in (6), the scope of $(\forall y)$ does not include $(Qx \rightarrow Rxy)$. Thus, none of the formulas (4), (5), or (6) are in prenex normal form.)

Theorem 10.7 *For each formula φ there is a formula ψ such that*

(i) *φ and ψ have the same free variables.*

(ii) *ψ is in prenex normal form.*

(iii) *φ is logically equivalent to ψ ($\vdash \varphi \leftrightarrow \psi$).*

(ψ is called a prenex normal form of φ.)

We shall delay the proof of this theorem until later. (See Exercise 15 E(23).) For now, we shall indicate with several examples how to find the formula ψ. We shall use logical theorems (1) through (6) if there is a negation sign in front of a quantifier, (7) and (8) to change the name of bound variables, and (12) through (19) to move quantifiers to the front. (See Section 10.6.)

Example 10.4

For each of the formulas φ given below, find a formula ψ that is a prenex normal form of φ (that is, ψ has the same free variables as φ; ψ is in prenex normal form; and ψ and φ are logically equivalent):

(a) $\neg(\exists x)[Ax \wedge (\forall y)(By \rightarrow Lxy)]$

(b) $(\forall x)(Px \rightarrow Qxy) \rightarrow [(\exists y)Py \rightarrow (\exists z)Qyz]$

(c) $(\exists x)(Rxy \rightarrow (\exists x)(Px \vee \neg(\exists y)Qxy)$

Solution: We shall construct a prenex normal form of each of the preceding formulas by giving a sequence of formulas, the first being the given formula, the last being its prenex normal form, and each formula in the sequence being logically equivalent to the preceding one. The number in parentheses following each formula is the number of the logical theorem (in the list at the end of the chapter) that is used to obtain that formula from the preceding one.

(a)

$$\neg(\exists x)[Ax \wedge (\forall y)(By \rightarrow Lxy)]$$
$$(\forall x)\neg[Ax \wedge (\forall y)(Py \rightarrow Lxy)] \qquad (1)$$
$$(\forall x)\neg(\forall y)[Ax \wedge (Py \rightarrow Lxy)] \qquad (12)$$
$$(\forall x)(\exists y)\neg(Ax \wedge (Py \rightarrow Lxy)] \qquad (2)$$

(A prenex normal form for a formula is not unique because, for example,

$$(\forall x)(\exists y)[Ax \rightarrow (Py \wedge \neg(Lxy)]$$

is also a prenex normal form for (a).)

(b)

$$(\forall x)(Px \rightarrow Qxy) \rightarrow [(\exists y)Py \rightarrow (\exists z)Qyz]$$
$$(\forall x)(Px \rightarrow Qxy) \rightarrow [(\exists u)Pu \rightarrow (\exists z)Qyz] \qquad (8)$$
$$(\forall x)(Px \rightarrow Qxy) \rightarrow (\forall u)[Pu \rightarrow (\exists z)Qyz] \qquad (18)$$
$$(\forall x)(Px \rightarrow Qxy) \rightarrow (\forall u)(\exists z)(Pu \rightarrow Qyz) \qquad (17)$$
$$(\exists x)[(Px \rightarrow Qxy) \rightarrow (\forall u)(\exists z)(Pu \rightarrow Qyz)] \qquad (19)$$
$$(\exists x)(\forall u)[(Px \rightarrow Qxy) \rightarrow (\exists z)(Pu \rightarrow Qyz)] \qquad (16)$$
$$(\exists x)(\forall u)(\exists z)[(Px \rightarrow Qxy) \rightarrow (Pu \rightarrow Qyz)] \qquad (17)$$

Notice that we had to change the name of the bound variable y on line 2 because, otherwise, logical theorem (18) would not be applicable on line 3 and (16) would not be applicable on line 6. (Why?) Also notice that y has two free occurrences on the first line, and it has the same two free occurrences on the last line.

(c)

$$(\exists x)Rxy \rightarrow (\exists x)(Px \vee \neg(\exists y)Qxy)$$
$$(\exists x)Rxy \rightarrow (\exists z)(Pz \vee \neg(\exists y)Qzy) \qquad (8)$$
$$(\exists x)Rxy \rightarrow (\exists z)(Pz \vee \neg(\exists u)Qzu) \qquad (8)$$
$$(\exists x)Rxy \rightarrow (\exists z)(Pz \vee (\forall u)\neg Qzu) \qquad (1)$$
$$(\exists x)Rxy \rightarrow (\exists z)(\forall u)(Pz \vee \neg Qzu) \qquad (14)$$

	$(\forall x)[Rxy \to (\exists z)(\forall u)(Pz \vee \neg Qzu)]$	(18)
	$(\forall x)(\exists z)[Rxy \to (\forall u)(Pz \vee \neg Qzu)]$	(17)
(*)	$(\forall x)(\exists z)(\forall u)[Rxy \to (Pz \vee \neg Qzu)]$	(16)

(What was the purpose of line 3 in the preceding sequence? That is, why did we change the name of the bound variable y in $\neg(\exists y)Qxy$?)

Alternatively, the last three lines could be replaced by the following:

	$(\exists z)[(\exists x)Rxy \to (\forall u)(Pz \vee \neg Qzu)]$	(17)
	$(\exists z)(\forall u)[(\exists x)Rxy \to (Pz \vee \neg Qzu)]$	(16)
(**)	$(\exists z)(\forall u)(\forall x)[Rxy \to (Pz \vee \neg Qzu)]$	(18)

Both (*) and (**) are prenex normal forms of (c).

10.6 USEFUL LOGICAL THEOREMS

Note that φ and ψ denote arbitrary formulas, and x and y denote arbitrary variables.

1. $\neg(\exists x)\varphi \leftrightarrow (\forall x)\neg\varphi$
2. $\neg(\forall x)\varphi \leftrightarrow (\exists x)\neg\varphi$
3. $\neg(\exists x)\neg\varphi \leftrightarrow (\forall x)\varphi$
4. $\neg(\forall x)\neg\varphi \leftrightarrow (\exists x)\varphi$
5. $\neg(\forall x)(\varphi \to \psi) \leftrightarrow (\exists x)(\varphi \wedge \neg\psi)$
6. $\neg(\exists x)(\varphi \wedge \psi) \leftrightarrow (\forall x)(\varphi \to \neg\psi)$

In (7) and (8), we assume that y is not free in φ and that the substitution of y for x in φ is valid.

7. $(\forall x)\varphi \leftrightarrow (\forall y)\varphi[y|x]$
8. $(\exists x)\varphi \leftrightarrow (\exists y)\varphi[y|x]$
9. $(\forall x)(\forall y)\varphi \leftrightarrow (\forall y)(\forall x)\varphi$
10. $(\exists x)(\exists y)\varphi \leftrightarrow (\exists y)(\exists x)\varphi$
11. $(\exists x)(\forall y)\varphi \to (\forall y)(\exists x)\varphi$ (but not conversely)

In (12) through (19), we assume that x is not free in φ.

12. $(\forall x)(\varphi \wedge \psi) \leftrightarrow (\varphi \wedge (\forall x)\psi)$
13. $(\exists x)(\varphi \wedge \psi) \leftrightarrow (\varphi \wedge (\exists x)\psi)$
14. $(\forall x)(\varphi \vee \psi) \leftrightarrow (\varphi \vee (\forall x)\psi)$

15. $(\exists x)(\varphi \vee \psi) \leftrightarrow (\varphi \vee (\exists x)\psi)$
16. $(\forall x)(\varphi \rightarrow \psi) \leftrightarrow (\varphi \rightarrow (\forall x)\psi)$
17. $(\exists x)(\varphi \rightarrow \psi) \leftrightarrow (\varphi \rightarrow (\exists x)\psi)$
18. $(\forall x)(\psi \rightarrow \varphi) \leftrightarrow ((\exists x)\psi \rightarrow \varphi)$
19. $(\exists x)(\psi \rightarrow \varphi) \leftrightarrow ((\forall x)\psi \rightarrow \varphi)$

EXERCISES 10

A. Suppose that φ and ψ are formulas and x is any variable that is not free in φ. Show that each of the following formulas is a logical theorem by giving a formal proof of each from the empty set of premises. (You may *not* use any of the logical theorems (12) through (19) in your proof. This exercise completes the proof of Theorem 10.6.) (Answer to 1.)

(1) $(\forall x)(\varphi \wedge \psi) \leftrightarrow [\varphi \wedge (\forall x)\psi]$
(2) $(\exists x)(\varphi \vee \psi) \leftrightarrow [\varphi \vee (\exists x)\psi]$
(3) $(\forall x)(\varphi \rightarrow \psi) \leftrightarrow [\varphi \rightarrow (\forall x)\psi]$
(4) $(\exists x)(\psi \rightarrow \varphi) \leftrightarrow [(\forall x)\psi \rightarrow \varphi]$

B. For each of the following formulas, determine whether it is a logical theorem. If it is a logical theorem, give a formal proof (from the empty set of premises); if not, give a counter-example. (You may use any of the logical theorems (1) through (19) on the list of logical theorems in your proof.) (Answers to 1, 3, 8, 12, 17, and 22.)

(1) $(\exists x)(Px \wedge Qx) \rightarrow [(\exists y)Py \wedge (\exists z)Qz]$
(2) $[(\forall x)Px \vee (\forall y)Qy] \rightarrow (\forall z)(Pz \vee Qz)$
(3) $[(\forall x)Px \rightarrow (\forall x)Qx] \rightarrow (\forall x)(Px \rightarrow Qx)$
(4) $(\forall x)(Px \leftrightarrow Qx) \rightarrow [(\forall x)Px \leftrightarrow (\forall x)Qx]$
(5) $[(\exists x)(Px \wedge Qx) \wedge (\exists x)(\neg Px \wedge Rx)] \rightarrow (\exists x)(Qx \wedge Rx)$
(6) $(\exists x)((\exists y)Py \rightarrow Qx) \rightarrow (\forall x)(Px \rightarrow (\exists y)Qy)$
(7) $[(\forall x)Px \rightarrow (\exists y)Qy] \rightarrow (\forall x)(\exists y)(Px \rightarrow Qy)$
(8) $(\forall x)(\exists y)(Px \wedge Qy) \rightarrow (\exists y)(\forall x)(Px \wedge Qy)$
(9) $(\exists y)(\forall x)(Px \wedge Qy) \rightarrow (\forall x)(\exists y)(Px \wedge Qy)$
(10) $(\forall x)(\forall y)(\forall z)(Rxy \wedge Ryz \rightarrow Rxz)$
(11) $(\forall x)(Px \leftrightarrow \neg Px) \rightarrow (\exists x)Qx$
(12) $[(\forall x)Px \rightarrow (\exists y)\neg Qy] \rightarrow [(\forall x)Qx \rightarrow (\exists y)\neg Py]$
(13) $(\forall x)(Px \rightarrow Qx) \wedge (\exists x)(Qx \rightarrow Pa) \rightarrow [(\exists x)Px \rightarrow Pa]$
(14) $(\forall x)[Px \rightarrow (\forall y)(Qy \rightarrow Rxy)] \wedge Qa \wedge \neg Rba \rightarrow \neg Pb$
(15) $(\forall x)(\exists y)Rxy \vee \neg(\forall x)(\exists y)Ryx$
(16) $(\forall x)(\exists y)(Rxy \vee \neg Ryx)$
(17) $(\forall x)(\exists y)Rxy \vee (\forall x)(\exists y)\neg Rxy$
(18) $(\forall x)(\exists y)Rxy \vee (\exists y)(\forall x)\neg Ryx$

(19) $(\exists x)[Px \wedge (\forall y)(Qy \to Rxy)] \wedge (\forall x)(\forall y)(Px \wedge Rxy \to Sy) \wedge (\forall x)(Sx \to Qx) \to (\forall x)(Qx \leftrightarrow Sx)$
(20) $(\forall x)(\forall y)(Px \wedge Qy \to Rxy) \wedge (\forall x)(\forall y)(Px \wedge Rxy \to Sy) \to (\forall x)(Qx \to Sx)$
(21) $(\forall x)(Px \to (\exists y)Qxy) \wedge (\exists x)[Px \wedge (\forall y)(Qxy \to Rxy)] \to (\exists x)(\exists y)Rxy$
(22) $[(\forall x)Px \to \neg(\forall x)Qx] \leftrightarrow (\exists x)(\forall y)(Px \to \neg Qy)$
(23) $[\neg(\exists x)(Px \wedge Qx) \vee (\exists x)Rx] \leftrightarrow (\forall x)(\forall y)(\neg Px \vee \neg Qx \vee Ry)$
(24) $(\forall x)(Fx \to \neg(\exists y)Gxy) \leftrightarrow (\forall x)(\forall y)(Fx \to \neg Gxy)$
(25) $\neg(\exists x)[(\exists y)Py \wedge Qx] \leftrightarrow (\forall x)(\forall y)(Py \to \neg Qx)$

C. For each of the following sentences,

(i) Translate into the Predicate Calculus, using the indicated predicate symbols and constants.
(ii) Negate each sentence.
(iii) Find a formula logically equivalent to the negation in which no negation sign precedes either a quantifier or another logical connective.
(iv) Translate the sentence found in part (c) back into idiomatic English.

(Example: All birds lay eggs. (Bx, Lx)

(i) $(\forall x)(Bx \to Lx)$
(ii) $\neg(\forall x)(Bx \to Lx)$
(iii) $(\exists x)(Bx \wedge \neg Lx)$
(iv) Some birds do not lay eggs.)

(Answers to 1, 5, 9, 13, 17, 20, and 23.)

(1) Some animals are pets. (Ax, Px)
(2) All students who have a college degree will get a job. (Sx, Dx, Jx)
(3) If Oscar graduates from high school, then he will go to some college. (o, Hx, Cx, Gxy, where Hx : x graduates from high school; Cx : x is a college; and Gxy : x goes to y.)
(4) All politicians and some attorneys are liars. (Px, Ax, Lx)
(5) If all rational numbers are real numbers, then all integers are real numbers. (Qx, Rx, Ix)
(6) Lions and tigers are prey animals. (Lx, Tx, Px)
(7) Only students who like physics like mathematics. (Sx, Px, Mx)
(8) Among the students in college, all those who like physics also like mathematics. (Sx, Cx, Px, Mx)
(9) If someone is older than George, then everyone is older than George. (Px, Oxy, e)
(10) Every student likes some professor. (Sx, Px, Lxy)
(11) Some student likes every professor. (Sx, Px, Lxy)
(12) Some student likes some professor. (Sx, Px, Lxy)
(13) All students like all professors. (Sx, Px, Lxy)

(14) If some professor is older than all students, then there is a student who is not older than some professor. (Px, Sx, Oxy)
(15) Either some student is not older than all professors or all professors are older than some student. (Sx, Px, Oxy)
(16) Given any two real numbers, there is a real number between them. (Rx, $Bxyz$)
(17) There are two integers that have at least one integer between them. (Ix, $Bxyz$)

In problems (18) through (24), use standard mathematical symbols. (See Exercises 5B.) Assume that the domain of all variables is the set of integers.

(18) Every integer is either positive or negative.
(19) Some integer is a multiple of seven.
(20) For each pair of distinct integers, one is even and the other is odd. (Use Ex : x is even; and Ox: x is odd. You may assume Ex is the same as $\neg Ox$.)
(21) Every even integer is a multiple of some odd integer. (Ex, Ox)
(22) There are even and odd integers such that one is a multiple of the other. (Ex, Ox)
(23) The sum of an odd and an even integer is odd. (Ex, Ox)
(24) The product of two even integers may be either a multiple of four or a multiple of six. (Ex)

In problems (25) and (26), assume all variables (including ϵ and δ) range over the set of real numbers and f is a function. In (25), we give the definition of "the function f is continuous on the interval $[a, b]$" and (26), "f is uniformly continuous on $[a, b]$."

*(25) For each positive ϵ and for each x in $[a, b]$, there is a δ such that for all y in $[a, b]$, if $|x - y| < \delta$, then $|f(x) - f(y)| < \epsilon$.
*(26) For each positive ϵ, there is a δ such that for all x and y in $[a, b]$ if $|x - y| < \delta$, then $|f(x) - f(y)| < \epsilon$.

D. Give a formal proof of each of the following valid arguments. (Answers to 1, 5, 9, 12, and 16.)

(1) **Premises:** $(\forall x)(Ax \to Bx)$, $\neg(\exists x)(Ax \wedge Bx)$
Conclusion: $(\forall x)\neg Ax$

(2) **Premises:** $(\forall x)(Ax \to Bx \vee Cx)$, $\neg(\exists x)(Bx \wedge Cx)$, $(\exists x)(Ax \wedge Cx)$
Conclusion: $(\exists x)(Ax \wedge \neg Bx)$

(3) **Premises:** Pa, $\neg(\exists x)(Px \wedge Qx)$, $(\forall x)(\neg Rxa \to Qx)$
Conclusion: $(\exists x)Rxx$

(4) **Premise:** $\neg(\exists x)(\forall y)(Py \wedge \neg Qx)$
Conclusion: $(\forall x)Px \to (\forall x)Qx$

(5) **Premise:** $\neg(\exists x)(Px \vee Qx) \vee (\exists x)Rx$
Conclusion: $(\forall x)(\exists y)(Px \vee Qx \to Ry)$

(6) **Premises:** $(\forall x)(Ax \to Bx)$, $(\exists x)[Cx \wedge \neg(\exists y)(By \wedge Dxy)]$
Conclusion: $(\exists x)[Cx \wedge \neg(\exists y)(Ay \wedge Dxy)]$

(7) **Premises:** $(\forall x)[\neg Qx \to (\exists y)(Rxy \vee Qx)]$, $\neg(\forall x)(Px \to Qx)$
Conclusion: $(\exists x)(\exists y)(Px \wedge Rxy)$

(8) **Premises:** $(\exists x)Px$, Qa, $\neg(\exists x)(Px \wedge Rxa)$,
$(\forall x)[Px \to (\forall y)(Sy \wedge Qy \to Rxy)]$
Conclusion: $\neg Sa$

(9) **Premises:** $(\exists x)(Ax \wedge Bx \wedge Cx)$, $\neg(\exists x)(Ax \wedge Bx \wedge Dx)$
Conclusion: $(\exists x)(Bx \wedge Cx \wedge \neg Dx)$

(10) **Premises:** $(\exists x)[Fx \wedge (\forall y)(Gy \to Hxy)]$, $\neg(\exists x)(\exists y)(Fx \wedge Hxy \wedge Iy)$
Conclusion: $\neg(\exists x)(Gx \wedge Ix)$

(11) **Premises:** $(\forall x)(Fx \to Gx) \to (\exists x)(Hx \wedge \neg Ix)$, $(\forall x)(Hx \to Ix)$
Conclusion: $(\exists x)(Fx \wedge \neg Gx)$

(12) **Premises:** $(\forall x)(Jx \to Kx) \to (\exists x)Lx \vee (\exists x)Mx$, $\neg(\exists x)Mx$
Conclusion: $(\exists x)(Lx \vee (\exists x)(Jx \wedge \neg Kx)$

(13) **Premises:** $(\forall x)(\forall y)[Px \wedge (\exists z)(Qz \wedge Ryz) \to Sxy]$,
$(\forall x)[(\exists y)(Py \wedge Syx) \to Tx]$, $\neg Ta$
Conclusion: $(\exists x)Px \to (\forall y)(Qy \to \neg Ray)$

(14) **Premises:** $(\forall x)(Ax \to Bx)$, $(\forall x)(\neg Cxx \wedge Bx \to Ax)$,
$(\exists x)(Dx \wedge \neg(\exists y)(By \wedge Cxy))$, $(\forall x)[Ax \to (\forall y)(Ay \to Cxy)]$
Conclusion: $(\forall x)(Dx \to Ax) \leftrightarrow \neg(\forall x)(Bx \to Cxx)$

(15) **Premises:** $(\forall x)[Jx \wedge Kx \to (Lx \leftrightarrow \neg(\exists y)(Myx \wedge \neg Ny))]$,
$\neg(\forall x)(Kx \to Jx)$, $(\forall x)(Jx \wedge Kx \wedge Lx \to Na)$
Conclusion: $(\forall x)[Jx \wedge Kx \wedge (\forall y)(Myx \to Ny) \to Na]$

(16) **Premises:** $(\forall x)[Px \wedge (\exists y)(Qy \wedge Rxy) \to (\exists z)(Sz \wedge Txz)]$,
$(\forall x)(\forall y)(Px \to Rxy)$, $(\exists x)Px$, $(\forall x)(\forall y)(Sy \to \neg Txy)$
Conclusion: $(\exists x)(\forall y)(Qy \to \neg Rxy)$

(17) **Premises:** $(\exists x)(Px \wedge Rxa)$, Sa, $(\forall x)[Px \wedge \neg(\exists y)(Qy \wedge Rxy) \to$
$\neg(\exists z)(Sz \wedge Rxz)]$
Conclusion: $(\exists x)(\exists y)(Px \wedge Qy \wedge Rxy)$

E. Translate each of the following arguments into the language of the Predicate Calculus, using the given predicate symbols and constants. If the argument is valid, give a formal proof; otherwise, give a counter-example. (Answers to 1, 4, 8, and 11.)

(1) All geniuses are intelligent. It is not the case that no student is intelligent. Therefore, some students are geniuses. (*Gx, Ix, Sx*)

(2) It is not the case that vegetables are fruit. All peas are vegetables. Therefore, it is not the case that peas are fruit. (*Vx, Fx, Px*)

(3) It is not the case that there is a snake that flies. It is not the case that all birds fly. Therefore, some birds are not snakes. (Sx, Fx, Bx)
(4) It is not the case that some vegetables are fruit. All peas are vegetables. Therefore, it is not the case that some peas are fruit. (Vx, Fx, Px)
(5) It is not the case that all men and women are bad drivers. Therefore, there is a bad driver who is neither a man nor a woman. (Mx, Wx, Bx)
(6) It is not the case that all men are bad drivers, and it is not the case that all women are bad drivers. Therefore, there is a bad driver who is neither a man nor a woman. (Mx, Wx, Bx)
(7) It is not the case that there is a man or a woman who is a bad driver. Therefore, no man or woman is a bad driver. (Mx, Wx, Bx)
(8) Every student who attends some college is happy. Oscar is a student who is not happy. Therefore, there does not exist a college. (Sx, Cx, Hx, Axy, o)
(9) Some teacher who likes every student is promoted. There is a teacher who is not promoted. Therefore, there is a teacher who does not like some student. (Tx, Sx, Px, Lxy)
(10) All planes that are inspected by every FAA official are safe. Not all planes are safe. There is a plane that was not inspected by some FAA official. (Px, Ox, Sx, Ixy)
(11) Every student pilot who flies to some city is accompanied by an instructor. Oscar is a student pilot who is not accompanied by any instructor. Therefore, Oscar does not fly to any city. (Sx, Cx, Ix, Fxy, Axy, o)
(12) Lions attack anyone who touches their cubs. Anyone who is attacked by a lion is killed. George is not killed. Therefore, if there is a lion, then George did not touch any cubs. (Lx, Cx, Kx, Txy, Axy, e)

***F.** Find a prenex normal form (as described in Theorem 10.7) for each of the following formulas. (Answers to 1, 4, 8, and 11.)

(1) $(\exists x)Px \rightarrow (\exists x)Qx$
(2) $\neg(\exists x)(Fx \vee Gx) \vee (\exists x)Hx$
(3) $(\forall x)(Fx \rightarrow Gxy) \rightarrow \neg(\exists y)Hy$
(4) $(\exists x)(Ax \wedge Bxy) \wedge (\exists y)(Cx \wedge Dxy)$
(5) $(\forall x)(Ax \rightarrow Bxy) \vee (\exists y)(Ay \wedge \neg(\exists z)Byz)$
(6) $(\exists x)(Px \wedge Rxy) \rightarrow (\forall y)(Qy \rightarrow Rxy)$
(7) $\neg(\forall x)(Px \rightarrow Qx) \wedge \neg(\exists y)(\exists z)(Pz \wedge Rxz \wedge Syx)$
(8) $(\forall x)(\exists y)Rxy \rightarrow (\exists y)(\forall x)Rxy$
(9) $\neg(\forall x)(\exists y)(Rxy \wedge Sxy) \vee \neg(\exists y)(\forall x)(Rxy \wedge Sxy)$
(10) $(\forall x)(Px \rightarrow (\exists y)(Qy \wedge Rxy)) \rightarrow \neg(\exists x)(\exists y)Rxy$
(11) $(\forall x)Jx \leftrightarrow (\exists x)Kx$
(12) $(\forall x)(Px \vee Qxy) \leftrightarrow (\exists x)(\exists y)(Qxy \wedge \neg Rxy)$
(13) $(\exists x)(Ax \wedge Bxy) \leftrightarrow \neg(\exists y)(Cy \vee Dxy)$
(14) $\neg[(Ex)Px \leftrightarrow (\forall y)Qxy]$

Chapter 11

Logic of Identity

In this chapter, we discuss both the semantics of identity and its formal structure. That is, if identity, =, is added to the language of the Predicate Calculus, then certain rules of inference must also be added if = is to be interpreted as identity (see Section 11.2). We start this chapter with a discussion of the use of identity in the English language.

11.1 IDENTITY IN ENGLISH

In English, we frequently put some form of the verb "to be" between two terms to indicate that they are identical. For example, we say

(1) John is the tallest person on our basketball team

to mean that John and the tallest person on our basketball team are identical – the same person. Thus, if we use the symbol = in our language to mean "identity," and if we let j and t represent John and the tallest person on our basketball team, respectively, then (1) is translated as $j = t$.

Sometimes the verb "to be" may be used in the past tense. For example, the sentence

(2) The best movie last year was King Kong

would be translated as $m = k$, because the best movie last year is said to be identical with King Kong.

On the other hand, if we say

(3) Five is prime,

we do *not* mean that five and prime are identical. Sentence (3) would be translated as $P5$, where the predicate Px is translated as "x is a prime." Similarly, if we say

(4) Lions are mammals,

we do *not* mean that lions and mammals are identical. Rather, (4) would be translated as $(\forall x)(Lx \to Mx)$, or every lion is a mammal.

It is usually fairly obvious when the verb "to be" is to be translated as identity and when it is not. See Exercises 11 A for some more examples.

With the identity symbol in our language, each of the following two expressions would be translated as $t_1 = t_2$, where t_1 and t_2 are two terms:

(5) t_1 and t_2 are (were) identical.

(6) t_1 is (was) identical to t_2.

Sometimes we use the word "equal" to replace "identical."

The expression

(7) t_1 and t_2 are distinct

means the same as

(8) t_1 and t_2 are not identical.

Thus (7) and (8) would each be translated as $\neg(t_1 = t_2)$, which we shall abbreviate as $t_1 \neq t_2$. The sentence "There exist three distinct elements in the set" is translated as

$$(\exists x)(\exists y)(\exists z)(Sx \wedge Sy \wedge Sz \wedge x \neq y \wedge x \neq z \wedge y \neq z).$$

The quantifier "there exists exactly one" can also be expressed in the language of the Predicate Calculus with Identity. Consider the sentence

(9) There is a problem that has exactly one solution.

Let $Px : x$ is a problem and $Sxy : y$ is a solution of x. The expression "exactly one solution" means "at least one solution and at most one solution." Then

(10) x has at least one solution

is translated as

(11) $(\exists y)Sxy$,

and

(12) x has at most one solution

is translated as

(13) $(\forall u)(\forall v)(Sxu \wedge Sxv \to u = v)$.

Therefore, (9) can be translated as

(14) $(\exists x)(Px \wedge (\exists y)Sxy \wedge (\forall u)(\forall v)(Sxu \wedge Sxv \rightarrow u = v))$.

Thus, to show that there is exactly one term with a given property involves two steps. First show that there is at least one such term, and then show that if there were two, say t_1 and t_2, then $t_1 = t_2$.

We shall show later (see Example 11.5) that (14) is logically equivalent to

(15) $(\exists x)[Px \wedge (\exists y)(Sxy \wedge (\forall z)(Sxz \rightarrow z = y))]$.

(We can now prove that $\vdash$ (14) $\rightarrow$ (15), but to prove the converse requires rules for =.) The translation of (15) is

(16) There is a problem that has a solution, and every solution is identical to the first solution.

The translation of (14) is

(17) There is a problem that has a solution, and each pair of solutions are identical.

Sentences (16) and (17) mean the same as (9) in English.

The expression "the only one" is similar to "exactly one." For example, the sentence

(18) Alice is the only one who passed

is translated as

(19) $Pa \wedge (\forall x)(Px \rightarrow x = a)$.

That is, Alice passed, and if anyone passed, that person and Alice are identical.

The sentence

(20) Alice likes only Norbert

is translated as

(21) $Lan \wedge (\forall x)(Lax \rightarrow x = n)$.

On the other hand, the sentence

(22) Only Alice likes Norbert

is translated as

(23) $Lan \wedge (\forall x)(Lxn \rightarrow x = a)$.

The position of the word "only" in a sentence frequently affects the meaning of the sentence.

11.2 RULES GOVERNING IDENTITY

In order for = to be interpreted as identity, we must state as logical theorems and rules the properties we want it to have. The first rule is that everything is identical to itself.

Ia : For each variable x, $(\forall x)(x = x)$ is a logical theorem; that is, $\vdash (\forall x)(x = x)$.

Thus, for example, any formula of the form "$t = t$" for any term t may be written as a line of a proof without any premise numbers and justified by *Ia* and $US[t|x]$.

The second rule is the Rule of Substitution for Identity. Before stating this rule, it is convenient to define another symbol.

Definition 11.1 If x is a variable, t a term, and φ a formula, $\varphi[x, t|x]$ is the formula obtained from φ by substituting t for some free occurrences of x. This substitution is called a *valid substitution* as long as no occurrence of a variable in t is bound in $\varphi[x, t|x]$.

Remember that $\varphi[t|x]$ is the formula obtained from φ by substituting t for *every* free occurrence of x, whereas $\varphi[x, t|x]$ is the formula obtained from φ by substituting t for *some* free occurrences of x.

The Rule of Substitution for Identity is stated as follows:

Ib : If x is a variable, t a term, and φ a formula, then

$$(\forall x)[x = t \rightarrow (\varphi \leftrightarrow \varphi[x, t|x])]$$

is a logical theorem, as long as the substitution of t for x in φ is valid.

To illustrate the use of *Ia* and *Ib*, we shall prove the following derived rules:

Ic : $\vdash (\forall x)(\forall y)(x = y \rightarrow y = x)$

Id : $\vdash (\forall x)(\forall y)(\forall z)(x = y \wedge y = z \rightarrow x = z)$

Proof of Ic:

{1}	1.	$x = y$	x, y *P* (for *CP*)
	2.	$x = x$	*Ia*
	3.	$x = y \rightarrow (x = x \leftrightarrow y = x)$	*Ib*
{1}	4.	$y = x$	x, y 1, 2, 3 *T*
	5.	$x = y \rightarrow y = x$	1, 4 *CP*
	6.	$(\forall x)(\forall y)(x = y \rightarrow y = x)$	5 *UG*(2)

In using the Rules for Identity, we shall suppress the use of Rules *LT* and *US*. Formally, line 2 should have been replaced by two lines:

2a.	$(\forall x)(x = x)$	*LT* (*Ia*)
2b.	$x = x$	2a *US* $[x\|x]$

Similarly for line 3. When we used *Ib* on line 3, the variable is x, the term t is y, and the formula φ is $x = x$. The formula $\varphi[x, t|x]$ is $y = x$. We substituted y for the first occurrence of x in $x = x$.

■

Proof of Id:

{1}	1.	$x = y \wedge y = z$	x, y, z *P* (for *CP*)
	2.	$x = y \rightarrow y = x$	*Ic*
{1}	3.	$y = x$	x, y 1, 2 *T*
{1}	4.	$y = z$	y, z, 1 *T*
	5.	$y = x \rightarrow (y = z \leftrightarrow x = z)$	*Ib*
{1}	6.	$x = z$	x, z 3, 4, 5 *T*
	7.	$x = y \wedge y = z \rightarrow x = z$	1, 6 *CP*
	8.	$(\forall x)(\forall y)(\forall z)(x = y \wedge y = z \rightarrow x = z)$	7 *UG* (3)

On line 2, we used *Ic* without any premise numbers because we had proved that *Ic* is a logical theorem. Then on line 5, where we used *Ib*, the variable is y, the term x, and the formula φ is $y = z$. $\varphi[y, x|y]$ is the formula $x = z$. We substituted x for y in $y = z$.

■

The Rules *Ia*, *b*, *c*, and *d* give us all the standard properties of identity. In the future, when we document a proof in which one of these rules is used, we shall just use the symbol *I* and not specify which part of the Rule *I* we are using.

11.3 PROOFS

The proofs of *Ic* and *Id* above illustrate how the rules for identity are used in proofs. We shall give additional examples in this section.

Example 11.1

Give a formal proof of the following valid argument:

Premises: $(\forall x)(Px \wedge Qx \rightarrow x = a \vee x = b)$, $\neg Ra$, $\neg(\exists x)(Qx \wedge Rx \wedge \neg Px)$

Conclusion: $(\forall x)(Qx \wedge Rx \rightarrow x = b)$

Solution:

{1}	1.	$(\forall x)(Px \wedge Qx \rightarrow x = a \vee x = b)$	P
{2}	2.	$\neg Ra$	P
{3}	3.	$\neg(\exists x)(Qx \wedge Rx \wedge \neg Px)$	P
{4}	4.	$Qx \wedge Rx$	x P (for CP)
{3}	5.	$(\forall x)(Qx \wedge Rx \rightarrow Px)$	3 LT
{3}	6.	$Qx \wedge Rx \rightarrow Px$	5 US $[x\|x]$
{3, 4}	7.	Px	x, 4, 6 T
{1}	8.	$Px \wedge Qx \rightarrow x = a \vee x = b$	1 US $[x\|x]$
{1, 3, 4}	9.	$x = a \vee x = b$	x, 4, 7, 8 T
{10}	10.	$x = a$	x P (for IP)
{2, 10}	11.	$\neg Rx$	x, 2 10 I
{2, 4, 10}	12.	$Rx \wedge \neg Rx$	x, 4, 11 T
{2, 4}	13.	$x \neq a$	x, 10, 12 IP
{1, 2, 3, 4}	14.	$x = b$	x, 9, 13 T
{1, 2, 3}	15.	$Qx \wedge Rx \rightarrow x = b$	4, 14 CP
{1, 2, 3}	16.	$(\forall x)(Qx \wedge Rx \rightarrow x = b)$	15 UG

On line 9, we derived $x = a \vee x = b$. We want to derive $x = b$, so we take $x = a$ as a premise and derive a contradiction. We conclude $x \neq a$; therefore, $x = b$. On line 11, we used the Rule of Substitution for Identity. We assumed $x = a$, line 10. Then, to obtain line 11, we substituted x for a on line 2, obtaining $\neg Rx$. This is an application of Ib.

Example 11.2 Show that the following set of premises is inconsistent by giving a formal proof of a contradiction:

$$Ja, Kb, (\forall x)(Jx \rightarrow Lx), (\forall x)(Kx \rightarrow \neg Lx), a = b$$

Solution:

{1}	1.	Ja	P
{2}	2.	Kb	P
{3}	3.	$(\forall x)(Jx \rightarrow Lx)$	P
{4}	4.	$(\forall x)(Kx \rightarrow \neg Lx)$	P
{5}	5.	$a = b$	P
{3}	6.	$Ja \rightarrow La$	3 US $[a\|x]$
{1, 3}	7.	La	1, 6 T
{4}	8.	$Kb \rightarrow \neg Lb$	4 US $[b\|x]$
{2, 4}	9.	$\neg Lb$	2, 8 T
{2, 4, 5}	10.	$\neg La$	5, 9 I
{1, 2, 3, 4, 5}	11.	$La \wedge \neg La$	7, 10 T

To obtain line 10, we substituted *a* for *b* on line 9. This is another application of *Ib*, the Rule of Substitution for Identity.

11.4 INTERPRETATION STRUCTURES

We have shown how the Rules for Identity are used in proofs, but we have not as yet explained how identity is interpreted in an interpretation structure. As we have seen in earlier chapters, an arbitrary n-place predicate may be interpreted as any set of ordered n-tuples of elements of the universe. This does not apply to $=$, however. In every interpretation structure, $=$ must be interpreted as identity; that is, $a = b$ is true if and only if a and b are the same elements in the universe. Otherwise, the logical theorems given in *Ia, b, c,* and *d* might not be true in every interpretation structure. (If *Ia, b, c,* and *d* are true in an interpretation structure, then $=$ is interpreted as an *equivalence relation*. See Definition 7.3 and Section 13.5.)

Example 11.3

Show that the following argument is not valid by finding a counter-example:

Premises: $(\exists x)(Fx \wedge Gx)$, Fa, $a = b$

Conclusion: Gb

Solution: We first try to prove it. Take $Fy \wedge Gy$ as a premise for *EP*. Because y does not have to be the same as a, there is no way to derive Gb. To show this, we take an interpretation structure with two elements (see Fig. 11.1). (We use the convention that **a** is the interpretation of a; **b**, the interpretation of b; and **c**, the interpretation of c.)

U = {a, c}

b = a

	F	G
a	*T*	*F*
c	*T*	*T*

Figure 11.1

In this interpretation structure, the first premise is true because **Fc** and **Gc** are true; the second premise is true because **Fa** is true; the third premise is true because **a** and **b** are the same (**a** = **b**); but the conclusion is false because **Gb**, which is the same thing as **Ga**, is false.

Example 11.4

Show that the following premises are consistent:

$$(\forall x)(Ff(x) \rightarrow \neg Gf(x)), a \neq b, a = c, f(c) \neq a, Fa \wedge Gb$$

Solution: The universe must have at least two elements because the second premise says $a \neq b$. The third premise says that a and c must have the same interpretation, and the fourth premise implies that $f(c)$ and a must have different interpretations (f is a function symbol). However, because a and c have the same interpretation, it follows that $f(a)$ and a must have different interpretations. We shall try an interpretation structure whose universe has two elements, **a** and **b**, and in which **c** and **a** are identical and **f** permutes **a** and **b**; that is, **f**(**a**) = **b** and **f**(**b**) = **a**. Then we have to find interpretations for F and G so that premises 1 and 5 are true (see Fig. 11.2).

U = {a, b}

c = a

	F	G
a	*T*	*F*
b	*F*	*T*

f(a) = b

f(b) = a

Figure 11.2

It follows now that premises 2 through 5 are true in this interpretation. Let us check premise 1. We have to check that for each **x** ∈ **U**, the following is true:

(1) **Ff**(**x**) → ¬**Gf**(**x**).

Replace **x** by **a**, and, using the definition of **f**, (1) becomes

(2) **Fb** → ¬**Gb**,

which is true because **Fb** is false. If we replace **x** by **b** in (1), we obtain

(3) **Fa** → ¬**Ga**.

Because **Ga** is false, (3) is also true. Thus, all the premises are true in the interpretation given in Figure 11.2, so the premises are consistent.

Remember that in every place a symbol appears in a premise, it has to have the same interpretation in the interpretation structure. Also, in an interpretation structure, a function must be defined on the whole universe, and its range must be a subset of the universe.

Example 11.5

Show that the following two formulas are logically equivalent:

(i) $(\exists x)Ax \wedge (\forall y)(\forall z)(Ay \wedge Az \rightarrow y = z)$

and

(ii) $$(\exists x)[Ax \wedge (\forall y)(Ay \rightarrow x = y)].$$

Solution: Each of these formulas is translated into English as

There is exactly one *x* such that *Ax*.

(See formulas (14) and (15) in the preceding section.)

We shall first give a formal proof of $\vdash$ (i) $\rightarrow$ (ii). This does not require any use of the rules for identity.

{1}	1.	$(\exists x)Ax \wedge (\forall y)(\forall z)(Ay \wedge Az \rightarrow y = z)$	*P* (for *CP*)
{1}	2.	$(\exists x)Ax$	1 *T*
{1}	3.	$(\forall y)(\forall z)(Ay \wedge Az \rightarrow y = z)$	1 *T*
{4}	4.	Ay	*y P* (for *CP*)
{5}	5.	Au	*u P* (for *EP*)
{1}	6.	$Au \wedge Ay \rightarrow u = y$	3 *US* (2) [*u*|*y*][*y*|*z*]
{1, 4, 5}	7.	$u = y$	*u*, *y* 4, 5, 6 *T*
{1, 5}	8.	$Ay \rightarrow u = y$	*u* 4, 7 *CP*
{1, 5}	9.	$(\forall y)(Ay \rightarrow u = y)$	*u* 8 *UG*
{1, 5}	10.	$Au \wedge (\forall y)(Ay \rightarrow u = y)$	*u* 5, 9 *T*
{1, 5}	11.	$(\exists x)[Ax \wedge (\forall y)(Ay \rightarrow x = y)]$	10 *EG* [*u*|*x*]
{1}	12.	$(\exists x)[Ax \wedge (\forall y)(Ay \rightarrow x = y)]$	2, 5, 11 *EP* [*u*|*x*]
	13.	(i) $\rightarrow$ (ii)	1, 12 *CP*

Next we shall prove the converse. In this proof, we do use Rule *I* on line 9.

{1}	1.	$(\exists x)[Ax \wedge (\forall y)(Ay \rightarrow x = y)]$	*P* (for *CP*)
{2}	2.	$Au \wedge (\forall y)(Ay \rightarrow u = y)$	*u P* (for *EP*)
{3}	3.	$Ay \wedge Az$	*y*, *z P* (for *CP*)
{2}	4.	$(\forall y)(Ay \rightarrow u = y)$	*u* 2 *T*
{2}	5.	$Ay \rightarrow u = y$	*u* 4 *US* [*y*|*y*]
{2}	6.	$Az \rightarrow u = z$	*u* 4 *US* [*z*|*y*]
{2, 3}	7.	$u = y$	*u*, *y* 3, 5 *T*
{2, 3}	8.	$u = z$	*u*, *z* 3, 6 *T*
{2, 3}	9.	$y = z$	*y*, *z* 7, 8 *I*
{2}	10.	$Ay \wedge Az \rightarrow y = z$	3, 9 *CP*
{2}	11.	$(\forall y)(\forall z)(Ay \wedge Az \rightarrow y = z)$	10 *UG* (2)
{2}	12.	Au	*u* 2 *T*
{2}	13.	$(\exists x)Ax$	12 *EG* [*u*|*x*]
{2}	14.	$(\exists x)Ax \wedge (\forall y)(\forall z)(Ay \wedge Az \rightarrow y = z)$	11, 13 *T*
{1}	15.	$(\exists x)Ax \wedge (\forall y)(\forall z)(Ay \wedge Az \rightarrow y = z)$	1, 2, 14 *EP* [*u*|*x*]
	16.	(ii) $\rightarrow$ (i)	1, 15 *CP*

11.5 IDENTITY IN MATHEMATICS

In mathematics, we frequently have to deal with formulas that contain both operations and relations. An *operation* is another name for a function. The function symbol for 2-place operations like addition, subtraction, multiplication, and division is normally written between the two arguments rather than in front of them. We write "1 + 2" rather than "+(1, 2)." In our formal language, we shall use a small circle, $\circ$, or an asterisk, $*$, between two terms to indicate a 2-place operation. For example, an interpretation structure for the formula

(1) $(a \circ b) * c = (a * c) \circ (b * c)$

must have interpretations for the two operation symbols, $\circ$ and $*$, and the three constants, *a, b, c*. We shall give two interpretation structures for formula (1). In the first, (1) is true, and in the second, (1) is false. In both structures,

(2) $\mathbf{U}$ = Set of positive integers.

(3) $\mathbf{a} = \mathbf{1}, \mathbf{b} = \mathbf{2}, \mathbf{c} = \mathbf{3}$.

In the first structure,

(4) $\mathbf{x} \circ \mathbf{y} = \mathbf{x} + \mathbf{y}$; $\quad \mathbf{x} * \mathbf{y} = \mathbf{x} \cdot \mathbf{y}$,

and in the second,

(5) $\mathbf{x} \circ \mathbf{y} = \mathbf{x} \cdot \mathbf{y}$; $\quad \mathbf{x} * \mathbf{y} = \mathbf{x} + \mathbf{y}$.

In the first structure, the interpretation of (1) is

(6) $(\mathbf{1} + \mathbf{2}) \cdot \mathbf{3} = (\mathbf{1} \cdot \mathbf{3}) + (\mathbf{2} \cdot \mathbf{3})$,

which is true, because the left side is **9** and so is the right side. In the second structure, the interpretation of (1) is

(7) $(\mathbf{1} \cdot \mathbf{2}) + \mathbf{3} = (\mathbf{1} + \mathbf{3}) \cdot (\mathbf{2} + \mathbf{3})$,

which is false, because the left side of equation (7) is **5** but the right side is **20**.

Note that it would be incorrect, for example, to take $\mathbf{U} = \{\mathbf{1}, \mathbf{2}, \mathbf{3}\}$ and then interpret $\circ$ and $*$ to be $+$ and $\cdot$, respectively. (Why?)

Example 11.6

Show that the following set of premises is inconsistent by giving a formal proof of a contradiction:

$$(\forall x)(\forall y)(x \circ y = x),\ (\forall x)(\forall y)(x \circ y = y),\ (\exists x)(\exists y)(x \neq y).$$

Solution:

{1}	1.	$(\forall x)(\forall y)(x \circ y = x)$	P
{2}	2.	$(\forall x)(\forall y)(x \circ y = y)$	P
{3}	3.	$(\exists x)(\exists y)(x \neq y)$	P
{4}	4.	$(\exists y)(u \neq y)$	u P (for EP)
{5}	5.	$u \neq v$	u, v P (for EP (v))
{1}	6.	$u \circ v = u$	1 US (2) $[u\|x][v\|y]$
{2}	7.	$u \circ v = v$	2 US (2) $[u\|x][v\|y]$
{1, 2}	8.	$u = v$	6, 7 I
{1, 2, 5}	9.	$u = v \wedge u \neq v$	u, v 5, 8 T
{1, 2, 5}	10.	$P \wedge \neg P$	9 T
{1, 2, 4}	11.	$P \wedge \neg P$	4, 5, 10 EP $[v\|y]$
{1, 2, 3}	12.	$P \wedge \neg P$	3, 4, 11 EP $[u\|x]$

(Explain the documentation on lines 6 and 7. Explain how Rule I was used on line 8. Explain the purpose of line 10. Could we have used Rule EP for u before using Rule EP for v? Explain.)

Example 11.7

Show that the following argument is not valid by constructing a counterexample:

Premises: $(\forall x)(\forall y)(\forall z)(x \circ (y \circ z) = (x \circ y) \circ z)$, $(\forall x)(x \circ a = x)$, $(\exists x)(\exists y)(x \neq y)$

Conclusion: $(\forall x)(\exists y)(x \circ y = a)$

Solution: It is natural in a problem like this to look for a numeric interpretation structure. Premise 1 says that $\circ$ is an associative operation; premise 2 says that a behaves like an identity element (**0** if $\circ$ is $+$; **1** if $\circ$ is $\cdot$); and premise 3 says that there are at least two elements. The conclusion says that every element has an inverse. (In our interpretation structure, the conclusion is to be false.) Let us try the following structure:

U = Set of non-negative integers.

$$\mathbf{x \circ y = x + y}, \qquad \mathbf{a = 0}.$$

Then premise 1 is true because $+$ is an associative operation on **U**; premise 2 is true because for all $\mathbf{x} \in \mathbf{U}$, $\mathbf{x} + \mathbf{0} = \mathbf{x}$; and premise **3** is true because there are at least two elements in **U**—say, **0** and **1**—such that $\mathbf{0} \neq \mathbf{1}$. However, the conclusion is false because, for example, there is no $\mathbf{y} \in \mathbf{U}$ such that $\mathbf{1} + \mathbf{y} = \mathbf{0}$.

Remember, the conclusion is false if we can find some $\mathbf{x} \in \mathbf{U}$ such that

(8) $(\exists \mathbf{y})(\mathbf{x} + \mathbf{y} = \mathbf{0})$

is false. The **x** we chose was **1**. Thus, to show that the conclusion is false, we must show that there is no $\mathbf{y} \in \mathbf{U}$ such that

(9) $\mathbf{1} + \mathbf{y} = \mathbf{0}$.

The only **y** that makes (9) true is $-\mathbf{1}$, and $-\mathbf{1}$ is not an element of **U**. Thus, the conclusion is false.

When we use an arithmetic interpretation, we assume that the arithmetic operations, constants, and relations have their normal interpretations. Thus, in Example 11.7, we used the fact that addition is associative on the set of non-negative integers, that **0** is the identity element for addition, that $\mathbf{0} \neq \mathbf{1}$, and that the only solution to equation (9) is $\mathbf{y} = -\mathbf{1}$.

Example 11.8

Give a formal proof from the empty set of premises to show that the following formula is a logical theorem:

(*) $$(\forall x)(x \leq x) \wedge (\forall x)[f(x) = f(f(x))] \rightarrow (\exists y)(y \leq f(y)).$$

Solution:

{1}	1.	$(\forall x)(x \leq x) \wedge (\forall x)[f(x) = f(f(x))]$	P (for CP)
{1}	2.	$(\forall x)(x \leq x)$	1 T
{1}	3.	$(\forall x)[f(x) = f(f(x))]$	1 T
{1}	4.	$f(x) \leq f(x)$	2 US $[f(x)\|x]$
{1}	5.	$f(x) = f(f(x))$	3 US $[x\|x]$
{1}	6.	$f(x) \leq f(f(x))$	4, 5 I
{1}	7.	$(\exists y)(y \leq f(y))$	6 EG $[f(x)\|y]$
	8.	(*)	1, 7 CP

After taking the premise for Rule *CP* on line 1, our aim is to find some term t so that $t \leq f(t)$. The term t we find, which has the desired property (line 6), is $f(x)$.

11.6 HIDDEN PREMISES

In normal discourse, we generally do not state the obvious. For example, we might say that

> Maine is north of New York, and New York is north of Florida.
> Therefore, Maine is north of Florida.

Symbolizing this argument in our language, we would have

Premise: $Nmn \wedge Nnl$

Conclusion: Nml

This is not a valid argument. However, an obvious statement was omitted—namely, the statement that the relation "north of" is transitive. Thus, there was a *hidden premise*:

(1) $(\forall x)(\forall y)(\forall z)(Nxy \wedge Nyz \rightarrow Nxz)$.

If we add (1) as a premise, the argument is easily shown to be valid.

Arguments that contain hidden premises (unstated premises that are obviously true) are called *enthymemes*. We shall give additional examples of enthymemes in the examples that follow and in the exercises.

Example 11.9

Each of the following arguments is an enthymeme. Add a hidden premise, and give a formal proof of each argument.

(a) The person who committed the crime was in Chicago at 6:00 PM. John was in New York at 6:00 PM. Therefore, John is not the person who committed the crime. (p, Cx, j, Nx)

(b) Alfred is on the swimming team. In every race, Alfred swims faster than every other member of the swimming team. Some member of the swimming team wins a race. Therefore, Alfred wins a race. (a; Tx; Rx; $Fxyz$: x swims faster than y in z; Wxy: x wins y)

Solution:

(a) We can translate the given argument as follows:

Premises: Cp, Nj

Conclusion: $j \neq p$

An obvious premise that was omitted was

If anyone is in New York at 6:00 PM, then he/she is not in Chicago at 6:00 PM.

This sentence is translated as

(*) $(\forall x)(Nx \rightarrow \neg Cx)$.

If we add (*) to the set of premises, the argument is valid.

{1}	1.	Cp	P
{2}	2.	Nj	P
{3}	3.	$(\forall x)(Nx \rightarrow \neg Cx)$	P (H)
{4}	4.	$j = p$	P (for IP)
{3}	5.	$Nj \rightarrow \neg Cj$	3 US $[j \vert x]$
{2, 3}	6.	$\neg Cj$	2, 5 T

{2, 3, 4}	7.	$\neg Cp$	4, 6 *I*
{1, 2, 3, 4}	8.	$Cp \wedge \neg Cp$	1, 7 *T*
{1, 2, 3}	9.	$j \neq p$	4, 8 *IP*

We put an "*H*" in parentheses on line 3 to indicate that this is the hidden premise.

(b) Premises: Ta, $(\forall x)(\forall y)(Tx \wedge Ry \wedge x \neq a \rightarrow Faxy)$,
$(\exists x)(\exists y)(Tx \wedge Ry \wedge Wxy)$

Conclusion: $(\exists x)(Rx \wedge Wax)$

Notice first in the translation of premise 2, $x \neq a$ is in the antecedent because of the word "other" in the premise. If we omitted $x \neq a$, it would imply also that Alfred swims faster than himself, which is false.

As it stands, this argument is not valid. (We leave the construction of a counter-example as an exercise.) However, there is an obviously true premise that is omitted:

For any race *z*, if *x* swims faster than *y* in *z*, then *y* does not win *z*.

In symbols,

(*) $(\forall x)(\forall y)(\forall z)(Rz \wedge Fxyz \rightarrow \neg Wyz)$.

We show below that if we add (*) as a premise, then the argument is valid.

(If you do not know what hidden premise to add, try to prove the argument. At the point where the proof breaks down, the premise or premises that are missing should be more apparent.)

{1}	1.	Ta	*P*
{2}	2.	$(\forall x)(\forall y)(Tx \wedge Ry \wedge x \neq a \rightarrow Faxy)$	*P*
{3}	3.	$(\exists x)(\exists y)(Tx \wedge Ry \wedge Wxy)$	*P*
{4}	4.	$(\forall x)(\forall y)(\forall z)(Rz \wedge Fxyz \rightarrow \neg Wyz)$	*P* (*H*)
{5}	5.	$(\exists y)(Tu \wedge Ry \wedge Wuy)$	*u P* (for *EP*)
{6}	6.	$Tu \wedge Rv \wedge Wuv$	*u*, *v P* (for *EP*(*v*))
{2}	7.	$Tu \wedge Rv \wedge u \neq a \rightarrow Fauv$	2 *US* $[u\|x][v\|y]$
{4}	8.	$Rv \wedge Fauv \rightarrow \neg Wuv$	4 *US* $[a\|x][u\|y][v\|z]$
{4, 6}	9.	$\neg Fauv$	*u*, *v* 6, 8 *T*
{2, 4, 6}	10.	$u = a$	*u* 6, 7, 9 *T*
{6}	11.	$Rv \wedge Wuv$	*u*, *v* 6 *T*
{2, 4, 6}	12.	$Rv \wedge Wav$	*v* 10, 11 *I*
{2, 4, 6}	13.	$(\exists x)(Rx \wedge Wax)$	12 *EG* $[v\|x]$
{2, 4, 5}	14.	$(\exists x)(Rx \wedge Wax)$	5, 6, 13 *EP* $[v\|y]$
{2, 3, 4}	15.	$(\exists x)(Rx \wedge Wax)$	3, 5, 14 *EP* $[u\|x]$

Notice that premise 1 is redundant. It is not used in this proof.

In an enthymeme, the hidden premise (or premises) is not necessarily unique. In Example 11.9(a), for example, we could just as well have added the premise

$$Nj \to \neg Cj,$$

and in (b),

$$(\forall y)(\forall z)(Rz \wedge Fayz \to \neg Wyz)$$

or

$$(\forall x)(\forall y)(\forall z)(Tx \wedge Rz \wedge Fxyz \to \neg Wyz).$$

If we had added the latter formula as the hidden premise in (b), then premise 1 is no longer redundant.

EXERCISES 11

A. Choose appropriate predicate symbols, constants, and function symbols, and translate each of the following sentences into the language of the Predicate Calculus with Identity. (Answers to 1, 5, 9, 13, 18, and 24.)

(1) Dr. Stakem is a radiologist.
(2) Oscar is not the person who committed the crime.
(3) Senator Jones was not the 37th president of the United States.
(4) Senator Jones was an American.
(5) Dr. Fillmore is either a dentist or a physician.
(6) George is neither a good typist nor a good programmer.
(7) Sally is the newest member of the secretarial staff.
(8) Members of Congress are Americans.
(9) No birds are mammals.
(10) Everyone, except George, passed the course.
(11) Only Norbert passed the course.
(12) There was exactly one person who passed the course.
(13) Arthur is Betsey's only husband.
(14) Professor Iten is the only distinguished professor at Wallington University.
(15) Alice is the only person in the class who received an A.
(16) Oscar is happy if everyone else is happy.
(17) There is someone who likes only Betty.
(18) Betty likes exactly one person in the class.
(19) If Norbert passes the course, then everyone else passes the course.
(20) Everyone, except Judy, solved the last problem on every exam.
(21) The person who committed the crime is the tallest person on some college basketball team.

(22) Oscar was the only student pilot who flew to Chicago with some instructor.
(23) Only Betty studied every subject with a professor.
(24) Student pilots can only fly to Chicago with an instructor.
(25) There is exactly one student who studied some subject with every professor.
(26) Only problems assigned by a teacher have solutions.

B. Redo Exercises 5 B.

C. Translate each of the following sentences into the language of the Predicate Calculus with Identity, using standard mathematical notation and only predicate symbols, constants and function symbols from the lists given below. (Answers to 1, 4, 7, 12, 17, and 22.)
For Exercises (1) through (10),

Ix: x is an integer.

Rx: x is a real number.

Bxyz: y is between x and z.

Px: x is a prime.

(x is a *factor* of y iff $(\exists z)(Iz \wedge y = x \cdot z)$.

(1) One and two are distinct positive integers.
(2) One is a positive integer, which is less than every other positive integer.
(3) There is at least one positive integer between one and seven.
(4) There are two distinct integers that have at most one integer between them.
(5) There are exactly two integers between one and four.
(6) There are only two positive integers less than three.
(7) Each positive integer greater than one is either a prime or has a prime factor.
(8) A positive integer is a prime if and only if its only factors are one and itself.
(9) Two is a prime, and it is less than every other prime.
(10) For each pair of positive integers x and y, if y is less than x, then there exist unique positive integers u and v such that $x = u \cdot y + v$.

For Exercises (11) through (14):

Px: x is a point.

Lx: x is a line.

Oxy: x is on y.

(11) Every point is on at least one line.
(12) For each pair of distinct points, there is a line on which they both lie.
(13) For each pair of distinct lines, there is at most one point on both lines.
(14) Given any two distinct points, there is exactly one line on which they both lie.

For Exercises (15) through (23):

Px: x is a polynomial equation with real coefficients.

Dxy: x is the degree of y.

Sxy: x is a solution of y.

Cx: x is a complex number.

Rx: x is a real number.

$f(x)$ is the discriminant of x.

(15) There is a polynomial equation with real coefficients of degree two such that none of its solutions are real.

(16) Every polynomial equation with real coefficients of degree three has a real solution and at most two solutions that are not real.

(17) There is a polynomial equation with real coefficients of degree two that has exactly one real solution.

(18) There is a polynomial equation with real coefficients of degree two that has exactly two solutions that are not real.

(19) There are polynomial equations with real coefficients that have at least three complex solutions.

(20) Every polynomial equation with real coefficients of degree three has at most three real solutions.

(21) A polynomial equation with real coefficients of degree two has exactly one solution iff its discriminant is zero.

(22) A polynomial equation with real coefficients of degree two has at least two distinct real solutions iff its discriminant is positive.

(23) If the discriminant of a polynomial equation with real coefficients of degree two is negative, then the equation has exactly two solutions that are not real.

D. Determine whether each of the following sets of formulas are consistent, and give reasons for your answer. That is, if the set is consistent, find an interpretation structure in which the formulas are all true; otherwise, give a formal proof of a contradiction. (Answers to 1, 5, 9, 12, and 16.)

(1) $(\exists x)(Ax \wedge \neg Bx)$, $Aa \wedge Bb$, $a = b$

(2) $\neg(\exists x)(Px \wedge \neg Rx)$, $\neg(\exists x)(Qx \wedge Rx)$, Pa, Qb, $a = b$

(3) $(\exists x)[Jx \wedge (\forall y)(Jy \rightarrow x = y)]$, Ja, Jb, $a \neq b$

(4) $(\forall x)(\forall y)(Px \wedge Qx \wedge Py \wedge Qy \rightarrow x = y)$, Pa, Qb, $a \neq b$

(5) $Pa \vee (\forall x)Qx$, $\neg Pb$, $a = b \wedge \neg Qc$

(6) $(\exists x)(\exists y)(x = y \rightarrow Px)$, $\neg(\exists x)Px$

(7) $(\forall x)(\exists y)(x \neq y)$, $(\exists x)(\exists y)(\forall z)(z = x \vee z = y)$

(8) $(\exists x)(\forall y)(x = y)$, $(\exists x)(\exists y)(x \neq y)$

(9) $(\forall x)(\exists y)(Rxy \rightarrow Ryx)$, $(\exists x)(Px \wedge \neg Rxx)$,
$(\exists x)(\forall z)(x \neq z \rightarrow \neg Px \wedge Rzx \wedge Rxz)$

(10) $(\forall x)(\forall y)(x \circ y = y \circ x)$, $(\exists z)(\forall x)(\forall y)(x \circ y = z)$

(11) $(\exists x)(\exists y)(Px \wedge Qy)$, $(\forall x)(\forall y)(Px \wedge Qy \rightarrow x \circ y \neq a)$,
$(\exists x)(\exists y)(x \circ y \neq a)$

(12) $(\forall x)(\exists y)(f(x) = f(y))$, $(\forall x)(\exists y)(f(x) \neq f(y))$
(13) Pa, $(\forall x)(Px \rightarrow (\exists y)Rxy)$, $(\forall x)(Rax \rightarrow x = b)$, $\neg Rab$
(14) $(\forall x)(\exists y)Rxy$, $(\forall x)(\forall y)(\forall z)(\forall w)(Rxy \wedge Rzw \rightarrow y = w)$, $(\forall y)(\exists x)\neg Rxy$
(15) $(\forall x)(\exists y)R(x \circ y)a$, $(\forall x)(\exists y)R(y \circ x)a$, $(\exists x)(\exists y)\neg R(x \circ y)(y \circ x)$
(16) $(\forall x)(\forall y)[x \circ y = a \rightarrow y = f(x)]$, $(\forall x)(x \circ x = a)$, $(\exists x)(x \neq f(x))$
(17) $(\exists x)Px$, $(\forall x)(Px \rightarrow Qx)$, $(\exists x)(\exists y)(x \neq y)$, $(\forall x)(\forall y)(Px \wedge Qy \rightarrow Rxy)$, $\neg(\exists x)(\exists y)(Rxy \wedge x \neq y)$
(18) $(\forall x)(\forall y)(\forall z)(x \circ (y \circ z) = (x \circ y) \circ z)$, $(\forall x)(x \circ a = x)$, $(\forall x)(\exists y)(x \circ y = a)$, $(\exists x)(\exists y)(x \neq y)$

E. For each of the following arguments, if it is valid, give a formal proof, if not, give a counter-example. (Answers to 1, 5, 8, 12, 16, and 21.)

(1) **Premises:** $(\exists x)(Jx \wedge Kx)$, $(\forall x)(Jx \rightarrow x = c)$
Conclusion: $Ka \wedge a = c$

(2) **Premises:** $(\forall x)(Px \vee x = b)$, $\neg(\forall x)(x = b)$
Conclusion: $(\exists x)Px$

(3) **Premises:** $(\exists y)(\forall x)(x = y \rightarrow Qx)$
Conclusion: $(\exists x)Qx$

(4) **Premises:** $Fa \wedge Gb$, $(\forall x)(Fx \rightarrow \neg Hx)$, $(\forall x)(\neg Gx \rightarrow Hx)$
Conclusion: $a = b$

(5) **Premises:** $(\forall x)(Rax \rightarrow a = x \vee a = b)$, $(\exists x)Rax$, $Sa \wedge \neg Sb$
Conclusion: Raa

(6) **Premises:** $(\exists y)(\forall x)(Fx \leftrightarrow x = y)$
Conclusion: $(\forall x)(Fx \rightarrow Gx) \leftrightarrow (\exists x)(Fx \wedge Gx)$

(7) **Premises:** $(\forall x)(\forall y)(Rxy \rightarrow x = y)$, $(\exists x)Px$, $(\exists x)\neg Px$
Conclusion: $(\exists x)(\exists y)\neg Rxy$

(8) **Premises:** $(\exists x)[Px \wedge (\forall y)(Py \rightarrow x = y)]$
Conclusion: $(\exists x)(\forall y)(Py \leftrightarrow x = y)$

(9) **Premises:** $(\exists x)(\exists y)(x \neq y)$, $(\exists x)Ax$, $(\forall x)(Ax \rightarrow Bx)$, $(\forall x)(\forall y)(Ax \wedge By \rightarrow Cxy)$
Conclusion: $(\exists x)(\exists y)(Cxy \wedge x \neq y)$

(10) **Premises:** $(\forall x)[Ax \rightarrow (\exists y)Bxy]$, $(\exists x)(Ax \wedge \neg Bxa)$
Conclusion: $(\exists x)(\exists y)(Bxy \wedge y \neq a)$

(11) **Premises:** $(\forall x)(\forall y)(Ax \wedge By \rightarrow x = y)$, $(\exists x)(Ax \wedge Cx)$, $(\exists x)(Bx \wedge Dx)$
Conclusion: $(\exists x)(Cx \wedge Dx)$

(12) **Premises:** $(\forall x)(\forall y)(Pxy \rightarrow Pyx)$, $(\forall x)Pxf(x)$
Conclusion: $(\forall x)Pf(f(x))f(x)$

(13) **Premises:** $(\forall x)(f(x) \neq x)$, $(\exists x)Qx$, $(\exists x)Qf(x)$, $(\exists x)Qf(x)$
Conclusion: $(\exists x)(\exists y)(Qx \wedge Qy \wedge x \neq y)$

(14) Premises: $(\forall x)Rxg(x)$, $(\exists x)(f(x) \neq g(f(x)))$
Conclusion: $(\exists x)(\exists y)(Rf(x)y \land f(x) \neq y)$

(15) Premises: $(\forall x)(\forall y)(Sxf(x, y)$, $(\exists x)[Sxf(x, x) \rightarrow (\forall y)(\forall z)(f(y, z) \neq x)]$
Conclusion: $(\exists x)(\exists y)(Sxy \land x \neq y)$

(16) Premises: $(\forall x)(\forall y)(\forall z)(Rxy \land Ryz \rightarrow Rxz)$, $(\forall x)(\forall y)(\exists z)(Rxz \land Ryz)$
Conclusion: $(\forall x)(\forall y)(Rxy \lor Ryx)$

(17) Premises: $(\exists x)[(\forall y)Fxy \lor (\forall y)Fyx]$, $(\forall x)(\neg Fxx \lor Gxa)$, $(\forall x)(x \neq a \rightarrow \neg Gxa)$, $(\forall x)(\forall y)(Gxa \rightarrow Gya)$
Conclusion: $(\exists y)(\forall x)(x = y)$

(18) Premises: $(\exists x)Rxx$, $(\forall x)(\forall y)(Sxxx \rightarrow Rxy)$, $(\forall x)(\forall y)(Sxyx \rightarrow x = y)$, $(\forall x)(\forall y)(Rxy \rightarrow (\exists z)Sxzy)$
Conclusion: $(\exists x)(\forall y)Rxy$

(19) Premises: $(\forall x)(\forall y)(x \circ y = y \circ x)$, $(\forall x)(\exists y)(x \circ y = a)$
Conclusion: $(\forall x)(x \circ a = x)$

(20) Premises: $(\forall x)(\forall y)(\forall z)[x \circ (y \circ z) = (x \circ y) \circ z]$, $(\forall x)(x \circ a = x)$, $(\forall x)(\exists y)(x \circ y = a)$, $(\exists x)(x \neq a)$
Conclusion: $(\exists y)(\forall x)(x \circ y = a)$

(21) Premises: $(\forall x)(\forall y)(x \circ y = y \circ x)$, $(\forall x)(\forall y)[f(x) \circ y = f(x \circ y)]$
Conclusion: $(\forall x)(\forall y)[f(x) \circ y = x \circ f(y)]$

(22) Premises: $(\forall x)(\forall y)(\forall z)[x \circ (y \circ z) = (x \circ y) \circ z]$, $(\forall x)(\forall y)(\exists z)(x = y \circ z)$, $(\forall x)(\forall z)(\exists y)(x = y \circ z)$
Conclusion: $(\forall y)(\forall z)[(\forall x)(x \circ y = x \land x \circ z = x) \rightarrow y = z]$

(23) Premises: $(\forall x)(\forall y)(Px \land Py \rightarrow Px \circ y)$, $(\exists x)[Px \land (\forall y)(Py \rightarrow Qy \circ x)]$
Conclusion: $(\exists x)(Px \land Qx)$

(24) Premises: $(\forall x)(\forall y)(Rxy \rightarrow \neg Ryx)$, $(\forall x)(\forall y)(Rxy \rightarrow x \neq y)$, $(\forall x)(\forall y)(\forall z)(Rxy \land Ryz \rightarrow Rxz)$, $(\forall x)(\forall z)(x \neq z \rightarrow (\exists y)Bxyz)$, $(\forall x)(\forall y)(\forall z)[Bxyz \rightarrow (Rxy \land Ryz) \lor (Rzy \land Ryx)]$
Conclusion: $(\forall x)(\forall z)[Rxz \rightarrow (\exists y)(Rxy \land Ryz)]$

F. For each of the following formulas, if it is a logical theorem, give a formal proof from the empty set of formulas; if not, give a counter-example. (Answers to 1, 5, and 9.)

(1) $(\forall x)[Px \leftrightarrow (\exists y)(Px \land Py)]$
(2) $(\exists x)(Ax \land Bx) \land (\exists x)(Ax \land x = a) \rightarrow (\exists x)(Bx \land x = a)$
(3) $(\exists y)(\forall x)(Fx \leftrightarrow x = y) \land (\exists x)(Fx \land Gx) \rightarrow (\forall x)(Fx \rightarrow Gx)$
(4) $(\forall x)(\forall y)(f(x) = y) \rightarrow (\forall x)(\forall y)(x = y)$
(5) $(\forall x)Rxx \land (\forall y)(g(f(y)) = f(y)) \rightarrow (\exists z)Rzg(z)$
(6) $(\exists x)(\exists y)(x \neq y) \land (\exists x)Px \land (\exists x)\neg Qx \rightarrow$
$(\exists x)(\exists y)\ (Px \land \neg Qy \land x \neq y)$
(7) $(\exists x)[Px \land (\forall y)(Py \leftrightarrow x = y)] \leftrightarrow (\exists x)(\forall y)(Py \leftrightarrow x = y)$

(8) $(\forall x)[Ax \rightarrow (Bx \leftrightarrow (\exists y)(x = y \wedge By))]$

(9) $(\forall x)(\forall y)(x \circ y = y \circ x) \wedge (\forall x)(f(x) \circ x = a) \rightarrow (\forall x)(\exists y)(x \circ y = a)$

(10) $(\forall x)(\forall y)(x \circ y = b \rightarrow y = g(x)) \rightarrow (\forall x)(\exists y)(x \circ y = b)$

G. Each of the following arguments is an enthymeme. Translate each into the language of the Predicate Calculus with Identity, add a suitable hidden premise or premises, and give a formal proof. (Answers to 1, 4, 7, 11, and 15.)

(1) Someone committed a burglary in Atlanta. Only Nancy or Judy could have committed the burglary. Nancy was in San Francisco. Therefore, Judy committed the burglary in Atlanta. (Bx, Ax, Sx, n, j).

(2) If someone likes Harry, then there exists a purple unicorn. Therefore, no one likes Harry. (Lxy, Px, Ux, a)

(3) If some student does not pass the course, then Professor Smith is happy. If every student passes the course, then some cow jumps over the moon. Therefore, Professor Smith is happy. (Sx, Px, Hx, Cx, Jx, s)

(4) Every person enjoys a good time. Therefore, some student enjoys a good time. (Px, Ex, Sx)

(5) Anyone older than Jeremy is also older than Ophelia. Jeremy is not older than Ophelia. Therefore, Ophelia is not older than Jeremy. (Oxy, j, o)

(6) Joan is Miriam's mother, and Miriam is Theresa's mother. Therefore, Joan is not Theresa's mother. (Mxy, j, m, t)

(7) There is someone in the first row who is in front of everyone else in the first row. Therefore, there is at most one person in the first row who is in front of everyone else in the first row. (Rx, Fxy)

(8) Therefore, any cat that has at most one scratched ear also has at least one ear that is not scratched. (Cx, Sx, Exy)

(9) Any horse that breaks a front leg is shot. Silver is a horse that has at most one front leg that is not broken. Therefore, Silver is shot. (Hx, Bx, Fxy, Sx, s)

(10) Every Harvard student, except Oscar, studies mathematics with a Harvard professor. Therefore, there is a Harvard professor. (Sx, Mxy, Px, o)

(11) Anyone who does not have the opportunity to study mathematics with Professor Smith is unfortunate. Alex did not have the opportunity to study any course with Professor Smith. Therefore, Alex is unfortunate. ($Oxyz$, Mx, Ux, Cx, a, s)

(12) Only Oscar flies to Middletown with an instructor. Therefore, there is a town to which Oscar flies. (o; m; $Fxyz$: x flies to y with z; Ix; Tx; Lxy: x flies to y)

(13) George is the only student who drew a polygon. Therefore, if any student drew a triangle, that student is George. (e, Sx, Dxy, Px, Tx)

(14) Only deer and lions live on the animal preserve. Lions kill only prey animals when they are hungry. Therefore, if a hungry lion kills an animal on the animal preserve, that animal is a deer. (Dx, Lx, Ax, Kxy, Px, Hx)

(15) Every student in the class who received a grade of A or B is happy. Some students in the class are not happy. Therefore, some students in the class received a grade lower than a B. (*Sx*: x is a student in the class; $g(x)$ is x's grade; *Hx*, *Lxy*, a, b)

(16) Every student in the class whose grade was higher than Elizabeth's grade received an A. The only grades given in the class were A, B, or C, and each was given to some student. Elizabeth's grade was not an A. Therefore, if Elizabeth is a student in the class, her grade is B. (*Sx*: x is a student in the class; *Hxy*: x is higher than y, $g(x)$ is x's grade; e: Elizabeth; a, b, c)

Chapter 12

Summary of the Rules of Inference

We shall put all the important definitions and rules of inference together in this chapter for easy reference.

12.1 LANGUAGE OF THE PREDICATE CALCULUS WITH IDENTITY

The symbols of the language are as follows:

Variables: x, y, z, u, v, w, and these letters with subscripts.

Connectives: $\neg, \wedge, \vee, \rightarrow, \leftrightarrow$.

Quantifiers: $\forall, \exists$.

Identity: $=$.

Constants: $a, b, \ldots, e, i, j, \ldots t$, and these letters with subscripts.

Function Symbols: f, g, h, and these letters with subscripts.

Predicate Symbols: $A, B, C, \ldots$, and these letters with subscripts.

(Parentheses are used for grouping. We associate the number of arguments with each function and predicate symbol.)

Definition 12.1 *Terms*

1. Constants and variables are terms.
2. If f is a function symbol of n variables and $t_1, t_2, \ldots, t_n$ are n terms, then $f(t_1, t_2, \ldots, t_n)$ is a term.
3. No other expressions are terms.

Definition 12.2

Atomic Formulas

1. If t_1 and t_2 are terms, then $t_1 = t_2$ is an atomic formula.
2. If P is an n-place predicate symbol and $t_1, t_2, \ldots, t_n$ are n terms, then $Pt_1 \ldots t_n$ is an atomic formula.
3. No other expressions are atomic formulas.

Definition 12.3

Formulas

1. An atomic formula is a formula.
2. If φ and ψ are formulas, then so are $(\neg\varphi)$, $(\varphi \wedge \psi)$, $(\varphi \vee \psi)$, $(\varphi \rightarrow \psi)$, and $(\varphi \leftrightarrow \psi)$.
3. If v is a variable and φ is a formula, then $((\forall v)\varphi)$ and $((\exists v)\varphi)$ are formulas.
4. No other expressions are formulas.

(Definitions 12.1 and 12.3 are examples of *inductive* definitions. This type of definition is discussed in more detail in Chapter 15. For most applications, it is sufficient if you are able to distinguish terms and formulas from meaningless strings of symbols.)

Definition 12.4

1. The *scope of a quantifier* in a formula is the quantifier together with the shortest formula following the quantifier.
2. An occurrence of a variable in a formula is said to be *bound* if it is within the scope of a quantifier. Otherwise, the occurrence of the variable is said to be *free*. A variable is said to be *bound (free) in a formula* if it has at least one bound (free) occurrence.
3. A formula with no free variables is called a *sentence.*
4. If φ is a formula, t is a term, and v is a variable, then $\varphi[t|v]$ denotes the formula obtained from φ by replacing every free occurrence of v by t. The substitution of t for v in φ is *valid* if no occurrence of a variable in t is bound in $\varphi[t|v]$.
5. If φ is a formula, t is a term, and v is a variable, then $\varphi[v,t|v]$ denotes the formula obtained from φ by replacing some free occurrences of v by t. The substitution of t for v in φ is *valid* if no occurrence of a variable in t is bound in $\varphi[v,t|v]$.

Definition 12.5

Proof

A *proof of* χ *from the premises* $\varphi_1, \varphi_2, \ldots, \varphi_n$ is a finite sequence of ordered pairs $\langle\alpha_1, \psi_1\rangle, \langle\alpha_2, \psi_2\rangle, \ldots, \langle\alpha_m, \psi_m\rangle$ where each α_i is a set of formulas and each ψ_i is a formula, $i = 1, 2, \ldots, m$, such that

1. $\alpha_m \subseteq \{\varphi_1, \varphi_2, \ldots, \varphi_n\}$ and ψ_m is χ.
2. Each ψ_i is either a premise or is deducible from preceding ψ_j's in the sequence using a rule of inference given below.

3. If ψ_i is a premise, then $\alpha_i = \{\psi_i\}$.
4. If ψ_i is deducible from preceding ψ_j's in the sequence using a rule of inference given below, then α_i is defined by the rule.

If a proof of χ from $\varphi_1, \varphi_2, \ldots, \varphi_n$ exists, then we say χ is *provable from* $\varphi_1, \varphi_2, \ldots, \varphi_n$ and write $\{\varphi_1, \varphi_2, \ldots, \varphi_n\} \vdash \chi$ or $\varphi_1, \varphi_2, \ldots, \varphi_n \vdash \chi$.
(In our proofs, we abbreviate the α's by replacing each premise by the line number where the premise occurs.)

Definition 12.6

If φ and ψ are formulas,

1. φ is a *logical theorem* if there is a proof of φ from the empty set of formulas. In symbols, $\vdash \varphi$.
2. φ *logically implies* ψ if $\vdash \varphi \rightarrow \psi$.
3. φ and ψ are *logically equivalent* if $\vdash \varphi \leftrightarrow \psi$.

Definition 12.7

Flagged Variable
If φ is a formula in a proof, a variable v is said to be *flagged* in φ if either

1. φ is a premise and v is free in φ, or
2. there is a premise ψ in which v is free, v is free in φ, and ψ is used to derive φ.

12.2 RULES OF INFERENCE

In what follows, φ, ψ, and χ and these letters with subscripts are formulas; u, v, and w are variables; t is a term; and the α and α_i's are sets of formulas.

P **(Rule for Premises).** $\langle\alpha, \varphi\rangle$ may occur in any proof as long as $\alpha = \{\varphi\}$. (φ is called a *premise*.)

T **(Rule for Tautologies).** $\langle\alpha, \varphi\rangle$ may occur in a proof if φ is a tautology and α is the empty set or if $\langle\alpha_1, \psi_1\rangle, \langle\alpha_2, \psi_2\rangle, \ldots, \langle\alpha_k, \psi_k\rangle$ precede $\langle\alpha, \varphi\rangle$ in the proof,

$$\psi_1 \wedge \psi_2 \wedge \cdots \wedge \psi_k \rightarrow \varphi$$

is a tautology and

$$\alpha = \alpha_1 \cup \alpha_2 \cup \cdots \cup \alpha_k.$$

CP **(Rule of Conditional Proof).** If $\langle\alpha \cup \{\varphi\}, \psi\rangle$ occurs in a proof, then $\langle\alpha, \varphi \rightarrow \psi\rangle$ may also occur in the proof. (In other words, if $\alpha \cup \{\varphi\} \vdash \psi$, then $\alpha \vdash \varphi \rightarrow \psi$.)

IP **(Rule of Indirect Proof).** If $\langle \alpha \cup \{\neg\varphi\}, \psi \wedge \neg\psi \rangle$ occurs in a proof, then $\langle \alpha, \varphi \rangle$ may also occur in the proof. (If $\alpha \cup \{\neg\varphi\} \vdash \psi \wedge \neg\psi$, then $\alpha \vdash \varphi$.)

US **(Rule of Universal Specification).** If $\langle \alpha, (\forall v)\varphi \rangle$ occurs in a proof, then $\langle \alpha, \varphi[t|v] \rangle$ may also occur in the proof, if the substitution of t for v in φ is valid. ($(\forall v)\varphi \vdash \varphi[t|v]$, if the substitution of t for v in φ is valid.)

UG **(Rule of Universal Generalization).** If $\langle \alpha, \varphi \rangle$ occurs in a proof, then $\langle \alpha, (\forall v)\varphi \rangle$ may also occur in the proof, if v is not free in α. ($\varphi \vdash (\forall v)\varphi$, as long as v is not flagged in φ.)

EG **(Rule of Existential Generalization).** If $\langle \alpha, \varphi[t|v] \rangle$ occurs in a proof, then $\langle \alpha, (\exists v)\varphi \rangle$ may also occur in the proof, if the substitution of t for v in φ is valid. ($\varphi[t|v] \vdash (\exists v)\varphi$, if the substitution of t for v in φ is valid.)

EP **(Rule of Existential Proof).** If $\langle \alpha \cup \{\varphi[u|v]\}, \psi \rangle$ occurs in a proof, then $\langle \alpha \cup \{(\exists v)\varphi\}, \psi \rangle$ may also occur in the proof, as long as the substitution of u for v in φ is valid, u is not free in α, and u is not free in ψ. (If $\alpha \cup \{\varphi[u|v]\} \vdash \psi$, then $\alpha \cup \{(\exists v)\varphi\} \vdash \psi$, as long as the substitution of u for v in φ is valid, u is not free in α, and u is not free in ψ.)

LT **(Rule for Logical Theorems).** If $\vdash \varphi$, the proof of φ is constructive, and α is the empty set, then $\langle \alpha, \varphi \rangle$ may occur in a proof. Or, if $\langle \alpha_1, \psi_1 \rangle, \langle \alpha_2, \psi_2 \rangle, \ldots, \langle \alpha_k, \psi_k \rangle$ occur in a proof, $\vdash \psi_1 \wedge \psi_2 \wedge \cdots \wedge \psi_k \rightarrow \varphi$, the proof of $\psi_1, \wedge \psi_2 \wedge \cdots \wedge \psi_k \rightarrow \varphi$ is constructive and $\alpha = \alpha_1 \cup \alpha_2 \cup \cdots \cup \alpha_k$, then $\langle \alpha, \varphi \rangle$ may also occur in the proof.

LE **(Rule for Logically Equivalent Formulas).** If $\langle \alpha, \varphi \rangle$ occurs in a proof, ψ is obtained from φ by replacing some occurrence of χ_1 by χ_2 and if $\vdash \chi_1 \leftrightarrow \chi_2$, then $\langle \alpha, \psi \rangle$ may also occur in the proof. (In fact, $\vdash \varphi \leftrightarrow \psi$.)

I **(Rules for Identity).**

(a) $\vdash (\forall v)(v = v)$.

(b) $\vdash (\forall v)[v = t \rightarrow (\varphi \leftrightarrow \varphi[v, t|v])]$, if the substitution of t for v is valid.

(c) $\vdash (\forall u)(\forall v)(u = v \rightarrow v = u)$.

(d) $\vdash (\forall u)(\forall v)(\forall w)(u = v \wedge v = w \rightarrow u = w)$.

12.3 GUIDELINES

1. To prove a conjunction, $\varphi \wedge \psi$, prove each of the conjuncts, φ and ψ, then use the rule $\{\varphi, \psi\} \vdash \varphi \wedge \psi$.
2. To prove a conditional $\varphi \rightarrow \psi$, take φ as a premise, derive ψ, then use the Rule *CP*.
3. To prove a disjunction, $\varphi \vee \psi$,
   (a) Prove either φ or ψ, then use the rule $\varphi \vdash \varphi \vee \psi$ or $\psi \vdash \varphi \vee \psi$.
   (b) Take $\neg\varphi$ as a premise, derive ψ, use Rule *CP* to derive $\neg\varphi \rightarrow \psi$, and, finally, use the rule $\neg\varphi \rightarrow \psi \vdash \varphi \vee \psi$.

4. To prove a biconditional, $\varphi \leftrightarrow \psi$, prove $\varphi \rightarrow \psi$ and $\psi \rightarrow \varphi$, then use the rule $\{\varphi \rightarrow \psi, \psi \rightarrow \varphi\} \vdash \varphi \leftrightarrow \psi$.
5. To prove an existential formula, $(\exists v)\varphi$, prove $\varphi[t|v]$ for some term t such that the substitution of t for v in φ is valid, then use Rule *EG*.
6. To prove a universal formula, $(\forall v)\varphi$, try one of the following:
   (a) If φ is a conditional, $\psi \rightarrow \chi$, take ψ as a premise and derive χ, use Rule *CP* to derive $\psi \rightarrow \chi$, and then use Rule *UG*.
   (b) If v is not free in any premise, take $\neg\varphi$ as a premise and derive a contradiction, use Rule *IP* to derive φ, and then use Rule *UG*.
7. Drop existential quantifiers before dropping universal quantifiers. That is, if any of the premises are existential formulas, choose *EP* premises and *EP* variables before using Rule *US* to drop the universal quantifiers.

   If none of the above methods work, try

8. An indirect proof. To prove φ, take $\neg\varphi$ as a premise and derive a contradiction. Then use Rule *IP* to derive φ.
9. An alternative proof.
   (a) To prove φ, find some formula ψ such that $\vdash \psi \rightarrow \varphi$ and $\vdash \neg\psi \rightarrow \varphi$, then use the rule $\{\psi \rightarrow \varphi, \neg\psi \rightarrow \varphi\} \vdash \varphi$.
   (b) To prove $\varphi \vee \psi \rightarrow \chi$, prove $\varphi \rightarrow \chi$ and $\psi \rightarrow \chi$, then use the rule $\{\varphi \rightarrow \chi, \psi \rightarrow \chi\} \vdash \varphi \vee \psi \rightarrow \chi$.

Definition 12.8

Valid and Invalid Arguments

An argument with premises $\varphi_1, \varphi_2, \ldots, \varphi_n$ and conclusion ψ is *valid* if for each interpretation structure **S** for $\{\varphi_1, \varphi_2, \ldots, \varphi_n, \psi\}$ and each sequence of elements **s** in **S**, if **s** satisfies each of $\varphi_1, \varphi_2, \ldots, \varphi_n$, then **s** satisfies ψ. In symbols,

$$\{\varphi_1, \varphi_2, \ldots, \varphi_n\} \models \psi.$$

Otherwise, the argument is *invalid*. (That is, an argument with premises $\varphi_1, \varphi_2, \ldots, \varphi_n$ and conclusion ψ is *invalid* if there is an interpretation structure **S** for $\{\varphi_1, \varphi_2, \ldots, \varphi_n, \psi\}$ and a sequence **s** in **S** such that **s** satisfies $\varphi_1, \varphi_2, \ldots$ and φ_n, but **s** does not satisfy ψ.) In the case that all the formulas involved in an argument are sentences, then the argument is *valid* if for each interpretation structure **S** for the sentences, whenever the premises are true in **S**, the conclusion is also true in **S**.

Completeness Theorem. If Γ is a set of sentences and φ is a sentence, then if $\Gamma \models \varphi$, then $\Gamma \vdash \varphi$.

Soundness Theorem. If Γ is a set of sentences and φ is a sentence, then if $\Gamma \vdash \varphi$, then $\Gamma \models \varphi$.

Compactness Theorem I. If Γ is a set of sentences and φ is a sentence, then if $\Gamma \models \varphi$, then there is a finite subset $\Delta \subseteq \Gamma$ such that $\Delta \models \varphi$.

Compactness Theorem II. If Γ is a set of sentences such that every finite subset of Γ has a model than Γ has a model.

Theorem 12.1 *An argument is valid if and only if it has a proof.*

Definition 12.9 *Consistent and Inconsistent sets of Formulas*
A set of formulas Γ is *consistent* if there is no formula φ such that $\Gamma \vdash \varphi \wedge \neg\varphi$. Otherwise, Γ is said to be *inconsistent.*

Theorem 12.2 *A set of formulas Γ is consistent if and only if there is an interpretation structure* **S** *for Γ and a sequence* **s** *in* **S** *such that* **s** *satisfies each formula in Γ.* (*If all the formulas in Γ are sentences then Γ is consistent if there is an interpretation structure for Γ in which each of the sentences in Γ is true.*)

12.4 ADDITIONAL GUIDELINES

1. To show an argument is valid, give a proof.
2. To show an argument is invalid, find an interpretation structure **S** and a sequence **s** in **S** such that **s** satisfies the premises but not the conclusion. (In the case that all the formulas in the argument are sentences, find an interpretation structure in which the premises are true and the conclusion is false.)
3. To show a set of formulas Γ is consistent, find an interpretation structure **S** and a sequence **s** in **S** such that **s** satisfies all the formulas in Γ. (In the case that all the formulas in Γ are sentences, find an interpretation structure in which each of the formulas in Γ is true.)
4. To show a set of formulas Γ is inconsistent give a proof of a contradiction from Γ.
5. To show an existential formula $(\exists v)\varphi$ is
   (a) true in a structure, show that φ is true for *some* element v in the universe.
   (b) false in a structure, show that φ is false for *all* elements v in the universe.
6. To show a universal formula $(\forall v)\varphi$ is
   (a) true in a structure, show that φ is true for *all* elements v in the universe.
   (b) false in a structure, show that φ is false for *some* element v in the universe.

12.5. USEFUL TAUTOLOGIES

("φ," "ψ," and "χ" denote arbitrary formulas.)

1. $\varphi \wedge (\varphi \rightarrow \psi) \rightarrow \psi$ (Rule of Detachment, Modus Ponens)
2. $\neg\psi \wedge (\varphi \rightarrow \psi) \rightarrow \neg\varphi$ (Modus Tollendo Tollens)

3. $\neg\varphi \wedge (\varphi \vee \psi) \rightarrow \psi$ (Modus Tollendo Ponens)
4. $\varphi \wedge \psi \rightarrow \varphi$ (Rule of Simplification)
5. $\varphi \rightarrow \varphi \vee \psi$ (Rule of Addition)
6. $\left.\begin{array}{l}\varphi \rightarrow (\psi \rightarrow \varphi \wedge \psi)\\ (\varphi \rightarrow \psi) \wedge (\varphi \rightarrow \chi) \rightarrow (\varphi \rightarrow \psi \wedge \chi)\end{array}\right\}$ (Rule of Adjunction)
7. $(\varphi \rightarrow \psi) \wedge (\psi \rightarrow \chi) \rightarrow (\varphi \rightarrow \chi)$ (Rule of Hypothetical Syllogism)
8. $\left.\begin{array}{l}(\varphi \rightarrow \psi) \wedge (\neg\varphi \rightarrow \psi) \rightarrow \psi\\ (\varphi \vee \psi \rightarrow \chi) \leftrightarrow (\varphi \rightarrow \chi) \wedge (\psi \rightarrow \chi)\end{array}\right\}$ (Rules of Alternative Proof)
9. $\left.\begin{array}{l}(\varphi \rightarrow \psi \wedge \neg\psi) \rightarrow \neg\varphi\\ \varphi \wedge \neg\varphi \rightarrow \psi\end{array}\right\}$ (Rule of Absurdity)
10. $\varphi \vee \neg\varphi$ (Rule of the Excluded Middle)
11. $\neg(\varphi \wedge \neg\varphi)$ (Rule of Contradiction)
12. $\left.\begin{array}{l}\varphi \vee \psi \leftrightarrow \psi \vee \varphi\\ \varphi \wedge \psi \leftrightarrow \psi \wedge \varphi\end{array}\right\}$ (Commutative Rules)
13. $\left.\begin{array}{l}\varphi \vee (\psi \vee \chi) \leftrightarrow (\varphi \vee \psi) \vee \chi\\ \varphi \wedge (\psi \wedge \chi) \leftrightarrow (\varphi \wedge \psi) \wedge \chi\end{array}\right\}$ (Associative Rules)
14. $\left.\begin{array}{l}\varphi \vee (\psi \wedge \chi) \leftrightarrow (\varphi \vee \psi) \wedge (\varphi \vee \chi)\\ \varphi \wedge (\psi \vee \chi) \leftrightarrow (\varphi \wedge \psi) \vee (\varphi \wedge \chi)\end{array}\right\}$ (Distributive Rules)
15. $\left.\begin{array}{l}\neg(\varphi \vee \psi) \leftrightarrow \neg\varphi \wedge \neg\psi\\ \neg(\varphi \wedge \psi) \leftrightarrow \neg\varphi \vee \neg\psi\end{array}\right\}$ (De Morgan's Rules)
16. $\neg(\neg\varphi) \leftrightarrow \varphi$ (Rule of Double Negation)
17. $\left.\begin{array}{l}(\varphi \rightarrow \psi) \leftrightarrow \neg\varphi \vee \psi\\ \neg(\varphi \rightarrow \psi) \leftrightarrow \varphi \wedge \neg\psi\end{array}\right\}$ (Rules for the Conditional)
18. $\left.\begin{array}{l}(\varphi \leftrightarrow \psi) \leftrightarrow (\varphi \rightarrow \psi) \wedge (\psi \rightarrow \varphi)\\ (\varphi \leftrightarrow \psi) \leftrightarrow (\varphi \wedge \psi) \vee (\neg\varphi \wedge \neg\psi)\\ (\varphi \leftrightarrow \psi) \leftrightarrow (\neg\varphi \vee \psi) \wedge (\varphi \vee \neg\psi)\end{array}\right\}$ (Rules for the Biconditional)
19. $\left.\begin{array}{l}\varphi \vee \varphi \leftrightarrow \varphi\\ \varphi \wedge \varphi \leftrightarrow \varphi\end{array}\right\}$ (Idempotency Rules)
20. $\left.\begin{array}{l}(\varphi \rightarrow \psi) \leftrightarrow (\neg\psi \rightarrow \neg\varphi)\\ (\varphi \rightarrow \neg\psi) \leftrightarrow (\psi \rightarrow \neg\varphi)\\ (\neg\varphi \rightarrow \psi) \leftrightarrow (\neg\psi \rightarrow \varphi)\end{array}\right\}$ (Rules of Contraposition)
21. $(\varphi \rightarrow (\psi \rightarrow \chi)) \leftrightarrow (\varphi \wedge \psi \rightarrow \chi)$ (Rule of Exportation-Importation)
22. $\left.\begin{array}{l}\varphi \vee (\varphi \wedge \psi) \leftrightarrow \varphi\\ \varphi \wedge (\varphi \vee \psi) \leftrightarrow \varphi\end{array}\right\}$ (Rules of Absorption)

12.6. USEFUL LOGICAL THEOREMS

("φ" and "ψ" denote arbitrary formulas, and "x" and "y" denote arbitrary variables.)

1. $\neg(\exists x)\varphi \leftrightarrow (\forall x)\neg\varphi$
2. $\neg(\forall x)\varphi \leftrightarrow (\exists x)\neg\varphi$
3. $\neg(\exists x)\neg\varphi \leftrightarrow (\forall x)\varphi$
4. $\neg(\forall x)\neg\varphi \leftrightarrow (\exists x)\varphi$
5. $\neg(\forall x)(\varphi \rightarrow \psi) \leftrightarrow (\exists x)(\varphi \wedge \neg\psi)$
6. $\neg(\exists x)(\varphi \wedge \psi) \leftrightarrow (\forall x)(\varphi \rightarrow \neg\psi)$

In theorems 7 and 8, we assume that y is not free in φ and that the substitution of y for x in φ is valid.

7. $(\forall x)\varphi \leftrightarrow (\forall y)\varphi[y|x]$
8. $(\exists x)\varphi \leftrightarrow (\exists y)\varphi[y|x]$
9. $(\forall x)(\forall y)\varphi \leftrightarrow (\forall y)(\forall x)\varphi$
10. $(\exists x)(\exists y)\varphi \leftrightarrow (\exists y)(\exists x)\varphi$
11. $(\exists x)(\forall y)\varphi \rightarrow (\forall y)(\exists x)\varphi$ (but not conversely)

In theorems 12 through 19, we assume that x is not free in φ.

12. $(\forall x)(\varphi \wedge \psi) \leftrightarrow (\varphi \wedge (\forall x)\psi)$
13. $(\exists x)(\varphi \wedge \psi) \leftrightarrow (\varphi \wedge (\exists x)\psi)$
14. $(\forall x)(\varphi \vee \psi) \leftrightarrow (\varphi \vee (\forall x)\psi)$
15. $(\exists x)(\varphi \vee \psi) \leftrightarrow (\varphi \vee (\exists x)\psi)$
16. $(\forall x)(\varphi \rightarrow \psi) \leftrightarrow (\varphi \rightarrow (\forall x)\psi)$
17. $(\exists x)(\varphi \rightarrow \psi) \leftrightarrow (\varphi \rightarrow (\exists x)\psi)$
18. $(\forall x)(\psi \rightarrow \varphi) \leftrightarrow ((\exists x)\psi \rightarrow \varphi)$
19. $(\exists x)(\psi \rightarrow \varphi) \leftrightarrow ((\forall x)\psi \rightarrow \varphi)$

Chapter 13

Mathematical Theories

A *theory* is a set of sentences Γ that is closed under logical implication. That is, if φ is a sentence and $\Gamma \vdash \varphi$, then $\varphi \in \Gamma$. (If Γ is inconsistent and is a theory, then every sentence is in Γ. Why?) The *language* of a theory Γ is the language of the Predicate Calculus with Identity along with the constants, function, and/or relation symbols that occur in Γ. Therefore, every such language contains variables, connectives, quantifiers, identity, and a list (finite or infinite) of constants, function, and/or predicate symbols. We shall call the variables, connectives, quantifiers, and identity *logical symbols* and the remaining symbols (constants, function, and predicate symbols) *nonlogical symbols*. A theory is called a *mathematical theory* if the nonlogical symbols have natural mathematical interpretations. We shall use *Group Theory* as an example of a mathematical theory.

13.1 GROUP THEORY

The first step in describing a theory is to specify the *nonlogical symbols*. Our theory will have three nonlogical symbols—a constant symbol, a unary operation symbol (a function symbol of one variable), and a binary operation symbol (a function symbol of two variables).

Nonlogical Symbols:
Constant: e.

Unary Operation Symbol: $*$. (We write 'x^*' instead of '$*x$.')

Binary Operation Symbol: $\circ$. (We write '$x \circ y$' instead of '$\circ xy$.')

The next step is to give a set (finite or infinite) of sentences that describes how the nonlogical symbols are supposed to behave. These sentences are called *axioms*.

Axioms:

$A1$: $(\forall x)(\forall y)(\forall z)[x \circ (y \circ z) = (x \circ y) \circ z]$.

$A2$: $(\forall x)(x \circ e = x)$.

$A3$: $(\forall x)(x \circ x^* = e)$.

Axiom $A1$ is the associative rule for the operation $\circ$; $A2$ describes what property e is to have (e is called an *identity* element); and $A3$ describes what property $*$ is to have ($*$ is called an *inverse* operation, and we call x^* the *inverse* of x).

Group Theory is the set of all sentences φ so that

(1) $$\{A1, A2, A3\} \vdash \varphi.$$

We shall call a sentence φ that satisfies (1), a *theorem of Group Theory* or just a *theorem*. Thus, when deriving a theorem of Group Theory, the axioms behave like premises. The Predicate Calculus with Identity serves as the *syntax* of the theory, whereas the nonlogical symbols and the axioms give the theory its *content*.

Before giving examples of theorems of Group Theory, we shall give another property of identity that is frequently useful in a mathematical theory.

Theorem 13.1 *Each of the following is a logical theorem:*

(a) $(\forall x)(\forall y)(x = y \rightarrow x^* = y^*)$.

(b) $(\forall x)(\forall y)(\forall z)(x = y \rightarrow x \circ z = y \circ z)$.

(c) $(\forall x)(\forall y)(\forall z)(x = y \rightarrow z \circ x = z \circ y)$.

Proof: We shall prove (b) and leave the proof of (a) and (c) as an exercise.

{1}	1.	$x = y$	x, y P (for CP)
	2.	$x \circ z = x \circ z$	I
{1}	3.	$x \circ z = y \circ z$	x, y, 1, 2 I
	4.	$x = y \rightarrow x \circ z = y \circ z$	1, 3 CP
	5.	$(\forall x)(\forall y)(\forall z)(x = y \rightarrow x \circ z = y \circ z)$	4 UG (3)

■

When we use one of the logical theorems given in Theorem 13.1 in a proof, we shall document it with the letter "*I*." Each is a property of identity.

A sentence φ is a theorem of Group Theory if there is a proof of an argument of the form:

Premises: $A1$, $A2$, $A3$

Conclusion: φ

Thus, in each of our formal proofs, the three axioms $A1$, $A2$, and $A3$ will be the first three lines. We shall call the nth theorem, which we prove in Group Theory, "Theorem n." Thus, if φ is Theorem n, for any n, then

(2) $$\vdash A1 \wedge A2 \wedge A3 \rightarrow \varphi.$$

That is, $A1 \wedge A2 \wedge A3 \rightarrow \varphi$ is a logical theorem.

Theorem 1 *Right Cancellation Law*:

$$(\forall x)(\forall y)(\forall z)(x \circ z = y \circ z \rightarrow x = y).$$

Proof: The idea of the proof is to assume

(i) $x \circ z = y \circ z.$

Then use Theorem 13.1(b) and "multiply" both sides of (i) on the right by z^*, obtaining,

(ii) $(x \circ z) \circ z^* = (y \circ z) \circ z^*.$

Apply the associative rule, $A1$,

(iii) $x \circ (z \circ z^*) = y \circ (z \circ z^*).$

Apply $A3$,

(iv) $x \circ e = y \circ e.$

Finally, apply $A2$, and obtain

(v) $x = y.$

The formal proof follows:

{1}	1.	$(\forall x)(\forall y)(\forall z)[x \circ (y \circ z) = (x \circ y) \circ z]$	P
{2}	2.	$(\forall x)(x \circ e = x)$	P
{3}	3.	$(\forall x)(x \circ x^* = e)$	P
{4}	4.	$x \circ z = y \circ z$	x, y, z P (for CP)
{4}	5.	$(x \circ z) \circ z^* = (y \circ z) \circ z^*$	x, y, z 4 I
{1}	6.	$x \circ (z \circ z^*) = (x \circ z) \circ z^*$	1 US (3) $[x\mid x][z\mid y][z^*\mid z]$
{1}	7.	$y \circ (z \circ z^*) = (y \circ z) \circ z^*$	1 US (3) $[y\mid x][z\mid y][z^*\mid z]$
{1, 4}	8.	$x \circ (z \circ z^*) = y \circ (z \circ z^*)$	x, y, z 5, 6, 7 I
{3}	9.	$z \circ z^* = e$	3 US $[z\mid x]$
{1, 3, 4}	10.	$x \circ e = y \circ e$	x, y 8, 9 I
{2}	11.	$x \circ e = x$	2 US $[x\mid x]$
{2}	12.	$y \circ e = y$	2 US $[y\mid x]$
{1, 2, 3, 4}	13.	$x = y$	x, y, 10, 11, 12 I
{1, 2, 3}	14.	$x \circ z = y \circ z \rightarrow x = y$	4, 13 CP
{1, 2, 3}	15.	$(\forall x)(\forall y)(\forall z)\ (x \circ z = y \circ z \rightarrow x = y)$	14 UG (3)

On line 5, we used Theorem 13.1 (b). Notice the use of Rule *US* on lines 6 and 7. Formally, they are incorrect uses of Rule *US*. (Why?) Explain how to correct it.

■

Theorem 2 *Commutative Law for e*:

$$(\forall x)(x \circ e = e \circ x).$$

Proof: Informally, it follows from *A*2 that $x \circ e = x$. Thus, we want to show that $e \circ x = x$. The idea is to find some element z so that

(3) $$(e \circ x) \circ z = x \circ z,$$

then apply Theorem 1, to cancel the z. We shall show that if we take z to be x^* in (3), then (3) is true. The formal proof follows:

{1}	1.	$(\forall x)(\forall y)(\forall z)[x \circ (y \circ z) = (x \circ y) \circ z]$	*P*
{2}	2.	$(\forall x)(x \circ e = x)$	*P*
{3}	3.	$(\forall x)(x \circ x^* = e)$	*P*
{2}	4.	$x \circ e = x$	2 *US* $[x\|x]$
{3}	5.	$x \circ x^* = e$	3 *US* $[x\|x]$
{1}	6.	$e \circ (x \circ x^*) = (e \circ x) \circ x^*$	1 *US* $[e\|x][x\|y][x^*\|z]$
{1}	7.	$(e \circ x) \circ x^* = e \circ (x \circ x^*)$	6 *I*
{1, 3}	8.	$(e \circ x) \circ x^* = e \circ e$	5, 7 *I*
{2}	9.	$e \circ e = e$	2 *US* $[e\|x]$
{1, 2, 3}	10.	$(e \circ x) \circ x^* = e$	8, 9 *I*
{1, 2, 3}	11.	$(e \circ x) \circ x^* = x \circ x^*$	5, 10 *I*
{1, 2, 3}	12.	$(e \circ x) \circ x^* = x \circ x^* \rightarrow e \circ x = x$	Theorem 1, *US* $[e \circ x\|x][x\|y][x^*\|z]$
{1, 2, 3}	13.	$e \circ x = x$	11, 12 *T*
{1, 2, 3}	14.	$x \circ e = e \circ x$	4, 13 *I*
{1, 2, 3}	15.	$(\forall x)(x \circ e = e \circ x)$	14 *UG*

To justify line 12 of this proof, we should have inserted two lines, 11.1 and 11.2, between lines 11 and 12 as follows:

	11.1	$A1 \wedge A2 \wedge A3 \rightarrow$ Theorem 1	*LT*
{1, 2, 3}	11.2	Theorem 1	1, 2, 3, 11.1 *T*

Then line 12 becomes

{1, 2, 3} 12. $(e \circ x) \circ x^* = x \circ x^* \rightarrow e \circ x = x$ 11.2 *US* $[e \circ x|x][x|y][x^*|z]$.

However, we shall usually omit lines like 11.1 and 11.2. The device used on line 12 allows us to use past results to obtain new theorems of Group Theory.

■

Theorem 3 *Uniqueness of the Identity*:

$$(\forall y)[(\forall x)(x \circ y = x) \rightarrow y = e].$$

Proof: Theorem 3 says that any y that satisfies $A2$ must be e.

{1}	1.	$(\forall x)(\forall y)(\forall z)[x \circ (y \circ z) = (x \circ y) \circ z]$	P
{2}	2.	$(\forall x)(x \circ e = x)$	P
{3}	3.	$(\forall x)(x \circ x^* = e)$	P
{4}	4.	$(\forall x)(x \circ y = x)$	y P (for CP)
{4}	5.	$e \circ y = e$	y 4 US $[e\|x]$
{1, 2, 3}	6.	$y \circ e = e \circ y$	Theorem 2, US $[y\|x]$
{1, 2, 3, 4}	7.	$y \circ e = e$	y 5, 6 I
{2}	8.	$y \circ e = y$	2 US $[y\|x]$
{1, 2, 3, 4}	9.	$y = e$	y 7, 8 I
{1, 2, 3}	10.	$(\forall x)(x \circ y = x) \rightarrow y = e$	4, 9 CP
{1, 2, 3}	11.	$(\forall y)[(\forall x)(x \circ y = x) \rightarrow y = e]$	10 UG

■

At this point, we can see how many details are necessary in a formal proof, even in this simple theory. Preceding many of the formal proofs we have given above, we have started with an "informal proof." We shall now discuss in more detail what we mean by an informal proof.

13.2 INFORMAL PROOFS

At the very least, an informal proof must contain enough details so that a logician who does not know anything about the content of the proof can convert it to a formal proof. Thus, whenever an axiom (or a premise) is used, it has to be explicitly cited. Let us look at some examples.

First let us look at the informal proof we gave above for Theorem 1. We shall repeat it here in a little more concise form.

We want to prove the following:

Theorem 1 $(\forall x)(\forall y)(\forall z)(x \circ z = y \circ z \rightarrow x = y)$.

Informal Proof:

Assume as a premise

(i) $x \circ z = y \circ z$.

(ii) $(x \circ z) \circ z^* = (y \circ z) \circ z^*$. (by (i) and I)

(iii) $x \circ (z \circ z^*) = y \circ (z \circ z^*)$. (by (ii) and $A1$)

(iv) $x \circ e = y \circ e$. (by (iii) and $A3$)

(v) $x = y$. (by (iv) and $A2$)

Therefore, Theorem 1 follows.

Now compare this informal proof with the formal proof on page 259.

Informal Proof	Formal Proof
(i)	line 4
(ii)	line 5
(iii)	line 8
(iv)	line 10
(v)	line 13

Starting with the informal proof, lines 1 through 5 of the formal proof are self-explanatory. Formula (iii) of the informal proof is obtained from (ii) and $A1$ using Rules *US* and *I*. The documentation in the informal proof "(by (ii) and $A1$)" means "(by (ii), $A1$, and the Rules of Logic)." Similarly, (iv) in the informal proof is obtained by (iii), $A3$, and the Rules of Logic; and (v), by (iv), $A2$, and the Rules of Logic. Thus, given the informal proof, a logician could construct the formal proof, because the structure of the informal proof is clearly indicated, and every time an axiom is used, it is specifically cited.

On the other hand, the discussion preceding the formal proof of Theorem 2 may be sufficient for a logician who knows something about the content to construct a formal proof, but it does not contain enough information for any logician to construct a formal proof because it does not, in all cases, cite the axioms that are used. We shall give an adequate informal proof below.

We wish to prove the following:

Theorem 2 $(\forall x)(x \circ e = e \circ x)$.

Informal Proof:

(i) $x \circ e = x$ (by $A2$)

(Next, we should like to find a z so that $(e \circ x) \circ z = x \circ z$. Then apply Theorem 1. We shall show that if z is x^*, then we get the desired result.)

(ii) $x \circ x^* = e$ (by $A3$)

(iii) $(e \circ x) \circ x^* = e \circ (x \circ x^*)$ (by $A1$)
$= e \circ e$ (by preceding line and $A3$)
$= e$ (by preceding line and $A2$)

It follows from (ii) and (iii) that

(iv) $(e \circ x) \circ x^* = x \circ x^*$

(v) $e \circ x = x$ (by (iv) and Theorem 1)

Therefore, it follows from (i) and (v) that

(vi) $x \circ e = e \circ x$.

This proves the theorem.

Comparing the informal proof with the formal proof,

Informal Proof	Formal Proof
(i)	line 4
(ii)	line 5
(iii)	lines 6–10
(iv)	line 11
(v)	line 13
(vi)	line 14

The expression in parentheses between (i) and (ii) explains briefly what is to follow. It's the plan of attack. It is important for the person who is proving the theorem, but it is not necessary for the logician who is going to construct the formal proof from the informal proof.

As another example, we shall give an informal proof of Theorem 3:

Theorem 3 $(\forall y)[(\forall x)(x \circ y = x) \rightarrow y = e]$.

Informal Proof:

Assume as a premise

(i) $x \circ y = x$ for every x.

It follows from (i) that

(ii) $e \circ y = e$.

(iii) $e \circ y = y \circ e$ (by Theorem 2)
$= y$ (by preceding line and $A2$)

It follows from (ii) and (iii) that

(iv) $y = e$.

■

As an exercise, the reader should construct a formal proof of Theorem 3, given the above informal proof.

In an informal proof, we have to be especially careful of flagged variables because they are usually not documented as they are in formal proofs. For example, in the informal proof of Theorem 3 above, in formula (i), y is flagged. It is free in a premise. Thus, we *cannot* deduce from (iv), $(\forall y)(y = e)$.

13.3 MODELS

A *model* for a theory Γ is an interpretation structure **S** for Γ such that all the formulas in Γ are true in **S**. In symbols, $\mathbf{S} \models \Gamma$. (See Definition 9.3.) If a theory Γ has a set of axioms Δ, then if $\mathbf{S} \models \Delta$, then $\mathbf{S} \models \Gamma$. (Why?)

A model of Group Theory is called a *group*. If **S** is an interpretation structure for Group Theory and if $\mathbf{S} \models \{A1, A2, A3\}$, then **S** is a model of Group Theory. Each of the following interpretation structures is an example of a group.

$$\mathbf{S} = \langle \mathbf{U}, \circ, ^*, \mathbf{e} \rangle.$$

(1) $\mathbf{U} = \{\mathbf{e}\}$, $\mathbf{e} \circ \mathbf{e} = \mathbf{e}$, $\mathbf{e}^* = \mathbf{e}$.

(2) $\mathbf{U} = \mathbb{Z}$ = set of all integers,
$\mathbf{e} = \mathbf{0}$, $\mathbf{x} \circ \mathbf{y} = \mathbf{x} + \mathbf{y}$, $\mathbf{x}^* = -\mathbf{x}$.

(3) $\mathbf{U} = \mathbb{R}^+$ = set of positive real numbers,
$\mathbf{e} = \mathbf{1}$, $\mathbf{x} \circ \mathbf{y} = \mathbf{x} \cdot \mathbf{y}$, $\mathbf{x}^* = \mathbf{1}/\mathbf{x}$.

(4) $\mathbf{U} = \{\mathbf{a}, \mathbf{b}, \mathbf{e}\}$, $\mathbf{e}^* = \mathbf{e}$, $\mathbf{a}^* = \mathbf{b}$, $\mathbf{b}^* = \mathbf{a}$.

∘	e	a	b
e	e	a	b
a	a	b	e
b	b	e	a

$\mathbf{x} \circ \mathbf{y}$

(5) $\mathbf{U} = \{\mathbf{e}, \mathbf{a}_1, \mathbf{a}_2, \mathbf{a}_3, \mathbf{a}_4, \mathbf{a}_5\}$, $\mathbf{e}^* = \mathbf{e}$, $\mathbf{a}_1^* = \mathbf{a}_1$, $\mathbf{a}_2^* = \mathbf{a}_2$, $\mathbf{a}_3^* = \mathbf{a}_3$, $\mathbf{a}_4^* = \mathbf{a}_5$, $\mathbf{a}_5^* = \mathbf{a}_4$.

∘	e	a_1	a_2	a_3	a_4	a_5
e	e	a_1	a_2	a_3	a_4	a_5
a_1	a_1	e	a_4	a_5	a_2	a_3
a_2	a_2	a_5	e	a_4	a_3	a_1
a_3	a_3	a_4	a_5	e	a_1	a_2
a_4	a_4	a_3	a_1	a_2	a_5	e
a_5	a_5	a_2	a_3	a_1	e	a_4

$\mathbf{x} \circ \mathbf{y}$

To show that each of the five interpretation structures is a group, show that each of the three axioms is true. Structure (1) is called a *trivial group*, and it is trivial to show that the axioms are true. To show that the axioms are true in (2) and (3), use the standard properties of arithmetic. For structures (4) and (5), it is easy to see from the tables that *A*2 and *A*3 are true. However, to see that *A*1 is true requires some work. We have to check that

$$\mathbf{x} \circ (\mathbf{y} \circ \mathbf{z}) = (\mathbf{x} \circ \mathbf{y}) \circ \mathbf{z}$$

for all $\mathbf{x}, \mathbf{y}, \mathbf{z} \in \mathbf{U}$ and this is quite tedious. For those who know a little algebra, however, it is quite simple, because model (4) has the same structure as the integers modulo **3** where $\mathbf{a} = \mathbf{1}$, $\mathbf{b} = \mathbf{2}$, $\mathbf{e} = \mathbf{0}$ and $\mathbf{x} \circ \mathbf{y} = \mathbf{x} + \mathbf{y}$ (modulo

3). (We say, **x** = **y** (modulo **3**) if **x** − **y** is divisible by **3**.) It is well known that modulo addition is associative.

The universe of model (5) can be interpreted as the following nonsingular 2 × 2 matrices:

$$\mathbf{e} = \begin{pmatrix} 1 & 0 \\ 0 & 1 \end{pmatrix}, \; \mathbf{a}_1 = \begin{pmatrix} 1 & 1 \\ 0 & 1 \end{pmatrix}, \; \mathbf{a}_2 = \begin{pmatrix} 1 & 0 \\ 1 & 1 \end{pmatrix},$$

$$\mathbf{a}_3 = \begin{pmatrix} 0 & 1 \\ 1 & 0 \end{pmatrix}, \; \mathbf{a}_4 = \begin{pmatrix} 0 & 1 \\ 1 & 1 \end{pmatrix}, \; \mathbf{a}_5 = \begin{pmatrix} 1 & 1 \\ 1 & 0 \end{pmatrix}.$$

The operation ∘ is matrix multiplication over the integers modulo **2**, and $\mathbf{x}^*$ is the inverse of the matrix **x**. It is a well-known fact of algebra that matrix multiplication is associative.

In models (1) through (4) above, the operation ∘ is commutative. That is,

(6) $$(\forall x)(\forall y)(x \circ y = y \circ x)$$

is true in each of these models. However, the commutative law is false in model (5) because, for example, $\mathbf{a}_1 \circ \mathbf{a}_2 = \mathbf{a}_4$, but $\mathbf{a}_2 \circ \mathbf{a}_1 = \mathbf{a}_5$. It follows that (6) is *not* a theorem of Group Theory because a theorem has the property that it is true in every model.

Thus, to show that a formula φ in the language of Group Theory is a theorem, show that

$$\{A1, A2, A3\} \vdash \varphi.$$

To show φ is not a theorem, find a model of Group Theory in which φ is false.

Example 13.1

Show that if φ is the sentence

$$(\exists x)(\forall y)(y \circ x = e),$$

then neither φ nor ¬φ is a theorem of Group Theory.

Solution: First, to show φ is not a theorem, we shall find a model of Group Theory in which φ is false. Translating φ into English, we obtain:

There is an element that serves as an inverse for every element.

Actually, this sentence is false in each of the models (2) through (5) given above. We shall show it is false in model (2). The interpretation of φ in (2) is

(7) $$(\exists \mathrm{x})(\forall \mathrm{y})(\mathrm{y} + \mathrm{x} = 0).$$

Suppose (7) were true in model (2). Then for some fixed **x**, we would have, for example,

$$\mathbf{1} + \mathbf{x} = \mathbf{0} \quad \text{and} \quad \mathbf{2} + \mathbf{x} = \mathbf{0}.$$

Using Theorem 1 of Group Theory, we obtain the contradiction **1** = **2**. Thus, (7) is false in model (2), so φ is not a theorem of Group Theory.

To show ¬φ is not a theorem, we must find a model of Group Theory in which ¬φ is false or, equivalently, φ is true. The trivial model, model (1) above, is such a model because there is only one element in the universe. Thus, the element that exists to make φ true is **e**.

Example 13.2

Let **S** = ⟨**U**, ∘, *, **e**⟩. Determine whether **S** is a group (a model of Group Theory) under the following conditions:

(a) **U** = {**−1/2, 0, 1/2**}, **x** ∘ **y** = **x** + **y**, **x*** = −**x**, **e** = **0**.

(b) **U** = {**1, 2**}, **x** ∘ **y** = min {**x, y**}, **x*** = **x**, **e** = **1**.

Solution:

(a) **S** is not a group in this case because it is not even an interpretation structure. One of the conditions on an interpretation structure is that for each function symbol, f, of n variables in the language, $\mathbf{f} : \mathbf{U^n} \to \mathbf{U}$. In case (a), + does not have this property because, for example, **1/2** + **1/2** = **1** ∉ **U**, so + is not a function from $\mathbf{U}^2$ to **U**.

(b) **S** is not a group in this case either because **2*** = **2**, **e** = **1** and **2** ∘ **2** = **2** ≠ **e**. Thus, $A3$ is not true in **S**.

13.4 DEFINITIONS

There are occasions, when a theory is being developed, that certain expressions occur frequently and it is convenient to give them names. For example, in Group Theory, we may want to talk about elements x and y that commute. That is,

$$x \circ y = y \circ x.$$

Let us define a new binary relation symbol C so that Cxy is an abbreviation for $x \circ y = y \circ x$.

In order to prove theorems about C, we must add C as a new nonlogical symbol to our language and also add a new axiom:

$A4$: $(\forall x)(\forall y)[Cxy \leftrightarrow (x \circ y = y \circ x)]$.

Axiom $A4$ serves as the *definition* of the symbol C.

With this additional machinery, we can prove, for example, the following theorem:

Theorem 4 $(\forall x)(x \circ x = e) \rightarrow (\forall x)(\forall y)Cxy.$

Proof: We shall give an informal proof. Suppose

(i) $x \circ x = e,$ for all x.

(It follows from $A4$ that we must prove $x \circ y = y \circ x$, for all x and y. We shall show that there is a z so that $(x \circ y) \circ z = (y \circ x) \circ z$, then apply Theorem 1.)

It follows from (i) that

(ii) $(x \circ y) \circ (x \circ y) = e.$

(iii)
$$\begin{aligned}(y \circ x) \circ (x \circ y) &= y \circ [x \circ (x \circ y)] && \text{(by } A1\text{)}\\ &= y \circ [(x \circ x) \circ y] && \text{(by } A1\text{)}\\ &= y \circ (e \circ y) && \text{(by (i))}\\ &= y \circ (y \circ e) && \text{(by Theorem 2)}\\ &= y \circ y && \text{(by } A2\text{)}\\ &= e && \text{(by (i))}\end{aligned}$$

(The documentation "by . . ." on each of the lines, except the first, in (iii) above, means "by the preceding line and . . . .") Thus it follows from (ii) and (iii) that

(iv) $(x \circ y) \circ (x \circ y) = (y \circ x) \circ (x \circ y).$

Now, applying Theorem 1, we obtain

(v) $x \circ y = y \circ x.$

So by $A4$,

(vi) $Cxy.$

■

Notice that by adding the new 2-place predicate symbol C and the new axiom $A4$, we really cannot prove any "new" theorems of Group Theory. Some theorems may now contain the symbol C, but whenever an expression of the form "Cxy" occurs in a theorem, if it is replaced by "$x \circ y = y \circ x$," then the resulting formula is provable from $A1$, $A2$, and $A3$. Thus, $A4$ serves as a definition of C.

On the other hand, if we added a new nonlogical symbol S and an axiom

(1) $$(\forall x)(Sx \leftrightarrow \neg Sx),$$

we can certainly prove additional theorems. Formula (1) logically implies a contradiction and a contradiction logically implies every formula φ. Thus, to define a new n-place relation symbol R, we add an axiom of the form

$$(\forall x_1)(\forall x_2) \ldots (\forall x_n)(Rx_1x_2 \ldots x_n \leftrightarrow \varphi),$$

where φ is a formula in the language of Group Theory that does not contain the symbol R and contains no free variables other than $x_1, x_2, \ldots, x_n$.

Similarly, for new constant or function symbols. For example, suppose we add a new function symbol f and its defining axiom is

(2) $$(\forall x)(\forall y)(\forall z)[f(x,y) = z \leftrightarrow z = (x \circ y) \circ (x^* \circ y^*)].$$

Formula (2) is logically equivalent to

(3) $$(\forall x)(\forall y)[f(x,y) = (x \circ y) \circ (x^* \circ y^*)].$$

Thus, it is clear that "$f(x, y)$" is just an abbreviation for "$(x \circ y) \circ (x^* \circ y^*)$." On the other hand, if we added a new function symbol g and an axiom

(4) $$(\forall x)(\forall y)(\forall z)[g(x,y) = z \leftrightarrow (x \circ z = y \vee y \circ z = x)],$$

we can prove new theorems in Group Theory. For example, using (4), we can prove

(5) $$(\forall x)(x = x^*),$$

which is not a theorem of Group Theory. (Formula (5) is false, for example, in model (2) in the previous section.)

An informal proof that (4) logically implies (5) is as follows:

(i) $x \circ x^* = e$	(by $A3$)
(ii) $g(x, e) = x^*$	(by (i) and (4))
(iii) $e \circ x = x \circ e$	(by Theorem 2)
$= x$	(by preceding line and $A2$)
(iv) $g(x, e) = x$	(by (4))
(v) $x = x^*$	(by (ii) and (iv)).

(As an exercise, change the preceding informal proof to a formal proof.)

The problem with formula (4) is that the formula

(6) $$x \circ z = y \vee y \circ z = x$$

does not define a function of x and y. Given x and y, there may be more than one z satisfying (6). Formula (4) defines a 3-place predicate symbol rather than a function symbol of two variables.

In order to define a function symbol f of n variables, we must have an axiom of the form

(7) $$(\forall x_1)(\forall x_2)\ldots(\forall x_n)(\forall y)[f(x_1, x_2, \ldots, x_n) = y \leftrightarrow \varphi],$$

where the formula φ satisfies the following properties:

1. The symbol f does not occur in φ.
2. The formula φ contains no free variables except $x_1, x_2, \ldots, x_n$ and y.
3. $\{A1, A2, A3\} \vdash (\forall x_1)(\forall x_2) \ldots (\forall x_n)(\exists y)[\varphi \wedge (\forall z)(\varphi[z|y] \rightarrow y = z)]$.

(We are assuming that the substitution of z for y in φ is valid). That is, condition 3 says that it is a theorem of Group Theory that for each $x_1, x_2, \ldots, x_n$, there is exactly one y such that φ. In this case, y is a function of $x_1, x_2, \ldots, x_n$ and f is the name of that function.

Example 13.3 Let Γ be a theory that has two nonlogical symbols—$\circ$, a binary operation symbol, and e, a constant—and two axioms:

$B1$: $(\forall x)(\exists y)(x \circ y) = e)$.

$B2$: $(\forall x)(\forall y)(\forall z)(z \circ x = z \circ y \rightarrow x = y)$.

Prove that we can define a new function symbol f in Γ so that

$B3$: $(\forall x)(\forall y)(f(x) = y \leftrightarrow x \circ y = e)$.

Solution: We must show that if φ is the formula

(8) $$x \circ y = e,$$

then φ satisfies conditions 1, 2, and 3 above. $B3$ corresponds to formula (7) above. It is clear that (8) satisfies conditions 1 and 2 because f does not occur in (8), and x and y are the only free variables in (8). To prove condition 3, we must prove

(9) $$\{B1, B2\} \vdash (\forall x)(\exists y)[x \circ y = e \wedge (\forall z)(x \circ z = e \rightarrow y = z)].$$

We shall give an informal proof. It follows from $B1$ that

$$(\forall x)(\exists y)(x \circ y = e).$$

(This proves existence.)

Suppose

$$x \circ y = e \quad \text{and} \quad x \circ z = e.$$

Then it follows from $B2$ that $y = z$. (This proves uniqueness.)

In this proof, we are using the fact that the following two formulas are logically equivalent:

$$(\forall x)[(\exists y)(x \circ y = e) \wedge (\forall y)(\forall z)(x \circ y = e \wedge x \circ z = e \rightarrow y = z)]$$

and

$$(\forall x)(\exists y)[x \circ y = e \wedge (\forall z)(x \circ z = e \rightarrow y = z)].$$

(See Example 11.5).

Because a constant is a function of 0 variables, its definition is similar to that of a function. For example, in a theory that has a binary operation symbol $\circ$ and a constant symbol a, we could define a new constant symbol b by the formula

$$(\forall x)(b = x \leftrightarrow x = a \circ a)$$

or, equivalently,

$$b = a \circ a.$$

The symbol b is just an abbreviation for $a \circ a$.

Basically, the purpose of a definition is either to give an abbreviation for some formula (a new predicate symbol) or a name for some term (a new constant or function symbol).

13.5 THEORY OF AN EQUIVALENCE RELATION

We complete this chapter with another example of a mathematical theory—namely, that of an equivalence relation. (See Definition 7.3). In this theory, which we shall call Θ, there is one nonlogical symbol, E, which is a 2-place predicate symbol. The axioms are as follows:

$E1$: $(\forall x)Exx$.

$E2$: $(\forall x)(\forall y)(Exy \rightarrow Eyx)$.

$E3$: $(\forall x)(\forall y)(\forall z)(Exy \wedge Eyz \rightarrow Exz)$.

Thus, E is reflexive ($E1$), symmetric ($E2$), and transitive ($E3$).

One of the interesting things about the theory Θ is that we can characterize its models.

Theorem 13.2 *Suppose* $\mathbf{S} = \langle \mathbf{U}, \mathbf{E} \rangle$ *is an interpretation structure for* Θ. *Then* $\mathbf{S}$ *is a model of* Θ *if and only if the following three conditions are satisfied:*

(1) $\mathbf{U}$ *is a union of sets* $\mathbf{A}_i$ *for* $i \in K$. *That is,*

$$\mathbf{U} = \bigcup_{i \in K} \mathbf{A}_i.$$

(2) *For each* $i, j \in K$, $i \neq j$, $\mathbf{A}_i \cap \mathbf{A}_j = \emptyset$, *where* $\emptyset$ *is the empty set. (That is,* $\mathbf{U}$ *is a union of pairwise disjoint sets).*

(3) *For all* $\mathbf{x}, \mathbf{y} \in \mathbf{U}$,

$$\mathbf{Exy} \text{ iff } (\exists i)(\mathbf{x} \in \mathbf{A}_i \wedge \mathbf{y} \in \mathbf{A}_i)$$

Proof: First suppose $\mathbf{S}$ satisfies (1), (2), and (3). We shall show that the three axioms of Θ, $E1$, $E2$, and $E3$ are true in $\mathbf{S}$.

(i) Axiom $E1$: Suppose $\mathbf{x} \in \mathbf{U}$. Then, by (1), there is an i so that $\mathbf{x} \in \mathbf{A}_i$. Clearly, if $\mathbf{x} \in \mathbf{A}_i$, then $\mathbf{x} \in \mathbf{A}_i$ and $\mathbf{x} \in \mathbf{A}_i$, so by (3) $\mathbf{Exx}$. Thus, $\mathbf{E}$ is reflexive.

(ii) Axiom *E*2: Suppose $\mathbf{x}, \mathbf{y} \in \mathbf{U}$ and $\mathbf{Exy}$. Then, by (3), there is an i so that $\mathbf{x} \in \mathbf{A}_i$ and $\mathbf{y} \in \mathbf{A}_i$, which implies $\mathbf{y} \in \mathbf{A}_i$ and $\mathbf{x} \in \mathbf{A}_i$. Thus, by (3), $\mathbf{Eyx}$. Therefore, $\mathbf{E}$ is symmetric.

(iii) Axiom *E*3: Suppose $\mathbf{x}, \mathbf{y}, \mathbf{z} \in \mathbf{U}$, $\mathbf{Exy}$, and $\mathbf{Eyz}$. By (3), there is an i and a j such that

$$\mathbf{x}, \mathbf{y} \in \mathbf{A}_i \text{ and } \mathbf{y}, \mathbf{z} \in \mathbf{A}_j.$$

Thus, $\mathbf{y} \in \mathbf{A}_i \cap \mathbf{A}_j$. But, if $i \neq j$, by (2), $\mathbf{A}_i \cap \mathbf{A}_j = \varnothing$, so it follows that $i = j$. Therefore, $\mathbf{x}, \mathbf{z} \in \mathbf{A}_i = \mathbf{A}_j$, so by (3), $\mathbf{Exz}$, which implies $\mathbf{E}$ is transitive.

It follows from (i), (ii), and (iii) that $\mathbf{S}$ is a model of Θ.

Conversely, suppose $\mathbf{S}$ is a model of Θ. We shall show that $\mathbf{S}$ must satisfy (1), (2), and (3). First, we shall define the $\mathbf{A}_i$'s. For each $\mathbf{u} \in \mathbf{U}$, define

(4) $$[\mathbf{u}] = \{\mathbf{x} \in \mathbf{U} : \mathbf{Exu}\}.$$

For each $\mathbf{u} \in \mathbf{U}$, $[\mathbf{u}]$ is called an *equivalence class* of $\mathbf{E}$. We shall show that the equivalence classes serve as the $\mathbf{A}_i$'s.

(iv) (1) is true in $\mathbf{S}$: Because $\mathbf{E}$ is reflexive (axiom *E*1 is true in $\mathbf{S}$), for all $\mathbf{u} \in \mathbf{U}$, $\mathbf{Euu}$. Thus, by (4), $\mathbf{u} \in [\mathbf{u}]$. Because every $\mathbf{u} \in \mathbf{U}$ belongs to some equivalence class,

$$\mathbf{U} = \cup\{[\mathbf{u}] : \mathbf{u} \in \mathbf{U}\}.$$

Thus, (1) is true.

(v) (2) is true in $\mathbf{S}$: We shall show that for all $\mathbf{u}, \mathbf{v} \in \mathbf{U}$, either $[\mathbf{u}] = [\mathbf{v}]$ or $[\mathbf{u}] \cap [\mathbf{v}] = \varnothing$. (This will prove (2)). We shall give an indirect proof. Suppose

(5) $$[\mathbf{u}] \neq [\mathbf{v}]$$

and

(6) $$[\mathbf{u}] \cap [\mathbf{v}] \neq \varnothing.$$

By (6), there is a $\mathbf{w} \in \mathbf{U}$ such that

(7) $$\mathbf{w} \in [\mathbf{u}] \cap [\mathbf{v}].$$

That is,

(8) $$\mathbf{w} \in [\mathbf{u}] \quad \text{and} \quad \mathbf{w} \in [\mathbf{v}].$$

By (5), there is a $\mathbf{z} \in \mathbf{U}$, such that

(9) $$\mathbf{z} \in [\mathbf{u}] \quad \text{and} \quad \mathbf{z} \notin [\mathbf{v}],$$

or

(10) $$\mathbf{z} \notin [\mathbf{u}] \quad \text{and} \quad \mathbf{z} \in [\mathbf{v}].$$

We shall assume (9). The proof is analogous if (10) holds.

It follows from (4) and (8) that **Ewu** and **Ewv**; and it follows from (4) and (9) that **Ezu**. Because **E** is symmetric (axiom $E2$ is true in **S**), **Ewu** implies **Euw**. Because **E** is transitive (axiom $E3$ is true in **S**),

(11) $$\mathbf{Ezu} \wedge \mathbf{Euw} \rightarrow \mathbf{Ezw}$$

and

(12) $$\mathbf{Ezw} \wedge \mathbf{Ewv} \rightarrow \mathbf{Ezv}.$$

Thus, we deduce, **Ezv**. But by (4), **Ezv** implies $\mathbf{z} \in [\mathbf{v}]$. This contradicts (9). We have shown that for all $\mathbf{u}, \mathbf{v} \in \mathbf{U}$, either $[\mathbf{u}] = [\mathbf{v}]$ or $[\mathbf{u}] \cap [\mathbf{v}] = \varnothing$, and this proves that (2) is true in **S**.

(vi) (3) is true in **S**: Suppose $\mathbf{x}, \mathbf{y} \in \mathbf{U}$ and **Exy**. Because **E** is reflexive, **Eyy**, so $\mathbf{y} \in [\mathbf{y}]$ by (4). Also by (4), **Exy** implies $\mathbf{x} \in [\mathbf{y}]$. Thus, both $\mathbf{x} \in [\mathbf{y}]$ and $\mathbf{y} \in [\mathbf{y}]$.

Conversely, suppose $\mathbf{x}, \mathbf{y} \in [\mathbf{z}]$. By (4), **Exz** and **Eyz**. Because **E** is symmetric, **Eyz** implies **Ezy**. Because **E** is transitive,

$$\mathbf{Exz} \wedge \mathbf{Ezy} \rightarrow \mathbf{Exy}.$$

This proves that (3) is true in **S**, and completes the proof of the theorem. ■

Suppose, for example, we have a model $\mathbf{S} = \langle \mathbf{U}, \mathbf{E} \rangle$ of Θ that has five equivalence classes: $\mathbf{A}_1$, $\mathbf{A}_2$, $\mathbf{A}_3$, $\mathbf{A}_4$, and $\mathbf{A}_5$. Then the universe, **U**, is partitioned by **E** into five pairwise disjoint sets, as shown in Figure 13.1. The partitions mean that if $\mathbf{x}, \mathbf{y} \in \mathbf{A}_i$, then **Exy**, but if $\mathbf{x} \in \mathbf{A}_i$, $\mathbf{y} \in \mathbf{A}_j$, $i \neq j$, then $\neg$**Exy**.

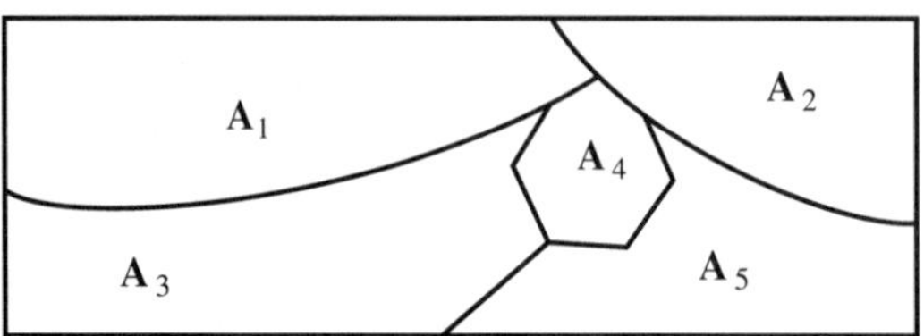

Figure 13.1

Example 13.4 Construct a theory for an equivalence relation so that every model has exactly two equivalence classes.

Solution: The nonlogical symbols of our theory, Θ_1, will be as follows:

E: a binary relation symbol.

a: a constant.

b: a constant.

The axioms are $E1$, $E2$, and $E3$, which were given previously, so that E is an equivalence relation, and also the following:

$E4$: $a \neq b$.

$E5$: $(\forall x)(Exa \vee Exb)$.

$E6$: $\neg Eab$.

Axiom $E4$ says that the two constants are distinct; $E5$, that there are at most two equivalence classes; and $E6$, that there are at least two equivalence classes because a and b do not belong to the same equivalence class.

The reader should check that, indeed, if $\mathbf{S} = \langle \mathbf{U}, \mathbf{E}, \mathbf{a}, \mathbf{b} \rangle$ is any model of Θ_1, then $\mathbf{S}$ must have exactly two equivalence classes.

Example 13.5. Determine whether the following two sentences are consistent with Θ, the theory of an equivalence relation:

$$P_1: (\forall x)(x = a \leftrightarrow \neg Exb),\ P_2: (\exists x)(Eax \wedge Ebx).$$

Solution: We must determine if the following set of five formulas is consistent: $E1$, $E2$, $E3$, $P1$, $P2$.

We shall show that they are inconsistent by giving an informal proof of a contradiction.

Suppose, by $P2$,

(i) $Eax \wedge Ebx$, for some x.

Then

(ii)	$Eax \wedge Exb$	(by (i) and $E2$)
(iii)	Eab	(by (ii) and $E3$)
(iv)	$a = a \leftrightarrow \neg Eab$	(by $P1$)

Thus, from (iii) and (iv), we obtain the contradiction

(v) $Eab \wedge \neg Eab$.

As an exercise, give a formal proof using the preceding informal proof.

Theorem 13.2 and Examples 13.4 and 13.5 illustrate some of the properties of theories in which logicians are interested: characteristics of models of theories, models that have specified properties, and the consistency of theories.

13.6 GROUP THEORY (SUMMARY)

Nonlogical symbols:

e: a constant.

$*$: a unary operation symbol.

$\circ$: a binary operation symbol.

Axioms:

$A1$: $(\forall x)(\forall y)(\forall z)(x \circ (y \circ z) = (x \circ y) \circ z)$

$A2$: $(\forall x)(x \circ e = x)$

$A3$: $(\forall x)(x \circ x^* = e)$

Theorem 1. $(\forall x)(\forall y)(\forall z)(x \circ z = y \circ z \rightarrow x = y)$

Theorem 2. $(\forall x)(x \circ e = e \circ x)$

Theorem 3. $(\forall y)[(\forall x)(x \circ y = x) \rightarrow y = e]$

Theorem 4. $(\forall x)(x \circ x = e) \rightarrow (\forall x)(\forall y)(x \circ y = y \circ x)$

Theorem 5. $(\forall x)(x \circ x^* = x^* \circ x)$

Theorem 6. $(\forall x)(\forall y)(\forall z)(z \circ x = z \circ y \rightarrow x = y)$

Theorem 7. $(\forall x)(\forall y)(x \circ y = e \rightarrow y = x^*)$

Theorem 8. $(\forall x)[(x^*)^* = x]$

Theorem 9.

(a) $(\forall x)(\forall y)(\exists z)(x = y \circ z)$

(b) $(\forall x)(\forall y)(\exists z)(x = z \circ y)$

Theorem 10. $(\forall x)(x \circ x = e) \rightarrow (\forall x)(x = x^*)$

Theorem 11. $(\forall x)(\forall y)((x \circ y)^* = y^* \circ x^*)$

Theorem 12. $(\exists y)(\forall x)(x \circ y = y \circ x)$

EXERCISES 13

A. Give a formal proof for each of the following informal proofs. (Answer to 3.)

(1) **Premises:** $(\exists x)Px$, $(\exists x)Qx$, $(\forall x)[Px \rightarrow (\forall y)(Qy \rightarrow Rxy)]$
Conclusion: $(\exists x)(\exists y)Rxy$

Informal Proof: Suppose

(i) Pu and (ii) Qv.

It follows from premise 3 that

(iii) $Pu \to (\forall y)(Qy \to Ruy)$.

From (i) and (iii), we deduce

(iv) $(\forall y)(Qy \to Ruy)$.

Then (iv) implies

(v) $Qv \to Ruv$.

Using (ii) and (v), we can deduce

(vi) Ruv,

from which the conclusion follows.

(2) **Premises:** $(\exists x)[Ax \wedge (\forall y)(By \to Cxy)]$, $(\forall x)(\forall y)(Ax \wedge Cxy \to Dy)$, $(\forall x)(Dx \to Bx)$

Conclusion: $(\forall x)(Bx \leftrightarrow Dx)$

Informal Proof: First assume

(i) Bx.

We can deduce from premise 1 that

(ii) $Au \wedge (\forall y)(By \to Cuy)$.

We can deduce from (ii) that

(iii) $Bx \to Cux$.

From (i) and (iii), deduce

(iv) Cux.

From premise 2, we obtain

(v) $Au \wedge Cux \to Dx$.

Thus from (ii), (iv), and (v), we can deduce Dx.

Conversely, we obtain

(vi) $Dx \to Bx$

from premise 3.

(3) **Premises:** $(\forall x)(\forall y)(\forall z)(Rxy \wedge Ryz \to Rxz)$, $(\forall x)(\forall y)(Rxy \to Ryx)$, $(\forall x)(\exists y)Rxy$

Conclusion: $(\forall x)Rxx$

Informal Proof: From premise 3, we can deduce

(i) Rxu, for some u.

By premise 2,

(ii) $Rxu \to Rux$.

Thus, from (i) and (ii), we deduce

(iii) Rux.

By premise 1,

(iv) $Rxu \land Rux \to Rxx$.

Thus from (i), (iii), and (iv), we deduce Rxx.

(4) Premises: $(\forall x)(\forall y)(x \circ y = y \circ x)$, $(\forall x)(\forall y)(x^* \circ y = (x \circ y)^*)$
Conclusion: $(\forall x)(\forall y)(x^* \circ y = x \circ y^*)$

Informal Proof: By premise 2,

$$
\begin{aligned}
x^* \circ y &= (x \circ y)^* \\
&= (y \circ x)^* && \text{(by preceding line and premise 1)} \\
&= y^* \circ x && \text{(by preceding line and premise 2)} \\
&= x \circ y^* && \text{(by preceding line and premise 1)}
\end{aligned}
$$

(5) Premises: $(\forall x)(\forall y)(\forall z)[x \circ (y \circ z) = (x \circ y) \circ z]$, $(\forall x)(\forall y)(\exists z)\ (x = y \circ z)$, $(\forall x)(\forall z)(\exists y)(x = y \circ z)$
Conclusion: $(\forall y)(\forall z)[(\forall x)(x \circ y = x \land x \circ z = x) \to y = z]$

Informal Proof: Suppose

(i) $x \circ y = x \land x \circ z = x$, for every x.

By premise 2, there is a u so that

(ii) $z = y \circ u$.

(iii) $y \circ z = y$. (by (i))

Therefore, by (ii) and (iii),

$$
\begin{aligned}
\text{(iv)}\quad y &= y \circ (y \circ u) \\
&= (y \circ y) \circ u && \text{(by preceding line and premise 1)} \\
&= y \circ u && \text{(by preceding line and (i))} \\
&= z && \text{(by preceding line and (ii))}
\end{aligned}
$$

B. Give (a) an informal proof and (b) a formal proof of each of the following theorems of Group Theory. You may use preceding theorems in the proof of later ones. (See the list of theorems of Group Theory at the end of Chapter 13.) (Answers to 1 and 5.)

(1) *Theorem 4: Sufficient Condition for Commutative Law*:

$$(\forall x)(x \circ x = e) \rightarrow (\forall x)(\forall y)(x \circ y = y \circ x).$$

(An informal proof was given in the text.)

(2) *Theorem 5: Commutative Law for Inverses*:

$$(\forall x)(x \circ x^* = x^* \circ x).$$

(3) *Theorem 6: Left Cancellation Law*:

$$(\forall x)(\forall y)(\forall z)(z \circ x = z \circ y \rightarrow x = y).$$

(4) *Theorem 7: Uniqueness of the Inverse*:

$$(\forall x)(\forall y)(x \circ y = e \rightarrow y = x^*).$$

(5) *Theorem 8: Inverse of the Inverse is the Original Element*:

$$(\forall x)[(x^*)^* = x].$$

(6) *Theorem 9: Solution of Equations*:
 (a) $(\forall x)(\forall y)(\exists z)(x = y \circ z)$
 (b) $(\forall x)(\forall y)(\exists z)(x = z \circ y)$

(7) *Theorem 10: Sufficient Condition for* $x = x^*$:

$$(\forall x)(x \circ x = e) \rightarrow (\forall x)(x = x^*).$$

(8) *Theorem 11: Inverse of a Product*:

$$(\forall x)(\forall y)((x \circ y)^* = y^* \circ x^*).$$

(9) *Theorem 12: Existence of an Element that Commutes with Every Element*:

$$(\exists y)(\forall x)(x \circ y = y \circ x).$$

C. For each of the following sentences, show that neither it nor its negation is a theorem of Group Theory. (Answers to 1, 5, and 10.)

(1) $(\exists x)(\forall y)(x = y)$.
(2) $(\exists x)(\exists y)[x \neq y \wedge (\forall z)(z = x \vee z = y)]$.
(3) $(\forall x)(\forall y)(x = x \circ y)$.
(4) $(\exists y)(\forall x)(x \circ y = y)$.
(5) $(\forall x)(\exists y)(x \circ y = y)$.
(6) $(\exists x)(x \neq x^*) \rightarrow (\exists x)(\exists y)(x \circ y \neq y \circ x)$.
(7) $(\forall x)(\forall y)((x \circ y)^* = x^* \circ y^*)$.
(8) $(\forall x)(\forall y)(x \circ y = y \circ x \rightarrow y = x^* \vee y = e)$.
(9) $(\exists y)(\forall x)(x \circ y = y \circ x^*)$.
(10) $(\forall x)(\forall y)(\exists z)(x = y \circ z \wedge x = z \circ y)$.
(11) $(\forall x)(\forall y)[(\exists z)(x \circ z = z \circ y) \rightarrow x \circ y = y \circ x]$.

D. Let $\mathbf{S} = \langle \mathbf{U}, \circ, {}^*, \mathbf{e} \rangle$. Determine in which of the following **S** is a group (a model of Group Theory). ($\mathbb{N}$ = set of non-negative integers; $\mathbb{Z}$ = set of all integers; $\mathbb{R}^+$ = the set of all positive real numbers; $\mathbb{R}$ = set of all real numbers.) (Answers to 1, 7, 12, and 15.)

(1) $\mathbf{U} = \{-1, 1\}$, $\mathbf{x} \circ \mathbf{y} = \mathbf{x} \cdot \mathbf{y}$, $\mathbf{x}^* = \mathbf{x}$, $\mathbf{e} = 1$.
(2) $\mathbf{U} = \{0, 1\}$, $\mathbf{x} \circ \mathbf{y} = \mathbf{x} + \mathbf{y}$, $\mathbf{x}^* = \mathbf{x}$, $\mathbf{e} = 0$.
(3) $\mathbf{U} = \{-1, 0, 1\}$, $\mathbf{x} \circ \mathbf{y} = \mathbf{x} + \mathbf{y}$, $\mathbf{x}^* = -\mathbf{x}$, $\mathbf{e} = 0$.
(4) $\mathbf{U} = \{2^z : z \in \mathbb{Z}\}$, $\mathbf{x} \circ \mathbf{y} = \mathbf{x} \cdot \mathbf{y}$, $\mathbf{x}^* = 1/\mathbf{x}$, $\mathbf{e} = 1$.
(5) $\mathbf{U} = \mathbb{R}^+$, $\mathbf{x} \circ \mathbf{y} = \mathbf{x} \cdot \mathbf{y}$, $\mathbf{x}^* = 1/\mathbf{x}$, $\mathbf{e} = 1$.
(6) $\mathbf{U} = \mathbb{R}$, $\mathbf{x} \circ \mathbf{y} = \mathbf{x} + \mathbf{y}$, $\mathbf{x}^* = -\mathbf{x}$, $\mathbf{e} = 0$.
(7) $\mathbf{U} = \{1/2, 1, 2\}$, $\mathbf{x} \circ \mathbf{y} = \mathbf{x} \cdot \mathbf{y}$, $\mathbf{x}^* = 1/\mathbf{x}$, $\mathbf{e} = 1$.
(8) $\mathbf{U} = \{a, b, c\}$, $\mathbf{x}^* - \mathbf{x}$, $\mathbf{e} - \mathbf{a}$.

∘	a	b	c
a	a	b	c
b	b	a	b
c	c	b	a

$\mathbf{x} \circ \mathbf{y}$

(9) $\mathbf{U} = \{0, 1, 2, 3\}$, $\mathbf{x} \circ \mathbf{y} = \mathbf{x} + \mathbf{y}$ (modulo 4), $0^* = 0$, $1^* = 3$, $2^* = 2$, $3^* = 1$, $\mathbf{e} = 0$.

In exercises (10) through (16), determine whether * and **e** can be defined so that **S** is a group.

(10) $\mathbf{U} = \mathbb{Z}$, $\mathbf{x} \circ \mathbf{y} = \mathbf{x}$.
(11) $\mathbf{U} = \mathbb{Z}$, $\mathbf{x} \circ \mathbf{y} = \mathbf{x} - \mathbf{y}$.
(12) $\mathbf{U} = \mathbb{Z}$, $\mathbf{x} \circ \mathbf{y} = \mathbf{x} + 2\mathbf{y}$.
(13) $\mathbf{U} = \mathbb{N}$, $\mathbf{x} \circ \mathbf{y} = \mathbf{x} + \mathbf{y}$.
(14) $\mathbf{U} = \{-1, 0, 1\}$.

∘	−1	0	1
−1	1	0	−1
0	0	0	0
1	−1	0	1

$\mathbf{x} \circ \mathbf{y}$

(15) $\mathbf{U} = \{a, b, c\}$.

∘	a	b	c
a	c	a	b
b	a	b	c
c	b	c	a

$\mathbf{x} \circ \mathbf{y}$

(16) $\mathbf{U} = \{\mathbf{a}, \mathbf{b}, \mathbf{c}\}$.

$\circ$	a	b	c
a	a	b	c
b	b	b	a
c	c	c	a

$\mathbf{x} \circ \mathbf{y}$

E. Determine whether each of the following sets of sentences is consistent with Θ, the theory of an equivalence relation. If the set is consistent with Θ, give a model of Θ in which the sentences are true; if not, give (a) an informal proof and (b) a formal proof of a contradiction. (Answers to 1, 6, and 10.)

(1) $(\forall x)(\exists y)\neg Exy$.
(2) $(\forall x)(\forall y)(Exy \leftrightarrow f(x) \neq f(y))$.
(3) $(\forall x)(\forall y)(Px \wedge Py \leftrightarrow Exy)$, $(\exists x)(\exists y)(x \neq y)$.
(4) $(\exists y)(\forall x)(f(x) = f(y))$, $(\forall x)(\forall y)(f(x) = f(y) \rightarrow x = y)$, $(\forall x)(\exists y)\neg Exy$.
(5) $(\forall x)(\exists y)(Exy \wedge x \neq y)$, $(\forall x)(\exists y)\neg Exy$.
(6) $(\exists x)(\exists y)(\exists z)(x \neq y \wedge Exz \wedge Eyz \wedge \neg Exy)$.
(7) $(\forall x)(\forall y)Exy$, $(\exists x)(\exists y)(\exists z)(x \neq y \wedge x \neq z \wedge y \neq z)$.
(8) $(\exists y)(\exists z)(\forall x)(x \neq y \leftrightarrow Exz)$, $(\forall y)(\forall z)(\exists z)(Eyx \wedge Ezx)$.
(9) $(\forall x)(\exists y)(Exy \wedge x \neq y)$, $(\forall x)(\forall y)\neg Exf(y)$, $(\forall x)(\exists y)(f(y) = x)$.
(10) $(\forall x)(Exa \vee Exb \vee Exc)$, $(\forall x)(\forall y)(x \neq y \rightarrow \neg Exy)$, $a \neq b$, $a \neq c$, $b \neq c$.

F. Let Γ be the following theory:

Nonlogical symbols: $\circ$, a binary operation symbol; e, a constant.

Axioms:

$B1$: $(\forall x)(\forall y)(\forall z)[x \circ (y \circ z) = (x \circ y) \circ z]$.

$B2$: $(\forall x)(x \circ e = x)$.

$B3$: $(\forall x)(\forall y)(\exists z)(x \circ z = y \wedge z \circ x = y)$.

(Answers to 1(i), 4(ii), and 5(i).)

(1) Give (a) an informal proof and (b) a formal proof of each of the following theorems of Γ:

(i) $(\forall x)(\forall y)(\forall z)(x \circ z = y \circ z \rightarrow x = y)$.
(ii) $(\forall x)(\forall y)(\forall z)(z \circ x = z \circ y \rightarrow x = y)$.
(iii) $(\forall x)(x \circ e = e \circ x)$.
(iv) $(\forall x)(\exists y)(x \circ y = e)$.
(v) $(\forall x)(\forall y)(\forall z)(x \circ y = e \wedge x \circ z = e \rightarrow y = z)$.

(2) Introduce a new function symbol of one variable, h, and a defining axiom $B4$ so that h is the inverse operation.

(3) Show that each of the following are logical consequences of $\{B1, B2, B3, B4\}$ by giving (a) an informal proof and (b) a formal proof:

(i) $(\forall x)(x \circ h(x) = e)$.

(ii) $(\forall x)(x \circ h(x) = h(x) \circ x)$.

(iii) $(\forall x)(h(h(x)) = x)$.

(iv) $(\forall x)(\forall y)(x \circ y = y \circ x)$.

(v) $(\forall x)(\forall y)(h(x \circ y) = h(y) \circ h(x))$.

(4) Give a model $\mathbf{S} = \langle \mathbf{U}, \circ, \mathbf{e} \rangle$ of Γ such that

(i) **U** has two elements.

(ii) **U** has three elements.

(iii) **U** has an infinite number of elements.

(iv) $\mathbf{S} \models (\exists x)(\exists y)(x \neq y) \wedge (\forall x)(x \circ x = e)$.

(v) $\mathbf{S} \models (\forall x)(\forall y)(x \circ y = x \vee x \circ y = y)$.

(5) Show by giving (a) an informal proof and (b) a formal proof of a contradiction that each of the following sets of formulas is inconsistent with Γ. (Assume "h" is defined as in F (2). You may use any of the theorems given in F (1) or (3) in your proof.)

(i) $(\exists x)(\exists y)(x \neq y)$, $(\forall x)(x \circ x = x)$.

(ii) $(\exists x)(\exists y)(x \neq y)$, $(\forall x)(\forall y)(x \circ y = x \vee x \circ y = y)$.

(iii) $(\forall x)(\forall y)(\forall z)(x \circ y = x \circ z \rightarrow y \circ z = e)$, $(\exists x)(\exists y)(x \neq y \wedge y = h(x))$.

(iv) $(\forall x)(x \circ x = e)$, $(\exists x)(x \neq h(x))$.

(v) $(\exists x)(\exists y)(h(x) = h(y) \wedge x \neq y)$.

G. What axioms must be added to the axioms $A1$, $A2$, and $A3$ of Group Theory to ensure that every model

$$\mathbf{S} = \langle \mathbf{U}, \circ, *, \mathbf{e} \rangle$$

satisfies the following:

(1) **U** has exactly two elements.

(2) **U** has at least three elements.

(3) $\circ$ is a commutative operation.

(4) **S** is a linearly ordered group.

(5) There is an equivalence relation that has exactly three equivalence classes.

(For (4) and (5), you must add new nonlogical symbols).

H. Suppose Γ is Group Theory (axioms $A1$, $A2$, and $A3$) and we add a new 2-place predicate symbol, R. In each of the following, assume the given sentences are added as axioms of Γ. In each case, determine if R is an equivalence relation. Give reasons for your answer. (Answers to 2 and 5.)

(1) $(\forall x)(\forall y)(Rxy \leftrightarrow x^* = y^*)$.

(2) $(\forall x)(\forall y)(Rxy \leftrightarrow x = y^*)$.

(3) $(\forall x)(x \circ x = e)$, $(\forall x)(\forall y)(Rxy \leftrightarrow x = y^*)$.

(4) $(\forall x)(\forall y)(Rxy \leftrightarrow x \circ y = y \circ x)$.

(5) $(\forall x)(x = x^*)$, $(\forall x)(\forall y)(Rxy \leftrightarrow x \circ y = y \circ x)$.

(6) $(\forall x)(\forall y)[Rxy \leftrightarrow (\exists z)(x \circ z = z \circ y)]$.

(7) $(\forall x)(\forall y)(x \circ y = y \circ x)$, $(\forall x)(\forall y)[Rxy \leftrightarrow (\exists z)(x \circ z = y \wedge y \circ z = x)]$.

Chapter 14

Peano Arithmetic and Mathematical Induction

In this chapter, we are going to formalize a theory that every school child learns in elementary school—namely, the theory of arithmetic. We shall use a set of axioms that is a modification of a system that was given by Richard Dedekind in 1879. In 1889, however, Giuseppe Peano published a formulation of the axioms that was widely read; consequently, these axioms are called *Peano's Axioms* and the resulting theory is called *Peano Arithmetic.* The intended model of the system is

$$\mathbf{S}_{\mathbb{N}} = \langle \mathbb{N}, \mathbf{+}, \boldsymbol{\cdot}, \mathbf{<}, \mathbf{0}, \mathbf{1} \rangle$$

where $\mathbb{N}$ is the set of non-negative integers (natural numbers).

14.1 THE AXIOM SYSTEM

The nonlogical symbols of our language are as follows:

Binary operation symbols: $+$ and $\cdot$.

Constants: 0 and 1.

2-place relation symbol: $<$.

Notice that $+$, $\cdot$, $<$, 0 and 1 are symbols in our formal language. In the standard interpretation in the structure $\mathbf{S}_{\mathbb{N}}$ above we use boldface type. Thus, in $\mathbf{S}_{\mathbb{N}}$, $\mathbf{+}$ is ordinary addition; $\boldsymbol{\cdot}$ is ordinary multiplication; $\mathbf{<}$ is less than; $\mathbf{0}$ is the zero element of $\mathbb{N}$; and $\mathbf{1}$ is the unit element of $\mathbb{N}$. It should be clear from the context whether we are in the formal language or in an interpretation.

The following will be the axioms of our system:

$P1$: $(\forall x)(\forall y)(x + y = y + x)$	(Commutative Rule for $+$)
$P2$: $(\forall x)(\forall y)(x \cdot y = y \cdot x)$	(Commutative Rule for $\cdot$)
$P3$: $(\forall x)(\forall y)(\forall z)[x + (y + z) = (x + y) = z]$	(Associative Rule for $+$)
$P4$: $(\forall x)(\forall y)(\forall z)[x \cdot (y \cdot z) = (x \cdot y) \cdot z]$	(Associative Rule for $\cdot$)
$P5$: $(\forall x)(\forall y)(\forall z)[x \cdot (y + z) = (x \cdot y) + (x \cdot z)]$	(Distributive Rule)
$P6$: $(\forall x)(x + 0 = x)$	(Property of 0)
$P7$: $(\forall x)(x \cdot 1 = x)$	(Property of 1)
(For each x, $x + 1$ is called the *successor* of x)	
$P8$: $(\forall x)(x + 1 \neq 0)$	(0 is not a successor)
$P9$: $(\forall x)(\forall y)(x + 1 = y + 1 \rightarrow x = y)$	(Successor function is 1–1)
$P10$: $(\forall x)(\forall y)(x < y \leftrightarrow (\exists z)(z \neq 0 \wedge y = x + z))$	(Property of $<$)

The last axiom, $P_\varphi 11$, is actually an *axiom schema*—a rule for producing axioms.

If φ is any formula in the language of Peano Arithmetic with x as a free variable, and if y is a variable that does not occur in φ, then the following formula is an axiom:

$P_\varphi 11$: $\varphi[0|x] \wedge (\forall y)(\varphi[y|x] \rightarrow \varphi[y + 1|x]) \rightarrow (\forall x)\varphi$.

The axiom schema $P_\varphi 11$ is called *mathematical induction.* It is an abbreviation for an infinite number of axioms because there is an axiom for each formula φ that has at least one free variable and there are an infinite number of such formulas in our language. (Similarly, the Rule of Substitution for Identity, Rule *Ib,* is a schema for producing an infinite number of logical theorems.) We shall call axiom schema $P_\varphi 11$, *MI* and call the theory of Peano Arithmetic, Π.

14.2 MATHEMATICAL INDUCTION

Before continuing our discussion of the theory Π, we shall try to supply motivation for the principle of mathematical induction. The discussion will be quite informal.

The word "induction" means generalizing from particular instances—that is, obtaining a general rule from particular instances of the rule. For example, let S_n be the sum of the first n powers of 2:

$$S_n = 2^0 + 2^1 + 2^2 + \cdots + 2^n.$$

Let us try to find a formula for S_n. A natural procedure would be to calculate S_n for some values of n and see if some pattern emerges:

n	0	1	2	3	4	$\cdots$
S_n	1	3	7	15	31	$\cdots$

It appears that each value of S_n is one less than some power of 2, so we make the educated guess that

(*) $$S_n = 2^{n+1} - 1.$$

However, we have only shown with our table that (*) is true for $n = 0, 1, 2, 3$, and 4. No matter how much we extend the table, we cannot prove that (*) holds for all non-negative values of n because there are an infinite number of cases to consider.

This is where mathematical induction (*MI*) comes to the rescue. The formula φ in *MI* is $S_n = 2^{n+1} - 1$. (Strictly speaking, this is not a formula in the language of Π because neither sums like S_n nor exponentiation has been defined. However, because we are just using this notation for examples, we shall use it here and define it in Chapter 15. Also, we shall use n instead of x as the free variable in φ. It is customary to use the letters i, j, k, m, and n as variables that range over the set of natural numbers. We shall discuss formulas like (*) in more detail in the next chapter.)

Our purpose is to prove $(\forall n)\varphi$. If we are to use *MI*, we must prove two things:

1. $\varphi[0|n]$ (The base step)
2. $(\forall m)(\varphi[m|n] \to \varphi[m + 1|n])$ (The induction step)

Then, $(\forall n)\varphi$ follows by *MI*. The base step allows us to assert

(i) $\varphi[0|n]$.

Using the induction step, we can derive

(ii) $\varphi[0|n] \to \varphi[1|n]$.

From (i) and (ii), we can deduce

(iii) $\varphi[1|n]$.

Then, using the induction step again, we obtain

(iv) $\varphi[1|n] \to \varphi[2|n]$,

so from (iii) and (iv) we can deduce

(v) $\varphi[2|n]$.

Using steps 1 and 2 above, we can continue this process as long as we like and deduce $\varphi[k|n]$ for any natural number k. This is the intuitive idea behind *MI* and explains why it is a reasonable assumption to make.

In the informal proofs that we give in this section, we shall assume that the reader is familiar with elementary algebra. Frequently, we shall just state "algebra" as justification for a step in an informal proof.

The informal proof of (*) using *MI* follows:

First we shall prove $\varphi[0|n]$, that is, $2^0 = 2^{0+1} - 1$. Using algebra we obtain $2^0 = 1$ and $2^{0+1} - 1 = 2 - 1 = 1$. Thus, $\varphi[0|n]$ holds.

To prove the induction step, we must prove

(1) $$\varphi[m|n] \rightarrow \varphi[m + 1|n],$$

for each $m \geq 0$. Formula (1) is a conditional. Normally, to prove a conditional, we take the antecedent, $\varphi[m|n]$, as a premise, derive the consequent, $\varphi[m + 1|n]$, and then use Rule *CP*. The premise $\varphi[m|n]$ is called the *induction hypotheses* (*IH*). Notice that when $\varphi[m|n]$ is taken as a premise, m is flagged, so we *cannot* deduce $(\forall m)\varphi[m|n]$. This is an error that is easy to make in an informal proof, so *be careful.*

In our example, to prove the induction step, we take as a premise the induction hypotheses

(2) $$S_m = 2^{m+1} - 1, \qquad (IH)$$

where m is a non-negative integer. We must prove

(3) $$S_{m+1} = 2^{(m+1)+1} - 1.$$

The informal proof of (3) is as follows:

(4)
$$\begin{aligned} S_{m+1} &= S_m + 2^{m+1} && \text{(Definition of } S_{m+1}\text{)} \\ &= (2^{m+1} - 1) + 2^{m+1} && (IH) \\ &= 2 \cdot 2^{m+1} - 1 && \text{(Algebra)} \\ &= 2^{(m+1)+1} - 1 && \text{(Algebra)} \end{aligned}$$

Now, it follows by *MI* that

$$S_n = 2^{n+1} - 1, \text{ for all } n \geq 0.$$

On line (4) above, we use the fact that the sum of the first $m + 1$ powers of 2 is obtained from the sum of the first m powers of 2 by adding 2^{m+1}. This is an example of *inductive reasoning,* and it plays an important role in the use of mathematical induction. Other examples of inductive reasoning include the following:

Calculate 2^{n+1} by multiplying 2^n by 2.

Calculate $(n + 1)!$ by multiplying $n!$ by $(n + 1)$.

We shall give additional examples later in illustrative examples.

In practice, we do not have to start the base step of *MI* at $n = 0$. We can start at any integer k. Formally, if we want to prove ψ for all $n \geq k$, let φ be the formula

$$\psi \vee n = 0 \vee n = 1 \vee \cdots \vee n = k - 1$$

and then apply *MI* to φ.

Example 14.1

Give an informal proof using *MI* that

$$2^n > n^2 + 19, \quad \text{for } n \geq 6.$$

Solution: Here φ is the formula

(5) $$2^n > n^2 + 19.$$

Base Step: The base step is to prove $\varphi[6|n]$, $2^6 > 6^2 + 19$. Calculating the left and right sides of the inequality, we see that

$$2^6 = 64 \qquad \text{and} \qquad 6^2 + 19 = 55.$$

This proves $\varphi[6|n]$ because $64 > 55$.

Induction Step: Suppose that for $m \geq 6$ we have,

(6) $$2^m > m^2 + 19. \qquad (IH)$$

(Now we want to prove

(7) $$2^{m+1} > (m + 1)^2 + 19.)$$

To prove (7), we start with the fact that

(8)
$$\begin{aligned}
2^{m+1} &= 2 \cdot 2^m \\
&> 2(m^2 + 19) && (IH) \\
&= m^2 + m^2 + 38 && \text{(Algebra)} \\
&> m^2 + 2m + 38 && \text{(Algebra, } m^2 > 2m \text{ because } m \geq 6) \\
&> m^2 + 2m + 1 + 19 && \text{(Algebra } 38 > 20) \\
&= (m + 1)^2 + 19 && \text{(Algebra)}
\end{aligned}$$

Thus, it follows by *MI* that (5) is true for all $n \geq 6$. (Actually, to prove the induction step, all that is required is that $m > 2$, but the base step is false if $m < 6$.)

Another way to prove the induction step in Example 14.1 is to start with line (6), the induction hypotheses, as we did, but then multiply both sides of line (6) by 2 and continue, using steps similar to those in (8) above. However, it is always *wrong* to assume what you are trying to prove. (Why?) Yet, we frequently see students doing just that when they are trying to prove the induction step, so *be careful.*

Example 14.2

Find a simple formula for the product

$$P_n = (1 - x)(1 + x)(1 + x^2)(1 + x^4) \cdots (1 + x^{2^n}),$$

and give an informal proof using mathematical induction that your formula is correct.

Solution: We shall make a table and see if we can find a pattern.

n	$\mathbf{P_n}$
0	$(1 - x)(1 + x) = 1 - x^2$
1	$(1 - x^2)(1 + x^2) = 1 - x^4$
2	$(1 - x^4)(1 + x^4) = 1 - x^8$
3	$(1 - x^8)(1 + x^8) = 1 - x^{16}$

Notice that we are using inductive reasoning to calculate the table:

$$P_{n+1} = P_n \cdot (1 + x^{2^{n+1}}).$$

We see from the table that a pattern does appear, and we guess that

(9) $$P_n = 1 - x^{2^{n+1}}.$$

We shall try to prove (9) for all $n \geq 0$ using *MI*.

Base Step: When $n = 0$, formula (9) becomes

(10) $$(1 - x)(1 + x) = 1 - x^{2^{0+1}}.$$

Using algebra, both the left and right side of (10) is equal to $1 - x^2$; thus, (10) holds.

Induction Step: Suppose, for $m \geq 0$,

(11) $$P_m = 1 - x^{2^{m+1}}. \qquad (IH)$$

Using inductive reasoning,

$$\begin{aligned} P_{m+1} &= P_m(1 + x^{2^{m+1}}) \\ &= (1 - x^{2^{m+1}})(1 + x^{2^{m+1}}) && (IH) \\ &= 1 - x^{2^{m+2}} && \text{(Algebra)} \end{aligned}$$

Thus, it follows by *MI* that (9) holds for all $n \geq 0$.

We do have to be careful about jumping to conclusions. For example, the values of the polynomial

$$f(n) = n^2 + n + 41$$

for $n = 0, 1, 2, \ldots, 39$ are given in the following table:

$f(0) = 41$	$f(10) = 151$	$f(20) = 461$	$f(30) = 971$
$f(1) = 43$	$f(11) = 173$	$f(21) = 503$	$f(31) = 1{,}033$
$f(2) = 47$	$f(12) = 197$	$f(22) = 547$	$f(32) = 1{,}097$
$f(3) = 53$	$f(13) = 223$	$f(23) = 593$	$f(33) = 1{,}163$
$f(4) = 61$	$f(14) = 251$	$f(24) = 641$	$f(34) = 1{,}231$
$f(5) = 71$	$f(15) = 281$	$f(25) = 691$	$f(35) = 1{,}301$
$f(6) = 83$	$f(16) = 313$	$f(26) = 743$	$f(36) = 1{,}373$
$f(7) = 97$	$f(17) = 347$	$f(27) = 797$	$f(37) = 1{,}447$
$f(8) = 113$	$f(18) = 383$	$f(28) = 853$	$f(38) = 1{,}523$
$f(9) = 131$	$f(19) = 421$	$f(29) = 911$	$f(39) = 1{,}601$

$$f(n) = n^2 + n + 41$$

It is not very difficult to check that the values of $f(n)$ for $n = 0, 1, 2, \ldots, 39$ are all primes. It would be easy to conclude that $f(n)$ is a prime for all $n \geq 0$; however, this conclusion would be false.

$$\begin{aligned} f(40) &= (40)^2 + 40 + 41 \\ &= (41) \cdot (41). \end{aligned}$$

Thus, $f(40)$ is clearly not a prime. The moral of this story is that even if your hypothesis is true for many, many values of n, it might not be true for all values of n. No finite number of special cases can prove a general result that holds for an infinite number of cases. Always prove your guess using mathematical induction or some other valid procedure.

We shall now give two examples to show that *both* the base step *and* the induction step in *MI* must be true in order to derive the conclusion of *MI* that φ holds for all n.

Example 14.3

(Base step false)

Let

$$S_n = 2 + 4 + 6 + \cdots + 2n.$$

Show that

(12) $$(S_m = m(m + 1) + 5) \rightarrow (S_{m+1} = (m + 1)(m + 2) + 5)$$

for all $m \geq 1$, but it is not true that

(13) $$S_n = n(n + 1) + 5$$

for any $n \geq 1$.

Solution: To prove (12), assume

(14) $$S_m = m(m + 1) + 5.$$

Using inductive reasoning,

$$\begin{aligned} S_{m+1} &= S_m + 2(m+1) \\ &= m(m+1) + 5 + 2(m+1) && (14) \\ &= (m+1)(m+2) + 5 && \text{(Algebra)} \end{aligned}$$

This proves (12). However, (13) is false for all n because S_n, being a sum of even integers, must be an even integer, but $n(n + 1) + 5$ is an odd integer for all n. (The product $n(n + 1)$ is even, and the sum of an even and an odd integer is odd.) Thus, (13) is false for all n.

Example 14.4

Find the error in the following argument:

Premise: All cats, in a set of n cats, have the same sex, $n \geq 1$.

Base Step: The premise is clearly true if $n = 1$.

Induction Step: Suppose the statement is true when there are m cats in a set, (*IH*). Then, whenever there are m cats in a set, they have the same sex. Suppose we have $m + 1$ cats, $C_1, C_2, C_3, C_4, \ldots, C_{m+1}$. If we take one cat away, it follows from the induction hypothesis that the remaining m cats all have the same sex. Thus, it follows that the cats

(*) $\qquad C_1, C_3, C_4, \ldots, C_{m+1}$

all have the same sex, and the cats

(**) $\qquad C_2, C_3, C_4, \ldots, C_{m+1}$

all have the same sex. Therefore, each of the $m - 1$ cats $C_3, C_4, \ldots, C_{m+1}$ has the same sex as C_3; by (*), C_1 has the same sex as C_3; and by (**), C_2 has the same sex as C_3. Thus, it follows that each of the $m + 1$ cats, $C_1, C_2, C_3, C_4, \ldots, C_{m+1}$, has the same sex as C_3, so they all have the same sex.

Solution: Clearly, the statement is false. However, if the base step and the induction step were true, it would follow by mathematical induction that the statement is true, and we would have a contradiction. The base step is true, so there must be an error in the induction step. Let us examine it more closely.

Let φ be the statement we are trying to prove. Notice the proof we gave in the induction step that $\varphi[m|n] \rightarrow \varphi[m + 1|n]$ is valid as long as $m \geq 2$. However, if $m = 1$ and we let C_1 and C_2 be two cats, our argument no longer holds. The fact that C_1 has the same sex as itself and C_2 has the same sex as itself does not imply that C_1 and C_2 have the same sex. Thus, we have a break in our argument, and this one small break is sufficient to destroy the whole argument.

Examples 14.3 and 14.4 illustrate that both the base step and the induction step must be true in order to deduce the conclusion of *MI*.

Our next example in this section is an application of *MI* in geometry.

Example 14.5

Find a formula for the number of lines connecting n points in a plane, no three of which are collinear, and give an informal proof using *MI* that your answer is correct.

Solution: We shall look at several special cases.

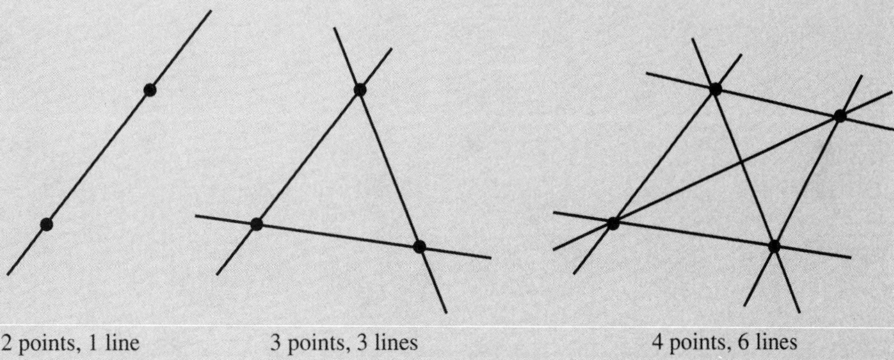

2 points, 1 line 3 points, 3 lines 4 points, 6 lines

Before long, our diagram will be too complicated to learn anything from it, so let's use some inductive reasoning. Let's look at the case in which there are four points (and six lines) and see what happens when we add a fifth point.

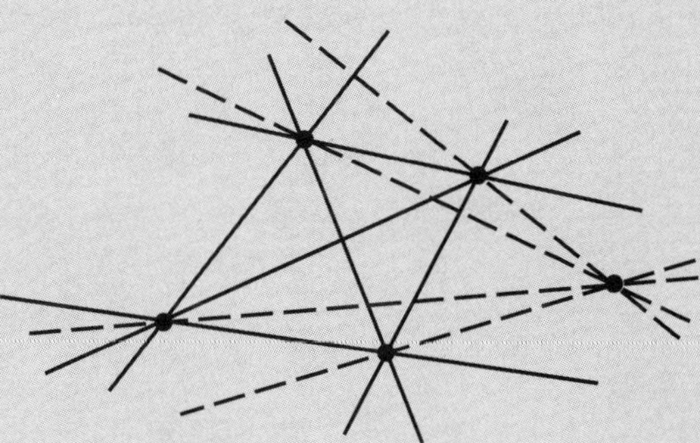

(The dotted lines are the additional lines connecting the fifth point to the other four.) When we add the fifth point, the number of new lines is the same as the number of points that were there previously (in this case, four).

Thus, when there are five points, there are $6 + 4 = 10$ lines. It is easy to check that this inductive principle is always true; that is, for each new point that is included, the number of lines is the number of lines there previously plus the number of points there previously.

The following pattern seems to be emerging:

Number of Points (n)	**Number of Lines**
2	1
3	1 + 2
4	1 + 2 + 3
5	1 + 2 + 3 + 4
6	1 + 2 + 3 + 4 + 5
⋮	⋮
n	S_n

where $S_n = 1 + 2 + 3 + \cdots + n - 1$. It can be shown that $S_n = \frac{(n-1)n}{2}$. (See Exercise 14 A (1).) Therefore, the statement we shall attempt to prove for $n \geq 2$ using *MI* is

φ: The numbers of lines connecting n points in a plane, no three of which are collinear, is $\frac{(n-1)n}{2}$.

Base Step: When $n = 2$, there is one line and $\frac{(2-1)2}{2} = 1$.

Induction Step: Suppose $m \geq 2$ and $\varphi[m|n]$. Suppose we have $m + 1$ points in a plane, no three of which are collinear. Then the number of lines connecting these $m + 1$ points, say L, is the number of lines connecting m of the points, say M, plus number of lines connecting the $m + 1$-st point to the first m points, m. Thus,

$$
\begin{aligned}
L &= M + m \\
&= \frac{(m-1)m}{2} + m && (IH) \\
&= \frac{m(m+1)}{2} && \text{(Algebra)}
\end{aligned}
$$

This proves the induction step, so it follows from *MI* that φ is true for all $n \geq 2$.

Example 14.6 Suppose we have an infinite supply of 3-cent stamps and 5-cent stamps. Give an informal proof using *MI* that it is possible to obtain exact postage of every denomination larger than 7 cents using just 3-cent and 5-cent stamps.

Solution: Base Step: In this problem, the base step is when $n = 8$, and 8-cent postage can be obtained using one 3-cent stamp and one 5-cent stamp.

Induction Step: The induction hypotheses is the premise that m cents postage can be obtained using 3-cent and 5-cent stamps, where $m \geq 8$. (We wish to show that we can obtain $m + 1$-cents postage using 3-cent and 5-cent stamps. We shall give an alternative proof.)

Case 1. Suppose the m-cents postage was obtained using at least one 5-cent stamp. In this case, replace one 5-cent stamp by two 3-cent stamps, thereby obtaining exactly $m + 1$-cents postage.

Case 2. Suppose only 3-cent stamps were used to obtain the m-cents postage. In this case, $m \geq 9$, so there must be at least three 3-cent stamps. Replace three 3-cent stamps by two 5-cent stamps, obtaining postage of exactly $m + 1$ cents.

Example 14.6 now follows by *MI*.

Note that it is possible to prove Example 14.6 without using induction. Give an alternative proof using the three cases $n = 3k$, $n = 3k + 1$, and $n = 3k + 2$. Show in each case how postage of n cents may be obtained using 3-cent and 5-cent stamps. Similarly, some of the exercises in A may be proved without using *MI*, but it will be good practice to use it.

(Some of the induction problems given in this section and in Exercises 14 A can be found in *Die Methode der Vollständigen Induktion* by I. S. Sominski, (VEB Deutscher Verlag der Wissenchaften, Berlin, 1960, and *Induction in Geometry* by L. I. Golovina and I. M. Yaglom, D. C. Heath and Co., Boston, 1963. See also *Induction and Analogy in Mathematics* by G. Polya, Princeton University Press, Princeton, 1954).)

14.3 THEOREMS OF Π

In this section, we give informal proofs of theorems in Peano Arithmetic, Π. It would be a good exercise to construct formal proofs from the informal ones. Remember that each formal proof must contain all the axioms of Π that are used in the proof. In particular, each instance of the induction schema $P_\varphi 11$ that is used must appear on a line of the proof. In the informal proofs given below, the parenthetic statements give the plan of attack.

Theorem 1 $(\forall x)(x + 1 \neq x)$.

Informal Proof: (The proof uses *MI*. The formula φ is $x + 1 \neq x$. Remember that a proof using *MI* has two steps. First prove $\varphi[0|x]$, or, in our case, $0 + 1 \neq 0$; and second, prove the conditional $\varphi[y|x] \to \varphi[y + 1|x]$ for each y, or, in our case,

$$y + 1 \neq y \to (y + 1) + 1 \neq y + 1, \text{ for each } y.$$

Then, it follows from *MI* that $(\forall x)\varphi$, or, in our case, Theorem 1.)
Base Step: It follows from *P8* that

(i) $0 + 1 \neq 0$,

so the base step is true.
Induction Step: Suppose

(ii) $y + 1 \neq y$. (*IH*)

(We want to prove $(y + 1) + 1 \neq y + 1$, so we shall give an indirect proof.) Suppose

(iii) $(y + 1) + 1 = y + 1$.

Then, by *P9*,

(iv) $y + 1 = y$.

However, (iv) contradicts (ii). Thus,

(v) $(y + 1) + 1 \neq y + 1$.

Now, Theorem 1 follows by *MI*. ■

The next theorem implies that the 0-element is unique.

Theorem 2 $(\forall x)(\forall y)(x + y = x \to y = 0)$.

Informal Proof: (The proof uses *MI*. The formula φ in *MI* is

$$(\forall y)(x + y = x \to y = 0).$$

First, we shall derive $\varphi[0|x]$ using a conditional proof.)
Suppose

(i) $0 + y = 0$.

Then

(ii) $y + 0 = 0$ (by (i) and *P1*)

(iii) $y = 0$ (by (ii) and *P6*)

This proves $\varphi[0|x]$,

(iv) $(\forall y)(0 + y = 0 \rightarrow y = 0)$.

Suppose

(v) $(\forall y)(z + y = z \rightarrow y = 0)$. (*IH*)

(We want to prove $(\forall y)((z + 1) + y = z + 1 \rightarrow y = 0)$.) Take as a premise

(vi)	$(z + 1) + y = z + 1$	(Premise for *CP*)
(vii)	$z + (1 + y) = z + 1$	(by (vi) and *P*3)
(viii)	$z + (y + 1) = z + 1$	(by (vii) and *P*1)
(ix)	$(z + y) + 1 = z + 1$	(by (viii) and *P*3)
(x)	$z + y = z$	(by (ix) and *P*9)
(xi)	$y = 0$	(by (x) and (v))

Thus, it follows from (v) through (xi) that

(xii) $(\forall y)(\varphi[z|x] \rightarrow \varphi[z + 1|x])$.

Theorem 2 follows from (iv), (xii), and *MI*.

■

Theorem 3 states that every element, except 0, has an immediate predecessor.

Theorem 3 $(\forall x)[x \neq 0 \rightarrow (\exists y)(x = y + 1)]$.

Informal Proof: (We shall again use *MI*. The formula φ is

$$x \neq 0 \rightarrow (\exists y)(x = y + 1).)$$

First, we obtain

(i) $0 \neq 0 \rightarrow (\exists y)(0 = y + 1)$.

(Formula (i) is $\varphi[0|x]$ and is true because the antecedent is false.)

Next, assume

(ii) $z \neq 0 \rightarrow (\exists y)(z = y + 1)$. (*IH*)

(It follows from *P*8 that the antecedent, $z + 1 \neq 0$, of $\varphi[z + 1|x]$ is true. Thus, to prove $\varphi[z + 1|x]$, we must prove

(iii) $(\exists y)(z + 1 = y + 1)$.

We shall prove (iii) without using the induction hypothesis.)

By the identity axiom, $z + 1 = z + 1$. Then, using Rule *EG* (with φ as the formula $z + 1 = y + 1$, z the term, and y the variable, so $\varphi[z|y]$ is $z + 1 = z + 1$), we obtain $(\exists y)(z + 1 = y + 1)$.

Theorem 3 follows by *MI*.

■

Theorem 4 $(\forall x)(\forall y)(\exists z)(x = y + z \vee y = x + z)$.

Informal Proof: (The proof is by *MI* on x. That is, φ is the formula

$$(\forall y)(\exists z)(x = y + z \vee y = x + z).)$$

First, we shall prove $\varphi[0|x]$.) It follows from the identity axiom and Rules *EG* and *US* that

(i) $(\forall y)(\exists z)(y = z)$.

By *P6*,

(ii) $z = z + 0$
$= 0 + z$ (by *P1*)

Consequently, from (i) and (ii), we can deduce $(\forall y)(\exists z)(y = 0 + z)$. It follows that

(iii) $(\forall y)(\exists z)(0 = y + z \vee y = 0 + z)$

and formula (iii) is $\varphi[0|x]$.

Next, suppose

(iv) $(\forall y)(\exists z)(u = y + z \vee y = u + z)$. (*IH*)

(We shall give an alternative proof of $\varphi[u + 1|x]$.)

First assume that

(v) $u = y + z$, for some z.

Then, adding one to both sides,

(vi) $u + 1 = (y + z) + 1$
$= y + (z + 1)$. (by *P3*)

Therefore,

(vii) $(\exists z)(u + 1 = y + z)$, for every y,

from whence it follows that

(viii) $(\forall y)(\exists z)(u + 1 = y + z \vee y = (u + 1) + z)$.

Next, assume that

(ix) $y = u + z$, for some z.

If $z = 0$, then, by *P6*, $u = y$. Thus,

(x) $u + 1 = y + 1$,

so there is a z such that

(xi) $u + 1 = y + z$

and (xi) implies (viii). If $z \neq 0$, then by Theorem 3, there is a v so that $z = v + 1$. Therefore, from (ix) we obtain

(xii) $y = u + (v + 1)$
$= u + (1 + v)$ (by $P1$)
$= (u + 1) + v.$ (by $P3$)

Consequently,

(xiii) $(\exists z)(y = (u + 1) + z)$,

and (xiii) again implies (viii).

We have now proved $\varphi[0|x]$ and $(\forall u)(\varphi[u|x] \rightarrow \varphi[u + 1|x])$. Thus, Theorem 4 follows by *MI*. (It follows from Theorem 4 and *P*10 that $<$ is a linear ordering relation. See Definition 7.3.)

■

Our next theorem is the cancellation rule for +. In Group Theory, we proved this by using the inverse operation. Here the proof is by *MI*.

Theorem 5 $(\forall x)(\forall y)(\forall z)(x + z = y + z \rightarrow x = y)$.

Informal Proof: (The proof is by induction on z. φ is the formula

$$(\forall x)(\forall y)(x + z = y + z \rightarrow x = y).)$$

Base Step: By *P*6, $x + 0 = x$ and $y + 0 = y$; therefore,

$$x + 0 = y + 0 \rightarrow x = y.$$

Induction step: The induction hypotheses is

$$x + z = y + z \rightarrow x = y. \qquad (IH)$$

(We want to prove

$$x + (z + 1) = y + (z + 1) \rightarrow x = y.)$$

Suppose

$$x + (z + 1) = y + (z + 1).$$

Then, by *P*3,

$$(x + z) + 1 = (y + z) + 1.$$

Applying *P*9,

$$x + z = y + z,$$

so by *IH*, we derive $x = y$. Now the theorem follows by *MI*.

■

Corollary $(\forall x)(\forall y)(\forall z)(z + x = z + y \rightarrow x = y)$.

The proof follows from Theorem 5, using *P*1.

The proof of the last theorem in this section will depend on preceding results. It will not require a proof by induction directly, but it does depend upon results that were proved by induction.

Theorem 6 $(\forall x)(x \cdot 0 = 0)$.

Informal Proof:

$$\begin{aligned} x \cdot 0 &= x \cdot (0 + 0) && \text{(by } P6\text{)} \\ &= x \cdot 0 + x \cdot 0 && \text{(by } P5\text{)} \\ x \cdot 0 + 0 &= x \cdot 0 + x \cdot 0 && \text{(by } P6\text{)} \\ 0 &= x \cdot 0 && \text{(by Theorem 5, Corollary)} \end{aligned}$$

■

In Exercises 14 B you are asked to give informal proofs of theorems in Peano Arithmetic. Not all the theorems require a proof using mathematical induction, and you may use preceding theorems to prove later ones. Theorems about the symbol $<$ will be found in the exercises. To aid you, we have prepared a list of all the axioms and theorems mentioned in this section and in the exercises. You will find this list in Section 14.5.

14.4 *MODELS AND THE COMPACTNESS THEOREM

The standard model of Peano Arithmetic (Π), as was mentioned earlier, is

$$\mathbf{S}_{\mathbb{N}} = \langle \mathbb{N}, +, \cdot, <, 0, 1 \rangle,$$

where $\mathbb{N}$ is the set of non-negative integers and the operations, relations, and constants are natural. In this section, we show how to use the Compactness Theorem to construct a "nonstandard" model of Π. (See Chapter 9 for a discussion of the Compactness Theorem.) The form of the Compactness Theorem we shall use here is the following:

Compactness Theorem: If Γ is a set of sentences such that each finite subset of Γ has a model, then Γ itself has a model. (See Theorem 9.4.)

We are going to extend Π to a new theory, which we shall call Γ, by adding a new constant, say ω, and an infinite number of axioms:

$G0$: $\omega \neq 0$

$G1$: $\omega \neq 1$

$G2$: $\omega \neq 1 + 1$

$G3$: $\omega \neq (1 + 1) + 1$

$G4$: $\omega \neq ((1 + 1) + 1) + 1$

$\vdots \qquad \vdots$

Theorem 14.1 *Each finite set of axioms of* Γ *has a model.*

Proof: Let Δ be a finite set of axioms of Γ, and let n be the largest integer such that $Gn \in \Delta$. Then, we claim that

$$\mathbf{S}_{\mathbb{N}}^{n} = \langle \mathbb{N}, +, \cdot, <, \mathbf{0}, \mathbf{1}, \mathbf{n+1} \rangle$$

is a model of Δ. It is assumed that $\mathbf{n} + \mathbf{1}$ is the interpretation of ω and that $\mathbf{2} = \mathbf{1} + \mathbf{1}, \mathbf{3} = (\mathbf{1} + \mathbf{1}) + \mathbf{1}, \mathbf{4} = ((\mathbf{1} + \mathbf{1}) + \mathbf{1}) + \mathbf{1},$ and so forth. Because $\mathbf{S}_{\mathbb{N}}^{n}$ is a model of $P1 - P_{\varphi}11$ and $G0 - Gn$ are true in $\mathbf{S}_{\mathbb{N}}^{n}$, it follows that $\mathbf{S}_{\mathbb{N}}^{n}$ is a model of Δ.

■

Now, it follows from the Compactness Theorem that Γ has a model,

$$\mathbf{S}_{\omega} = \langle \mathbb{N}_{\omega}, +, \cdot, <, \mathbf{0}, \mathbf{1}, \boldsymbol{\omega} \rangle.$$

By construction, $\mathbf{0}, \mathbf{1}, \boldsymbol{\omega} \in \mathbb{N}_{\omega}$. It is easy to prove each of the following properties of $\mathbf{S}_{\omega}$. (We leave the proofs for the exercises. See Exercises 14 D.)

(1) $$\mathbb{N} \subseteq \mathbb{N}_{\omega}$$

(Formally, $\mathbb{N}$ is not a subset of $\mathbb{N}_{\omega}$, but $\mathbb{N}_{\omega}$ has a subset with the same structure as $\mathbb{N}$.)

(2)
$$\boldsymbol{\omega} + \mathbf{1} \in \mathbb{N}_{\omega}$$
$$(\boldsymbol{\omega} + \mathbf{1}) + \mathbf{1} \in \mathbb{N}_{\omega}$$
$$((\boldsymbol{\omega} + \mathbf{1}) + \mathbf{1}) + \mathbf{1} \in \mathbb{N}_{\omega}$$
$$\vdots$$

(3)
$$(\exists \mathbf{x}_0)(\mathbf{x}_0 \in \mathbb{N}_{\omega} \wedge \mathbf{x}_0 + \mathbf{1} = \boldsymbol{\omega})$$
$$(\exists \mathbf{x}_1)(\mathbf{x}_1 \in \mathbb{N}_{\omega} \wedge (\mathbf{x}_1 + \mathbf{1}) + \mathbf{1} = \boldsymbol{\omega})$$
$$(\exists \mathbf{x}_2)(\mathbf{x}_2 \in \mathbb{N}_{\omega} \wedge ((\mathbf{x}_2 + \mathbf{1}) + \mathbf{1}) + \mathbf{1} = \boldsymbol{\omega})$$
$$\vdots$$

Let Ω be the set of all the predecessors and successors of $\boldsymbol{\omega}$. That is,

$$\Omega = \{\mathbf{x} : (\exists \mathbf{n})(\mathbf{n} \in \mathbb{N} \wedge (\mathbf{x} = \mathbf{n} + \boldsymbol{\omega} \vee \mathbf{x} + \mathbf{n} = \boldsymbol{\omega}))\}.$$

Then Ω looks like $\mathbb{Z}$, the set of all integers. Also,

(4) For each $\mathbf{n} \in \mathbb{N}$ and for each $\mathbf{x} \in \boldsymbol{\Omega}, \mathbf{n} < \mathbf{x}$.

Thus, with respect to the ordering $<$, $\mathbb{N}_{\omega}$ contains a set with the same structure as $\mathbb{N}$ followed by a set with the same structure as $\mathbb{Z}$. However, it is not as simple as this. Infinite integers like $\boldsymbol{\omega}+\boldsymbol{\omega}$, $(\boldsymbol{\omega}+\boldsymbol{\omega})+\boldsymbol{\omega}, \ldots, \boldsymbol{\omega} \cdot \boldsymbol{\omega}$, $(\boldsymbol{\omega} \cdot \boldsymbol{\omega}) \cdot \boldsymbol{\omega}, \ldots$ are also in $\mathbb{N}_{\omega}$, and these infinite integers are not in Ω. Each of these infinite numbers generates a copy of $\mathbb{Z}$, similar to Ω. In fact, it can be shown that there is no first or last copy of $\mathbb{Z}$ and that between any two of these copies of $\mathbb{Z}$, there is another copy. (Use the fact that for any infinite integer $\boldsymbol{\omega}'$, either $\boldsymbol{\omega}'$ or $\boldsymbol{\omega}'+\mathbf{1}$ is even and, therefore, divisible by **2**.)

Because of the nature of the Predicate Calculus, even though we intended to axiomatize $\mathbb{N}$, we also have to put up with other types of models. The model $\mathbf{S}_\omega$ is an example. For additional information about models of Π, see the references given below.

(References for this section include *A Mathematical Introduction to Logic* by H. B. Enderton, Academic Press, New York, 1972, and *Introduction to Mathematical Logic*, 3rd ed., by E. Mendelson, Wadsworth & Brooks/Cole, Monterey, CA, 1987.)

14.5 PEANO ARITHMETIC (Π)

Nonlogical Symbols:

Binary operation symbols: $+, \cdot$.

Constants: 0, 1.

2-place relation symbol: $<$.

Axioms:

*P*1: $(\forall x)(\forall y)(x + y = y + x)$.

*P*2: $(\forall x)(\forall y)(x \cdot y = y \cdot x)$.

*P*3: $(\forall x)(\forall y)(\forall z)[x + (y + z) = (x + y) + z]$.

*P*4: $(\forall x)(\forall y)(\forall z)[x \cdot (y \cdot z) = (x \cdot y) \cdot z]$.

*P*5: $(\forall x)(\forall y)(\forall z)[x \cdot (y + z) = x \cdot y + x \cdot z]$.

*P*6: $(\forall x)(x + 0 = x)$.

*P*7: $(\forall x)(x \cdot 1 = x)$.

*P*8: $(\forall x)(x + 1 \neq 0)$.

*P*9: $(\forall x)(\forall y)(x + 1 = y + 1 \rightarrow x = y)$.

*P*10: $(\forall x)(\forall y)[x < y \leftrightarrow (\exists z)(z \neq 0 \wedge y = x + z)]$.

$P_\varphi 11$: (*MI*) If φ is any formula in the language of Peano Arithmetic (Π) with x as a free variable, and if y is a variable that does not occur in φ, then the following is an axiom:

$$\varphi[0|x] \wedge (\forall y)(\varphi[y|x] \rightarrow \varphi[y + 1|x]) \rightarrow (\forall x)\varphi.$$

Theorems:

Theorem 1: $(\forall x)(x + 1 \neq x)$.

Theorem 2: $(\forall x)(\forall y)(x + y = x \rightarrow y = 0)$.

Theorem 3: $(\forall x)[x \neq 0 \rightarrow (\exists y)(x = y + 1)]$.

Theorem 4: $(\forall x)(\forall y)(\exists z)(x = y + z \vee y = x + z)$.

Theorem 5: $(\forall x)(\forall y)(\forall z)(x + z = y + z \rightarrow x = y)$.

Corollary: $(\forall x)(\forall y)(\forall z)(z + x = z + y \rightarrow x = y)$.

Theorem 6: $(\forall x)(x \cdot 0 = 0)$.

Theorem 7: $(\forall x)(\forall y)(x + y = 0 \rightarrow x = 0 \wedge y = 0)$.

Theorem 8: $(\forall x)(\forall y)(\forall z)[(x + y) \cdot z = x \cdot z + y \cdot z]$.

Theorem 9: $(\forall x)(\forall y)(x \cdot y = 0 \rightarrow x = 0 \vee y = 0)$.

Theorem 10: $(\forall y)[(\forall x)(x \cdot y = x) \rightarrow y = 1]$.

Theorem 11: $(\forall x)(\forall y)(\forall z)(z \neq 0 \wedge x \cdot z = y \cdot z \rightarrow x = y)$.

Theorem 12: $(\forall x)[x \neq 0 \wedge x \neq 1 \rightarrow (\exists y)(x = y + (1 + 1))]$.

Theorem 13: $(\forall x)(x \not< x)$.

Theorem 14: $(\forall x)(x < x + 1)$.

Theorem 15: $(\forall x)(x \not< 0)$.

Theorem 16: $(\forall x)(\forall y)(\forall z)(x < y \wedge y < z \rightarrow x < z)$.

Theorem 17: $(\forall x)(\forall y)(x < y \rightarrow x \neq y \wedge y \not< x)$.

Theorem 18: $(\forall x)(\forall y)(x = y \rightarrow x \not< y \wedge y \not< x)$.

Theorem 19: $(\forall x)(\forall y)(x = y \vee x < y \vee y < x)$.

Theorem 20: $(\forall x)(\forall y)(\forall z)(x < y \rightarrow x + z < y + z)$.

Theorem 21: $(\forall x)(\forall y)(\forall z)(z \neq 0 \wedge x < y \rightarrow x \cdot z < y \cdot z)$.

Theorem 22: $(\forall x)(\forall y)(\forall z)(x + z < y + z \rightarrow x < y)$.

Theorem 23: $(\forall x)(\forall y)(\forall z)(x \cdot z < y \cdot z \rightarrow x < y)$.

Theorem 24: $(\forall x)\neg(\exists y)(x < y \wedge y < x + 1)$.

Theorem 25: $(\forall x)(\forall y)(x < y + 1 \rightarrow x < y \vee x = y)$.

Theorem 26: $(\forall x)(\forall y)[y \neq 0 \rightarrow (\exists u)(\exists v)(x = y \cdot u + v \wedge v < y)]$.

Theorem 27: $(\forall x)(\forall y)[y \neq 0 \rightarrow (\forall u_1)(\forall u_2)(\forall v_1)(\forall v_2)(x = y \cdot u_1 + v_1 \wedge v_1 < y \wedge x = y \cdot u_2 + v_2 \wedge v_2 < y \rightarrow u_1 = u_2 \wedge v_1 = v_2)]$.

(In Theorems 28 through 31, "2" is an abbreviation for "1 + 1" and "3" is an abbreviation for "2 + 1.")

Theorem 28: $(\forall x)(\exists y)(x = 2 \cdot y \vee x = 2 \cdot y + 1)$.

Theorem 29: $(\forall x)(\exists y)(x = 3 \cdot y \vee x = 3 \cdot y + 1 \vee x = 3 \cdot y + 2)$.

Theorem 30: $(\forall x)(\forall x_1)(\forall y)(\forall y_1)(x < y \wedge x = 2 \cdot x_1 \wedge y = 2 \cdot y_1 \rightarrow x < x_1 + y_1 \wedge x_1 + y_1 < y)$.

Theorem 31: $(\forall x)(\forall x_1)(\forall y)(\forall y_1)(x < y \wedge x = 2 \cdot x_1 + 1 \wedge y = 2 \cdot y_1 + 1 \rightarrow x < (x_1 + y_1) + 1 \wedge (x_1 + y_1) + 1 < y)$.

EXERCISES 14

A. (Answers to 1, 8, 12, 16, 21, 24, 30, 36, 43, 47, 53, 57(a), and 58(a).) In these exercises, we use the following notation:

$$\sum_{i=k}^{n} f(i) = f(k) + f(k+1) + \cdots + f(n).$$

$$\prod_{i=k}^{n} f(i) = f(k) \cdot f(k+1) \cdot \ldots \cdot f(n).$$

For example, the sum of the first n powers of 2, S_n, in this notation is

$$S_n = \sum_{i=0}^{n} 2^i.$$

The Product P_n given in Example 14.2 is

$$P_n = (1-x) \cdot \prod_{i=0}^{n} (1 + x^{2^i}).$$

In Exercises (1) through (42), give an informal proof using mathematical induction.

(1) $\displaystyle\sum_{i=1}^{n-1} i = \frac{(n-1)n}{2}, n \geq 2.$

(2) $\displaystyle\sum_{i=1}^{n} i^2 = \frac{1}{6}\, n(n+1)(2n+1),\ n \geq 1.$

(3) $\displaystyle\sum_{i=1}^{n} i^3 = \frac{1}{4}\, n^2(n+1)^2,\ n \geq 1.$

(4) $\displaystyle\sum_{i=1}^{n} (2i-1)^2 = \frac{1}{3}\, n(2n-1)(2n+1),\ n \geq 1.$

(5) $\displaystyle\sum_{i=1}^{n} \frac{1}{(2i-1)(2i+1)} = \frac{n}{2n+1}, n \geq 1.$

(6) $\displaystyle\sum_{i=1}^{n} \frac{i^2}{(2i-1)(2i+1)} = \frac{n(n+1)}{2(2n+1)}, n \geq 1.$

(7) $\displaystyle\sum_{i=1}^{n} \frac{1}{i(i+2)} = \frac{3}{4} - \frac{2n+3}{2(n+1)(n+2)}, n \geq 1.$

(8) $\displaystyle\sum_{i=1}^{n} i(i+1) = \frac{1}{3}\, n(n+1)(n+2),\ n \geq 1.$

(9) $\displaystyle\sum_{i=1}^{n} i(i+1)(i+2) = \frac{1}{4}\, n(n+1)(n+2)(n+3),\ n \geq 1.$

(10) $\displaystyle\sum_{i=0}^{n} (2^i + 3^i) = 2^{n+1} + \frac{1}{2}3^{n+1} - \frac{3}{2},\ n \geq 0.$

(11) $\sum_{i=1}^{n} i^3 = \left(\sum_{i=1}^{n} i\right)^2, n \geq 1.$

(12) $\sum_{i=1}^{n} i \cdot i! = (n + 1)! - 1, n \geq 1.$

(13) $2^n > n, n \geq 0.$

(14) $2^n > n^2, n \geq 5.$

(15) $3^n > 2^n + n, n \geq 2.$

(16) $n! > 3^n, n \geq 7.$

(17) $2^{n^2} > n!, n \geq 1$. (You may use (13) in your proof.)

(18) $\sum_{i=1}^{n} i! < (n + 1)!, n \geq 1.$

(19) $\sum_{i=1}^{n} \frac{1}{\sqrt{i}} > \sqrt{n}, n \geq 2.$

(20) $\left(\sum_{i=1}^{n} i\right)\left(\sum_{i=1}^{n} \frac{1}{i}\right) \geq n^2, n \geq 1.$

(21) $\sum_{i=1}^{n} \frac{1}{n + i} > \frac{13}{24}, n \geq 2.$

(22) $n(n + 1)$ is divisible by 2, $n \geq 0$.

(23) $n^3 + 2n$ is divisible by 3, $n \geq 0$.

(24) $2^{2n} - 1$ is divisible by 3, $n \geq 0$.

(25) $n(n + 1)(n + 2)$ is divisible by 6, $n \geq 0$. (You may use (22) in your proof.)

(26) $12^n + 2 \cdot 5^{n-1}$ is divisible by 7, $n \geq 1$.

(27) $3^{4n} - 5^{2n}$ is divisible by 7, $n \geq 1$.

(28) $n^3 + (n + 1)^3 + (n + 2)^3$ is divisible by 9, $n \geq 0$.

(29) $(11)^{n+2} + (12)^{2n+1}$ is divisible by 133, $n \geq 0$.

(30) If x is any positive real number, then $(1 + x)^n \geq 1 + nx, n \geq 0$.

(31) If the terms in the sequence $a_1, a_2, a_3, \ldots$ are all positive real numbers, then

$$\prod_{i=1}^{n} (1 + a_i) \geq 1 + \sum_{i=1}^{n} a_i, n \geq 1.$$

(32) If $0 < a_i < 1$ for all i, then

$$\prod_{i=1}^{n} (1 - a_i) \geq 1 - \sum_{i=1}^{n} a_i, n \geq 1.$$

(33) Binomial Theorem: If a and b are real numbers, then

$$(a + b)^n = \sum_{i=0}^{n} \frac{n!}{(n - i)!i!} a^{n-i} b^i, n \geq 1.$$

(Assume $0! = 1$.)

(In Exercises (34) through (36), trigonometric identities must be used for some steps of the proof.)

(34) DeMoivre's Theorem:

$$(\cos x + i \sin x)^n = \cos(nx) + i \sin(nx), n \geq 1.$$

(In DeMoivre's Theorem, "i" is the complex number such that $i^2 = -1$.)

(35) $\sum_{i=1}^{n} \sin(ix) = \frac{\sin[(n+1)x/2]}{\sin(x/2)} \sin\left(\frac{nx}{2}\right), n \geq 1.$

(36) $\prod_{i=0}^{n} \cos(2^i x) = \frac{\sin(2^{n+1}x)}{2^{n+1}\sin x}, n \geq 0.$

(37) $\int x^n e^x dx = e^x \sum_{i=0}^{n} (-1)^i \frac{n!x^{n-i}}{(n-i)!} + C, n \geq 0.$ (Assume $0! = 1$).

(38) Do the postage stamp problem (Example 14.6) under the following conditions:
  (a) With an infinite supply of 2-cent and 3-cent stamps, obtain postage of every denomination greater than 1 cent.
  (b) With an infinite supply of 3-cent and 7-cent stamps, obtain postage of every denomination greater than 11 cents.
  (c) With an infinite supply of 2-cent and $2k + 1$-cent stamps, $(k \geq 1)$, obtain postage of every denomination greater than $2k - 1$ cents.

(39) If $n \geq 2$ people stand in line and if the first person is a man and the last person is a woman, then somewhere in line there must be a man directly in front of a woman.

(40) Suppose there are $2n$ participants in a tennis match, $(n \geq 1)$. If in the first round there are n games, each with two players, then the pairing for the first round can be arranged in exactly $\Pi_{i=1}^{n}(2i - 1)$ different ways.

(41) Suppose there are $n \geq 2$ intervals (finite or infinite) on a straight line such that each pair have a point in common. Then all n intervals have a point in common.

(42) If a positive integer is composed of 3^n identical digits, then it is divisible by 3^n, $(n \geq 1)$. (You may use the fact that each positive integer is uniquely expressible in the form $\Sigma_{i=0}^{n} a_i 10^{m_i}$, where for each i, a_i and m_i are non-negative integers such that $0 \leq a_i \leq 9$ and $m_0 > m_1 > \cdots > m_n$. You may also use other elementary properties of integers such as an integer is divisible by 3 iff the sum of its digits is divisible by 3.)

In Exercises (43) through (58), find a formula for the given expression and give an informal proof using mathematical induction that it is correct.

(43) $\sum_{i=1}^{n} (2i - 1), n \geq 1.$

(44) $\sum_{i=1}^{n} \frac{1}{i(i+1)}, n \geq 1.$

(45) $\sum_{i=1}^{n} (-1)^{i+1} i^2, n \geq 1.$

(46) $\sum_{i=1}^{n} \frac{1}{(4i^2 - 1)}, n \geq 1.$

(47) $\prod_{i=2}^{n} \left(1 - \frac{1}{i^2}\right), n \geq 2.$

(48) $\prod_{i=1}^{n} \left(1 - \frac{4}{(2i - 1)^2}\right), n \geq 1.$

(49) The number of subsets of an n-element set $n \geq 1$. (Include the empty set and the set itself as subsets.)

(50) The number of two-element subsets of a set with $n \geq 2$ elements.

(51) The number of ways $n \geq 1$ blocks, all of different colors, can be arranged on a straight line.

(52) The n-th derivative of $f(x) = 1/(1 + x), x \neq -1, n \geq 1$.

(53) The n-th derivative of $f(x) = 1/(4 - 5x), x \neq 4/5, n \geq 1$.

(54) The n-th derivative of $f(x) = 1/\sqrt{1 - x}, x < 1, n \geq 1$.

(55) The number of parts a plane is divided by $n \geq 1$ straight lines such that each pair of lines intersect, but no three lines intersect in a point.

(56) The number of parts a plane is divided by $n \geq 1$ circles such that each pair of circles intersect in two points and no three pass through the same point.

(57) A *convex* polygon is a polygon such that each (interior) angle is less than 180 degrees. A *diagonal* of a polygon is a straight line connecting two nonadjacent vertices.

(a) The sum of the angles of an n-sided convex polygon, $n \geq 3$.

(b) The number of diagonals of an n-sided convex polygon, $n \geq 3$.

(c) The number of nonintersecting diagonals needed to subdivide an n-sided convex polygon into triangles, $n \geq 3$.

(58) We shall construct a sequence of two dimensional figures so that the first is a square of side s and each subsequent figure is obtained from the preceding one by replacing the middle third of each side by a square. (______ is replaced by _⌐¬_.)

(a) The number of sides of the n-th figure, $n \geq 1$.

(b) The length of a side of the n-th figure, $n \geq 1$.

(c) The perimeter of the n-th figure, $n \geq 1$.

(d) The area of the n-th figure, $n \geq 1$.

(You may use the formula

$$\sum_{i=0}^{n} ar^i = \frac{a(r^{n+1} - 1)}{r - 1}, r \neq 1.$$

It is interesting to note that as $n \to \infty$, the perimeter of the n-th figure increases without limit, but the area of the n-th figure approaches $2s^2$.)

B. Give an informal proof of each of the following theorems of Peano Arithmetic. You may use preceding theorems to prove later ones. (See the list of axioms and theorems of Peano Arithmetic in Section 14.5.) (Answers to 1, 5, 10, 16, and 19.)

(1) Theorem 7: $(\forall x)(\forall y)(x + y = 0 \rightarrow x = 0 \wedge y = 0)$.
(2) Theorem 8: $(\forall x)(\forall y)(\forall z)[(x + y) \cdot z = x \cdot z + y \cdot z]$.
(3) Theorem 9: $(\forall x)(\forall y)(x \cdot y = 0 \rightarrow x = 0 \vee y = 0)$.
(4) Theorem 10: $(\forall y)[(\forall x)(x \cdot y = x) \rightarrow y = 1]$.
(5) Theorem 11: $(\forall x)(\forall y)(\forall z)(z \neq 0 \wedge x \cdot z = y \cdot z \rightarrow x = y)$.
(6) Theorem 12: $(\forall x)[x \neq 0 \wedge x \neq 1 \rightarrow (\exists y)(x = y + (1 + 1))]$.
(7) Theorem 13: $(\forall x)(x \not< x)$.
(8) Theorem 14: $(\forall x)(x < x + 1)$.
(9) Theorem 15: $(\forall x)(x \not< 0)$.
(10) Theorem 16: $(\forall x)(\forall y)(\forall z)(x < y \wedge y < z \rightarrow x < z)$.
(11) Theorem 17: $(\forall x)(\forall y)(x < y \rightarrow x \neq y \wedge y \not< x)$.
(12) Theorem 18: $(\forall x)(\forall y)(x = y \rightarrow x \not< y \wedge y \not< x)$.
(13) Theorem 19: $(\forall x)(\forall y)(x = y \vee x < y \vee y < x)$.
(14) Theorem 20: $(\forall x)(\forall y)(\forall z)(x < y \rightarrow x + z < y + z)$.
(15) Theorem 21: $(\forall x)(\forall y)(\forall z)(z \neq 0 \wedge x < y \rightarrow x \cdot z < y \cdot z)$.
(16) Theorem 22: $(\forall x)(\forall y)(\forall z)(x + z < y + z \rightarrow x < y)$.
(17) Theorem 23: $(\forall x)(\forall y)(\forall z)(x \cdot z < y \cdot z \rightarrow x < y)$.
(18) Theorem 24: $(\forall x)\neg(\exists y)(x < y \wedge y < x + 1)$.
(19) Theorem 25: $(\forall x)(\forall y)(x < y + 1 \rightarrow x < y \vee x = y)$.
(20) Theorem 26: $(\forall x)(\forall y)[y \neq 0 \rightarrow (\exists u)(\exists v)(x = y \cdot u + v \wedge v < y)]$.
(21) Theorem 27: $(\forall x)(\forall y)[y \neq 0 \rightarrow (\forall u_1)(\forall u_2)(\forall v_1)(\forall v_2)(x = y \cdot u_1 + v_1 \wedge v_1 < y \wedge x = y \cdot u_2 + v_2 \wedge v_2 < y \rightarrow u_1 = u_2 \wedge v_1 = v_2)]$.

(In Exercises (22) through (25), "2" is an abbreviation for "1 + 1" and "3" is an abbreviation for "2 + 1.")

(22) Theorem 28: $(\forall x)(\exists y)(x = 2 \cdot y \vee x = 2 \cdot y + 1)$.
(23) Theorem 29: $(\forall x)(\exists y)(x = 3 \cdot y \vee x = 3 \cdot y + 1 \vee x = 3 \cdot y + 2)$.
(24) Theorem 30: $(\forall x)(\forall x_1)(\forall y)(\forall y_1)(x < y \wedge x = 2 \cdot x_1 \wedge y = 2 \cdot y_1 \rightarrow x < x_1 + y_1 \wedge x_1 + y_1 < y)$.
(25) Theorem 31: $(\forall x)(\forall x_1)(\forall y)(\forall y_1)(x < y \wedge x = 2 \cdot x_1 + 1 \wedge y = 2 \cdot y_1 + 1 \rightarrow x < (x_1 + y_1) + 1 \wedge (x_1 + y_1) + 1 < y)$.

C. In this set of exercises, we introduce a new nonlogical 2-place relation symbol, $|$, ($x|y$ is interpreted to mean x divides y) and a new axiom

***P*12.** $(\forall x)(\forall y)[x|y \leftrightarrow (\exists z)(y = x \cdot z)]$.

Give an informal proof of each of the following using axioms *P*1 through *P*12. You may also use any of the preceding theorems of Π and you may use preceding theorems to prove later ones. (Answers to 1, 6, and 9.)

(1) $(\forall x)(x|x)$.
(2) $(\forall x)(1|x)$.
(3) $(\forall x)(x|0)$.
(4) $(\forall x)(\forall y)(\forall z)(x|y \wedge y|z \rightarrow x|z)$.
(5) $(\forall x)(\forall y)(y \neq 0 \wedge x|y \rightarrow x < y \vee x = y)$.

(6) $(\forall x)(\forall y)(x|y \wedge y|x \rightarrow x = y)$.
(7) $(\forall x)(\forall y)(\forall z)[x|y \rightarrow x|(y \cdot z)]$.
(8) $(\forall x)(\forall y)(\forall z)[x|y \wedge x|z \rightarrow x|(y + z)]$.
(9) $(\forall x)(x|1 \rightarrow x = 1)$.
(10) $(\forall x)(\forall y)(x|y \wedge x|(y + 1) \rightarrow x = 1)$.

*D. Give an informal proof of each of the following theorems about $\mathbf{S}_\omega$, models of Π, and other models.

(1) $\mathbb{N} \subseteq \mathbb{N}_\omega$. (Actually, this is only true if we assume that the operations, relations, and constants in $\mathbf{S}_\mathbb{N}$ are the same as the corresponding operations, relations, and constants in $\mathbf{S}_\omega$. We shall make these assumptions in these exercises.)

(2) For each $\mathbf{n} \in \mathbb{N}$, $\boldsymbol{\omega} + \mathbf{n} \in \mathbb{N}_\omega$.

(3) For each $\mathbf{n} \in \mathbb{N}$, there is an $\mathbf{x} \in \mathbb{N}_\omega$ such that $\mathbf{x} + \mathbf{n} = \boldsymbol{\omega}$.

(4) Let

$$\boldsymbol{\Omega} = \{\mathbf{x} \in \mathbb{N}_\omega : (\exists \mathbf{n})[\mathbf{n} \in \mathbb{N} \wedge (\mathbf{x} = \boldsymbol{\omega} + \mathbf{n} \vee \boldsymbol{\omega} = \mathbf{x} + \mathbf{n})]\}$$

Prove that for each $\mathbf{n} \in \mathbb{N}$ and for each $\mathbf{x} \in \Omega$, $\mathbf{n} < \mathbf{x}$.

(5) Prove that Π has no finite models.

(6) Define Ω_{ω_1} to be a $\mathbb{Z}$-chain in $\mathbb{N}_\omega$ if there is an infinite integer $\boldsymbol{\omega}_1 \in \mathbb{N}_\omega$ such that

$$\boldsymbol{\Omega}_{\omega_1} = \{\mathbf{x} \in \mathbb{N}_\omega : (\exists \mathbf{n})[\mathbf{n} \in \mathbb{N} \wedge (\mathbf{x} = \boldsymbol{\omega}_1 + \mathbf{n} \vee \boldsymbol{\omega}_1 = \mathbf{x} + \mathbf{n})]\}.$$

Prove that

(a) There is no smallest $\mathbb{Z}$-chain.
(b) There is no largest $\mathbb{Z}$-chain.
(c) There is a $\mathbb{Z}$-chain between any two distinct $\mathbb{Z}$-chains.

(7) An integer x is said to be a *prime* if $1 < x$ and the only divisors of x are 1 and x, itself. Prove that for each infinite set of primes there is a model of Π in which there is an infinite integer that has exactly those primes as its divisors. (Remark: Because there are an uncountable number of infinite sets of primes, it can be shown that (7) implies that Π has an uncountable number of countable models.)

(8) A set $\mathbf{A} \subseteq \mathbb{N}_\omega$ is said to be *definable* if there is a formula φ in the language of Π with one free variable $\mathbf{x}$ such that $\mathbf{A} = \{\mathbf{x}: \varphi(\mathbf{x})\}$.

(a) Prove that the empty set, $\emptyset$, and $\mathbb{N}_\omega$ are definable.
(b) If $\mathbf{A}, \mathbf{B} \subseteq \mathbb{N}_\omega$ are definable, show that $\mathbf{A} \cap \mathbf{B}$, $\mathbf{A} \cup \mathbf{B}$, and $\mathbf{A} - \mathbf{B}$ are definable.
(c) Prove that every nonempty definable subset of $\mathbb{N}_\omega$ has a smallest element. (*Hint:* Use *MI*. It follows that no $\mathbb{Z}$-chain is definable. See (6)).
(d) Prove that $\mathbb{N}$ is not definable.

(9) Suppose φ is a formula in the language of Π with one free variable x such that φ is true in $\mathbf{S}_\omega$ for arbitrarily large finite x. Prove that φ must be true in $\mathbf{S}_\omega$ for some infinite x.

(10) (A *prime* is defined in (7)). Prove that there is an infinite prime in $\mathbb{N}_\omega$. (*Hint:* There is no largest finite prime.)

(11) Let φ be a sentence in the language of Group Theory. (See Chapter 13.) Suppose that for each natural number n, there is a group with $\geq n$ elements in which φ is true. Prove that there must be an infinite group in which φ is true. (Recall that a "group" is a model of Group Theory.)

(12) Let Γ be the theory of a well-ordering relation; that is, the only non-logical symbol of Γ is a 2-place relation symbol R, and Γ has the following axioms:

$W1$: $(\forall x)\neg(xRx)$.

$W2$: $(\forall x)(\forall y)(\forall z)(xRy \wedge yRz \rightarrow xRz)$.

$W3$: $(\forall x)(\forall y)(xRy \vee yRx \vee x = y)$.

$W_\varphi 4$: If φ is any formula in the language of Γ with exactly one free variable x such that y does not occur in φ. Then the following is an axiom:

$$(\exists x)\varphi \rightarrow (\exists x)[\varphi \wedge (\forall y)(\varphi[y|x] \rightarrow \neg yRx)].$$

(a) Prove that there is a model of Γ that has an infinite descending R-chain

$$\ldots Rx_4\ Rx_3\ Rx_2\ Rx_1$$

(b) The result in (a) seems to be paradoxical because $W_\varphi 4$ seems to say that each set of elements (that is, $\{x{:}\varphi(x)\}$) has an R-smallest element. This implies there are no infinite descending R-chains. Can you explain this apparent paradox?

(13) A *graph* is a set of vertices (points) and edges (pairs of vertices or, geometrically, a line connecting two vertices). Two vertices are said to be *adjacent* if there is an edge connecting them. A graph can be *colored with n colors* if it is possible to assign each of the vertices exactly one color from the set of n colors in such a way that no pair of adjacent vertices are assigned the same color. A graph is said to be *finite* if it has a finite number of vertices.

(a) Using the sentential variables

V_i^k: vertex i has color k.

A_{ij}: vertices i and j are adjacent.

Write down formulas in the Sentential Calculus that would be interpreted to mean "A graph with m vertices can be colored with n colors."

(b) Prove that if G is a graph with a countably infinite number of vertices such that each finite subgraph can be colored with n colors, then G can be colored with n colors.

Chapter 15

Inductive Definitions, Complete Induction, and Automated Theorem-Proving

Inductive reasoning, as was mentioned in Chapter 14, plays an important role in proving theorems and in calculations. For example, suppose we want to calculate 2^{11} given that $2^{10} = 1{,}024$. We could ignore the given information and calculate 2^{11} by multiplying 2 by itself 11 times. However, the easy way to do the problem is to use the fact that 2^{11} can be obtained from 2^{10} by multiplying by 2. That is,

$$2^{11} = 2 \cdot 2^{10} = (2) \cdot (1{,}024) = 2{,}048.$$

In general, we are using the fact that

$$2^{n+1} = 2 \cdot 2^n, \qquad n \geq 0.$$

This idea leads to the notion of an *inductive* (or *recursive*) definition. (The justification of this kind of definition would take us beyond Peano Arithmetic, so we shall give an informal presentation.)

15.1 INDUCTIVE DEFINITIONS

We can define the exponential function 2^n for $n \geq 0$ by the following two equations.

1. $2^0 = 1$. (Initial step)
2. $2^{n+1} = 2 \cdot 2^n$, for $n \geq 0$. (Inductive step)

From these two equations, we can calculate 2^n for each $n \geq 0$. For example, suppose we wanted to calculate 2^4. We would proceed as follows:

$$
\begin{array}{lll}
& 2^0 = 1 & \text{(by 1)} \\
\text{(i)} & 2^1 = 2^{0+1} = 2 \cdot 2^0 & \text{(by 2 and algebra)} \\
& = 2 \cdot 1 = 2 & \text{(by 1 and algebra)} \\
\text{(ii)} & 2^2 = 2^{1+1} = 2 \cdot 2^1 & \text{(by 2 and algebra)} \\
& = 2 \cdot 2 = 4 & \text{(by (i) and algebra)} \\
\text{(iii)} & 2^3 = 2^{2+1} = 2 \cdot 2^2 & \text{(by 2 and algebra)} \\
& = 2 \cdot 4 = 8 & \text{(by (ii) and algebra)} \\
\text{(iv)} & 2^4 = 2^{3+1} = 2 \cdot 2^3 & \text{(by 2 and algebra)} \\
& = 2 \cdot 8 = 16 & \text{(by (iii) and algebra)}
\end{array}
$$

(As mentioned previously in Chapter 14, we use the term "algebra" to justify any well-known property of the natural numbers. For example, in (i) above, the algebra we are using is that $0 + 1 = 1$ and $2 \cdot 1 = 2$. Similarly for the other lines.)

The inital step, 1, gives us the initial value of the function and the inductive step, 2, gives a rule to calculate a value of the function if we know the preceding value. In general, the type of definition we are using has the following form.

Definition 15.1

Inductive Definition. For each integer k, for each real number r, and for each function $G : \mathbb{R}^2 \to \mathbb{R}$, there is a unique function F whose domain is the set of all integers $\geq k$ such that F satisfies the following two conditions:

1. $F(k) = r$. (Initial step)
2. $F(n + 1) = G(n, F(n))$ for all $n \geq k$. (Inductive step)

In the preceding example, $k = 0$, $r = 1$, $G(n, F(n)) = 2 \cdot F(n)$ and the function F we are defining is the exponential function $F(n) = 2^n$ for all $n \geq 0$.

In an inductive definition, the proof that the function F is unique is an easy proof by mathematical induction and will be left as an exercise. However, the proof that the function F exists uses set theoretical techniques that we shall not discuss here. Instead, we shall give several additional examples of inductive definitions.

Example 15.1

Suppose

1. $F(0) = 1$.
2. $F(n + 1) = (n + 1) \cdot F(n)$, for $n \geq 0$.

Show that F is the factorial function, $F(n) = n!$.

Solution: In the definition given above, $k = 0$, $r = 1$, and $G(n, F(n)) = (n + 1) \cdot F(n)$. We shall give an informal proof using mathematical induction that $F(n) = n!$.

Base Step: $F(0) = 1$ and $0!$ is defined to be 1, so $F(0) = 0!$.

Induction Step: Suppose for $m \geq 0$, $F(m) = m!$ (induction hypotheses, *IH*). Then, by 2 above,

$$\begin{aligned} F(m+1) &= (m+1) \cdot F(m) \\ &= (m+1) \cdot m! && (IH) \\ &= (m+1)! && \text{(Property of the factorial function)} \end{aligned}$$

Thus, it follows by *MI* (mathematical induction) that $F(n) = n!$ for all $n \geq 0$.

A *sequence* is a function whose domain is a set of integers. It follows that an inductive definition is a definition of a sequence. Frequently, the argument of a sequence is written as a subscript.

Sometimes, in an inductive definition, it may be very difficult to give an explicit formula for the function defined.

Example 15.2

Define a sequence S so that

1. $S_1 = \sqrt{2}$.
2. $S_{n+1} = \sqrt{2 + S_n}$, for $n \geq 1$.

The first few terms of this sequence are $\sqrt{2}$, $\sqrt{2 + \sqrt{2}}$, $\sqrt{2 + \sqrt{2 + \sqrt{2}}}, \ldots$. There is no simple formula for S_n. Even without a simple formula, we can prove several properties of this sequence S:

(a) $S_n < 2$, for $n \geq 1$.

(b) S_n is irrational, for $n \geq 1$.

(c) $S_n < S_{n+1}$, for $n \geq 1$.

(d) $\lim_{n \to \infty} S_n = 2$.

Solution: We shall prove (a) through (c) using *MI*.

(a) Base Step: $S_1 = \sqrt{2} = 1.4142 \cdots < 2$. Thus, the base step is true.

Induction Step: Suppose $S_m < 2$, *IH*.

$$\begin{aligned} S_{m+1} &= \sqrt{2 + S_m} && \text{(2, inductive step)} \\ &< \sqrt{2 + 2} && (IH \text{ and algebra}) \\ &= 2 && \text{(Algebra)} \end{aligned}$$

Thus, (a) follows by *MI*.

(b) Base Step: $S_1 = \sqrt{2}$. It is a well-known fact of number theory that $\sqrt{2}$ is irrational; that is, $\sqrt{2}$ is not expressible as the ratio of two integers.

Induction Step: Suppose S_m is irrational for $m \geq 1$ (*IH*). By 2, $S_{m+1} = \sqrt{2 + S_m}$. We shall give an indirect proof that S_{m+1} is irrational. Suppose S_{m+1} is rational. Then there are two integers p and q so that

$$\sqrt{2 + S_m} = \frac{p}{q}.$$

Squaring both sides, we obtain

$$2 + S_m = \frac{p^2}{q^2}.$$

Solving for S_m,

(*) $$S_m = \frac{p^2}{q^2} - 2 = \frac{p^2 - 2q^2}{q^2}.$$

Because p and q are integers, (*) implies that S_m is the ratio of two integers and, therefore, rational. However, this contradicts *IH*, where we assumed S_m was irrational. This proves the induction step, so the result follows by *MI*.

(c) Base step: $S_1 = \sqrt{2} = 1.4142\ldots$. By 2, $S_2 = \sqrt{2 + \sqrt{2}} = 1.8478\ldots$. It follows that $S_1 < S_2$.

Induction step: The induction hypothesis, *IH*, is $S_m < S_{m+1}$, for $m \geq 1$. (We wish to show that $S_{m+1} < S_{m+2}$.)

$$\begin{aligned} S_{m+1} &= \sqrt{2 + S_m} && \text{(2, inductive step)} \\ &< \sqrt{2 + S_{m+1}} && \text{(}IH\text{ and algebra)} \\ &= S_{m+2} && \text{(2, inductive step)} \end{aligned}$$

Thus, it follows by *MI* that $S_n < S_{n+1}$ for all $n \geq 1$.

(d) To prove that $\lim_{n\to\infty} S_n = 2$, we shall use several properties of limits that are known from analysis. First we must prove that the sequence S has a limit. It is a well-known theorem of calculus that every bounded, increasing (or decreasing) infinite sequence has a limit. It follows from part (a) that our sequence S is bounded. (All the terms are positive, so $0 < S_n < 2$ for all $n \geq 1$.) It follows from part (c) that S is increasing. Thus, $\lim_{n\to\infty} S_n$ exists.

Suppose $\lim_{n\to\infty} S_n = L$. Another well-known property of limits is that $\lim_{n\to\infty} S_n = \lim_{n\to\infty} S_{n+1}$. By the definition of S (inductive step), we have

(*) $$S_{n+1} = \sqrt{2 + S_n}.$$

Taking the limit of both sides of (*) and using properties of limits, we obtain

$$(**) \qquad L = \sqrt{2 + L}.$$

Squaring both sides of (**) and solving for L, we obtain

$$L^2 = 2 + L$$
$$L^2 - L - 2 = 0$$
$$L = 2 \quad \text{or} \quad L = -1$$

However, L cannot be negative because all the terms in the sequence are positive. Consequently, L must be 2, which imples $\lim_{n\to\infty} S_n = 2$. This completes the proof of (d).

We see from Example 15.2 that even though we could not find an explicit formula for S_n, we could still discover a lot of information about the sequence S. Note also that because of the form of an inductive definition, it is natural to use mathematical induction to prove properties of sequences defined by induction.

In Chapter 14, we used summations (Σ) and products (Π). These are functions that are defined by induction.

Definition 15.2

Definition of $\sum_{i=k}^{n} f(i) = S_n$.

1. $S_k = f(k)$.
2. $S_{n+1} = S_n + f(n + 1)$, for $n \geq k$.

Definition of $\prod_{i=k}^{n} f(i) = P_n$. $(k > 0)$.

1. $P_k = f(k)$.
2. $P_{n+1} = P_n \cdot f(n + 1)$, for $n \geq k$.

On occasion, it is useful in step 2 of an inductive definition to define a function (or sequence) at $n + 1$ as a function of some or all of its preceding values. The following is an example of this type of function.

Example 15.3

Define a sequence S as follows:

1. $S_1 = 2$.
2. $S_{n+1} = \prod_{i=1}^{n} S_i$, for $n \geq 1$.

We can calculate values of this sequence as follows: by 1, $S_1 = 2$. Then, using 2 and algebra, we obtain

$$S_2 = \prod_{i=1}^{1} S_i = S_1 = 2$$

$$S_3 = \prod_{i=1}^{2} S_i = S_1 \cdot S_2 = 4$$

$$S_4 = \prod_{i=1}^{3} S_i = S_1 \cdot S_2 \cdot S_3 = 16$$

$$S_5 = \prod_{i=1}^{4} S_i = S_1 \cdot S_2 \cdot S_3 \cdot S_4 = 256, \text{etc.}$$

It appears from the above calculations that

(*) $$S_{n+1} = (S_n)^2, \qquad \text{for } n \geq 2.$$

We shall prove (*) below.

Assume $n \geq 2$. Then, by 2 in Example 15.3,

(*) $$\begin{aligned} S_{n+1} &= \prod_{i=1}^{n} S_i \\ &= \left(\prod_{i=1}^{n-1} S_i\right) \cdot S_n && \text{(Definition 15.2(b))} \\ &= S_n \cdot S_n && \text{(Example 15.3, 2)} \\ &= (S_n)^2 && \text{(Algebra)} \end{aligned}$$

Thus, it follows that $S_{n+1} = (S_n)^2$ for $n \geq 2$.

Let us also try to guess an explicit formula for S_n. It appears from the table above that S_n is some power of 2. We guess that

(**) $$S_n = 2^{2^{n-2}}, \qquad \text{for } n \geq 2.$$

We shall prove (**) by *MI*.

Base Step: $S_2 = 2$ and $2^{2^{2-2}} = 2^{2^0} = 2^1 = 2$. So (**) holds when $n = 2$.

Induction Step: Suppose $S_m = 2^{2^{m-2}}$, for $m \geq 2$ (*IH*). Then, using (*), which we proved above,

$$\begin{aligned} S_{m+1} &= (S_m)^2 \\ &= (2^{2^{m-2}})^2 && (IH) \\ &= 2^{2^{(m+1)-2}} && \text{(Algebra)} \end{aligned}$$

Thus, (**) follows by *MI*.

Each inductive definition is an algorithm that describes a method for evaluating a function. The next example is called *Horner's method*. It is used to evaluate a polynomial in one variable. (Horner's method is actually synthetic division.)

Example 15.4

Horner's Method. Let

$$p(x) = a_0x^n + a_1x^{n-1} + \cdots + a_{n-1}x + a_n$$

be a polynomial in x. The coefficients $a_0, a_1, \ldots, a_n$ are real numbers, and n is a non-negative integer. For each real number r, define a sequence S^p for the polynomial $p(x)$ as follows:

1. $S_0^p = a_0$
2. $S_{m+1}^p = a_{m+1} + r \cdot S_m^p,\ 0 \le m \le n - 1.$

Show that $S_n^p = p(r)$, for $n \ge 0$. (Notice that the values of the sequence S^p form the bottom line in synthetic division.)

Solution: The proof is by *MI* on n, the degree of the polynomial.

Base Step: If $n = 0$, $p(x) = a_0$ for all x and $S_0^p = a_0$ by the initial step in the definition of S^p.

Induction Step: Suppose $p(x)$ is a polynomial of degree $m + 1$, $m \ge 0$. The induction hypothesis, *IH*, is that all polynomials $q(x)$ of degree m have the property that

$$q(r) = a_m + rS_m^q.$$

Suppose

$$p(x) = a_0x^{m+1} + a_1x^m + \cdots + a_mx + a_{m+1}.$$

Then

$$\begin{aligned} p(r) &= a_0r^{m+1} + a_1r^m + \cdots + a_mr + a_{m+1} \\ &= a_{m+1} + r(a_0r^m + a_1r^{m-1} + \cdots + a_m) && \text{(Algebra)} \\ &= a_{m+1} + r(a_m + rS_m^p) && (\mathit{IH}) \\ &= a_{m+1} + rS_m^p && \text{(2, inductive step)} \\ &= S_{m+1}^p && \text{(2, inductive step)} \end{aligned}$$

Thus, it follows by *MI* that $S_n^p = p(r)$, for $n \ge 0$.

Example 15.5 is an inductive definition in which there are two initial steps, and subsequent values of the sequence depend on two of the preceding values.

Example 15.5

We define a sequence f (called the *Fibonacci sequence*) as follows:

1. $f_1 = 1, f_2 = 1$.
2. $f_{n+1} = f_n + f_{n-1}$, for $n \geq 2$.

Each term, after the second, is the sum of the two preceding terms. The first eight terms of the Fibonacci sequence are 1, 1, 2, 3, 5, 8, 13, and 21. By using step 2, we can calculate as many of the terms as we like.

In Chapter 1 in Section 1.6, we described a formula of the Sentential Calculus. Actually, we gave an inductive definition of a formula. Recall that the language of the Sentential Calculus has the following symbols: sentential variables (A, B, . . .); connectives ($\neg$, $\wedge$, $\vee$, $\rightarrow$, $\leftrightarrow$); and parentheses for grouping. Some strings of these letters and symbols are formulas, and some are not. If S is a string of these symbols, we define the *length* of S as the number of sentential variables and connectives that occur in S.

Example 15.6

If S is a string of symbols of the Sentential Calculus, then S is a *formula* iff

1. If the length of S is 1, then S is a sentential variable (an atomic formula).
2. If the length of S is $n + 1$, then there are formulas P and Q of length less than $n + 1$ such that one of the following is true:
   (a) S is $(\neg P)$
   (b) S is $(P \wedge Q)$
   (c) S is $(P \vee Q)$
   (d) S is $(P \rightarrow Q)$
   (e) S is $(P \leftrightarrow Q)$

Thus, as illustrated in Chapter 1, each formula is constructed from atomic formulas using (a) through (e). The definition of a formula of length $n + 1$ depends upon the definitions of formulas of shorter length. We can think of Example 15.6 as defining a function F whose domain is the set of strings of symbols, S, in the Sentential Calculus such that

$$F(S) = \begin{cases} 1, & \text{if } S \text{ is a formula} \\ 0, & \text{otherwise.} \end{cases}$$

Examples 15.3, 15.5, and 15.6 illustrate a more general type of inductive definition than that given in Definition 15.1. In the general inductive definition, there may be more than one initial step, and the inductive step may depend on more than one earlier value. We shall not attempt to give a formal definition of a general inductive definition here. Intuitively, an inductive definition defines a "tree." The first level is all the initial values, and we can climb to higher levels using the inductive step. In Definition 15.1, all the vertices of the tree lie on a straight line. We illustrated a more general tree in Chapters 1 and 5. Definitions 5.3 and 5.5 are also examples of a general inductive definition.

Frequently, to prove properties of functions defined by general inductive definitions, it is convenient to use a stronger form of mathematical induction that is called *complete* (or *strong*) induction.

15.2 COMPLETE INDUCTION

First, we shall state the schema that we shall call "complete induction" (*CI*) and then show that it is a theorem of Peano Arithmetic.

Theorem 15.1 Complete Induction (*CI*) *If* φ *is any formula in the language of Peano Arithmetic with* x *as a free variable and if* y *and* z *are variables that do not occur in* φ*, then the following is a theorem of Peano Arithmetic:*

(*) $$\varphi[0|x] \wedge (\forall y)[(\forall z)(z<y+1 \rightarrow \varphi[z|x]) \rightarrow \varphi[y+1|x]] \rightarrow (\forall x)\varphi.$$

Before proving Theorem 15.1, let us examine it more closely. Just as for *MI*, *CI* has two steps—the base step and the induction step.
Base Step: $\varphi[0|x]$.
Induction Step: $(\forall y)[(\forall z)(z < y + 1 \rightarrow \varphi[z|x]) \rightarrow \varphi[y + 1|x]]$.

(Informally, we write the induction step as

$$(\forall y)[\varphi[0|x] \wedge \varphi[1|x] \wedge \cdots \wedge \varphi[y|x] \rightarrow \varphi[y+1|x]],$$

but this is not a formula in our language.) The corresponding induction step for *MI* is

$$(\forall y)(\varphi[y|x] \rightarrow \varphi[y+1|x]).$$

We see that the induction step for *CI* has a stronger hypothesis than that for *MI*. (It corresponds to the inductive step in the general inductive definition.) We shall illustrate *CI* below in the examples.

Proof of Theorem 15.1: We shall assume the formula φ satisfies the hypothesis of *CI*. We are trying to prove the formula (*), which is a conditional formula, so we shall start by taking the antecedent as a premise:

(i) $\varphi[0|x]$,

and

(ii) $(\forall y)[(\forall z)(z < y + 1 \rightarrow \varphi[z|x]) \rightarrow \varphi[y + 1|x]]$.

Now we want to derive $(\forall x)\varphi$. We shall define a formula ψ to be

(**) $$(\forall z)(z<y+1 \rightarrow \varphi[z|x]).$$

The formula ψ is a formula of Peano Arithmetic, which has y as a free variable. We shall show that ψ satisfies the base step and induction step of *MI*.

First note that $\psi[0|y]$ is logically equivalent to $\varphi[0|x]$. (Why?) Thus, by (i), ψ satisfies the base step of *MI*.

Next, suppose $\psi[u|y]$. (To prove ψ satisfies the induction step of *MI*, we must derive $\psi[u + 1|y]$.) From (ii) and (**), (using $US[u|y]$), we obtain

(iii) $\psi[u|y] \rightarrow \varphi[u + 1|y]$.

From $\psi[u|y]$ and (iii), we can derive $\varphi[u + 1|y]$. The formula

$$\psi[u|y] \wedge \varphi[u + 1|y]$$

is logically equivalent to $\psi[u + 1|y]$. (Why?) Thus, for all u,

$$\psi[u|y] \rightarrow \psi[u + 1|y],$$

which proves that ψ satisfies the induction step of *MI*. Consequently, $(\forall y)\psi$ follows from *MI*. But

$$(\forall y)\psi \rightarrow (\forall x)\varphi$$

is provable in Peano Arithmetic. (Why?). Thus, *CI* is a Theorem of Peano Arithmetic.

■

As our first illustration of *CI*, we shall return to Example 15.3 and give a proof by *CI* of the main result.

Example 15.7

We shall define a sequence S as follows:

1. $S_1 = 2$
2. $S_{n+1} = \prod_{i=1}^{n} S_i$, for $n \geq 1$.

Prove using *CI* that

(*) $$S_{n+1} = (S_n)^2, n \geq 2.$$

(See Example 15.3 for comparison. There we prove (*) using *MI*.)

Solution: The base step is the same as in Example 15.3.

Induction Step: The induction hypothesis, *IH*, is

(i) $S_{k+1} = (S_k)^2$ for all k, $1 < k < m + 1$.

That is,

(ii) $S_3 = (S_2)^2, S_4 = (S_3)^2, \ldots, S_{m+1} = (S_m)^2$.

(We wish to prove $S_{m+2} = (S_{m+1})^2$.) Using the inductive step in the definition of S,

$$\begin{aligned}
S_{m+2} &= \prod_{i=1}^{m+1} S_i \\
&= S_1 \cdot S_2 \cdot \prod_{i=3}^{m+1} S_i && \text{(Definition 15.2(b))} \\
&= (S_1)^2 \cdot \prod_{i=3}^{m+1} S_i && (S_2 = S_1) \\
&= (S_1)^2 \cdot \prod_{i=2}^{m} (S_i)^2 && (IH) \\
&= \prod_{i=1}^{m} (S_i)^2 && \text{(Definition 15.2(b))} \\
&= \left(\prod_{i=1}^{m} S_i\right)^2 && \text{(Algebra)} \\
&= (S_{m+1})^2 && \text{(2, inductive step)}
\end{aligned}$$

Thus, it follows by *CI* that we have proved (*).

Example 15.8 Prove the following about the Fibonacci sequence f. (See Example 15.5.)

$$f_n \le f_{n+1}, \text{for } n \ge 1.$$

Solution: The proof is by *CI*.

Base Step: In the base step, we shall include the first two cases because, as we shall see below, the induction step is only valid for $m \ge 2$. By the definition of f, $f_1 = f_2 = 1$ and $f_3 = f_1 + f_2 = 2$, therefore,

$$f_1 \le f_2 \quad \text{and} \quad f_2 \le f_3.$$

Induction Step: The induction hypothesis (*IH*) is

$$f_1 \le f_2, f_2 \le f_3, \ldots, f_m \le f_{m+1}, \quad m \ge 1.$$

(We want to prove $f_{m+1} \le f_{m+2}$.) Using 2 in the definition of f,

(i) $f_{m+2} = f_{m+1} + f_m$.

Assume $m \ge 2$. Then, by *IH*,

(ii) $f_{m-1} \le f_m$ and $f_m \le f_{m+1}$.

(Note that (ii) is true only if $m \geq 2$. Also note that we only use the last two inequalities in *IH*. However, if we had been using *MI* rather that *CI*, we could only have assumed $f_m \leq f_{m+1}$.)

Using (i), (ii), and algebra, we obtain

(iii) $f_m + f_{m-1} \leq f_{m+2}$.

However, from 2 in the definition of f,

(iv) $f_m + f_{m-1} = f_{m+1}$.

Thus, from (iii) and (iv), we obtained the desired result, for $m \geq 2$,

(v) $f_{m+1} \leq f_{m+2}$.

Because we had the foresight to prove $f_2 \leq f_3$ in the base step, (v) holds for all $m \geq 1$. The desired result follows by *CI*.

Example 15.9

Marriage Problem. Suppose that, given any set of n men, each subset of k of them, $1 \leq k \leq n$, knows at least k women. (That is, if M is any set of k men, then there is a set W of at least k women such that for each $w \in W$, there is an $m \in M$ such that m knows w. Several men in M may know the same woman in W and/or some men in M may know several women in W.) Prove that, under these conditions, it is possible for each of the n men to marry some woman he knows.

Solution: The proof is by complete induction on n.

Base Step: If there is only one man, then, by hypothesis, he knows at least one woman, so he can marry her.

Induction Step: Suppose we have a set M of $m + 1$ men such that each subset of k of them, $1 \leq k \leq m + 1$, knows at least k women. Let W be the set of all women that are known by some man in M. By hypothesis, there are at least $m + 1$ women in W. Our induction hypothesis, *IH*, is that if X is any set of j men, $1 \leq j < m + 1$, such that each subset of k of them, $1 \leq k \leq j$, knows at least k women, then it is possible for each man in X to marry some woman he knows. There are two cases to consider.

Case 1. Each subset of M with $k < m + 1$ men knows at least $k + 1$ women. Marry one man off to some woman he knows. Then there is a set of m men left that has the property that each subset of k of them knows at least k women. Thus, by *IH*, each of these m men can marry some woman whom he knows. Therefore, each of the original $m + 1$ men can marry some woman he knows.

Case 2. There is at least one subset $A \subseteq M$ with k members, $1 \leq k < m + 1$, who know exactly k women. Let W be the set of all women who are known by some man in M, and let $W_A \subseteq W$ be the set of k women whom are known by some man in A. The set A satisfies the hypothesis of IH (A has less than $m + 1$ members, and each subset of A with j members, $1 \leq j \leq k$, knows at least j women). Thus, it follows from IH that each man in A can marry some woman in W_A whom he knows. All the men in A and all the women in W_A can marry each other. Now, we claim that the men in $M - A$ and the women in $W - W_A$ also satisfy the hypothesis of IH. (Clearly, $M - A$ has $m + 1 - k < m + 1$ members, and each subset of $M - A$ with j members, $1 \leq j \leq m + 1 - k$, knows at least j members of $W - W_A$. If some subset $X \subseteq M - A$ with j members knew less than j members of $W - W_A$, then $A \cup X \subseteq M$ has $k + j$ members and knows less than $k + j$ members of W. This contradicts our assumption about M and W.) Therefore, it follows by IH that each of the men in $M - A$ can marry some woman he knows in $W - W_A$.

Consequently, we have proved that each man in M can marry some woman he knows in W. This completes the proof of the induction step. Thus, it follows by *CI* that every man can marry some woman he knows.

Our last two examples in this section depend on the definition of a formula of the Sentential Calculus given in Example 15.6.

Example 15.10

Using Example 15.6, give an informal proof using *CI* that, in every formula in the Sentential Calculus, the left and right parentheses occur in pairs.

Solution: The proof is by *CI* on the length of the formula. (The length of a formula is the number of sentential variables and connectives that occur in the formula.)

Base Step: If the formula is of length 1, then it is a sentential variable. (See 1 in Example 15.6.) There are no parentheses around sentential variables; thus, vacuously, the parentheses occur in pairs.

Induction Step: The induction hypothesis, *IH*, is that parentheses occur in pairs in all formulas of length $< m + 1$, where $m \geq 1$. Let φ be a formula of length $m + 1$. Then, by 2 in Example 15.6, there are formulas ψ_1 and ψ_2 of the Sentential Calculus such that φ is one of the following five formulas:

1. $(\neg\psi_1)$.
2. $(\psi_1 \wedge \psi_2)$.
3. $(\psi_1 \vee \psi_2)$.
4. $(\psi_1 \rightarrow \psi_2)$.
5. $(\psi_1 \leftrightarrow \psi_2)$.

The length of each of ψ_1 and ψ_2 is less than $m + 1$. Thus, by *IH*, the parentheses in ψ_1 and in ψ_2 occur in pairs. It follows from 1 through 5 above that the parentheses in φ also occur in pairs.

This proves the induction step, so it follows by *CI* that the parentheses occur in pairs in all formulas of the Sentential Calculus.

Before completing the last example in this section, we shall define a function w as follows.

Definition 15.3

We define a function w whose domain is the set of all finite strings of symbols of the Sentential Calculus.

1. $w(A) = -1$ if A is a sentential variable.
2. $w(\neg) = 0$.
3. $w(\vee) = 1$, $w(\wedge) = 1$, $w(\rightarrow) = 1$, $w(\leftrightarrow) = 1$.
4. If S is a string of symbols of the Sentential Calculus and S is $S's$, where s is either a sentential variable or one of the connectives $\neg$, $\vee$, $\wedge$, $\rightarrow$, or $\leftrightarrow$, then $w(S) = w(S') + w(s)$.

(In other words, the value of $w(S)$ is the sum of the values $w(s)$, where s is a symbol (excluding a parenthesis) in S. This is an inductive definition. Steps 1 through 3 are the initial steps, and step 4 is the inductive step.)

Example 15.11

Give an informal proof using *CI* that if φ is a formula of the Sentential Calculus, then $w(\varphi) = -1$.

Solution: The proof is by *CI* on the length of φ (using Example 15.6).

Base Step: If φ is of length 1 and is a formula, then, by Example 15.6, φ must be a sentential variable. By Definition 15.3, $w(\varphi) = -1$.

Induction Step: Suppose φ is a formula of length $m + 1$. The induction hypothesis, *IH*, is the statement that if ψ is a formula of length less than $m + 1$, then $w(\psi) = -1$.

Because φ is a formula of length greater than 1, it follows from Example 15.6 that there are formulas ψ_1 and ψ_2 such that φ is one of the following formulas:

1. $(\neg\psi_1)$.
2. $(\psi_1 \vee \psi_2)$.
3. $(\psi_1 \wedge \psi_2)$.
4. $(\psi_1 \rightarrow \psi_2)$.
5. $(\psi_1 \leftrightarrow \psi_2)$.

Because the length of φ is $m + 1$, it follows in each case that the length of ψ_1 and of ψ_2 is less than $m + 1$. It follows from *IH* that

$$(*) \qquad w(\psi_1) = -1, w(\psi_2) = -1.$$

There are five cases to consider, corresponding to the five formulas 1 through 5 above.

Case 1. φ is $(\neg\psi_1)$. Then

$$\begin{aligned} w(\varphi) &= w(\neg) + w(\psi_1) && \text{(Definition 15.3)} \\ &= 0 + w(\psi_1) && \text{(Definition 15.3)} \\ &= 0 + (-1) && (IH) \\ &= -1 && \text{(Algebra)} \end{aligned}$$

Case 2. φ is $(\psi_1 \vee \psi_2)$. Then

$$\begin{aligned} w(\varphi) &= w(\psi_1) + w(\vee) + w(\psi_2) && \text{(Definition 15.3)} \\ &= w(\psi_1) + 1 + w(\psi_2) && \text{(Definition 15.3)} \\ &= (-1) + 1 + (-1) && (IH) \\ &= -1 && \text{(Algebra)} \end{aligned}$$

Cases 3 through 5 are similar to Case 2 and will be omitted.
Thus, it follows by *CI* that if φ is a formula of the Sentential Calculus, $w(\varphi) = -1$.

15.3 AUTOMATED THEOREM-PROVING

There is an algorithm for deciding whether a formula of the Sentential Calculus is a tautology. One such algorithm is to make a truth table and check to see if the last column is all *T*'s. Thus, it follows that there is an algorithm for determining whether an argument in the Sentential Calculus is valid. (An argument with premises $\varphi_1, \varphi_2, \ldots, \varphi_n$ and conclusion ψ is valid iff $\varphi_1 \wedge \varphi_2 \wedge \cdots \wedge \varphi_n \rightarrow \psi$ is a tautology.) For the Predicate Calculus, however, there is no algorithm, like a truth table, for proving valid arguments that will work for every case. (This important result is due to Alonzo Church in 1936.[1] It is based on results of Kurt Gödel.) Even so, it is desirable to try to prove valid arguments in the Predicate Calculus using the computer. In this way, we approximate the way a rational being reasons using the computer. Such algorithms are called *automated*

[1] *Journal of Symbolic Logic*, Vol. 37, 1936, pp. 1–18. Correction, ibid, pp. 101–102.

reasoning or *automated theorem-proving algorithms*. This is part of an area of study that is called *artificial intelligence*.

Automated theorem-proving algorithms for the Predicate Calculus give rise to a theorem prover that will find a proof of any valid argument, but will not give a proof of an invalid argument. The main differences between this type of algorithm and, for example, the truth table algorithm for the Sentential Calculus are the following: (1) If an argument is valid, it normally cannot be determined in advance how long it will take the theorem-prover to prove it; and (2) if an argument is invalid, there is no guarantee that the theorem-prover will ever discover that fact. Although making a truth table for an argument that has many sentential variables may take a long time, one knows in advance that if an argument has n sentential variables, its truth table will have 2^n lines. There is no such rule for the Predicate Calculus.

In this section, we give some simple examples of one approach to automated theorem-proving. The language we shall use to describe the automated reasoning program is the language of the Predicate Calculus with Identity, with the following changes:

1. The sentential connectives are $\neg$ and $\vee$. (These connectives are sufficient. See Definition 16.1 ff.)
2. No quantifiers will be used. If a variable x is free in a formula, it is assumed it is within the scope of $(\forall x)$. (Existential quantifiers may be eliminated using "Skolem functions" (see Chapter 16). In this section, however, we shall restrict ourselves to formulas that do not contain existential quantifiers.)

The Rules of Inference include the following rules:

R1 (Double Negation): From $\neg\neg\varphi$, infer φ, and from φ, infer $\neg\neg\varphi$.

R2 (Commutative Rule): From $\varphi \vee \psi$, infer $\psi \vee \varphi$. (The associative rule for $\vee$ is implicitly assumed.)

R3 (Useful Tautologies): From $\neg(\varphi_1 \vee \varphi_2 \vee \cdots \vee \varphi_n)$, infer $\neg\varphi_i$ for each $i = 1, 2, \ldots, n$, and from φ_i, infer $\varphi_1 \vee \varphi_2 \vee \cdots \vee \varphi_n$ for each $i = 1, 2, \ldots, n$.

R4 (Unification—A combination of the Rules *US* and tautologies):
(a) From $\varphi \vee \psi$ and $\neg\varphi[a_1|x_1, a_2|x_2, \ldots, a_n|x_n]$, infer $\psi[a_1|x_1, a_2|x_2, \ldots, a_n|x_n]$.
(b) From $\neg\varphi \vee \psi$ and $\varphi[a_1|x_1, a_2|x_2, \ldots, a_n|x_n]$, infer $\psi\ [a_1|x_1, a_2|x_2, \ldots, a_n|x_n]$.

R5 (Resolution—A combination of tautologies):
(a) From $\varphi \vee \psi$ and $\neg\psi \vee \chi$, infer $\varphi \vee \chi$.
(b) From $\varphi \vee \neg\psi$ and $\psi \vee \chi$, infer $\varphi \vee \chi$.

R6 Rules for Identity (see *Ia, Ib, Ic,* and *Id* in Chapter 11)

Frequently, combinations of these rules are used. For example,

From $\neg Axy \vee \neg Bxy \vee Cxy$, $\neg Cab$ and Aab, infer $\neg Bab$.

Here, we use two applications of R4.

From $\neg Axy \vee \neg Bxy \vee Cxy$, $Aab \vee Dab$ and Bab, infer $Cab \vee Dab$.

In this argument, we used Commutative and Associative Rules (from $\neg Axy \vee \neg Bxy \vee Cxy$, infer $\neg Bxy \vee Cxy \vee \neg Axy$); R4 (from $\neg Bxy \vee Cxy \vee \neg Axy$ and Bab, infer $Cab \vee \neg Aab$); and R5 (from $Cab \vee \neg Aab$ and $Aab \vee Dab$, infer $Cab \vee Dab$). In what follows, we will normally not mention uses of Rules R1 and R2.

Let's see what happens if we submit the following argument to our automated reasoning machine:

All ghosts who live in the old mansion like Oscar. Oscar does not like anyone who likes him. Therefore, any ghost whom Oscar likes does not live in the old mansion. (Gx, Mx, Lxy, o)

Translating into the Predicate Calculus using the indicated predicate symbols and constants, we obtain the following:

Premises: $(\forall x)(Gx \wedge Mx \rightarrow Lxo)$, $(\forall x)(Lxo \rightarrow \neg Lox)$

Conclusion: $(\forall x)(Gx \wedge Lox \rightarrow \neg Mx)$

Next, drop universal quantifiers and rewrite the premises and conclusion using only the connectives $\neg$ and $\vee$.

$$Gx \wedge Mx \rightarrow Lxo \Leftrightarrow \neg(Gx \wedge Mx) \vee Lxo$$
$$\Leftrightarrow \neg Gx \vee \neg Mx \vee Lxo$$
$$Lxo \rightarrow \neg Lox \Leftrightarrow \neg Lxo \vee \neg Lox$$
$$Gx \wedge Lox \rightarrow \neg Mx \Leftrightarrow \neg(Gx \wedge Lox) \vee \neg Mx$$
$$\Leftrightarrow \neg Gx \vee \neg Lox \vee \neg Mx$$

We have to give the machine some strategy. We must tell it where to start, in what order to apply the rules, and so forth. (We shall not discuss strategy in any detail here.) In this example, the machine would start with the following two formulas:

(1) $\neg Gx \vee \neg Mx \vee \neg Lxo.$

(2) $\neg Lxo \vee \neg Lox.$

Then from (1), (2), and R5, the machine would infer

(3) $\neg Gx \vee \neg Mx \vee \neg Lox.$

Formula (3) is tautologically equivalent to the conclusion, so the proof is complete.

Let's try a more complicated problem.

Example 15.12 A man is supposed to bring a dog, a lamb, and a bag of hay from the left bank of a river to the right bank on a rowboat. The boat is very small, so the man can carry only one of these items on the boat at a time. In addition, he cannot leave the dog alone with the lamb, nor the lamb alone with the hay. Is there any way the man can accomplish his task and bring the dog, the lamb, and the hay from the left bank to the right bank of the river?

Solution: In order for our machine to solve the problem, we must translate it into the language that the machine understands, and we must give the machine *all* the relevant information. First, define the constants and predicate symbols:

$$d\text{: dog,}\qquad h\text{: hay,}\qquad l\text{: lamb,}\qquad m\text{: man.}$$

We shall also use terms to represent sets:

$$t_x : \{x\},\quad t_{xy} : \{x, y\},\quad t_{xyz} : \{x, y, z\},\quad t_{mxyz} : \{m, x, y, z\},\quad t_0 : \phi \text{ (the empty set)}$$

We shall use one predicate symbol:

Bxy: x is on the left bank and y is on the right bank.

Next, we must give the machine all the information that is given. At first, everything is on the left bank:

(1) $$Bt_{mdhl}t_0.$$

We must also tell the machine that two sets are identical if their elements are permuted. This leads to 11 equations.

(2)
$$\begin{aligned}
t_{xy} &= t_{yx}\\
t_{xyz} &= t_{xzy}\\
&= t_{yxz}\\
&= t_{yzx}\\
&= t_{zxy}\\
&= t_{zyx}\\
t_{mxyz} &= t_{mxzy}\\
&= t_{myxz}\\
&= t_{myzx}\\
&= t_{mzxy}\\
&= t_{mzyx}
\end{aligned}$$

The machine has to know which positions are not allowed:

(3) $Bt_{dl}t_{mh}$

(4) $Bt_{hl}t_{md}$

(5) $Bt_{mh}t_{dl}$

(6) $Bt_{md}t_{hl}$

(7) $Bt_{dhl}t_{m}$

(8) $Bt_{m}t_{dhl}$

There are other positions that are not allowed, but we do not have to state them explicitly because they are equivalent to (3) through (8) by (2) and R6. If the machine gets to any one of the positions (3) through (8), it goes back and tries another line of attack.

Finally, we have to tell the machine the possible ways of getting from one position to another using the fact that the boat can only carry the man and one other item.

(9) $\neg Bt_{mxyz}t_0 \vee Bt_{yz}t_{mx}.$

(From the starting position, the man rows one item to the right bank.)

(10) $\neg Bt_{yz}t_{mx} \vee Bt_{myz}t_{x}.$

(From the second position, the man rows back to the left bank alone. If he took x with him, he would be back in the starting position.)

(11) $\neg Bt_{myz}t_{x} \vee Bt_{z}t_{mxy}.$

(From the third position on the left bank, the man rows an item to the right bank.)

(12) $\neg Bt_{z}t_{mxy} \vee t_{xy} = t_{dl} \vee t_{xy} = t_{lh} \vee Bt_{mz}t_{xy}.$

(From the fourth position, if the man can leave x and y together, he rows back alone to the left bank.)

(13) $\neg Bt_{z}t_{mxy} \vee \neg(t_{xy} = t_{dl} \vee t_{xy} = t_{hl}) \vee Bt_{mzx}t_{y}.$

(From the fourth position, if the man cannot leave x and y together, he brings one back to the left bank.)

(14) $\neg Bt_{mz}t_{xy} \vee Bt_0 t_{mxyz}.$

(If the man and one item are on the left bank, the man rows that item to the right bank.)

Now the machine has all the necessary information, so we give it the strategy. This time, we shall try an indirect proof, and give the machine the negation of the position we are trying to achieve!

(15) $\neg Bt_0 t_{mdhl}$.

We start the machine to see if it can find a contradiction. From (1), (9), and R4, infer

(16) $Bt_{hl}t_{md}$.

Line (16) is the same as line (4), a position that is not allowed, so the machine will go back to (1) and try again.

(17)	$Bt_{mldh}t_0$	((1), (2), R6)
(18)	$Bt_{dh}t_{ml}$	((9), (17), R4)
(19)	$Bt_{mdh}t_l$	((10), (18), R4)
(20)	$Bt_h t_{mld}$	((11), (19), R4)
(21)	$Bt_{mhl}t_d$	((13), (20), R3, R4)
(22)	$Bt_l t_{mhd}$	((11), (21), R4)
(23)	$Bt_{ml}t_{hd}$	((12), (22), R4)
(24)	$Bt_0 t_{mhdl}$	((14), (23), R4)
(25)	$Bt_0 t_{mdhl}$	((2), (24), R6)

Line (25) contradicts line (15), so the machine has solved the problem and proved that it is possible for the man to bring all the items to the right bank. The solution found is given in the table below using lines (1) and (18) through (24).

Left Bank	Right Bank
mdhl	ϕ
dh	*ml*
mdh	*l*
h	*mdl*
mhl	*d*
l	*mdh*
ml	*dh*
ϕ	*mdhl*

Our last example in this section is Example 3.3 (a). In Chapter 3, we solved this problem in a few lines. Here, we shall show how our machine handles it.

Example 15.13 Four different people—Arthur (a), Betty (b), Charles (c), and Dorothy (d)—have four different positions (not necessarily in the given order): lawyer (L), pilot (P), teacher (T), and veterinarian (V). Each person holds exactly one position, and each position is held by exactly one person. In addition,

(∗) $\neg Va, \quad \neg Vc, \quad \neg Vb, \quad \neg Pb, \quad \neg Lc, \quad \neg Pc, \quad \neg Ld.$

Is the information consistent?

Solution: For our machine to solve this problem, we must give it all the information. We are given that no two people are the same:

(1) $a \neq b, a \neq c, a \neq d, b \neq c, b \neq d, c \neq d.$

We are also given that each person holds at most one position:

(2)
$$\neg Lx \vee \neg(Px \vee Tx \vee Vx)$$
$$\neg Px \vee \neg(Lx \vee Tx \vee Vx)$$
$$\neg Tx \vee \neg(Lx \vee Px \vee Vx)$$
$$\neg Vx \vee \neg(Lx \vee Px \vee Tx)$$

Each person holds at least one position:

(3) $Lx \vee Px \vee Tx \vee Vx.$

Each position is held by at least one person:

(4)
$$La \vee Lb \vee Lc \vee Ld$$
$$Pa \vee Pb \vee Pc \vee Pd$$
$$Ta \vee Tb \vee Tc \vee Td$$
$$Va \vee Vb \vee Vc \vee Vd$$

Finally, we give the machine the information given in (∗) above and tell it to stop if either it arrives at a contradiction or it arrives at a set of formulas of the form

$$Lx, \quad Py, \quad Tz, \quad Vw,$$

where x, y, z, and w are some permutation of the constants a, b, c, and d. Based on the strategy it is given, our machine might proceed as follows:

(5)	Vd	((∗), (4), R4)
(6)	Tc	((∗), (3), R4)
(7)	$La \vee Lb$	((∗), (4), R4)
(8)	$Pa \vee Pd$	((∗), (4), R4)
(9)	$\neg(Ld \vee Pd \vee Td)$	((2), (5), R4)

(10)	$\neg Pd$	((9), R3)
(11)	Pa	((8), (10), R4)
(12)	$\neg(La \vee Ta \vee Va)$	((2), (11), R4)
(13)	$\neg La$	((12), R3)
(14)	Lb	((7), (13), R3)

At this point, the machine would stop because, using (14), (11), (6), and (5), it has constructed the following set:

(**) Lb, Pa, Tc, Vd.

Thus, the information given in Example 15.13 is consistent. How would you find out from the machine whether the result found in (**) is unique?

The examples we have given here should give you the flavor of automated theorem-proving. (The proof of the Completeness Theorem in Chapter 16 also describes an algorithm for proving theorems.) It is a relatively new field, but it is important and is growing rapidly. Even though we have given a simplified version, it should be clear that machines are not intelligent. You cannot assume that they know anything. Every bit of information that we take for granted has to be told to the machine. We frequently hear students say "We do not have to know how to do . . . because the computer can do it." However, we see from these simple examples that problems may have to be carefully analyzed before they are submitted to a computer, and, in many cases, the programmer may have to give the machine the process to be used to solve the problem.

(For more details about automated theorem proving, see *Automated Reasoning: Introduction and Applications* by L. Wos, R. Overbeek, E. Lusk, and J. Boyle, Prentice-Hall, New York, 1984. This is an excellent introductory text on the subject.)

EXERCISES 15

A. Each of the following is an inductive definition of a sequence S. Express S_n as a function of n, and give an informal proof using induction (either mathematical induction or complete induction) that your formula is correct. (Answers to 1, 5, 10, 15, 20, and 25.)

(1) $S_0 = 1$, $S_{n+1} = \frac{1}{2}S_n$, $n \geq 0$.
(2) $S_0 = a$, $S_{n+1} = bS_n$, $n \geq 0$.
(3) $S_0 = 1$, $S_{n+1} = 2(n + 1)S_n$, $n \geq 0$.
(4) $S_0 = 1$, $S_{n+1} = (an + b)S_n$, $n \geq 0$.
(5) $S_0 = 1$, $S_{n+1} = S_n + 2^{n+1}$, $n \geq 0$.

(6) $S_0 = 1,\ S_{n+1} = S_n + 3^n,\ n \geq 0.$
(7) $S_0 = a,\ S_{n+1} = S_n + b,\ n \geq 0.$
(8) $S_0 = a,\ S_{n+1} = S_n + n,\ n \geq 0.$
(9) $S_0 = 1,\ S_{n+1} = 3S_n + 1,\ n \geq 0.$
(10) $S_0 = 1,\ S_{n+1} = 4S_n - 1,\ n \geq 0.$
(11) $S_0 = a,\ S_{n+1} = bS_n + c,\ b \neq 0,\ n \geq 0.$
(12) $S_0 = a,\ S_{n+1} = S_n + bn + c,\ n \geq 0.$
(13) $S_0 = 0,\ S_1 = 1,\ S_{n+1} = 2S_n - S_{n-1},\ n \geq 1.$
(14) $S_0 = a,\ S_1 = b,\ S_{n+1} = 2S_n - S_{n-1},\ n \geq 1.$
(15) $S_0 = 1,\ S_1 = 0,\ S_{n+1} = 3S_n - 2S_{n-1},\ n \geq 1.$
(16) $S_0 = 1,\ S_1 = 3,\ S_{n+1} = 4S_n - 4S_{n-1},\ n \geq 1.$
(17) $S_0 = 0,\ S_1 = 3,\ S_{n+1} = -S_n + 2S_{n-1},\ n \geq 1.$
(18) $S_0 = 0,\ S_1 = 2,\ S_{n+1} = 4S_n - 3S_{n-1},\ n \geq 1.$
(19) $S_0 = 0,\ S_1 = 4,\ S_{n+1} = -2S_n + 3S_{n-1},\ n \geq 1.$
(20) $S_0 = a,\ S_1 = b,\ S_{n+1} = 4S_n - 4S_{n-1},\ n \geq 1.$
(21) $S_0 = a,\ S_1 = b,\ S_{n+1} = 6S_n - 9S_{n-1},\ n \geq 1.$
(22) $S_0 = a,\ S_1 = b,\ S_{n+1} = cS_n - (c^2/4)\ S_{n-1},\ n \geq 1.$
(23) $S_0 = 1,\ S_{n+1} = \sum_{i=0}^{n} S_i,\ n \geq 0.$
(24) $S_0 = a,\ S_{n+1} = \sum_{i=0}^{n} S_i,\ n \geq 0.$
(25) $S_0 = 1,\ S_{n+1} = 2\sum_{i=0}^{n} S_i,\ n \geq 0.$
(26) $S_0 = 1,\ S_{n+1} = a\sum_{i=0}^{n} S_i,\ n \geq 0.$
(27) $S_0 = 3,\ S_{n+1} = 2 + \prod_{i=0}^{n} S_i,\ n \geq 0.$
(28) $S_0 = 4,\ S_{n+1} = \sqrt{S_n + \sqrt{S_{n-1} + \sqrt{S_{n-2} + \sqrt{\cdots}}}}$

B. Use Horner's method to calculate the sequence S^p for each of the polynomials $p(x)$ given below. (See Example 15.4.) (Answer to 1.)
(1) $p(x) = 3x^4 - 7x^3 + 2x^2 - 3x + 2.$
(2) $p(x) = 3x^5 - 4x^3 + x^2 + x - 3.$
(3) $p(x) = x^5 + 3x^4 - 7x - 1.$
(4) $p(x) = -x^6 + 4x^4 - x^2 + 4.$
(5) $p(x) = 3x^6 - 4x^5 - 4x^3 + 7x^2.$
(6) $p(x) = -x^6 - 3x^5 + 3x^3 - 6x.$

C. The Fibonacci sequence, f, was defined in Example 15.5 ($f_1 = f_2 = 1$; $f_{n+1} = f_n + f_{n-1}$, for $n \geq 2$). We shall also define a similar sequence called the "Lucas sequence," ℓ ($\ell_1 = 1$, $\ell_2 = 3$; $\ell_{n+1} = \ell_n + \ell_{n-1}$, for $n \geq 2$). Each of the following exercises refer to these sequences. Give an informal proof of each using induction (either mathematical induction or complete induction). (Answers to 1, 5, 7, 14, and 18.)

(1) $f_n + 2f_{n+1} = f_{n+3}, n \geq 1$.
(2) $f_{n+4} = 3f_{n+2} - f_n, n \geq 1$.
(3) $f_{n+6} = 4f_{n+3} + f_n, n \geq 1$.
(4) $f_{n+8} = 7f_{n+4} - f_n, n \geq 1$.
(5) $f_{n+10} = 11f_{n+5} + f_n, n \geq 1$.
(6) $f_n + \ell_n = 2f_{n+1}, n \geq 1$.
(7) $\sum_{i=1}^{n} \ell_i = \ell_{n+2} - 3, n \geq 1$.
(8) $\sum_{i=1}^{n} f_i = f_{n+2} - 1, n \geq 1$.
(9) $\sum_{i=1}^{n} \ell_{2i} = \ell_{2n+1} - 1, n \geq 1$.
(10) $\sum_{i=1}^{n} f_{2i} = f_{2n+1} - 1, n \geq 1$.
(11) $\sum_{i=1}^{n} \ell_{2i-1} = \ell_{2n} - 2, n \geq 1$.
(12) $\sum_{i=1}^{n} f_{2i-1} = f_{2n}, n \geq 1$.
(13) $\sum_{i=1}^{n} (f_i)^2 = f_n f_{n+1}, n \geq 1$.
(14) $\ell_n < 2^n, n \geq 1$.
(15) $f_n < 2^n, n \geq 1$.
(16) $f_{n-1}f_{n+1} - (f_n)^2 = (-1)^n, n \geq 2$.
(17) $\ell_{n-1}\ell_{n+1} - (\ell_n)^2 = 5(-1)^{n-1}, n \geq 2$.
(18) $1.614 < \dfrac{f_{n+1}}{f_n} < 1.620, n \geq 7$.
(19) $1.6169 < \dfrac{\ell_{n+1}}{\ell_n} < 1.621, n \geq 7$.
(20) $f_n = \dfrac{1}{\sqrt{5}}\left[\left(\dfrac{1+\sqrt{5}}{2}\right)^n - \left(\dfrac{1-\sqrt{5}}{2}\right)^n\right], n \geq 1$.
(21) $\ell_n = \left(\dfrac{1+\sqrt{5}}{2}\right)^n + \left(\dfrac{1-\sqrt{5}}{2}\right)^n, n \geq 1$.
(22) $f_m f_n + f_{m+1}f_{n+1} = f_{m+n+1}, m \geq 1, n \geq 1$.
(23) $f_n\ell_m + f_{n+1}\ell_{m+1} = \ell_{m+n+1}, m \geq 1, n \geq 1$.

D. We define a *word* to be a finite string of the digits 0, 1, 2, . . . , 9. A *binary* word only contains digits from the set {0, 1}, and a *ternary* word only contains digits from {0, 1, 2}. Give an inductive definition for each of the following. (Answers to 1, 5, and 8.)

(1) A binary word in which all the 0's precede all the 1's.
(2) A binary word that contains exactly one 0.
(3) A ternary word that does not contain any 1's.
(4) A ternary word whose first digit is 0 and last digit is 2.
(5) If w is a binary word, let w^* be the word obtained from w by replacing each 0 by a 1 and each 1 by a 0. Give an inductive definition of w^* for any binary word w.
(6) The *reverse* of a word w is a word $\overleftarrow{w}$, which has all the symbols of w in reverse order. (For example, $\overleftarrow{0012012} = 2102100$.) Give an inductive definition of $\overleftarrow{w}$ for any ternary word w. Prove that $\overleftarrow{w_1w_2} = \overleftarrow{w_2}\,\overleftarrow{w_1}$.
(7) The number of binary words of length n that do not contain the pattern 11.
(8) The number of binary words of length n that do not contain the pattern 110.
(9) The number of ternary words of length n that do not contain the pattern 11.
(10) The number of ternary words of length n that do not contain the pattern 012.

E. Give informal proofs of each of the following. (Answers to 1, 5, 7, 15, 18, 20, and 24(ai), (ciii).)

(1) A sequence S is defined by $S_1 = 1$, $S_2 = 2$; $S_{n+1} = 2S_n + S_{n-1}$, $n \geq 2$. Prove that $S_n \leq (\frac{5}{2})^n$, $n \geq 1$.
(2) A sequence S is defined by $S_1 = 1$, $S_2 = 5$; $S_{n+1} = 2S_n + 3S_{n-1}$, $n \geq 2$. Prove that $3^n \leq S_{n+1} \leq 2 \cdot 3^n$, $n \geq 1$.
(3) A sequence S is defined by: $S_1 = S_2 = S_3 = 1$; $S_{n+2} = S_{n+1} + S_n + S_{n-1}$, $n \geq 2$.
 (a) Prove that S_n is odd, $n \geq 1$.
 (b) Prove that $S_{n+1} \leq 2^n$, $n \geq 0$.
(4) A sequence S is defined by $S_1 = S_2 = S_3 = 1$; $S_{n+2} = S_{n+1} + S_n + S_{n-1}$, $n \geq 2$.
 (a) Prove that $S_{n+2} \geq 2S_n$, $n \geq 2$.
 (b) Prove that $S_{n+2} \geq (\sqrt{2})^n$, $n \geq 2$.
(5) A sequence S is defined by: $S_1 = 1$; $S_{n+1} = \dfrac{1}{1 + S_n}$, $n \geq 1$.
 (a) Prove that $S_n = \dfrac{f_n}{f_{n+1}}$, $n \geq 1$, where f is the Fibonacci sequence, $f_1 = f_2 = 1$; $f_{n+1} = f_{n-1} + f_n$, $n \geq 2$.
 (b) Assuming S has a limit, prove $\lim\limits_{n\to\infty} S_n = \dfrac{\sqrt{5} - 1}{2}$.
(6) A sequence S is defined by $S_1 = 3$, $S_{n+1} = (S_n)^2$, $n \geq 1$. Prove that $\lim\limits_{n\to\infty} S_n = \infty$. (Hint, express S_n as a function of n.)

(7) A sequence S is defined by: $S_1 = 2$; $S_{n+1} = \frac{1}{4}[(S_n)^2 + 3]$.
(a) Prove that $S_n > S_{n+1}$, $n \geq 1$.
(b) Prove that S has a limit, and find the limit.

(8) A sequence S is defined by $S_1 = 1$, $S_{n+1} = \frac{1}{5}[(S_n)^2 + 6]$.
(a) Prove that $S_n < S_{n+1}$, $n \geq 1$.
(b) Prove that S has a limit, and find the limit.

(9) A sequence S is defined by $S_0 = 2$; $S_{n+1} = \sqrt{S_n}$, $n \geq 0$.
(a) Prove that $S_n > S_{n+1}$, $n \geq 0$.
(b) Prove that S has a limit, and find the limit.

(10) A sequence S is defined by $S_0 = 0$; $S_{n+1} = \sqrt{a + S_n}$, $n \geq 0$, where $a > 0$.
(a) Find a positive real number M such that $S_n < M$, $n \geq 0$.
(b) Prove that $S_n < S_{n+1}$, $n \geq 0$.
(c) Prove that S has a limit, and find the limit.

(11) Pigeon-hole principle. If p pigeons are placed in n pigeon holes and $p > n$, then at least one of the pigeon holes must receive two pigeons.

(12) Prove that a positive integer $n \geq 2$ is either a prime or has a prime factor. (p is a *prime* if p is an integer, $p > 1$, and the only factors of p are 1 and p itself.)

(13) Jigsaw-puzzle problem. Suppose a jigsaw puzzle has n pieces. Each time a piece or a connected set of pieces is connected to another piece or connected set of pieces, it is called a *move*. Prove that there must be exactly $n - 1$ moves to complete the puzzle.

(14) (a) Suppose that there are 3^n ($n \geq 1$) coins and that all the coins are identical except for one that is heavier. Prove that the heavier coin can be found in n weighings on a balance.
(b) Same as part (a), except that you do not know whether the bad coin is heavier or lighter than the others. Prove that you can find the bad coin, and determine whether it is heavier or lighter than the others in $n + 1$ weighings on a balance.

(15) The product of two 2×2 matrices is defined as follows:

$$\begin{pmatrix} a_1 & a_2 \\ a_3 & a_4 \end{pmatrix}\begin{pmatrix} b_1 & b_2 \\ b_3 & b_4 \end{pmatrix} = \begin{pmatrix} a_1b_1 + a_2b_3 & a_1b_2 + a_2b_4 \\ a_3b_1 + a_4b_3 & a_3b_2 + a_4b_4 \end{pmatrix}.$$

If A is any matrix, then A^n means A multiplied by itself n times. Prove that if $M = \begin{pmatrix} 0 & 1 \\ 1 & 1 \end{pmatrix}$, then $M^n = \begin{pmatrix} f_{n-1} & f_n \\ f_n & f_{n+1} \end{pmatrix}$, $n \geq 2$, where f is the Fibonacci sequence: $f_1 = f_2 = 1$; $f_{n+1} = f_n + f_{n-1}$, $n \geq 2$.

(16) If R is a partial ordering on a set D, a set $A \subseteq D$ is said to be an *anti-chain* if for all $a, b \in A$, $a \neq b$, $\neg aRb$, and $\neg bRa$. (See Definition 7.3.) R is a *chain* if R is a linear ordering. Prove that if D is finite, R is a partial ordering on D, and n is the maximum size of an anti-chain in D, then D can be partitioned into n chains.

(17) If $\varphi_1, \varphi_2, \ldots, \varphi_n$ are n formulas in the Sentential Calculus, prove that for $n \geq 1$,

$$\neg(\varphi_1 \wedge \varphi_2 \wedge \cdots \wedge \varphi_n) \Leftrightarrow (\neg\varphi_1 \vee \neg\varphi_2 \vee \cdots \vee \neg\varphi_n).$$

(This is a generalization of De Morgan's rule.)

(18) If φ is a formula in the language of Peano Arithmetic with x as a free variable, and if y and z are variables that do not occur in φ, then the following is a theorem:

$$(\forall y)[(\forall z)(z < y \rightarrow \varphi[z|x]) \rightarrow \varphi[y|x]] \rightarrow (\forall x)\varphi.$$

(This implies that the base step is not necessary in complete induction.)

(19) Let φ be a formula in the Sentential Calculus. Let n be the number of places in φ at which a binary connective $\wedge$, $\vee$, $\rightarrow$, or $\leftrightarrow$ occurs in φ, and let m be the number of places at which a sentential variable occurs in φ. Prove that $m = n + 1$.

(20) A formula of the Sentential Calculus is called *positive* if it is built up from the sentential variables using only the connectives $\wedge$ and $\vee$. Prove that no positive formula is a tautology and that no positive formula is a contradiction.

(21) If φ is a formula of the Sentential Calculus that only contains the connectives $\neg$, $\wedge$, and $\vee$, and φ' is obtained from φ by replacing each $\wedge$ by $\vee$ and replacing each $\vee$ by $\wedge$, then prove that φ is a tautology if and only if $\neg\varphi'$ is a tautology.

(22) If φ is a formula of the Predicate Calculus (without identity) that only contains the connectives $\neg$, $\wedge$, and $\vee$, and φ' is obtained from φ by interchanging $\wedge$ and $\vee$ (as in (21)) and also interchanging $\forall$ and $\exists$, prove that $\vdash \varphi$ if and only if $\vdash \neg\varphi'$.

(23) Prove that each formula of the Predicate Calculus is logically equivalent to a formula in prenex normal form. (See Definition 5.5 and Theorem 10.7.)

(24) Suppose φ and ψ are formulas of the Sentential Calculus. If we write $\neg\varphi$ instead of $(\neg\varphi)$, $\wedge\varphi\psi$ instead of $(\varphi \wedge \psi)$, $\vee\varphi\psi$ instead of $(\varphi \vee \psi)$, $\rightarrow \varphi\psi$ instead of $(\varphi \rightarrow \psi)$, and $\leftrightarrow \varphi\psi$ instead of $(\varphi \leftrightarrow \psi)$, there is no longer any need for parentheses. (This way of writing a formula is called *Polish Notation* and may be familiar to you due to its use in some pocket calculators.) For example, the formula

$$(P \rightarrow ((\neg Q) \wedge (\neg R)))$$

in Polish Notation would be

$$\rightarrow P \wedge \neg Q \neg R.$$

(a) Write each of the following formulas in Polish Notation:

(i) $(\neg P \vee Q) \rightarrow \neg(P \wedge \neg R)$.

(ii) $(P \rightarrow \neg Q) \leftrightarrow \neg(R \wedge \neg P)$.

(iii) $(P \wedge Q) \wedge (R \vee \neg Q) \rightarrow \neg P$.

(iv) $P \rightarrow (Q \rightarrow R) \wedge (\neg P \rightarrow Q)$.

(v) $\neg(P \rightarrow \neg Q) \wedge (P \rightarrow \neg Q)$.

(vi) $\neg(P \wedge (Q \vee (\neg P \leftrightarrow R \wedge Q)))$.

(b) If φ is written in Polish Notation and $w(\varphi)$ is defined as in Definition 15.3, prove that φ is a formula of the Sentential Calculus if and only if the following two conditions are true:

(i) $w(\varphi) = -1$.

(ii) $w(\varphi') \geq 0$ for each proper initial segment φ' of φ. (Note that φ' is a string of symbols, but it is not necessarily a formula.)

(c) Apply the test given in part (b) to determine which of the following expressions are formulas of the Sentential Calculus written in Polish notation. Write those that are in the standard way.

(i) $\neg \leftrightarrow AB \wedge \neg C$.

(ii) $\rightarrow \wedge AB\neg C\vee \rightarrow AC$.

(iii) $\rightarrow A \rightarrow B \leftrightarrow CD$.

(iv) $\rightarrow \neg \rightarrow AB \rightarrow\rightarrow B\neg C \rightarrow \neg CB$.

(v) $\rightarrow \wedge AB\neg \vee A \vee BC$.

(vi) $\wedge A \vee B \rightarrow C \leftrightarrow D$.

(vii) $\wedge \vee \rightarrow\leftrightarrow ABCDE$.

(viii) $\neg A \wedge B \rightarrow C$.

(ix) $\rightarrow A \wedge \vee B\neg C \leftrightarrow DE$.

(x) $\neg\vee \rightarrow\leftrightarrow \neg \leftrightarrow ABCDE$.

(25) If ψ is a formula in the Sentential Calculus, give a definition of a *subformula* of ψ. (It then follows immediately that if ψ is a proper subformula of

(a) $\neg\varphi$, then ψ occurs in φ.

(b) $\varphi_1 \rightarrow \varphi_2$, then ψ occurs in φ_1 or in φ_2.)

(26) Let S_n be the set of all formulas of the Sentential Calculus that can be formed using only sentential variables from the list $A_1, A_2, \ldots, A_n$. Define a binary relation R_n on S_n so that if $\varphi, \psi \in S_n$,

$$\varphi R_n \psi \quad \text{iff} \quad \varphi \Leftrightarrow \psi.$$

(a) Prove that R_n is an equivalence relation, $n \geq 1$. (See Definition 7.3.)

(b) If $\varphi \in S_n$, $[\varphi] = \{\psi \in S_n : \varphi R_n \psi\}$ is called an *equivalence class* of R_n. Prove that R_n has 2^{2^n} equivalence classes, $n \geq 1$. (*Hint*: Use the disjunctive normal form. See Definition 4.1.)

F. For each of the arguments given in Exercises 7 B (except Exercise 12)[2],

---

[2]If some of the premises in Exercise 12 are written in prenex normal form, they contain existential quantifiers.

(i) write the premises and the conclusion in the language of the automated theorem-proving machine;

(ii) Describe how the machine might give a proof.

(Answers to 1, 7, and 11.)

G. Four different people—Arthur (*a*), Betty (*b*), Charles (*c*) and Dorothy (*d*)—have four different positions (not necessarily in the given order): lawyer (*L*), pilot (*P*), teacher (*T*), and veterinarian (*V*). Each person holds exactly one position, and each position is held by exactly one person. (See Example 15.13.) For each of the following situations, answer the questions as our automated theorem-proving machine might answer them:

(i) Are the conditions consistent?

(ii) If the conditions are consistent, is each person's position uniquely determined?

(Answers to 1 and 4.)

(1) $\neg Vb$, $\neg Vc$, $\neg Va$, $\neg Pa$, $\neg Tb$, $\neg Lc$.

(2) $\neg Pa$, $\neg Pb$, $\neg Ta$, $\neg Td$, $\neg Va$, $\neg Vc$, $\neg Vb$, $\neg Lb$.

(3) $\neg Ta$, $\neg Tb$, $\neg Va$, $\neg Vb$, $\neg Lb$, $\neg Pb$, $\neg Tc$, $\neg Vc$.

(4) $\neg La$, $\neg Pb$, $\neg Tc$, $\neg Vd$, $\neg Va$, $\neg Vb$, $\neg Lc$, $\neg Ld$, $\neg Ta$, $\neg Td$.

(5) $\neg La$, $\neg Ld$, $\neg Pb$, $\neg Pc$, $\neg Tb$, $\neg Tc$, $\neg Vc$, $\neg Vd$.

(6) $Va \vee Vc$, $\neg Lb$, $\neg Pb$, $Tc \vee Vc$, $\neg La$.

(7) $Va \vee Vc$, $\neg Lb$, $\neg Pb$, $\neg Lc$, $\neg Tc$, $\neg Td$, $\neg La$.

H. Solve each of the following problems as the automated theorem proving machine might solve it. (See Examples 15.12 and 15.13.) (Answers to 1 and 3(a).)

(1) Suppose there are 9 coins, all identical except for one that is heavier than the others. Can you determine the heavier coin in two weighings on a balance?

(2) Suppose there are three boxes, one contains red balls, the second contains blue balls, and the third contains green balls. However, all the boxes are mislabeled. You look in the first box, which is labeled red balls, and see it contains blue balls. Can you determine the contents of the other two boxes?

(3) We return again to TUFA Island where there are two races of people—the Tu's who always tell the truth, and the Fa's, who never tell the truth. (See Chapter 3.) We meet three natives—*U*, *V* and *W*—who make the statements given below. In each case, can it be determined to which race each person belongs?

(a) *U*: None of us are Tu's.
V: Exactly one of us is a Tu.

(b) *U*: We are all Tu's.
V: Exactly one of us is a Tu.
W: Exactly two of us are Tu's.

(c) *U*: *V* and *W* belong to the same race.
V: Both *W* and I are Tu's.
W: *V* is a Fa.

(d) *U*: *V* and *W* are both Fa's.
V: *U* is a Tu.
W: *U* is a Tu.

(e) *U*: *V* and *W* do not belong to the same race.
V: *U* is a Tu.
W: *U* is a Tu.

(4) Three married couples come to a river where they find a boat that can carry, at most, two persons at a time. However, no husband will permit his wife to be left without him, neither on a river bank nor in the boat, if there is another man present. Is there any way the three couples can cross from the left bank to the right bank of the river under these conditions?

(5) Three men and three women come to a river where they find a boat that can carry, at most, two persons at a time. The three men wish to cross the river, but they have a problem. No man wants to be on either bank of the river if the women outnumber the men on that bank. Is there a way for the three men to cross the river?

Chapter 16

*Soundness and Completeness

In this chapter, we prove that our rules of inference preserve truth (the Soundness Theorem) and that each sentence that is true in every interpretation structure is provable (the Completeness Theorem). First, in order to simplify some of the proofs, we show that some of the connectives and quantifiers in the language of the Predicate Calculus are, in a certain sense, redundant. (As a prerequisite for this chapter, the reader should be familiar with the disjunctive normal form for a formula in the Sentential Calculus, Section 4.1, and the prenex normal form for a formula in the Predicate Calculus, Section 10.5.)

16.1 COMPLETE SETS OF CONNECTIVES AND QUANTIFIERS

Definition 16.1 A set X of connectives is said to be *complete* if every formula of the Sentential Calculus is logically equivalent to a formula in which only connectives from X occur. Similarly, a set Y of connectives and quantifiers is said to be *complete* if every formula of the Predicate Calculus is logically equivalent to a formula in which only connectives and quantifiers from Y occur.

For example, the set $X_1 = \{\neg, \vee, \wedge, \forall\}$ is complete. First, note that the existential quantifier $(\exists x)$ can be replaced by $\neg(\forall x)\neg$ because for each formula φ

$$\vdash (\exists x)\varphi \leftrightarrow \neg(\forall x)\neg\varphi.$$

We know (Theorem 10.7) that every formula of the Predicate Calculus is logically equivalent to a formula in prenex normal form and that (Theorem 4.2) every formula in the Sentential Calculus is tautologically equivalent (and, therefore, logically equivalent) to a formula in disjunctive normal form. Thus, it follows that the set X_1 is complete.

(Alternatively, use the fact

$$(P \rightarrow Q) \Leftrightarrow (\neg P \vee Q)$$

and

$$(P \leftrightarrow Q) \Leftrightarrow (\neg P \vee Q) \wedge (\neg Q \vee P).$$

Then, the theorem follows using complete induction on the number of connectives in the formula.)

It is also easy to show that the following sets are complete sets of connectives for the Sentential Calculus:

$$X_2 = \{\neg, \vee\},$$
$$X_3 = \{\neg, \wedge\},$$
$$X_4 = \{\neg, \rightarrow\}.$$

Moreover, suppose we define new binary connectives $\downarrow$ and $\uparrow$, using the following truth tables:

P	Q	$P \downarrow Q$
T	T	F
T	F	F
F	T	F
F	F	T

P	Q	$P \uparrow Q$
T	T	F
T	F	T
F	T	T
F	F	T

Then, the following two sets are also complete:

$$X_5 = \{\downarrow\},$$
$$X_6 = \{\uparrow\}.$$

The proofs will be left for the exercises.

However, for example, the set $Y = \{\vee, \wedge\}$ is not complete. To prove this, it is sufficient to show that there is a formula in the Sentential Calculus that is not tautologically equivalent to a formula that just contains connectives from the set $\{\vee, \wedge\}$. We shall prove by complete induction that any formula of the Sentential Calculus that has only one sentential variable P and only contains connectives from the set $\{\vee, \wedge\}$ is true when P is true and false when P is false. Thus, it follows that the formula $\neg P$ is not tautologically equivalent to any such formula.

Theorem 16.1 *If φ is a formula of the Sentential Calculus that only has one sentential variable P and only contains connectives from the set $\{\vee, \wedge\}$, then φ is true when P is true, and false when P is false.*

Proof: The proof is by complete induction on the length of φ.
Base Step: The length of φ is 1. Then φ must be P. The statement is clearly true for P.

Induction Step: Assume the theorem is true for all formulas of length $\leq m$, and suppose φ is a formula of length $m + 1$. Then φ must be of either of the following forms:

(i) $(\psi_1 \wedge \psi_2)$

or

(ii) $(\psi_1 \vee \psi_2)$,

where ψ_1 and ψ_2 are formulas of length $\leq m$ whose only sentential variable is P. By the induction hypotheses, each of ψ_1 and ψ_2 is true when P is true, and false when P is false. Thus, from the truth tables for $\wedge$ and $\vee$, the same is true for φ.

Consequently, the theorem follows by complete induction.

■

16.2 SENTENTIAL CALCULUS (SOUNDNESS THEOREM)

The Soundness Theorem is the statement that the rules of inference preserve truth. That is,

$$\text{If } \Gamma \vdash \varphi, \text{ then } \Gamma \models \varphi.$$

We shall show that both the Rules P and T preserve truth and that the Rules CP and IP are consequences of Rules P and T.

Theorem 16.2 *If φ is a formula of the Sentential Calculus, and Γ is a set of formulas of the Sentential Calculus such that*

(i) Γ $\quad n.$ $\quad \varphi$ $\quad P$

or

(ii) Γ $\quad n.$ $\quad \varphi$ $\quad n_1, n_2, \ldots, n_k$ $\quad T$

are lines of a proof, then whenever all the formulas in Γ are true, φ must also be true.

Proof: In case (i), $\Gamma = \{\varphi\}$, so clearly if all the formulas in Γ are true, so is φ.

For case (ii), the proof is by complete induction on the length of the proof of φ from Γ.

Base Step: If the length of the proof is 1, then φ is a tautology. Thus, $\Gamma \models \varphi$ for every Γ.

Induction Step: Suppose that the theorem is true for all proofs of length $< m + 1$ and the length of the proof of φ from Γ is $m + 1$. Then there must be formulas $\psi_1, \psi_2, \ldots, \psi_k$ in the proof on lines $n_1, n_2, \ldots, n_k$ such that $n_i < m + 1$ for $i = 1, 2, \ldots, k$ and

Γ	n_i.	ψ_i	(Reason)

Because the following line is also in the proof

Γ	$m + 1$.	φ	$n_1, n_2, \ldots, n_k T$

it follows that

(*) $\psi_1 \land \psi_2 \land \cdots \land \psi_k \rightarrow \varphi$ is a tautology.

By the induction hypotheses, $\Gamma \models \psi_i$, $i = 1, 2, \ldots, k$, so it follows from (*) that $\Gamma \models \varphi$. Thus, (ii) follows by complete induction. ∎

Theorem 16.3 *If $\Gamma, \varphi \vdash \psi$, then $\Gamma \vdash \varphi \rightarrow \psi$. (Rule CP, or, as it is sometimes called, the Deduction Theorem.)*

Proof: The proof is by complete induction on the length of the proof of ψ from Γ and φ.
Base Step: Suppose $\Gamma, \varphi \vdash \psi$, and the length of the proof is 1.

Case 1. ψ is derived using Rule *P*. Then either ψ is φ or $\psi \in \Gamma$. If ψ is φ, because $\varphi \rightarrow \varphi$ is a tautology, it follows that $\Gamma \vdash \varphi \rightarrow \psi$. If $\psi \in \Gamma$, then the proof of $\varphi \rightarrow \psi$ is as follows:

	1.	$\psi \rightarrow (\varphi \rightarrow \psi)$	*T*
{2}	2.	ψ	*P*
{2}	3.	$\varphi \rightarrow \psi$	1, 2 *T*

Thus, $\Gamma \vdash \varphi \rightarrow \psi$ because $\{\psi\} \subseteq \Gamma$.

Case 2. ψ is derived using Rule *T*. Because the length of the proof of ψ is 1, ψ must be a tautology. Then, $\varphi \rightarrow \psi$ is also a tautology, so $\Gamma \vdash \varphi \rightarrow \psi$.
Induction Step: The induction hypothesis is the following: If χ is any formula such that $\Gamma, \varphi \vdash \chi$ and the length of the proof of χ from Γ and φ is less than $m + 1$, then $\Gamma \vdash \varphi \rightarrow \chi$.

Suppose $\Gamma, \varphi \vdash \psi$, and the length of the proof of ψ from Γ and φ is $m + 1$. If ψ is φ, $\psi \in \Gamma$, or ψ is a tautology, the proof that $\Gamma \vdash \varphi \rightarrow \psi$ is the same as in the base step. The only other possibility is that there are formulas $\psi_1, \psi_2, \ldots, \psi_k$ so that $\Gamma, \varphi \vdash \psi_i$, $i = 1, 2, \ldots, k$ and

(*) $\psi_1 \land \psi_2 \land \cdots \land \psi_k \rightarrow \psi$

is a tautology. Because the length of the proof of each ψ_i is less than $m + 1$, it follows from the induction hypothesis that

(**) $\Gamma \vdash \varphi \rightarrow \psi_i$, $i = 1, 2, \ldots, k$.

However, the formula

(***) $$(\varphi \to \psi_1) \wedge \cdots \wedge (\varphi \to \psi_k) \wedge (*) \to (\varphi \to \psi)$$

is a tautology. Thus, it follows from (**), (***) is a tautology, and from Rule *T* that $\Gamma \vdash \varphi \to \psi$. Now, using complete induction, we obtain the desired result. ■

Theorem 16.4 *If* $\Gamma, \neg\varphi \vdash \psi \wedge \neg\psi$, *then* $\Gamma \vdash \varphi$. *(Rule IP)*

Proof: Suppose $\Gamma, \neg\varphi \vdash \psi \wedge \neg\psi$. By Rule *CP*,

(*) $$\Gamma \vdash \neg\varphi \to \psi \wedge \neg\psi.$$

However,

(**) $$(\neg\varphi \to \psi \wedge \neg\psi) \to \varphi$$

is a tautology. Thus, using (*), (**) is a tautology, and Rule *T*, we obtain $\Gamma \vdash \varphi$.

■

16.3 SENTENTIAL CALCULUS (COMPLETENESS THEOREM)

The Completeness Theorem states

$$\text{If } \Gamma \models \varphi \text{ then } \Gamma \vdash \varphi.$$

In the Sentential Calculus, this is translated as "If Γ tautologically implies φ, then there is a proof of φ from Γ." Because of the way Rule *T* is given in our system, the Completeness Theorem is trivial. It follows immediately from Rule *T*. However, we shall show here that it is possible to restrict Rule *T* to a small number of tautologies and still obtain the Completeness Theorem.

We shall define another system that we shall call *SC*. Because the set of connectives $\{\neg, \to\}$ is a complete set (see Exercise 15 A(3)), we shall restrict the language of *SC* to these two connectives. We also have sentential variables and parentheses in the language of *SC*.

The rules of *SC* are Rule *P* (Premises), Rule *CP* (Conditional Proof), and the following five rules, which replace Rule *T*:

*R*1: $\neg\neg\varphi$ is derivable from φ.

*R*2: $\varphi \to \psi$ is derivable from $\neg\varphi$.

*R*3: $\psi \to \varphi$ is derivable from φ.

R4: $\neg(\varphi \rightarrow \psi)$ is derivable from φ and $\neg\psi$.

R5: ψ is derivable from $\varphi \rightarrow \psi$ and $\neg\varphi \rightarrow \psi$.

We shall use the symbol "$\Gamma \vdash_{SC} \varphi$" to mean "there is a proof of φ from Γ in *SC*." The Completeness Theorem for *SC* is

(*) $$\text{If } \Gamma \models \varphi, \text{ then } \Gamma \vdash_{SC} \varphi.$$

Because a proof is finite in length, it is sufficient to assume that Γ is finite, and in this case, it is easy to see that (*) is equivalent to

(**) $$\text{If} \models \varphi, \text{ then} \vdash_{SC} \varphi.$$

That is, if φ is a tautology, then there is a proof of φ in *SC* from the empty set of premises. Before proving (**), we shall prove a preliminary result.

Lemma 16.5 *Let φ be a formula in SC such that $A_1, A_2, \ldots, A_k$ is a list of all the sentential variables that occur in φ. For any assignment of truth values to $A_1, A_2, \ldots, A_k$, let A_i' be A_i, if A_i was assigned the value T; and let A_i' be $\neg A_i$ if A_i was assigned the value F. Also, let φ' be φ if φ takes the value T under the given assignment of truth values to the sentential variables in φ; and let φ' be $\neg\varphi$ if φ takes the value F. Then*

$$A_1', A_2', \ldots, A_k' \vdash_{SC} \varphi'.$$

As an example, suppose φ is $A_1 \rightarrow (\neg A_2 \rightarrow A_3)$. The truth table for φ is given in Figure 16.1:

A_1	A_2	A_3	φ
T	*T*	*T*	*T*
T	*T*	*F*	*T*
T	*F*	*T*	*T*
T	*F*	*F*	*F*
F	*T*	*T*	*T*
F	*T*	*F*	*T*
F	*F*	*T*	*T*
F	*F*	*F*	*T*

Figure 16.1

For the truth assignment on line 2 of the truth table in Figure 16.1, A_1' is A_1, A_2' is A_2, A_3' is $\neg A_3$ and φ' is φ. The lemma implies

$$A_1, A_2, \neg A_3 \vdash_{SC} A_1 \rightarrow (\neg A_2 \rightarrow A_3).$$

For the truth assignment on line 4, A_1' is A_1, A_2' is $\neg A_2$, A_3' is $\neg A_3$ and φ' is $\neg\varphi$. For this case, using the lemma, we obtain

$$A_1, \neg A_2, \neg A_3 \vdash_{SC} \neg[A_1 \rightarrow (\neg A_2 \rightarrow A_3)].$$

Proof of Lemma 16.5: The proof is by complete induction on the length of φ.
Base Step: The length of φ is 1. Then φ is A_1 and the lemma becomes

$$(*) \qquad A_1 \vdash_{SC} A_1 \quad \text{or} \quad \neg A_1 \vdash_{SC} \neg A_1,$$

depending upon whether A_1 is assigned T or A_1 is assigned F, respectively. It follows from Rule P that (*) is true.
Induction Step: The induction hypothesis is that the lemma holds for all formulas of SC of length $< m + 1$. Let φ be a formula of length $m + 1$. Suppose some truth assignment is given. There are two cases to consider.

Case 1. φ is $\neg\psi$. The length of ψ is less than $m + 1$. If ψ takes the value T under the given truth assignment, then ψ' is ψ and φ takes the value F, so φ' is $\neg\varphi$ or $\neg\neg\psi$. By the induction hypothesis,

$$A'_1, A'_2, \ldots, A'_k \vdash_{SC} \psi.$$

Then using $R1$,

$$A'_1, A'_2, \ldots, A'_k \vdash_{SC} \neg\neg\psi.$$

However, $\neg\neg\psi$ is φ', so this proves the case when ψ takes the truth value T.
If ψ takes the truth value F, then φ takes the value T. Thus, φ' is φ or $\neg\psi$, and ψ' is $\neg\psi$. By the induction hypothesis,

$$A'_1, A'_2, \ldots, A'_k \vdash_{SC} \neg\psi.$$

Because φ' is also $\neg\psi$, this completes the proof of Case 1.

Case 2. φ is $\psi_1 \to \psi_2$. The length of each of the formulas ψ_1 and ψ_2 is $< m + 1$, so by the induction hypotheses,

(i) $A'_1, A'_2, \ldots, A'_k \vdash_{SC} \psi'_1$

and

(ii) $A'_1, A'_2, \ldots, A'_k \vdash_{SC} \psi'_2$.

If ψ_1 takes the value F, then ψ'_1 is $\neg\psi_1$ and φ takes the value T, so φ' is $\psi_1 \to \psi_2$. Thus, using (i) and $R2$, we obtain

$$A'_1, A'_2, \ldots, A'_k \vdash_{SC} \varphi'.$$

On the other hand, if ψ_2 takes the value T, then φ also takes the value T. Therefore, ψ'_2 is ψ_2 and φ' is $\psi_1 \to \psi_2$. It follows from (ii) and $R3$ that

$$A'_1, A'_2, \ldots, A'_k \vdash_{SC} \varphi'.$$

The final possibility in Case 2 is that ψ_1 takes the value T and ψ_2 takes the value F. Then φ takes the value F. In this case, ψ'_1 is ψ_1, ψ'_2 is $\neg\psi_2$ and φ' is $\neg(\psi_1 \to \psi_2)$. Now use (i), (ii), and $R4$ to obtain the desired result.

We have completed the proof of the induction step, so Lemma 16.5 follows by complete induction.

■

Using Lemma 16.5, it is easy to prove the completeness theorem for *SC*.

Theorem 16.6 *The Completeness Theorem for SC is as follows:*

$$\textit{If} \models \varphi, \textit{ then} \vdash_{SC} \varphi.$$

Proof: Suppose φ is a tautology. Let $A_1, A_2, \ldots, A_k$ be a list of all the sentential variables that occur in φ. For any assignment of truth values to the A_i's, let A_i' and φ' be defined as in Lemma 16.5. Because φ is a tautology, for any assignment of truth values, φ' *is the same as* φ.

Let us make two assignments of truth values to the A_i's, the only difference between them being that in the first, A_k is T, so A_k' is A_k, and in the second, A_k is F, so A_k' is $\neg A_k$. Applying Lemma 16.5, we obtain

(i) $A_1', A_2', \ldots, A_{k-1}', A_k \vdash_{SC} \varphi$

and

(ii) $A_1', A_2', \ldots, A_{k-1}', \neg A_k \vdash_{SC} \varphi$.

Using Rule *CP*, it follows that

(iii) $A_1', A_2', \ldots, A_{k-1}' \vdash_{SC} A_k \rightarrow \varphi$.

and

(iv) $A_1', A_2', \ldots, A_{k-1}' \vdash_{SC} \neg A_k \rightarrow \varphi$.

Using (iii), (iv), and *R*5, we obtain

(v) $A_1', A_2', \ldots, A_{k-1}' \vdash_{SC} \varphi$.

Repeating this process $k - 1$ times, we obtain the desired result, $\vdash_{SC} \varphi$.

■

We have shown that in the Sentential Calculus, and also in *SC*, that a formula φ is provable if and only if φ is a tautology. It follows, therefore, that the Sentential Calculus and *SC* are *consistent*; that is, there is no formula φ such that both φ and $\neg\varphi$ are provable. For if φ is provable, then φ is a tautology. However, $\neg\varphi$ is not a tautology because it takes the truth value F for every assignment of truth values to its sentential variables. Consequently, $\neg\varphi$ cannot be provable.

16.4 PREDICATE CALCULUS (SOUNDNESS THEOREM)

We shall give an intuitive idea of the proof of the Soundness Theorem for the Predicate Calculus. (For a rigorous proof, see, for example, *Introduction to Mathematical Logic* by E. Mendelson, 3rd ed., Wadsworth & Brooks/Cole, Monterey, CA, 1987.) We wish to prove the following: If $\Gamma \vdash \varphi$, then $\Gamma \models \varphi$. That is, the Rules of Inference preserve truth. (See Definitions 12.5 and 12.8.) We shall prove that the Rules *US* and *UG* preserve truth and then show that the Rules *EG, EP, LT*, and *LE* are derived rules. (The Soundness Theorem for the Sentential Calculus was proved in Section 16.2.)

Theorem 16.7 *Suppose t is a term, u is a variable, φ is a formula and Γ is a set of formulas.*

(i) *Rule US preserves truth: If the substitution of t for u in φ is valid, then $(\forall u)\varphi \models \varphi[t|u]$.*

(ii) *Rule UG preserves truth: If $\Gamma \models \varphi$ and u is not free in Γ, then $\Gamma \models (\forall u)\varphi$.*

Proof: First we note that if **a** is a sequence of elements from an interpretation structure **S**, then **a** satisfies $(\forall u)\varphi$ in **S** iff **a** satisfies φ for every interpretation of u in **S**.

(i) The proof of (i) is by complete induction on the length of φ. It is a rather long and tedious proof, so we shall omit it.

(ii) Suppose **S** is an interpretation structure for $\Gamma \cup \{\varphi\}$, **a** is a sequence in **S**, and u is not free in Γ. By hypothesis, $\Gamma \models \varphi$. Thus, if **a** satisfies Γ, then **a** satisfies φ. Because u is not free in Γ, **a** satisfies Γ for every interpretation of u in **S**. Consequently, **a** satisfies φ for every interpretation of u in **S**, which implies that **a** satisfies $(\forall u)\varphi$ in **S**.

■

We shall use the fact that the set $\{\neg, \rightarrow, \forall\}$ is a complete set of connectives and quantifiers for the Predicate Calculus. Thus, the Rules *EG* and *EP* can be stated as follows:

Rule *EG*: $\varphi[t|u] \vdash \neg(\forall u)\neg\varphi$, if the substitution of t for u in φ is valid.

Rule *EP*: If $\Gamma, \varphi[u|v] \vdash \psi$, then $\Gamma, \neg(\forall v)\neg\varphi \vdash \psi$, if the substitution of u for v in φ is valid and v is not free in Γ and v is not free in ψ.

Theorem 16.8 *Rule EG is a derived rule.*

Proof: Suppose the substitution of t for u in φ is valid. Then the substitution of t for u in $\neg\varphi$ is also valid. Therefore, using Rule *US*,

$$(\forall u)\neg\varphi \vdash \neg\varphi[t|u].$$

Thus, by Rule *CP*,

$$\vdash (\forall u)\neg\varphi \rightarrow \neg\varphi[t|u].$$

Using Rule *T* (Rule of Contraposition),

$$\vdash \varphi[t|u] \rightarrow \neg(\forall y)\neg\varphi,$$

from which Rule *EG* follows.

■

Theorem 16.9 *Rule EP is a derived rule.*

Proof: We may assume without loss of generality that the *EP* variable u is not free in Γ and ψ. Then Rule *EP* becomes

$$\text{If } \Gamma, \varphi \vdash \psi, \text{ then } \Gamma, \neg(\forall u)\neg\varphi \vdash \psi, \text{ if } u \text{ is not free in } \Gamma \text{ or in } \psi.$$

Suppose Γ, $\varphi \vdash \psi$, and u is not free in Γ or in ψ, then by Rule *CP*,

$\Gamma \vdash \varphi \rightarrow \psi$	
$\Gamma \vdash \neg\psi \rightarrow \neg\varphi$	(by *T*)
$\Gamma, \neg\psi \vdash \neg\varphi$	(by *T*)
$\Gamma, \neg\psi \vdash (\forall u)\neg\varphi$	(by *UG*)
$\Gamma \vdash \neg\psi \rightarrow (\forall u)\neg\varphi$	(by *CP*)
$\Gamma \vdash \neg(\forall u)\neg\varphi \rightarrow \psi$	(by *T*)

This proves Rule *EP*.

■

The Rule *LT* is provable because it is just an abbreviation for a formal proof, sometimes combined with an application of Rule *T*. We shall prove that the Rule *LE* is also provable.

Theorem 16.10 *Rule LE is a derived rule. If ψ is obtained from φ by replacing some occurrence of χ_1 by χ_2 and $\vdash \chi_1 \leftrightarrow \chi_2$, then $\vdash \varphi \leftrightarrow \psi$.*

Proof: The proof is by complete induction on the number of connectives and quantifiers in φ.
Base Step: φ is an atomic formula, $Rt_1t_2 \ldots t_n$ (no connectives or quantifiers). Then χ_1 is φ and χ_2 is ψ, so $\vdash \chi_1 \leftrightarrow \chi_2$ implies $\vdash \varphi \leftrightarrow \psi$.

Induction Step: There are three cases to consider. In each case, we shall assume that χ_1 is not φ, otherwise, the proof is trivial.

Case 1. *φ is $\neg\varphi'$.* If ψ is obtained from φ by replacing some occurrence of χ_1 by χ_2, then ψ is $\neg\psi'$, where ψ' is obtained from φ' by replacing the same occurrence of χ_1 by χ_2. (See Exercise 15E (25)(a).) Thus, the induction hypothesis holds for φ', so $\vdash \psi_1 \leftrightarrow \psi_2$ implies $\vdash \varphi' \leftrightarrow \psi'$. It follows, using a tautology (the Rule of Contraposition), that $\vdash \varphi \leftrightarrow \psi$.

Case 2. φ is $\varphi' \rightarrow \varphi''$. Then if χ_2 replaces χ_1 in φ, it must do so in either φ' or φ''. (See Exercise 15E (25)(b).) Thus, ψ is of the form $\psi' \rightarrow \psi''$, where either

(*) ψ' is obtained from φ' by replacing some occurrence of χ_1 by χ_2 and χ'' is the same as φ''.

or

(**) ψ'' is obtained from φ'' by replacing some occurrence of χ_1 by χ_2 and ψ' is the same as φ'.

In either case, it follows from the induction hypothesis and a tautology ($\vdash P \leftrightarrow P$) that

$$\text{If} \vdash \psi_1 \leftrightarrow \psi_2, \text{then (i)} \vdash \varphi' \leftrightarrow \psi' \text{ and (ii)} \vdash \varphi'' \leftrightarrow \psi''.$$

We want to prove

$$(\varphi' \rightarrow \varphi'') \leftrightarrow (\psi' \rightarrow \psi'').$$

We shall use an indirect proof to prove

$$(\varphi' \rightarrow \varphi'') \rightarrow (\psi' \rightarrow \psi'').$$

(The proof of the converse is analogous.)

{1}	1.	$\varphi' \rightarrow \varphi''$	*P* (for *CP*)
{2}	2.	$\psi' \wedge \neg\psi''$	*P* (for *IP*)
	3.	$\neg\psi'' \rightarrow \neg\varphi''$	(ii) *T*
{2}	4.	$\neg\varphi''$	2, 3 *T*
{1, 2}	5.	$\neg\varphi'$	1, 4 *T*
	6.	$\psi' \rightarrow \varphi'$	(i) *T*
{2}	7.	φ'	2, 6 *T*
{1, 2}	8.	$\varphi' \wedge \neg\varphi'$	5, 7 *T*
{1}	9.	$\neg(\psi' \wedge \neg\psi'')$	2, 8 *IP*
{1}	10.	$\psi' \rightarrow \psi''$	9 *T*
	11.	$(\varphi' \rightarrow \varphi'') \rightarrow (\psi' \rightarrow \psi'')$	1, 10 *CP*

Case 3. φ is $(\forall u)\varphi'$. If we replace χ_1 by χ_2 to obtain ψ, then ψ must be of the form $(\forall u)\psi'$, where ψ' is obtained from φ' by replacing χ_1 by χ_2. (The reasoning is similar to that in Case 1.) By the induction hypotheses,

(*) $$\vdash \chi_1 \leftrightarrow \chi_2 \text{ implies} \vdash \varphi' \leftrightarrow \psi'.$$

We wish to prove

(**) $$\vdash \chi_1 \leftrightarrow \chi_2 \text{ implies} \vdash \varphi \leftrightarrow \psi.$$

The proof of (**) from (*) is an easy proof using *US* and *UG*. We leave the details as an exercise.

Theorem 16.10 follows by complete induction. ■

16.5 PREDICATE CALCULUS (COMPLETENESS THEOREM)

The proof we give here of the Completeness Theorem is based on an exposition by L. Lipshitz of a proof by J. Herbrand. We start by describing *Skolem functions.* Let φ be a sentence in prenex normal form. For example, suppose φ is $(\forall x)(\exists y)\psi(x, y)$, where $\psi(x, y)$ has no quantifiers. Let f be a function letter of one variable such that f does not occur in ψ, and let φ_s be the formula $(\forall x)\psi(x, f(x))$.

Let **S** be an interpretation structure for φ. Let **S*** result from **S** by defining some new function of one variable on the universe of **S**, but keeping the original functions and relations the same. We call **S*** an *expansion* of **S**, and **S** is called a *contraction* of **S***.

Suppose $\mathbf{S} \models \varphi$; that is, **S** is a model of φ. Then for each $\mathbf{a} \in \mathbf{U}$ (the universe of **S**), there is a $\mathbf{b} \in \mathbf{U}$ such that $\mathbf{S} \models \psi(\mathbf{a}, \mathbf{b})$. For each $\mathbf{a} \in \mathbf{U}$, choose such a $\mathbf{b} \in \mathbf{U}$ and define $\mathbf{f}(\mathbf{a}) = \mathbf{b}$. It follows that if **S*** is the expansion of **S** obtained by adjoining **f**, then $\mathbf{S}^* \models \varphi_S$. Conversely, if **T** is an interpretation structure for φ_S such that $\mathbf{T} \models \varphi_S$, then it is clear that $\mathbf{T} \models \varphi$. Thus, looking for a model of φ is equivalent to looking for a model of φ_S.

The new function **f** we constructed above is called a *Skolem function.* Skolem functions choose "witnesses" for the existential quantifiers and thereby allow us to deal with a formula that only contains universal quantifiers.

Definition 16.2 Skolem functions.

Let φ be any sentence of the Predicate Calculus in prenex normal form. Then φ_S results from φ by replacing each existentially quantified variable x by a new term $f(y_1, y_2, \ldots, y_k)$ where $y_1, y_2, \ldots, y_k$ are all the universally quantified variables preceding x, and f is a function letter of k variables that does not occur in φ. (A function of no variables is a constant.)

As another example, if

$$\varphi \text{ is } (\exists x_1)(\forall y_1)(\forall y_2)(\exists x_2)(\exists x_3)(\forall y_3)(\exists x_4)\psi(x_1, y_1, y_2, x_2, x_3, y_3, x_4)$$

then

$$\varphi_S \text{ is } (\forall y_1)(\forall y_2)(\forall y_3)\psi(c, y_1, y_2, f(y_1, y_2), g(y_1, y_2), y_3, h(y_1, y_2, y_3)).$$

You should be convinced by the discussion at the beginning of this section that the following theorem is true.

Theorem 16.11 *If φ is any sentence in the Predicate Calculus in prenex normal form, then any model of φ can be expanded to a model of φ_S, and any model of φ_S is a model of φ.*

(Actually, the proof is by induction on the number of quantifiers in φ.)

We proved in Chapter 9 (Theorem 9.1) that the Completeness Theorem is equivalent to the following:

(*) Every consistent set of sentences has a model.

The next question we shall address is "How do we go about constructing a model for a sentence φ?" First, we shall assume that φ is in prenex normal form. Then it follows from Theorem 16.11 that we may replace φ by φ_S. Thus, we may assume φ has the form

$$(\forall x_1)(\forall x_2)\ldots(\forall x_n)\psi(x_1, x_2, \ldots, x_n),$$

where ψ contains no quantifiers. We shall also assume, for simplicity, that ψ does not contain composition of function letters. (Otherwise, the models that we construct below would be more difficult to describe.)

A model of φ is an interpretation structure **S** for φ with universe **U** so that for all elements $\mathbf{c}_1, \mathbf{c}_2, \ldots, \mathbf{c}_n \in \mathbf{U}$,

$$\mathrm{S} \models \psi(\mathrm{c}_1, \mathrm{c}_2, \ldots, \mathrm{c}_n).$$

(That is, the interpretation of ψ in **S** is true for all $\mathbf{c}_1, \mathbf{c}_2, \ldots, \mathbf{c}_n \in \mathbf{U}$.)

We shall describe a procedure that will either produce a model of φ or will show that no such model exists. (The same procedure works for a finite number of sentences instead of just one sentence. This is illustrated in Examples 16.1, 16.2, and 16.3.)

We shall start our procedure with a set $\mathbf{U}_0$ which consists of interpretations for all the constants that occur in φ. (If φ has no constants, put some element in $\mathbf{U}_0$, say $\mathbf{c}_1 \in \mathbf{U}_0$, just so we will have some place to start.) Suppose

$$\mathrm{U}_0 = \{\mathrm{c}_1, \mathrm{c}_2, \ldots, \mathrm{c}_k\}.$$

Let $\mathbf{U}_1$ be $\mathbf{U}_0$, together with the set of interpretations of all terms that can be obtained by applying the function letters that occur in ψ just once to the constants whose interpretations are in $\mathbf{U}_0$. For example, suppose $\mathbf{U}_0 = \{\mathbf{c}_1, \mathbf{c}_2\}$ and the function letters f and g occur in ψ, where f is one-place and g is two-place. Then

$$\mathrm{U}_1 = \{\mathrm{c}_1, \mathrm{c}_2, \mathrm{f}(\mathrm{c}_1), \mathrm{f}(\mathrm{c}_2), \mathrm{g}(\mathrm{c}_1, \mathrm{c}_1), \mathrm{g}(\mathrm{c}_1, \mathrm{c}_2), \mathrm{g}(\mathrm{c}_2, \mathrm{c}_1), \mathrm{g}(\mathrm{c}_2, \mathrm{c}_2)\}$$

Next, let $\boldsymbol{\beta}_1$ be the conjunction of all the instances of $\boldsymbol{\psi}$ that can be obtained by substituting the elements of $\mathbf{U}_0$ for $x_1, x_2, \ldots, x_n$ in all possible ways. Now interpret the relation symbols in ψ on $\mathbf{U}_1$ in all possible ways and see if any satisfy $\boldsymbol{\beta}_1$. (Note that if $=$ is one of the relation symbols in ψ, then it has to be interpreted as identity. Also note that if we allowed composition of functions in ψ, the universe at level 1 would be $\mathbf{U}_k$, for some $k > 1$, instead of $\mathbf{U}_1$.) In this way, we may obtain several partial models, say $\mathbf{S}_{11}, \mathbf{S}_{12}, \ldots, \mathbf{S}_{1n_1}$, all with universe $\mathbf{U}_1$. (These may not be models because they may not be closed under the functions. That is, for example, we may have $\mathbf{f}(\mathbf{c}_1) \in \mathbf{U}_1$, but $\mathbf{f}(\mathbf{f}(\mathbf{c}_1)) \notin \mathbf{U}_1$.)

Next, let $\mathbf{U}_2$ be $\mathbf{U}_1$, together with the elements obtained by applying the functions to the elements of $\mathbf{U}_1$, just once. Let $\boldsymbol{\beta}_2$ be the conjunction of all the instances of $\boldsymbol{\psi}$ that we get by substituting the elements of $\mathbf{U}_1$ for the variables of $\boldsymbol{\psi}$ in all possible ways. Interpret the relations on $\mathbf{U}_2$ in all possible ways and see which of these satisfy $\boldsymbol{\beta}_2$. Call the interpretation structures that do satisfy $\boldsymbol{\beta}_2$, $\mathbf{S}_{21}$, $\mathbf{S}_{22}$, . . . , $\mathbf{S}_{2n_2}$. Notice that each $\mathbf{S}_{2j}$ is an extension of exactly one $\mathbf{S}_{1i}$. (Why?)

If we continue in this way, we get a tree. (See Fig. 16.2.) In the example in Figure 16.2, $\mathbf{S}_{11}$ extends in three different ways to the second level; $\mathbf{S}_{12}$ does not extend in any way; $\mathbf{S}_{13}$ extends in two ways; and so forth.

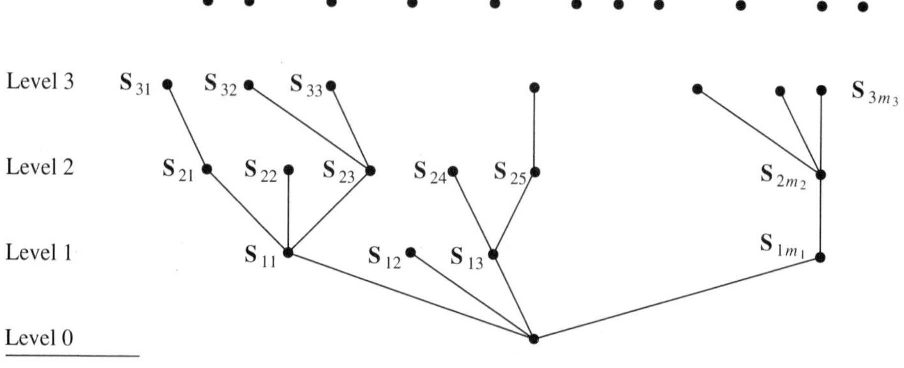

Figure 16.2

The important facts about the tree are that at each level there are only a finite number of nodes (partial models) and that we can (at least in principle, although it may be a lot of work) actually write down as much of the tree as we like.

One of two things will happen:

1. The process of building the tree will never end.
2. At some level, say level m, we cannot continue. There is no way to define the relations so that $\boldsymbol{\beta}_m$ is true.

We shall show below that in case 1, there is a model of φ, and in case 2, there is a proof of $\neg\varphi$. But first, let us look at some examples.

Example 16.1

Construct a model of the following sentences:

$$(\forall x)Rax,$$
$$(\forall x)(\forall y)(Rxy \wedge Ryx \rightarrow x = y),$$
$$(\forall x)(\exists y)(Rxy \wedge x \neq y).$$

(As was mentioned above, the procedure for any finite number of sentences is the same as that for one sentence.)

Solution: First we shall put in Skolem functions and obtain the three sentences:

$$(\forall x)Rax,$$
$$(\forall x)(\forall y)(Rxy \wedge Ryx \rightarrow x = y),$$
$$(\forall x)(Rxf(x) \wedge x \neq f(x)).$$

Then $\mathbf{U}_0 = \{\mathbf{a}\}$, $\mathbf{U}_1 = \{\mathbf{a}, \mathbf{f}(\mathbf{a})\}$, and

$$\beta_1 \text{ is } Raa \wedge (Raa \wedge Raa \rightarrow a = a) \wedge (Raf(a) \wedge a \neq f(a)).$$

We must define **R** on $\mathbf{U}_1$ in all possible ways so that the interpretation of β_1 is true. Thus, the interpretation of R and f must be such that each of the following is true:

$$Raa, \qquad Raf(a), \qquad \text{and} \qquad a \neq f(a).$$

It remains to choose the truth values of $Rf(a)a$ and $Rf(a)f(a)$. There are $2^2 = 4$ possibilities, and in each case, we will obtain a partial model of β_1.

$\mathbf{S}_{11}$: $\neg Rf(a)a$ and $Rf(a)f(a)$ are each true.

$\mathbf{S}_{12}$: $\neg Rf(a)a$ and $\neg Rf(a)f(a)$ are each true.

$\mathbf{S}_{13}$: $Rf(a)a$ and $Rf(a)f(a)$ are each true.

$\mathbf{S}_{14}$: $Rf(a)a$ and $\neg Rf(a)f(a)$ are each true.

(Remember that Raa, $Raf(a)$, and $a \neq f(a)$ each have true interpretations in $\mathbf{S}_{11} - \mathbf{S}_{14}$.)

Before continuing to level 2 of our tree, we shall introduce some notation. Define $f^{(0)}(a) = a$ and $f^{(n+1)}(a) = f(f^{(n)}(a))$, $n \geq 0$. Then, for example, $f(f(f(a)))$ is abbreviated as $f^{(3)}(a)$.

Now, for level 2, we set the universe $\mathbf{U}_2 = \{\mathbf{a}, \mathbf{f}(\mathbf{a}), \mathbf{f}^{(2)}(\mathbf{a})\}$ and the formula β_2 is

$$\beta_1 \wedge Raf(a) \wedge (Raf(a) \wedge Rf(a)a \rightarrow a = f(a)) \wedge$$
$$(Rf(a)a \wedge Raf(a) \rightarrow f(a) = a) \wedge (Rf(a)f(a) \wedge Rf(a)f(a) \rightarrow f(a) = f(a)) \wedge$$
$$(Rf(a)f^{(2)}(a) \wedge f(a) \neq f^{(2)}(a)).$$

To make β_2 true, we must interpret R and f on $\mathbf{U}_2$ so that all of the following are true:

$$Raa, Raf(a), \neg Rf(a)a, Rf(a)f^{(2)}(a), a \neq f(a), f(a) \neq f^{(2)}(a), a \neq f^{(2)}(a).$$

(Note that we cannot have that $a = f^{(2)}(a)$ is true, for then we would have the contradiction that $Rf(a)a$ and $\neg Rf(a)a$ both have true interpretations.)

It remains to choose the truth values of

$$Raf^{(2)}(a), Rf^{(2)}(a)a, Rf(a)f^{(2)}(a), Rf^{(2)}(a)f(a).$$

There are $2^4 = 16$ possible choices, so at level 2, we have 16 partial models. Let $\mathbf{S}_{21}$ be an extension of $\mathbf{S}_{11}$ and $\mathbf{S}_{22}$ an extension of $\mathbf{S}_{12}$ in which all of the following have true interpretations:

$$Raf^{(2)}(a), Rf(a)f^{(2)}(a), \neg Rf^{(2)}(a)a, \neg Rf^{(2)}(a)f(a).$$

Our tree, up to this point, is pictured in Figure 16.3.

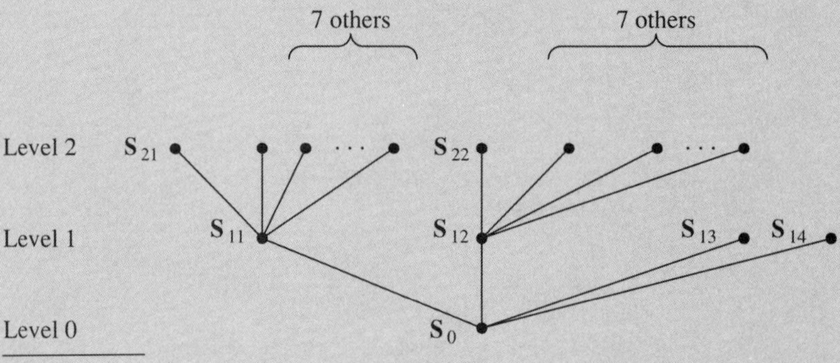

Figure 16.3

Note that $\mathbf{S}_{13}$ and $\mathbf{S}_{14}$ do not extend to level 2 because on level 2 all partial models must have $\neg Rf(a)a$ true but $Rf(a)a$ is true in $\mathbf{S}_{13}$ and $\mathbf{S}_{14}$.

There are many ways that our tree can be extended indefinitely, but we shall only indicate two of these alternatives. We shall show one way the tree can be extended through $\mathbf{S}_{11}$, $\mathbf{S}_{21}$, and one way through $\mathbf{S}_{12}$, $\mathbf{S}_{22}$. Such extensions are called *branches*. (A *branch* of a tree is a sequence of nodes N_0, N_1, N_2, . . . , such that for each i, there is a line connecting N_i and N_{i+1}.)

In each of these partial models, we shall assume $f^{(k)}(a) \neq f^{(\ell)}(a)$ is true if $k \neq \ell$. In $\mathbf{S}_{m1}$, the interpretation of $Rf^{(k)}(a)f^{(\ell)}(a)$ is true if $k \leq \ell$, but false otherwise; in $\mathbf{S}_{m2}$, the interpretation of $Rf^{(k)}(a)f^{(\ell)}(a)$ is true if $k < \ell$ and if $k = \ell = 0$, but false otherwise. Then the branches,

(*) $$\mathbf{S}_0, \mathbf{S}_{11}, \mathbf{S}_{21}, \ldots, \mathbf{S}_{m1}, \ldots$$

and

(**) $$\mathbf{S}_0, \mathbf{S}_{21}, \mathbf{S}_{22}, \ldots, \mathbf{S}_{m2}, \ldots$$

continue indefinitely.

Let $\langle \mathbf{U}, \mathbf{R}, \mathbf{f}, \mathbf{a} \rangle$ be an interpretation structure for the given sentences. The branch (*) gives us the model, which looks like

(***) $$\langle \mathbb{N}, \leq, \mathbf{s}, \mathbf{0} \rangle,$$

where $\mathbb{N}$ is the set of natural numbers; $\leq$ is the natural less-than-or-equal-to relation on $\mathbb{N}$; **s** is the successor function on $\mathbb{N}$, $\mathbf{s}(\mathbf{x}) = \mathbf{x} + \mathbf{1}$, for each $\mathbf{x} \in \mathbb{N}$; and **0** is the zero element of $\mathbb{N}$.

On the other hand, the branch (**) gives us a model that is the same as (***), except for the interpretation of R. Here the interpretation of R is a relation $<^*$, where for all $\mathbf{x}, \mathbf{y}, \in \mathbb{N}$, $\mathbf{x} <^* \mathbf{y}$ iff $\mathbf{x} = \mathbf{y} = \mathbf{0}$ or $\mathbf{x} < \mathbf{y}$. ($<$ is the natural less-than relation on $\mathbb{N}$.)

There are other branches that extend indefinitely, but the models they lead to are not as easy to describe as the preceding two models.

Example 16.2

Determine whether the sentences

$$(\forall x)(\exists y)[(Py \wedge Qy) \vee (Px \rightarrow Qx)]$$
$$(\forall x)(Px \rightarrow \neg Rx)$$
$$(\exists x)Px$$
$$(\forall x)(Px \rightarrow \neg Qx)$$

have a model.

Solution: First we put in Skolem functions

$$(\forall x)[(Pf(x) \wedge Qf(x)) \vee (Px \rightarrow Qx)]$$
$$(\forall x)(Px \rightarrow \neg Rx)$$
$$Pa$$
$$(\forall x)(Px \rightarrow \neg Qx)$$

Then

$$\mathbf{U}_0 = \{\mathbf{a}\},$$
$$\mathbf{U}_1 = \{\mathbf{a}, \mathbf{f(a)}\}.$$

β_1 **is** $[(Pf(a) \wedge Qf(a)) \vee (Pa \rightarrow Qa)] \wedge (Pa \rightarrow \neg Ra) \wedge Pa \wedge (Pa \rightarrow \neg Qa)$.

For β_1 to be true, each of the following must be true:

(i) $$Pa, \neg Qa, \neg Ra, Pf(a), Qf(a).$$

We still have a choice for $Rf(a)$, so there will be two partial models at level 1:

$\mathbf{S}_{11}$: $Rf(a)$ true.

$\mathbf{S}_{12}$: $\neg Rf(a)$ true.

Continuing,

$$U_2 = \{a, f(a), f^{(2)}(a)\}.$$

β_2 is $\beta_1 \wedge [(Pf^{(2)}(a) \wedge Qf^{(2)}(a)) \vee (Pf(a) \to Qf(a))] \wedge (Pf(a) \to \neg Rf(a)) \wedge$
$$(Pf(a) \to \neg Qf(a)).$$

We saw at level 1 that $Pf(a)$ and $Qf(a)$ must be true, but if β_2 is to be true, its last conjunct has to be true. Thus, if $Pf(a)$ is true, $\neg Qf(a)$ must also be true. We cannot have a model in which both $Qf(a)$ and $\neg Qf(a)$ are true. Therefore, there is no way to define a partial model at level 2. Our tree is given in Figure 16.4.

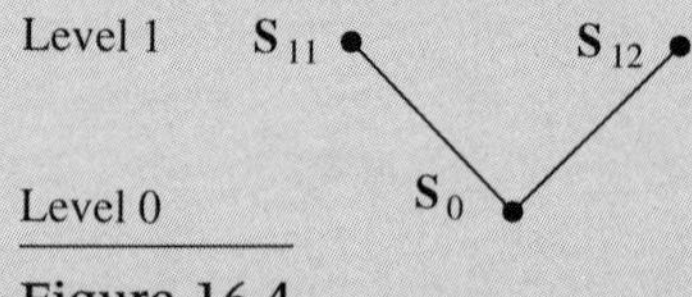

Figure 16.4

Because the procedure does not construct a model, we shall show that, indeed, the four sentences given in Example 16.2 do not have a model because they are inconsistent. The proof of inconsistency follows the preceding argument very closely:

{1}	1.	$(\forall y)(\exists x)[(Px \wedge Qx) \vee (Py \to Qy)]$	P
{2}	2.	$(\forall x)(Px \to \neg Rx)$	P
{3}	3.	$(\exists x)Px$	P
{4}	4.	$(\forall x)(Px \to \neg Qx)$	P
{5}	5.	Pz	z P (for EP)
{1}	6.	$(\exists x)[(Px \wedge Qx) \vee (Pz \to Qz)]$	1 US $[z\|y]$
{7}	7.	$(Pu \wedge Qu) \vee (Pz \to Qz)$	z, u P (for $EP(u)$)
{2}	8.	$Pz \to \neg Rz$	2 US $[z\|x]$
{4}	9.	$Pz \to \neg Qz$	4 US $[z\|x]$
{2, 5}	10.	$\neg Rz$	z, 5 8 T
{4, 5}	11.	$\neg Qz$	z, 5, 9 T
{4, 5, 7}	12.	$Pu \wedge Qu$	u 5, 7, 11 T
{4, 5, 7}	13.	Pu	u 12 T
{4, 5, 7}	14.	Qu	u 12 T
{1}	15.	$(\exists x)[(Px \wedge Qx) \vee (Pu \to Qu)]$	1 US $[u\|y]$
{16}	16.	$(Pv \wedge Qv) \vee (Pu \to Qu)$	u, v P (for $EP(v)$)
{2}	17.	$Pu \to \neg Ru$	2 US $[u\|x]$
{4}	18.	$Pu \to \neg Qu$	4 US $[u\|x]$
{4, 5, 7}	19.	$\neg Qu$	u, 13, 18 T
{4, 5, 7}	20.	$Qu \wedge \neg Qu$	u 14, 19 T
{4, 5, 7}	21.	$A \wedge \neg A$	20 T
{1, 4, 5}	22.	$A \wedge \neg A$	6, 7, 21 EP
{1, 3, 4}	23.	$A \wedge \neg A$	3, 5, 22 EP

To compare the formal proof with the preceding construction, the "z" in the proof corresponds to "a" in the construction; "u" to "$f(a)$"; and "v" to "$f^{(2)}(a)$." Lines 5, 7, 8, and 9 contain the conjuncts appearing in β_1, and lines 16, 17, and 18 contain the additional conjuncts appearing in β_2. In (i) above, we stated the formulas that had to have true interpretations, and these formulas appear on lines 5, 10, 11, 13, and 14 of the formal proof. Lines 19 and 20 mirror the informal proof of the contradiction. (Of course, lines 21 through 23 of the formal proof are needed to show that the contradiction was actually derived from the original premises.) Premise 2 was not used to derive the contradiction, so we could have shortened the proof by omitting all lines of the proof that depend on premise 2. We included them because we were following the informal argument given in the construction.

Summarizing, lines 1 through 14 of the formal proof were used to derive the formula β_1 and its consequences, particularly Qu (or $Qf(a)$ in the construction). Then lines 15 through 20 were used to show that β_2 implies $\neg Qu$ (or $\neg Qf(a)$) and thus we have a contradiction.

Before we give the proof of the Completeness Theorem, we prove a lemma.

Lemma 16.12 (König's Infinity Lemma). *Let T be a tree with an infinite number of nodes, but such that the number of nodes at each level is finite. Then there is an infinite branch through the tree.*

Proof: Define a sequence of nodes as follows: N_0 is the node on the bottom. On level 1, there are only a finite number of nodes, so at least one of them must have an infinite number of nodes above it. Let N_1 be such a node. Now look at the nodes on level 2, which are above N_1. At least one of these has an infinite number of nodes above it. Let N_2 be such a node. Continuing in this way, we obtain an infinite branch $N_0, N_1, N_2, \ldots$.

■

Theorem 16.13 The Completeness Theorem. *Every consistent set of sentences has a model.*

Proof: Let

$$\Gamma = \{\varphi_1, \varphi_2, \varphi_3, \ldots\}$$

be a consistent set of sentences. We construct a tree similarly to the method we used above, except at level 1 we take φ_1 as our formula; at level 2, $\varphi_1 \wedge \varphi_2$; at level 3, $\varphi_1 \wedge \varphi_2 \wedge \varphi_3$; and so forth.

First suppose that the tree of partial models is finite. That means that at some level, say m, there is no way to make the sentence β_m true. This implies that $\neg\beta_m$ is a tautology. Thus, $\beta_m \vdash A \wedge \neg A$. Because $\{\varphi_1, \varphi_2, \ldots, \varphi_m\} \vdash \beta_m$, we obtain the contradiction that Γ is not consistent.

Consequently, it follows that if Γ is consistent, the tree, T, of partial models is infinite. Because each level is finite, it follows from Lemma 16.12 that T must have an infinite branch. Let the partial models in an infinite branch be $\mathbf{S}_0$, $\mathbf{S}_1$, $\mathbf{S}_2$, . . . with corresponding universes $\mathbf{U}_0$, $\mathbf{U}_1$, $\mathbf{U}_2$, ... Because $\mathbf{S}_{i+1}$ is an extension of $\mathbf{S}_i$, for each $i = 1, 2, \ldots,$ all the definitions of the functions and relations made in each $\mathbf{S}_i$ are consistent. Define the interpretation structure $\mathbf{S}$ so that its universe is $\mathbf{U} = \cup_{i=0}^{\infty} \mathbf{U}_i$, and its functions, constants, and relations are defined as they are along the branch; that is, the elements of $\mathbf{U}$ are all the elements that can be built up from the interpretation of the constants by using the functions. The relations are defined as we proceed along the branch. It is not hard to see that $\mathbf{S}$ is a model of Γ because suppose some φ_i is false in $\mathbf{S}$. Then, because φ_i is a universal sentence, it would be false in one of the partial models, and that cannot happen.

This completes the proof of the Completeness Theorem.

■

(We have proved the Completeness Theorem here for the case that Γ is, at most, countably infinite. If Γ is uncountable, the proof requires more sophisticated set-theoretic techniques.)

The ideas used in the proof of the Completeness Theorem describe an algorithm for constructing a proof. (This is an example of another automated theorem-proving algorithm. See Section 15.3.) The algorithm was illustrated in Example 16.2. We give another example below.

Example 16.3

Use the procedure described in the proof of the Completeness Theorem to give a proof of the following valid argument:

Premises: $(\forall x)[Px \to (\forall y)(Qy \to \neg Rxy)]$, $(\forall x)[Px \to (\exists y)(Sy \wedge Rxy)]$, $(\exists x)Px$.
Conclusion: $(\exists x)(Sx \wedge \neg Qx)$

Solution: Try to construct a model for the premises and the negation of the conclusion. Then, when we get stuck, we use the procedure to construct an indirect proof of the argument. The first step is to write the premises and the negation of the conclusion in prenex normal form:

$$(\forall x)(\forall y)[Px \to (Qy \to \neg Rxy)]$$
$$(\forall x)(\exists y)[Px \to (Sy \wedge Rxy)]$$
$$(\exists x)Px$$
$$(\forall x)(Sx \to Qx).$$

Next, introduce Skolem functions:

$$(\forall x)(\forall y)[Px \to (Qy \to \neg Rxy)]$$
$$(\forall x)[Px \to (Sf(x) \wedge Rxf(x))]$$
$$Pa$$
$$(\forall x)(Sx \to Qx).$$

Take $\mathbf{U}_0 = \{\mathbf{a}\}$ and $\mathbf{U}_1 = \{\mathbf{a}, \mathbf{f}(\mathbf{a})\}$.

β_1 **is** $[Pa \to (Qa \to \neg Raa)] \wedge [Pa \to (Sf(a) \wedge Raf(a))] \wedge Pa \wedge (Sa \to Qa)$.

Thus, if β_1 is to be true, each of the following must have true interpretations:

(i) $\qquad Pa, \quad Sf(a), \quad Raf(a).$

It turns out that there are 32 partial models at level 2. (We leave the proof of this as an exercise. It is not needed in this example.) However, we cannot continue the tree because two of the conjuncts that appear in β_2 are

(ii) $\qquad [Pa \to (Qf(a) \to \neg Raf(a))], \quad (Sf(a) \to Qf(a)).$

It is easy to see using (i) and (ii) that we can derive the contradiction $Raf(a) \wedge \neg Raf(a)$.

A formal proof of the argument using this procedure follows:

{1}	1.	$(\forall x)[Px \to (\forall y)(Qy \to \neg Rxy)]$	P
{2}	2.	$(\forall x)[Px \to (\exists y)(Sy \wedge Rxy)]$	P
{3}	3.	$(\exists x)Px$	P
{4}	4.	$\neg(\exists x)(Sx \wedge \neg Qx)$	P (for IP)
{1}	5.	$(\forall x)(\forall y)[Px \to (Qy \to \neg Rxy)]$	1 LE
{2}	6.	$(\forall x)(\exists y)[Px \to (Sy \wedge Rxy)]$	2 LE
{4}	7.	$(\forall x)(Sx \to Qx)$	4 LT
{8}	8.	Pz	z P (for EP)
{1}	9.	$Pz \to (Qz \to \neg Rzz)$	5 US $[z \mid x][z \mid y]$
{2}	10.	$(\exists y)[Pz \to (Sy \wedge Rzy)]$	6 US $[z \mid x]$
{4}	11.	$Sz \to Qz$	7 US $[z \mid x]$
{12}	12.	$Pz \to (Su \wedge Rzu)$	z, u P (for $EP(u)$)
{8, 12}	13.	$Su \wedge Rzu$	z, u, 8, 12 T
{4}	14.	$Su \to Qu$	7 US $[u \mid x]$
{1}	15.	$Pz \to (Qu \to \neg Rzu)$	5 US $[z \mid x][u \mid y]$
{1, 8}	16.	$Qu \to \neg Rzu$	z, u, 8, 15 T
{4, 8, 12}	17.	Qu	u 13, 14 T
{1, 4, 8, 12}	18.	$\neg Rzu$	z, u 16, 17 T
{1, 4, 8, 12}	19.	$Rzu \wedge \neg Rzu$	z, u, 13, 18 T
{1, 4, 8, 12}	20.	$A \wedge \neg A$	19 T
{1, 2, 4, 8}	21.	$A \wedge \neg A$	2, 12, 20 EP
{1, 2, 3, 4}	22.	$A \wedge \neg A$	3, 8, 21 EP
{1, 2, 3}	23.	$(\exists x)(Sx \wedge \neg Qx)$	4, 22 IP

In the formal proof, "z" corresponds to "a" in the informal proof, and "u" corresponds to "$f(a)$." In the formal proof, the first thing we did was to find prenex normal forms for the three premises and the negation of the conclusion (lines 3, 5, 6, and 7). Then we construct the formula β_1. Its conjuncts appear on lines 8, 9, 11, and 12. Then, on lines 8 and 13, we have the formulas (i) above, which have to be true if β_1 is true. The two conjuncts of β_2 ((ii) above) that we need are given on lines 14 and 15. The contradiction is derived on line 19. Thus, we see the formal proof is very similar to the procedure used to try to construct a model. Additional examples are given in Exercises 16.

(For more information, consult *Introduction to Mathematical Logic* by E. Mendelson, 3rd ed., Wadsworth & Brooks/Cole, Monterey, CA, 1987, and *A Mathematical Introduction to Logic* by H. B. Enderton, Academic Press, New York, 1972.)

EXERCISES 16

A. Prove that each of the following sets of connectives is complete. In each case, use complete induction to show that the missing connectives can be replaced by the given connectives. You may use the fact that the set $\{\neg, \vee, \wedge\}$ is complete, and you may use preceding results to prove later ones. (Answer to 1, 4, and 8.)

(1) $\{\neg, \vee\}$.

(2) $\{\neg, \wedge\}$.

(3) $(\neg, \rightarrow\}$.

(4) $\{\downarrow\}$.

(5) $\{\uparrow\}$.

(6) $\{\wedge, \leftrightarrow, \oplus\}$, where $\oplus$ is defined by the following truth table:

P	Q	$P \oplus Q$
T	T	F
T	F	T
F	T	T
F	F	F

(7) $\{\perp, M^{\perp}\}$, where $\perp$ is a constant whose truth value is false and $M^{\perp}$ is a ternary (3-place) connective defined by the following truth table:

P	Q	R	$M^{\perp}(P, Q, R)$
T	T	T	F
T	T	F	F
T	F	T	F
T	F	F	T
F	T	T	F
F	T	F	T
F	F	T	T
F	F	F	T

(8) $\{\neg, \top, M^{\top}\}$, where $\top$ is a constant whose truth value is true and $M^{\top}$ is a ternary (3-place) connective defined by the following truth table:

P	Q	R	$M^{\top}(P, Q, R)$
T	T	T	T
T	T	F	T
T	F	T	T
T	F	F	F
F	T	T	T
F	T	F	F
F	F	T	F
F	F	F	F

B. Prove that none of the following sets are complete. (Answers to 1 and 3.)

(1) $\{\wedge, \rightarrow\}$.
(2) $\{\wedge, \leftrightarrow\}$.
(3) $\{\neg, \leftrightarrow\}$.
(4) $\{\neg, \oplus\}$, where the connective $\oplus$ is defined in $A(6)$.
(5) $\{M^{\perp}\}$, where the connective $M^{\perp}$ is defined in $A(7)$.
(6) $\{\neg, M^{\top}\}$, where the connective $M^{\top}$ is defined in $A(8)$.
(7) $\{\perp, \top, \neg, \leftrightarrow, \oplus\}$, where the connective $\oplus$ is defined in $A(6)$, and the constants $\perp$ and $\top$ are defined in $A(7)$ and (8).

C. Each of the following problems are related to the procedure used to prove the Completeness Theorem. (Answers to 1(a)(i), 2(a)(c), 3(a)(e) and 4(a)(d).)

(1) Introduce Skolem functions in each of the following sentences. (First write the sentence in prenex normal form, if it is not already in that form.)

(a) $(\forall x)(\exists y)(\exists z)(\forall u)(Pxy \wedge Qxz \rightarrow Ryzu)$.
(b) $(\forall x)(\exists y)(\forall z)(\exists u)(\exists v)[(Px \rightarrow Py) \wedge (Qxzu \rightarrow Pv)]$.
(c) $(\exists x)(\exists y)(\forall z)(\exists u)(Fx \wedge Gyu \leftrightarrow Hzu)$.
(d) $(\exists x)(\forall y)(\exists z)(\exists u)(\forall v)[Fxu \rightarrow (Fyz \rightarrow Gxzu \vee Hyuv)]$.
(e) $(\forall x)[Ax \rightarrow (\exists y)(\exists z)(Bxy \wedge Cyz)]$.
(f) $(\exists x)[Ax \wedge (\forall y)(Bxy \rightarrow (\exists z)Cxyz)]$.

(g) $(\forall x)Px \rightarrow (\exists x)Qx$.
(h) $(\forall x)(\forall y)(Px \wedge Qy \rightarrow Rxy) \rightarrow (\forall x)(\exists z)(Rxz \wedge Pz)$.
(i) $(\exists x)(\forall y)(Fxy \vee Gyz) \rightarrow (\forall x)(\exists y)(\forall z)(Fxy \wedge Gyz \rightarrow Hxyz)$.
(j) $(\exists x)(\forall y)Rxy \leftrightarrow (\forall x)Px$.
(k) $(\exists x)Px \wedge (\exists x)Qx \leftrightarrow (\forall x)Rx$.

(2) Using the procedure used to prove the Completeness Theorem, try to guess a model for each of the following sets of formulas by finding all partial models through level 3 (that is, find $\mathbf{U}_0$, $\mathbf{U}_1$, β_1, $\mathbf{U}_2$, β_2, $\mathbf{U}_3$, β_3, and all the related partial models), then try to find a branch through the tree. (First write each sentence in prenex normal form and then introduce Skolem functions. If there are no constants, take $\mathbf{U}_0 = \{\mathbf{a}\}$. See Example 16.1.)
(a) $(\exists x)Px$, $(\forall x)(\exists y)(x \neq y)$.
(b) $(\forall x)Rxx$, $(\forall x)Rxf(x)$, $(\forall x)(\exists y)\neg Rxy$.
(c) $(\forall x)(\exists y)(Rxy \wedge x \neq y)$, $(\forall x)(\forall y)(Rxy \rightarrow Ryz)$, $(\forall x)\neg Rxx$.
(d) $(\forall x)\neg Rxx$, $(\forall x)(\forall y)(\forall z)(Rxy \wedge Ryz \rightarrow Rxz)$, $(\forall x)(\exists y)Rxy$.
(e) $(\exists x)(\exists y)(Fx \wedge Gxy)$, $(\forall x)(\exists y)(\neg Gxy \wedge x \neq y)$.
(f) $(\forall x)(\forall y)(Fx \wedge Fy \rightarrow Gxy)$, $(\exists x)Fx$, $(\forall x)(\exists y)\neg Gxy$.
(g) $(\forall x)\neg Rxx$, $(\forall x)(\forall y)(Rxy \rightarrow \neg Ryx)$, $(\forall x)(\forall y)(\forall z)(Rxy \wedge Ryz \rightarrow Rxz)$, $(\forall x)(\forall y)(\exists z)(Rxy \rightarrow Rxz \wedge Rzy)$, $(\exists x)(\forall y)(x \neq y \rightarrow Rxy)$.
(h) $(\forall x)Fxx$, $(\forall x)(\forall y)(Fxy \wedge Fyx \rightarrow x = y)$,$(\forall x)(\forall y)(\forall z)(Fxy \wedge Fyz \rightarrow Fxz)$, $(\forall x)(\exists y)(x \neq y \wedge Fxy)$, $(\forall x)(\exists y)(x \neq y \wedge Fyx)$.

(3) Use the procedure illustrated in Example 16.3 to give a formal proof of each of the following valid arguments:

(a) **Premises:** $(\forall x)(\forall y)(Ax \wedge By \rightarrow \neg Lxy)$, $(\forall x)(\exists y)[(By \rightarrow \neg Lxy) \rightarrow Dx]$
Conclusion: $(\forall x)(Ax \rightarrow Dx)$

(b) **Premise:** $(\forall x)(Fx \rightarrow Gx)$
Conclusion: $\neg(\exists x)(\forall y)(\exists z)[Fx \wedge \mathrm{K}xz \wedge (Gx \rightarrow \neg Kxy)]$

(c) **Premises:** $(\forall x)[Px \rightarrow (\exists y)(Qy \wedge Rxy)]$, $(\exists x)[Px \wedge Sx \wedge (\forall y)(Rxy \rightarrow Sy)]$
Conclusion: $(\exists x)(Qx \wedge Sx)$

(d) **Premise:** $(\forall x)(\exists y)(Py \wedge \neg Qx)$
Conclusion: $\neg[(\exists x)Px \rightarrow (\exists \mathrm{x})Qx]$

(e) **Premises:** $(\forall x)(Ax \rightarrow Bx)$, $(\exists x)(Cx \wedge (\forall y)(By \rightarrow Dxy)]$
Conclusion: $(\exists x)[Cx \wedge (\forall y)(Ay \rightarrow Dxy)]$

(f) **Premise:** $(\exists x)[Px \wedge (\forall y)(Py \rightarrow x = y)]$
Conclusion: $(\exists x)(\forall y)(x = y \rightarrow Py)$

(g) **Premises:** $(\forall x)[Ax \rightarrow (\exists y)Bxy]$, $(\exists x)(Ax \wedge \neg Bxa)$
Conclusion: $(\exists x)(\exists y)(Bxy \wedge y \neq a)$

(h) **Premises:** $(\exists x)[(\forall y)Fxy \vee (\forall y)Fyx]$, $(\forall x)(\neg Fxx \vee Gxa)$, $(\forall x)(x \neq a \rightarrow \neg Gxa)$, $(\forall x)(\forall y)(Gxa \rightarrow Gya)$

Conclusion: $(\exists x)(\forall y)(x = y)$

(4) Use the procedure illustrated in Example 16.3 to give a formal proof of each of the following logical theorems:

(a) $(\forall x)[Ax \rightarrow (\exists y)(x = y \wedge Ay)]$.

(b) $(\forall x)(Px \rightarrow Qx) \rightarrow [(\forall x)Px \rightarrow (\forall x)Qx]$.

(c) $(\exists x)(\forall y)[Px \vee (Py \rightarrow Qy)]$.

(d) $(\forall x)(\exists y)(Px \wedge Qy) \rightarrow (\exists y)(\forall x)(Px \wedge Qy)$.

(e) $(\forall x)(Fx \rightarrow (\exists y)Gxy) \wedge (\exists x)[Fx \wedge (\forall y)(Gxy \rightarrow Hxy)] \rightarrow (\exists x)(\exists y)Hxy$.

(f) $(\exists x)(\forall y)(\forall z)[Ax \wedge (Ay \wedge Az \rightarrow y = z)] \rightarrow (\exists x)(\forall y)[Ax \wedge (Ay \rightarrow x = y)]$.

D. Prove each of the following. (Answers to 1, 3(a), and 5(a) and (b).)

(1) If Γ is a consistent set of sentences, then for every sentence φ either $\Gamma \cup \{\varphi\}$ or $\Gamma \cup \{\neg\varphi\}$ is consistent.

(2) Suppose φ is the sentence

$$(\forall x)(\forall y)(\forall z)[Rxx \wedge (Rxy \vee Ryx) \wedge (Rxy \wedge Ryz \rightarrow Rxz)] \rightarrow (\exists y)(\forall x)Ryx.$$

Prove that φ is true in every interpretation structure for φ that has a finite universe, but φ is not provable.

(3) Suppose ψ is a formula of the Predicate Calculus that contains no quantifiers, no constants, and no function symbols.

(a) Suppose $x_1, x_2, \ldots, x_n$ are the only free variables in ψ, and suppose φ is the sentence $(\forall x_1)(\forall x_2) \ldots (\forall x_n)\psi$. Prove that $\vdash \varphi$ iff φ is true in every interpretation structure for φ whose universe has $\leq n$ elements.

(b) Suppose $x_1, x_2, \ldots, x_n$ are the only free variables in ψ, and suppose φ is the sentence $(\exists x_1)(\exists x_2) \ldots (\exists x_n)\psi$. Prove that $\vdash \varphi$ iff φ is true in every interpretation structure for φ whose universe has one element.

(c) Suppose $x_1, x_2, \ldots, x_n, y_1, y_2, \ldots, y_m$ are the only free variables in ψ, and suppose φ is the sentence $(\forall x_1)(\forall_2) \ldots (\forall x_n)(\exists y_1)(\exists y_2) \ldots (\exists y_m)\psi$. Prove that $\vdash \varphi$ iff φ is true in every interpretation structure for φ whose universe has $\leq n$ elements.

(4) Suppose φ is a sentence of the Predicate Calculus such that φ contains no function symbols and the only predicate symbols it contains are k unary (1-place) predicate symbols. Prove that $\vdash \varphi$ iff φ is true in every interpretation structure for φ that has $\leq 2^k$ elements in its universe.

(5) In Chapter 13, we defined a theory to be a set of sentences that is closed under logical implication. A theory Γ is said to be *complete* if for any sentence φ in the language of Γ either $\varphi \in \Gamma$ or $\neg\varphi \in \Gamma$.

(a) Show that the set of all logical theorems of the Predicate Calculus is not a complete theory.

(b) Let **S** be an interpretation structure, and let Γ be the set of all sentences that are true in **S**. Show that Γ is a complete theory.
(c) Prove that Group Theory is not complete. (See Chapter 13.)
(d) Prove that a theory Γ is complete iff for all sentences φ and ψ in the language of Γ, if $\Gamma \vdash \varphi \vee \psi$, then $\Gamma \vdash \varphi$ or $\Gamma \vdash \psi$.

(6) Let Γ_1 and Γ_2 be two theories in the same language L such that for every interpretation structure **S** for L, $\mathbf{S} \models \Gamma_1$ iff $\mathbf{S} \not\models \Gamma_2$. Prove that there are sentences $\varphi_1 \in \Gamma_1$ and $\varphi_2 \in \Gamma_2$ such that for all $\psi_1 \in \Gamma_1$ and all $\psi_2 \in \Gamma_2$,

$$\varphi_1 \vdash \psi_1 \quad \text{and} \quad \varphi_2 \vdash \psi_2.$$

References

Bundy, A. *The Computer Modelling of Mathematical Reasoning*, Academic Press, New York, 1983.

Chang, C., and Lee, C. *Symbolic Logic and Mechanical Theorem Proving*, Academic Press, New York, 1973.

Copi, I. M. *Symbolic Logic*, 5th ed., MacMillan, New York, 1979.

Enderton, H. B. *A Mathematical Introduction to Logic*, Academic Press, New York, 1972.

Gersting, J. L. *Mathematical Structures for Computer Science*, 2nd ed., W. H. Freeman, San Francisco, 1982.

Golovina, L. I., and Yaglom, I. M. *Induction in Geometry*, D. C. Heath, Boston, 1963.

Herbrand, J. *Recherches sur la theorie de la demonstration*, Ph.D. thesis, University of Paris, 1930.

Hill, F. J., and Peterson, G. R. Introduction to Switching Theory and Logical Design, Wiley, New York, 1968.

Hillman, A. P., Alexanderson, G. L. and Grassl, R. M. *Discrete and Combinatorial Mathematics*, Dellen-MacMillan, New York, 1987.

Liu, C. L. *Elements of Discrete Mathematics*, 2nd ed., McGraw-Hill, New York, 1985.

Mendelson, E. *Boolean Algebra & Switching Circuits*, Schaum's Outline Series, McGraw-Hill, New York, 1970.

Mendelson, E. *Introduction to Mathematical Logic*, 3rd ed., Wadsworth & Brooks/Cole, Monterey, CA, 1987.

Meyer, D. G. *Little Bits of Digital Wisdom*, 10th ed., A Lecture-Workbook for EE 266, Purdue Universtiy, West Lafeyette, IN, 1984.

Polya, G. *Induction and Analogy in Mathematics*, Princeton University Press, Princeton, NJ, 1954.

Ross, K. A., and Wright, C. R. B. *Discrete Mathematics*, Prentice-Hall, New York, 1985.

Smullyan, R. *What Is the Name of This Book?*, Prentice-Hall, New York, 1978.

Sominski, I. S. *Die Methode der Vollständigen Induction*, Veb Deutscher Verlag der Wissenschaften, 1960.

Summers, G. J. *New Puzzles in Logical Deduction*, Dover, New York, 1968.

Suppes, P. *Introduction to Logic*, Van Nostrand Reinhold, New York, 1957.

Wos, L., Overbeek, R., Lusk, E., and Boyle, J. *Automated Reasoning: Introduction and Applications*, Prentice-Hall, New York, 1984.

Answers to Selected Exercises

Chapter 1

A. 5. $(O \vee G) \wedge \neg(O \wedge G)$.
7. $\neg(H \wedge S) \rightarrow \neg P$ or $\neg P \vee (H \wedge S)$.
11. $\neg(S \wedge H) \rightarrow \neg P$.
15. $D \rightarrow (O \wedge V \rightarrow S)$.
17. $Q \rightarrow (P \leftrightarrow E \wedge L)$.
20. $D \rightarrow (I \rightarrow P)$.
22. $P \rightarrow (I \wedge A \rightarrow G)$.
24. $S_3 \wedge S_4 \rightarrow I$.

B. 2. A: Virginia passed algebra; G: Virginia passed geometry; H: Virginia would be admitted to Harvard; Y: Virginia would be admitted to Yale.
$\neg(A \wedge G) \rightarrow \neg(H \vee Y)$
5. O: Oscar will graduate from college; V: Virginia will graduate from college; L: Oscar passes the logic course; P: Virginia passes the logic course.
$(O \rightarrow L) \wedge (V \rightarrow P)$
9. X: x is an integer; I: $x + 1$ is an even integer; E: x is an even integer.
$X \rightarrow (I \vee E) \wedge \neg(I \wedge E)$
11. A: A is a true sentence ($\neg A$: A is a false sentence); C: The conditional sentence $A \rightarrow B$ is of the form $F \rightarrow T$; D: The conditional sentence $A \rightarrow B$ is of the form $F \rightarrow F$; B: B is true ($\neg B$: B is false).
$\neg A \rightarrow (B \rightarrow C) \wedge (\neg B \rightarrow D)$

C. 2. Converse: Two lines are not parallel, if they intersect.
Inverse: Two lines will not intersect, if they are parallel.
Contrapositive: Two lines are parallel, if they do not intersect.
7. Converse: If a triangle satisfies the Pythagorean Theorem, then it is a right triangle.
Inverse: If a triangle is not a right triangle, then it does not satisfy the Pythagorean Theorem.
Contrapositive: If a triangle does not satisfy the Pythagorean Theorem, then it is not a right triangle.

D. **3.**

A	B	C	$\neg B$	$A \vee \neg B$	$\neg A$	$C \wedge \neg A$	$A \vee \neg B \rightarrow C \wedge \neg A$
T	T	T	F	T	F	F	F
T	T	F	F	T	F	F	F
T	F	T	T	T	F	F	F
T	F	F	T	T	F	F	F
F	T	T	F	F	T	T	T
F	T	F	F	F	T	F	T
F	F	T	T	T	T	T	T
F	F	F	T	T	T	F	F

Chapter 2

A. (Truth tables are omitted.)

1. (*iv*); **7.** (*i*), (*ii*), (*iii*); **9.** (*i*); **10.** (*ii*).

B.

1. (14) Distributive Rules; $P = A \vee B$, $Q = C$, $R = D$.

5. (9) Rule of Absurdity; $P = A \wedge B \wedge C$, $Q = D$.

9. (15) De Morgan's Rules; $P = A$, $Q = \neg B \vee C$.

C. (Truth tables are omitted, but counter-examples are included.)

2. Premises: $(O \rightarrow M) \wedge (M \rightarrow G)$, $O \vee G$

Conclusion: $\neg M$

Argument is not valid.

O	M	G	$(O \rightarrow M) \wedge (M \rightarrow G)$	$O \vee G$	$\neg M$
T	T	T	T	T	F

4. Premises: $O \wedge G \rightarrow V$, G

Conclusion: $V \vee \neg O$

Argument is valid.

7. Premises: $O \rightarrow M \vee V$, $G \rightarrow V$, $(\neg G \vee M) \wedge (V \rightarrow O)$

Conclusion: $V \leftrightarrow O$

Argument is not valid.

O	M	V	G	$O \rightarrow M \vee V$	$G \rightarrow V$	$(\neg(G \vee M) \wedge (V \rightarrow O)$	$V \leftrightarrow O$
T	T	F	F	T	T	T	F

D. **4.** Argument is valid.

{1}	1.	$O \wedge G \rightarrow V$	P
{2}	2.	G	P
{3}	3.	O	P (for CP)
{1, 2, 3}	4.	V	1, 2, 3 T
{1, 2}	5.	$O \rightarrow V$	3, 4 CP
{1, 2}	6.	$\neg O \vee V$	5 T

8. Argument is valid.

{1}	1.	$O \to \neg(M \vee V)$	P
{2}	2.	$(\neg V \to \neg G) \wedge (\neg M \to G)$	P
{3}	3.	O	P (for IP)
{1, 3}	4.	$\neg(M \vee V)$	1, 3 T
{1, 3}	5.	$\neg M \wedge \neg V$	4 T
{2}	6.	$\neg V \to \neg G$	2 T
{2}	7.	$\neg M \to G$	2 T
{1, 2, 3}	8.	$\neg G$	5, 6 T
{1, 2, 3}	9.	G	5, 7 T
{1, 2, 3}	10.	$G \wedge \neg G$	8, 9 T
{1, 2}	11.	$\neg O$	3, 10 IP

E. (Truth tables are omitted, but counter-examples are included.)

1. Argument is not valid.

O	V	M	$O \to M \vee V$	M	$V \to O$	O
F	F	T	T	T	T	F

4. Argument is valid.

8. Argument is valid.

F. 4.

{1}	1.	$O \to V$	P
{2}	2.	$G \to M$	P
{3}	3.	$G \to O \vee \neg M$	P
{4}	4.	$G \vee \neg M$	P
{5}	5.	$M \vee \neg V$	P
{5}	6.	$V \to M$	5 T
{4}	7.	$M \to G$	4 T
{1, 4, 5}	8.	$O \to G$	1, 6, 7 T
{9}	9.	G	P (for CP)
{2, 9}	10.	M	2, 9 T
{3, 9}	11.	$O \vee \neg M$	3, 9 T
{2, 3, 9}	12.	O	10, 11 T
{2, 3}	13.	$G \to O$	9, 12 CP
{1, 2, 3, 4, 5}	14.	$O \leftrightarrow G$	8, 13 T

G. 1. A: $a < b$; B: $b < c$; C: $a < c$.

Premises: $A \wedge B \to C, \quad A \wedge \neg C$

Conclusions: $\neg B$

Argument is valid.

{1}	1.	$A \wedge B \to C$	P
{2}	2.	$A \wedge \neg C$	P
{1, 2}	3.	$\neg(A \wedge B)$	1, 2 T
{1, 2}	4.	$\neg A \vee \neg B$	3 T
{1, 2}	5.	$\neg B$	2, 4 T

5. P: $b < c$; N: $-c < -b$; A: $a + (-c) < a + (-b)$.

Premises: $P \leftrightarrow N$, $\neg A \rightarrow \neg P$
Conclusion: $P \leftrightarrow A$

Argument is not valid.

P	N	A	$P \leftrightarrow N$	$\neg A \rightarrow \neg P$	$P \leftrightarrow A$
F	F	T	T	T	F

8. (b). Z: $x < 0$; T: $x < 2$; X: $g(x) = x$; F: $g(x) = 5$; S: $g(x) = 6$.

Premises: $T \rightarrow (Z \rightarrow X)$, $T \rightarrow (\neg Z \rightarrow F)$, $\neg T \rightarrow S$, $Z \vee \neg T$
Conclusion: $X \vee S$

Argument is valid.

{1}	1.	$T \rightarrow (Z \rightarrow X)$	P
{2}	2.	$T \rightarrow (\neg Z \rightarrow F)$	P
{3}	3.	$\neg T \rightarrow S$	P
{4}	4.	$Z \vee \neg T$	P
{5}	5.	$\neg(X \vee S)$	P (for IP)
{5}	6.	$\neg X \wedge \neg S$	5 T
{3, 5}	7.	T	3, 6 T
{3, 4, 5}	8.	Z	4, 7 T
{1, 3, 5}	9.	$Z \rightarrow X$	1, 7 T
{1, 3, 4, 5}	10.	X	8, 9 T
{1, 3, 4, 5}	11.	$X \wedge \neg X$	6, 10 T
{1, 3, 4}	12.	$X \vee S$	5, 11 IP

11. **Premises:** $(A \rightarrow C) \wedge (B \rightarrow D)$, $\neg C \wedge D$
Conclusion: $\neg A \wedge B$

Argument is not valid.

A	B	C	D	$(A \rightarrow C) \wedge (B \rightarrow D)$	$\neg C \wedge D$	$\neg A \wedge B$
F	F	F	T	T	T	F

H. 1.

{1}	1.	$A \rightarrow (B \rightarrow C)$	P
{2}	2.	$C \rightarrow \neg D$	P
{3}	3.	$\neg E \rightarrow D$	P
{4}	4.	B	P (for CP)
{5}	5.	A	P (for CP)
{1, 5}	6.	$B \rightarrow C$	1, 5 T
{1, 4, 5}	7.	C	4, 6 T
{1, 2, 4, 5}	8.	$\neg D$	2, 7 T

	$\{1, 2, 3, 4, 5\}$	9.	E	3, 8 T
	$\{1, 2, 3, 4\}$	10.	$A \rightarrow E$	5, 9 CP
	$\{1, 2, 3\}$	11.	$B \rightarrow (A \rightarrow E)$	4, 10 CP

4.	$\{1\}$	1.	$(A \rightarrow B) \vee (A \wedge \neg C)$	P
	$\{2\}$	2.	A	P
	$\{3\}$	3.	$C \rightarrow \neg B$	P
	$\{4\}$	4.	$\neg C \rightarrow B$	P
	$\{5\}$	5.	C	P (for IP)
	$\{3, 5\}$	6.	$\neg B$	3, 5 T
	$\{2, 3, 5\}$	7.	$A \wedge \neg B$	2, 6 T
	$\{2, 3, 5\}$	8.	$\neg(A \rightarrow B)$	7 T
	$\{1, 2, 3, 5\}$	9.	$A \wedge \neg C$	1, 8 T
	$\{1, 2, 3, 5\}$	10.	$C \wedge \neg C$	5, 9 T
	$\{1, 2, 3\}$	11.	$\neg C$	5, 10 IP

8.	$\{1\}$	1.	$A \wedge B \rightarrow C$	P
	$\{2\}$	2.	$A \vee E$	P
	$\{3\}$	3.	$G \rightarrow (\neg C \wedge \neg D)$	P
	$\{4\}$	4.	$A \leftrightarrow B$	P
	$\{5\}$	5.	$A \rightarrow C$	P
	$\{6\}$	6.	$C \rightarrow \neg D$	P
	$\{7\}$	7.	$B \rightarrow D$	P
	$\{8\}$	8.	A	P (for IP)
	$\{5, 8\}$	9.	C	5, 8 T
	$\{4, 8\}$	10.	B	4, 8 T
	$\{4, 7, 8\}$	11.	D	7, 10 T
	$\{4, 6, 7, 8\}$	12.	$\neg C$	6, 11 T
	$\{4, 5, 6, 7, 8\}$	13.	$C \wedge \neg C$	9, 12 T
	$\{4, 5, 6, 7\}$	14.	$\neg A$	8, 13 IP
	$\{2, 4, 5, 6, 7\}$	15.	E	2, 14 T
	$\{2, 4, 5, 6, 7\}$	16.	$E \vee \neg G$	15 T
	$\{2, 4, 5, 6, 7\}$	17.	$G \rightarrow E$	16 T
	$\{2, 4, 5, 6, 7\}$	18.	$\neg A \wedge (G \rightarrow E)$	14, 17 T

10.	$\{1\}$	1.	$A \rightarrow B$	P
	$\{2\}$	2.	$C \rightarrow \neg B$	P
	$\{3\}$	3.	$\neg C \rightarrow D$	P
	$\{4\}$	4.	$A \wedge \neg D$	P
	$\{1, 4\}$	5.	B	1, 4 T
	$\{1, 2, 4\}$	6.	$\neg C$	2, 5 T
	$\{1, 2, 3, 4\}$	7.	D	3, 6 T
	$\{1, 2, 3, 4\}$	8.	$D \wedge \neg D$	4, 7 T
		9.	$(D \wedge \neg D) \rightarrow (E \rightarrow F)$	T
	$\{1, 2, 3, 4\}$	10.	$E \rightarrow F$	8, 9 T

Chapter 3

A. 1. S: Oscar studies.
L: Oscar will pass his logic course.
Inconsistent.

{1}	1.	$S \to L$	P
{2}	2.	$\neg S \to L$	P
{3}	3.	$\neg L$	P
{1, 3}	4.	$\neg S$	1, 3 T
{1, 2, 3}	5.	L	2, 4 T
{1, 2, 3}	6.	$L \wedge \neg L$	3, 5 T

6. G: Miriam will graduate.
M: Miriam gets married.
H: Miriam helps Virginia with her homework.
V: Virginia will graduate.
W: Miriam will go to work.
Consistent.

G	M	W	H	V	$G \vee M$	$W \leftrightarrow M$	$W \to \neg H$	$\neg H \to \neg V$	$G \leftrightarrow \neg V$
T	T	T	F	F	T	T	T	T	T

9: O: Oscar graduates.
G: George graduates.
S: Oscar gets a job.
E: George gets a job.
Consistent.

O	G	S	E	$O \wedge G \to S \vee E$	$S \to \neg E$	$G \to \neg O$	$\neg(O \vee G) \wedge (S \vee G)$
F	F	T	F	T	T	T	T

11. A: $a < b$
B: $b < c$.
C: $a < c$.
Consistent.

A	B	C	$A \wedge B \to C$	$A \wedge \neg B \to \neg C$	$\neg A \wedge B \to C$	$C \leftrightarrow (A \wedge B) \vee \neg A$
F	F	T	T	T	T	T

15. A: A is true.
B: B is true.
D: $A \to B$ is of the form $F \to T$.
E: $A \to B$ is of the form $F \to F$.
Consistent.

A	B	D	E	$\neg A \to (B \to D) \wedge (\neg B \to E)$	$B \wedge \neg A \to D \wedge \neg(B \vee E)$
T	F	T	F	T	T

19. C: f is continuous on (a, b).
M: f has a maximum on $[a, b]$.
N: f has a minimum on $[a, b]$.
A: f is continuous at a.
B: f is continuous at b.
Inconsistent.

{1}	1.	$C \wedge A \wedge B \wedge M \wedge \neg N$	P
{2}	2.	$C \rightarrow (M \wedge N) \vee \neg(A \wedge B)$	P
{1, 2}	3.	$(M \wedge N) \vee \neg(A \wedge B)$	1, 2 T
{1, 2}	4.	$M \wedge N$	1, 3 T
{1, 2}	5.	$N \wedge \neg N$	1, 4 T

B. **3.** Inconsistent.

{1}	1.	$A \vee B \rightarrow \neg C$	P
{2}	2.	$A \wedge C$	P
{1, 2}	3.	$\neg(A \wedge B)$	1, 2 T
{1, 2}	4.	$\neg A \wedge \neg B$	3 T
{1, 2}	5.	$A \wedge \neg A$	2, 4 T

5. Inconsistent.

{1}	1.	$A \rightarrow B$	P
{2}	2.	$C \rightarrow \neg B$	P
{3}	3.	$\neg C \rightarrow D$	P
{4}	4.	$A \wedge \neg D$	P
{1, 4}	5.	B	1, 4 T
{1, 2, 4}	6.	$\neg C$	2, 5 T
{1, 2, 3, 4}	7.	D	3, 6 T
{1, 2, 3, 4}	8.	$D \wedge \neg D$	4, 7 T

9. Consistent.

A	B	C	D	$A \leftrightarrow B \vee C$	$A \vee \neg D$	$\neg B \vee D$	$C \vee \neg D$	$B \leftrightarrow \neg A$
T	F	T	F	T	T	T	T	T

13. Inconsistent.

{1}	1.	$A \rightarrow C$	P
{2}	2.	$A \vee \neg B$	P
{3}	3.	$C \rightarrow D \vee E$	P
{4}	4.	$D \rightarrow F \wedge \neg B$	P
{5}	5.	$E \rightarrow \neg A \wedge F$	P
{6}	6.	$A \vee \neg F$	P
{7}	7.	$\neg B \vee G \rightarrow H \wedge I$	P
{8}	8.	$H \rightarrow B$	P
{9}	9.	$\neg A$	P (for IP)
{2, 9}	10.	$\neg B$	2, 9 T

{2, 8, 9}	11.	$\neg H$	8, 10 *T*
{2, 9}	12.	$\neg B \vee G$	10 *T*
{2, 7, 9}	13.	$H \wedge I$	7, 12 *T*
{2, 7, 8, 9}	14.	$H \wedge \neg H$	11, 13 *T*
{2, 7, 8}	15.	A	9, 14 *IP*
{2, 7, 8}	16.	$A \vee \neg F$	15 *T*
{2, 7, 8}	17.	$\neg(\neg A \wedge F)$	16 *T*
{2, 5, 7, 8}	18.	$\neg E$	5, 17 *T*
{1, 3}	19.	$A \rightarrow D \vee E$	1, 3 *T*
{1, 2, 3, 7, 8}	20.	$D \vee E$	15, 19 *T*
{1, 2, 3, 5, 7, 8}	21.	D	18, 20 *T*
{1, 2, 3, 4, 5, 7, 8}	22.	$F \wedge \neg B$	4, 21 *T*
{1, 2, 3, 4, 5, 7, 8}	23.	$\neg H$	8, 22 *T*
{1, 2, 3, 4, 5, 7, 8}	24.	$\neg B \vee G$	22 *T*
{1, 2, 3, 4, 5, 7, 8}	25.	$H \wedge I$	7, 24 *T*
{1, 2, 3, 4, 5, 7, 8}	26.	$H \wedge \neg H$	23, 25 *T*

C. **1.** (i) Consistent.
(ii) All guilty, or *A, B, D* guilty and *C* innocent, or *A* guilty and *B, C, D* innocent.

4. (i) Consistent.
(ii) Cannot happen.

D. **1.** (i) Inconsistent.

3. (i) Consistent.
(ii) Arthur is the lawyer, Betty is the pilot, Charles is the teacher, and Dorothy is the vet. Or Arthur is the pilot, Betty is the lawyer, Charles is the teacher, and Dorothy is the vet.

E. **1.** (a) *R* and *T* are Fa's, and *S* is a Tu or a Fa.

2. (a) Yes. *V* is a Fa, and *W* is a Tu.

3. (a) Yes. *X* is a Fa, *Y* is a Tu, and *Z* is Smith Jr.

3. (d) ϵ_1's grade was 60, ϵ_2's grade was 70, and ϵ_3's grade was 90.

Chapter 4

A. **1.** (d) $(A \wedge B \wedge \neg C) \vee (A \wedge \neg B \wedge C) \vee (\neg A \wedge B \wedge C) \vee (\neg A \wedge \neg B \wedge C)$.

2. (a) $(A \wedge B \wedge \neg C \wedge D) \vee (\neg A \wedge B \wedge C \wedge D) \vee (\neg A \wedge B \wedge C \wedge \neg D) \vee (\neg A \wedge B \wedge \neg C \wedge D)$.

3. (b) $(A \wedge B) \vee (A \wedge \neg B) \vee (\neg A \wedge \neg B)$.
(h) $(A \wedge \neg B \wedge C) \vee (\neg A \wedge \neg B \wedge C)$.
(l) $(A \wedge B \wedge C \wedge \neg D) \vee (A \wedge \neg B \wedge C \wedge D) \vee (A \wedge \neg B \wedge \neg C \wedge D) \vee (A \wedge \neg B \wedge \neg C \wedge \neg D)$.

B. **1.** $P \vee Q$.

4. PQ.

8. R.
11. $\bar{P}\bar{Q}R \vee \bar{R}(P \vee Q)$.
13. $R \vee \bar{P}\bar{Q}$.
15. $\bar{P}Q \vee PR$.
17. $\bot$.
21. $\bar{Q}(RS \vee \bar{R}\bar{S}) \vee \bar{P}(R\bar{S} \vee \bar{R}S)$.
25. $\bar{Q}(P \vee \bar{S}) \vee \bar{P}R\bar{S} \vee P\bar{R}\bar{S}$.

C. 2. $A \vee B$.
5. $\bar{B}$.
8. $A \vee BC \vee CD$.

D. 1. $\overline{PQ}(\bar{P} \vee Q)$.
4. $\overline{P\bar{Q}}(Q \vee R)$.
7. $P \vee (\bar{P} \vee Q)Q$ and $(P \vee (\bar{P} \vee Q)Q)\bar{R}$.
11. $PQ\bar{R} \vee P\bar{Q}(R \vee S) \vee \overline{R \vee S}$.
13. $\bar{P}QR\bar{S} \vee P\bar{Q}RS \vee \bar{P}\bar{Q}RS$.

E. 1. $A \vee B \vee C \vee D$.
3. $A(B \vee C) \vee BC$.
6. $A\bar{B}\bar{C} \vee B(\bar{A} \vee C)$.

F. 1. $PQ\bar{R} \vee \bar{P}\bar{Q}R$.
3. $P\bar{R} \vee \bar{P}Q \vee \bar{Q}R$ or $P\bar{Q} \vee \bar{P}R \vee Q\bar{R}$.
4. (c) $\bar{P}_3\bar{P}_2(\bar{P}_1\bar{P}_0 \vee P_1P_0) \vee \bar{P}_3P_2P_1\bar{P}_0 \vee P_3\bar{P}_2\bar{P}_1P_0$.
7. (b) $P \Leftrightarrow QD$.
$C \Leftrightarrow Q \vee D$.
$I \Leftrightarrow \bar{Q} \vee \bar{D}$.
8. (a) $P \Leftrightarrow Q \vee D \vee N$.
$C \Leftrightarrow Q \vee D$.
$I \Leftrightarrow \overline{Q \vee D \vee N} \Leftrightarrow \bar{Q}\bar{D}\bar{N}$.
9. (c) $P \Leftrightarrow Q \vee D_1D_2$.
$I \Leftrightarrow \bar{P}$.
$C \Leftrightarrow \overline{\bar{Q}\bar{N}(D_1D_2 \vee \bar{D}_1\bar{D}_2)}$.
10. (b) $C(AD \vee B)$.
12. (a) $G_1 \Leftrightarrow A$, $G_2 \Leftrightarrow A(BC \vee BD \vee CD) \vee BCD$.

Chapter 5

A. 2. $(\exists x)(Cx \wedge \neg Tx)$.
4. $(\forall x)(Sx \wedge Cx \rightarrow \neg Px)$ or $\neg(\exists x)(Sx \wedge Cx \wedge Px)$.
7. $So \wedge Sm \wedge Mo \wedge Mm$.

10. $(\forall x)(Sx \wedge Mx \rightarrow (Ax \rightarrow Hx))$.
13. $(\forall x)(Sx \wedge Oxm \rightarrow Oxi)$.
15. $(\forall x)(Px \wedge Oxo \rightarrow (\exists y)(Sy \wedge Oxy))$.
19. $(\forall x)(Sx \wedge (\forall y)(Py \rightarrow Lxy) \rightarrow Lxx)$.
21. $(\forall x)(Sx \wedge Mx \rightarrow (\forall y)(Fy \rightarrow \neg Axy))$ or $(\forall x)(Sx \wedge Mx \rightarrow \neg(\exists y)(Fy \wedge Axy))$ or $\neg(\exists x)(Sx \wedge Mx \wedge (\exists y)(Fy \wedge Axy))$.
23. $(\forall x)(Fx \rightarrow Mx) \rightarrow (\exists y)(\exists z)(Sy \wedge Fz \wedge Ayz)$.
25. $(\forall x)(Fx \wedge Axe \rightarrow \neg Mx)$ or $\neg(\exists x)(Fx \wedge Axe \wedge Mx)$.
28. $(\exists x)(\exists y)(\exists z)(Sx \wedge Cy \wedge Iz \wedge Fxyz)$.
30. $(\forall x)[(\exists y)(\exists z)(Cy \wedge Iz \wedge Fxyz) \rightarrow Sx]$ or $(\forall x)(\forall y)(\forall z)(Cy \wedge Iz \wedge Fxyz \rightarrow Sx)$.
34. $(\exists x)(Ix \wedge (\forall y)(Sy \rightarrow (\exists z)(Cz \wedge Fyzx))]$.
37. $(\forall x)[Sx \wedge (\forall y)(\forall z)(Cy \wedge Iz \rightarrow \neg Fxyz) \rightarrow \neg Ix]$ or $\neg(\exists x)[Sx \wedge \neg(\exists y)(\exists z)(Cy \wedge Iz \wedge Fxyz) \wedge Ix]$.
39. $(\forall x)(Dx \rightarrow \neg Sxf(1, 3))$ or $\neg(\exists x)(Dx \wedge Sxf(1, 3))$.
41. $(\exists x)(Dx \wedge Sxh(2)) \rightarrow Snh(2) \wedge Sah(2))$.
45. $(\forall x)(Dx \rightarrow (\forall y)(Ey \rightarrow (\exists z)(Pzy \wedge Sxz)))$.
47. $(\forall x)[(\exists y)(Ey \wedge (\forall z)(Pzy \rightarrow Sxz)) \rightarrow \neg Dx]$ or $\neg(\exists x)(Dx \wedge (\exists y)(Ey \wedge (\forall z)(Pzy \rightarrow Sxz)))$.

B.
1. $(\forall x)(\forall y)(x + y = y + x)$.
3. $(\exists x)(\exists y)(x = 3 \cdot y)$.
6. $(\forall x)[(\exists y)(y > 0 \wedge x = 7 \cdot y) \rightarrow x > 5]$.
9. $(\exists x)[x > 0 \wedge (\exists y)(x = 2 \cdot y + 1) \wedge (\forall z)(z > 0 \wedge (\exists w)(z = 2 \cdot w + 1) \rightarrow z \geq x)]$.
11. $(\forall x)(\exists y)(x + y = 0)$.
15. $(\exists x)[(\exists y)(x = 2 \cdot y + 1 \wedge (\forall z)((\exists u)(z = 2 \cdot u) \rightarrow (\exists v)(z = x \cdot v))]$.
18. $(\exists x)(\exists y)[((\exists u_1)(\exists v_1)(x = 2 \cdot u_1 \wedge y = 2 \cdot v_1) \vee (\exists u_2)(\exists v_2)(x = 2 \cdot u_2 + 1 \wedge y = 2 \cdot v_2 + 1)) \wedge (\neg(\exists z_1)(x = 3 \cdot z_1) \vee \neg(\exists z_2)(y = 3 \cdot z_2))]$.
21. $(\exists u)(\exists v)(\exists x)(u \cdot x + v = 0)$.
23. $(\forall u)(\forall v)(\forall x)(\forall y)(u \cdot x + v = 0 \wedge u \cdot y + v = 0 \rightarrow x = y)$.

C.
1. Some cat does not have a tail.
4. No cat likes any dog.
7. Only cats bite all dogs.
10. Some cat either bites or dislikes every dog without a tail.
13. If Nancy solves a problem on some exam, then so does Alex.
18. If Nancy solves some problem on every exam, so does every woman.
20. Some man solves the same problem on every exam.
22. Between every pair of distinct real numbers, there is a real number.
25. Every nonzero real number has a unique multiplicative inverse.
27. There exists a prime.
29. No odd multiple of 5 is even.

D.
3. $(\forall x)(Px \rightarrow (\exists y)(Dy \wedge Syx))$, $(\exists x)(Px \wedge Sox)$, $(\exists x)(Dx \wedge (\forall y)(Py \rightarrow Sxy))$, $(\forall x)(Dx \rightarrow (\exists y)(Py \wedge Sxy))$.

7. $(\forall x)(Ix \rightarrow (\exists y)(Iy \wedge Mxy))$, $(\exists x)(Ix \wedge (\forall y)(Iy \rightarrow (Mxy \rightarrow x = y)))$, $(\forall x)(Ix \rightarrow Mxx)$.

F. 2. (i); 8. (i); 13. (iii)(iv); 15. (v); 20. (ii)(iv); 24. (v); 30. (iii).

Chapter 6

A. 1. **S** = ⟨**U**, **P**, **Q**, **R**⟩.

True: **U** = {**a**}.

	P	**Q**	**R**
a	*T*	*T*	*T*

False: **U** = {**a**}.

	P	**Q**	**R**
a	*T*	*T*	*F*

4. **S** = ⟨**U**, **P**, **Q**, **a**, **b**⟩.

True: **U** = {**a**, **b**}.

	P
a	*T*

Q	**b**
a	*F*

False: **U** = {**a**, **b**}.

	P
a	*F*

7. **S** = ⟨**U**, **P**, **Q**, **R**, **S**⟩.

True: **U** = {**a**}.

	P	**Q**
a	*T*	*F*

R	**a**
a	*F*

False: **U** = {**a**}.

	P
a	*F*

9. **S** = ⟨**U**, **P**, **Q**, **R**, **a**⟩.

True: **U** = {**a**}.

	P
a	*T*

Q	**a**
a	*F*

False: **U** = {**a**}.

	P
a	*F*

12. **S** = ⟨**U**, **P**, **Q**, **R**, **f**⟩.

True: **U** = {**a**}.

	P
a	*F*

False: **U** = {**a**}. **f**(**a**) = **a**.

	P
a	*T*

Q	**a**
a	*T*

15. **S** = ⟨**U**, **R**⟩.

True: **U** = $\mathbb{R}$. **Rxy** : **x** ≤ **y**.

False: **U** = {**a**}.

R	**a**
a	*F*

B. 1. **S** = ⟨**U**, **A**, **B**⟩.

U = {**a**}.

	A
a	*F*

4. **S** = ⟨**U**, **P**, **Q**, **a**⟩.

U = {**a**}.

	P	**Q**
a	*F*	*T*

8. **S** = ⟨**U**, **L**, **a**, **b**⟩.

U = {**a**, **b**}.

L	**a**	**b**
a	*F*	*T*
b	*F*	*T*

12. **S** = ⟨**U**, **R**⟩.

U = $\mathbb{R}$. **Rxy** : **x** ≤ **y**.

15. **S** = ⟨**U**, **P**, **Q**, **f**⟩.

U = {**a**, **b**}. **f**(**x**, **y**) = **x**.

	P	**Q**
a	*T*	*T*
b	*T*	*F*

Chapter 7

A. 1.

{1}	1.	$(\forall x)(Ax \to Bx)$	*P*
{2}	2.	$(\forall x)(Cx \to \neg Bx)$	*P*
{1}	3.	$Ax \to Bx$	1 *US* $[x \vert x]$
{2}	4.	$Cx \to \neg Bx$	2 *US* $[x \vert x]$
{2}	5.	$Bx \to \neg Cx$	4 *T*
{1, 2}	6.	$Ax \to \neg Cx$	3, 5 *T*
{1, 2}	7.	$(\forall x)(Ax \to \neg Cx)$	6 *UG*

5.

{1}	1.	$(\forall x)(Ax \to Bx \vee \neg Cx)$	*P*
{2}	2.	$(\forall x)(Ax \wedge Bx \to Dx)$	*P*
{3}	3.	$(\forall x)(Ax \wedge \neg Cx \to \neg Ex)$	*P*
{4}	4.	$Ax \wedge Ex$	*x P* (for *CP*)
{1}	5.	$Ax \to Bx \vee \neg Cx$	1 *US* $[x \vert x]$
{1, 4}	6.	$Bx \vee \neg Cx$	*x* 4, 5 *T*
{3}	7.	$Ax \wedge \neg Cx \to \neg Ex$	3 *US* $[x \vert x]$

{3, 4}	8.	$\neg(Ax \wedge \neg Cx)$	x, 4, 7 T
{3, 4}	9.	$\neg Ax \vee Cx$	x 8 T
{3, 4}	10.	Cx	x 4, 9 T
{1, 3, 4}	11.	Bx	x, 6, 10 T
{2}	12.	$Ax \wedge Bx \rightarrow Dx$	2 US $[x\|x]$
{1, 2, 3, 4}	13.	Dx	x, 4, 11, 12 T
{1, 2, 3}	14.	$Ax \wedge Ex \rightarrow Dx$	4, 13 CP
{1, 2, 3}	15.	$(\forall x)(Ax \wedge Ex \rightarrow Dx)$	14 UG

9.

{1}	1.	$(\forall x)(\forall y)(\forall z)(Rxy \wedge Ryz \rightarrow Rxz)$	P
{2}	2.	$(\forall x)(\forall y)(Rxy \rightarrow Ryx)$	P
{3}	3.	Rxy	x, y P (for CP)
{2}	4.	$Rxy \rightarrow Ryz$	2 US $[x\|x][y\|y]$
{2, 3}	5.	Ryx	x, y 3, 4 T
{1}	6.	$Rxy \wedge Ryx \rightarrow Rxx$	1 US $[x\|x][y\|y][x\|z]$
{1, 2, 3}	7.	Rxx	x, 3, 5, 6 T
{1, 2}	8.	$Rxy \rightarrow Rxx$	3, 7 CP
{1, 2}	9.	$(\forall x)(\forall y)(Rxy \rightarrow Rxx)$	8 UG (2)

12.

{1}	1.	$(\forall x)(Ax \rightarrow \neg Rxa)$	P
{2}	2.	$Ab \wedge Ba$	P
{3}	3.	$(\forall x)(Cx \rightarrow Bx)$	P
{4}	4.	$(\forall x)(Ax \rightarrow (\forall y)(By \wedge Cy \rightarrow Rxy))$	P
{1}	5.	$Ab \rightarrow \neg Rba$	1 US $[b\|x]$
{1, 2}	6.	$\neg Rba$	2, 5 T
{4}	7.	$Ab \rightarrow (\forall y)(By \wedge Cy \rightarrow Rby)$	4 US $[b\|x]$
{2, 4}	8.	$(\forall y)(By \wedge Cy \rightarrow Rby)$	2, 7 T
{2, 4}	9.	$Ba \wedge Ca \rightarrow Rba$	8 US $[a\|y]$
{1, 2, 4}	10.	$\neg(Ba \wedge Ca)$	6, 9 T
{1, 2, 4}	11.	$\neg Ba \vee \neg Ca$	10 T
{1, 2, 4}	12.	$\neg Ca$	2, 11 T

B. 1. $(\forall x)(Bx \rightarrow Wx)$, $(\forall x)(Fx \rightarrow \neg Wx)$ $\therefore$ $(\forall x)(Fx \rightarrow \neg Bx)$.

{1}	1.	$(\forall x)(Bx \rightarrow Wx)$	P
{2}	2.	$(\forall x)(Fx \rightarrow \neg Wx)$	P
{1}	3.	$Bx \rightarrow Wx$	1 US $[x\|x]$
{2}	4.	$Fx \rightarrow \neg Wx$	2 US $[x\|x]$
{1}	5.	$\neg Wx \rightarrow \neg Bx$	3 T
{1, 2}	6.	$Fx \rightarrow \neg Bx$	4, 5 T
{1, 2}	7.	$(\forall x)(Fx \rightarrow \neg Bx)$	6 UG

7. $(\forall x)(Lx \rightarrow Gx)$, $(\forall x)(Gx \wedge Hx \rightarrow \neg Ix)$, $(\forall x)(Gx \rightarrow \neg Hx)$ $\therefore$ $(\forall x)(Lx \rightarrow Ix \vee \neg Hx)$.

{1}	1.	$(\forall x)(Lx \rightarrow Gx)$	P
{2}	2.	$(\forall x)(Gx \wedge Hx \rightarrow \neg Ix)$	P
{3}	3.	$(\forall x)(Gx \rightarrow \neg Hx)$	P

{4}	4.	Lx	x P (for CP)
{1}	5.	$Lx \to Gx$	1 US $[x\vert x]$
{1, 4}	6.	Gx	x, 4, 5 T
{3}	7.	$Gx \to \neg Hx$	3 US $[x\vert x]$
{1, 3, 4}	8.	$\neg Hx$	x 6, 7 T
{1, 3, 4}	9.	$Ix \vee \neg Hx$	x 8 T
{1, 3}	10.	$Lx \to Ix \vee \neg Hx$	4, 9 CP
{1, 3}	11.	$(\forall x)(Lx \to Ix \vee \neg Hx)$	10 UG

9. $(\forall x)(\forall y)(Px \wedge Sy \to Oxy)$, $\neg Oot \wedge St$ $\therefore$ $\neg Po$.

{1}	1.	$(\forall x)(\forall y)(Px \wedge Sy \to Oxy)$	P
{2}	2.	$\neg Oot \wedge St$	P
{1}	3.	$Po \wedge St \to Oot$	1 US $[o\vert x][t\vert y]$
{1, 2}	4.	$\neg(Po \wedge St)$	2, 3 T
{1, 2}	5.	$\neg Po \vee \neg St$	4 T
{1, 2}	6.	$\neg Po$	2, 5 T

11. $(\forall x)(Fx \vee Mx \to (\forall y)(Cy \to Lxy))$, $Fn \wedge \neg Lnp$, $(\forall x)(Mx \to \neg Lxs)$, Mr $\therefore$ $\neg Cp \wedge \neg Cs$.

{1}	1.	$(\forall x)(Fx \vee Mx \to (\forall y)(Cy \to Lxy))$	P
{2}	2.	$Fn \wedge \neg Lnp$	P
{3}	3.	$(\forall x)(Mx \to \neg Lxs)$	P
{4}	4.	Mr	P
{3}	5.	$Mr \to \neg Lrs$	3 US $[r\vert x]$
{3, 4}	6.	$\neg Lrs$	4, 5 T
{1}	7.	$Fr \vee Mr \to (\forall y)(Cy \to Lry)$	1 US $[r\vert x]$
{4}	8.	$Fr \vee Mr$	4 T
{1, 4}	9.	$(\forall y)(Cy \to Lry)$	7, 8 T
{1, 4}	10.	$Cs \to Lrs$	9 US $[s\vert y]$
{1, 3, 4}	11.	$\neg Cs$	6, 10 T
{1}	12.	$Fn \vee Mn \to (\forall y)(Cy \to Lny)$	1 US $[n\vert x]$
{2}	13.	$Fn \vee Mn$	2 T
{1, 2}	14.	$(\forall y)(Cy \to Lny)$	12, 13 T
{1, 2}	15.	$Cp \to Lnp$	14 US $[p\vert y]$
{1, 2}	16.	$\neg Cp$	2, 15 T
{1, 2, 3, 4}	17.	$\neg Cp \wedge \neg Cs$	11, 16 T

C. 1.

{1}	1.	$(\forall x)(\forall y)(Dx \wedge Dy \to (Rxy \to \neg Ryx))$	P
{2}	2.	Dx	x P (for CP)
{1}	3.	$Dx \wedge Dx \to (Rxx \to \neg Rxx)$	1 US $[x\vert x][x\vert y]$
{1, 2}	4.	$Rxx \to \neg Rxx$	x 2, 3 T
{1, 2}	5.	$\neg Rxx$	x T
{1}	6.	$Dx \to \neg Rxx$	2, 5 CP
{1}	7.	$(\forall x)(Dx \to \neg Rxx)$	6 UG

5.

{1}	1.	$(\forall x)Exx$	P
{2}	2.	$(\forall x)(\forall y)(Exy \to Eyx)$	P
{3}	3.	$(\forall x)(\forall y)(\forall z)(Exy \wedge Eyz \to Exz)$	P
{4}	4.	$Exy \wedge Exz$	x, y, z P (for CP)
{2}	5.	$Exy \to Eyx$	2 US $[x\|x][y\|y]$
{2, 4}	6.	Eyx	x, y 4, 5 T
{3}	7.	$(\forall z)(Euv \wedge Evz \to Euz)$	3 US $[u\|x][v\|y]$
{3}	8.	$(\forall u)(\forall v)(\forall z)(Euv \wedge Evz \to Euz)$	7 UG (2)
{3}	9.	$Eyx \wedge Exz \to Eyz$	8 US $[y\|u][x\|v][z\|z]$
{2, 3, 4}	10.	Eyz	y, z 4, 6, 9 T
{2, 3}	11.	$Exy \wedge Exz \to Eyz$	4, 10 CP
{2, 3}	12.	$(\forall x)(\forall y)(\forall z)(Exy \wedge Exz \to Eyz)$	11 UG (3)

D. 1.

{1}	1.	$(\forall x)(Px \to Qx)$	P
{2}	2.	$(\forall x)(Px \to \neg Qx)$	P
{3}	3.	Pa	P
{1}	4.	$Pa \to Qa$	1 US $[a\|x]$
{1, 3}	5.	Qa	3, 4 T
{2}	6.	$Pa \to \neg Qa$	2 US $[a\|x]$
{2, 3}	7.	$\neg Qa$	3, 6 T
{1, 2, 3}	8.	$Qa \wedge \neg Qa$	5, 7 T

5.

{1}	1.	$(\forall x)(Px \to Qx \wedge Rx)$	P
{2}	2.	$(\forall x)(Sx \to \neg Rx)$	P
{3}	3.	$\neg(\forall x)(Px \to \neg Sx)$	P
{1}	4.	$Px \to Qx \wedge Rx$	1 US $[x\|x]$
{2}	5.	$Sx \to \neg Rx$	2 US $[x\|x]$
{2}	6.	$Rx \to \neg Sx$	5 T
	7.	$Qx \wedge Rx \to Rx$	T
{1, 2}	8.	$Px \to \neg Sx$	4, 5, 7 T
{1, 2}	9.	$(\forall x)(Px \to \neg Sx)$	8 UG
{1, 2, 3}	10.	$(\forall x)(Px \to \neg Sx) \wedge \neg(\forall x)(Px \to \neg Sx)$	3, 9 T

10.

{1}	1.	$(\forall x)(\forall y)(\forall z)(Rxy \wedge Ryz \to Rxz)$	P
{2}	2.	$(\forall x)(\forall y)(Rxy \to Ryx)$	P
{3}	3.	Rab	P
{4}	4.	$\neg Rbb$	P
{2}	5.	$Rab \to Rba$	2 US $[a\|x][b\|y]$
{2, 3}	6.	Rba	3, 5 T
{1}	7.	$Rba \wedge Rab \to Rbb$	1 US $[b\|x][a\|y][b\|z]$
{1, 2, 3}	8.	Rbb	3, 6, 7 T
{1, 2, 3, 4}	9.	$Rbb \wedge \neg Rbb$	4, 8 T

16.

{1}	1.	Tab	*P*
{2}	2.	$(\forall x)(\forall y)(Txy \rightarrow Qx \land Ry)$	*P*
{3}	3.	$(\forall x)(\forall y)(\forall z)(Sxyz \rightarrow Px)$	*P*
{4}	4.	$(\forall x)(\forall y)(Qx \land Ry \rightarrow Scyx)$	*P*
{5}	5.	$\neg Pc$	*P*
{2}	6.	$Tab \rightarrow Qa \land Rb$	2 *US* $[a\|x][b\|y]$
{1, 2}	7.	$Qa \land Rb$	1, 6 *T*
{4}	8.	$Qa \land Rb \rightarrow Scba$	4 *US* $[a\|x][b\|y]$
{1, 2, 4}	9.	$Scba$	7, 8 *T*
{3}	10.	$Scba \rightarrow Pc$	3 *US* $[c\|x][b\|y][a\|z]$
{1, 2, 3, 4}	11.	Pc	9, 10 *T*
{1, 2, 3, 4, 5}	12.	$Pc \land \neg Pc$	5, 11 *T*

E. 1.

{1}	1.	$(\forall x)(Ax \rightarrow Bx)$	*P*
{2}	2.	$(\forall y)(Cy \rightarrow \neg By)$	*P*
{3}	3.	Az	*z P* (for *CP*)
{1}	4.	$Az \rightarrow Bz$	1 *US* $[z\|x]$
{1, 3}	5.	Bz	*z* 3, 4 *T*
{2}	6.	$Cz \rightarrow \neg Bz$	2 *US* $[z\|y]$
{1, 2, 3}	7.	$\neg Cz$	*z* 5, 6 *T*
{1, 2}	8.	$Az \rightarrow \neg Cz$	3, 7 *CP*
{1, 2}	9.	$(\forall z)(Az \rightarrow \neg Cz)$	8 *UG*

3.

{1}	1.	$(\forall x)(\forall y)(Axy \land Ayx \rightarrow Bxy)$	*P*
{2}	2.	$(\forall z)(Axz \land Ayz)$	*x, y P* (for *CP*)
{2}	3.	$Axy \land Ayy$	*x, y* 2 *US* $[y\|z]$
{2}	4.	$Axx \land Ayx$	*x, y* 2 *US* $[x\|z]$
{2}	5.	$Axy \land Ayx$	*x, y* 3, 4 *T*
{1}	6.	$Axy \land Ayx \rightarrow Bxy$	1 *US* $[x\|x][y\|y]$
{1, 2}	7.	Bxy	*x, y* 5, 6 *T*
{1}	8.	$(\forall z)(Axz \land Ayz) \rightarrow Bxy$	2, 7 *CP*
{1}	9.	$(\forall y)[(\forall z)(Axz \land Ayz) \rightarrow Bxy]$	8 *UG*
{1}	10.	$(\forall x)(\forall y)[(\forall z)(Axz \land Ayz) \rightarrow Bxy]$	9 *UG*

F. 1. Line 9: Incorrect use of Rule *UG*. (Correct usage: $(\forall x)(Px \rightarrow Sa \land Ta)$.)

5. x is flagged on lines 3, 6, and 8.
Line 4: Incorrect use of Rule *US*.

9. Line 6: Incorrect use of Rule *US*.
Line 7: Incorrect use of Rule *T*.

Chapter 8

A. 1.

{1}	1.	$(\forall x)(Ax \land Bx \rightarrow Cx)$	*P*
{2}	2.	$(\exists x)(Bx \land \neg Cx)$	*P*

	{3}	3.	$By \land \neg Cy$	y P (for EP)
	{1}	4.	$Ay \land By \to Cy$	1 US $[y\vert x]$
	{1, 3}	5.	$\neg(Ay \land By)$	y 3, 4 T
	{1, 3}	6.	$\neg Ay \lor \neg By$	y 5 T
	{1, 3}	7.	$\neg Ay$	y 3, 6 T
	{1, 3}	8.	$(\exists x)(\neg Ax)$	7 EG $[y\vert x]$
	{1, 2}	9.	$(\exists x)(\neg Ax)$	2, 3, 8 EP $[y\vert x]$

4.	{1}	1.	$(\forall x)(Px \to Qx)$	P
	{2}	2.	$(\exists x)(Rx \land \neg Qx)$	P
	{3}	3.	$(\forall x)(Rx \to Px \lor Sx)$	P
	{4}	4.	$Rz \land \neg Qz$	z P (for EP)
	{1}	5.	$Pz \to Qz$	1 US $[z\vert x]$
	{1, 4}	6.	$\neg Pz$	z, 4, 5 T
	{3}	7.	$Rz \to Pz \lor Sz$	3 US $[z\vert x]$
	{3, 4}	8.	$Pz \lor Sz$	z 4, 7 T
	{1, 3, 4}	9.	Sz	z 6, 8 T
	{1, 3, 4}	10.	$Rz \land Sz$	z 4, 9 T
	{1, 3, 4}	11.	$(\exists x)(Rx \land Sx)$	10 EG $[z\vert x]$
	{1, 2, 3}	12.	$(\exists x)(Rx \land Sx)$	2, 4, 11 EP $[z\vert x]$

8.	{1}	1.	$(\exists y)(\forall x)Rxy$	P
	{2}	2.	$(\forall x)Rxz$	z P (for EP)
	{2}	3.	Rxz	z 2 US $[x\vert x]$
	{2}	4.	$(\exists y)(Rxy)$	3 EG $[z\vert y]$
	{1}	5.	$(\exists y)(Rxy)$	1, 2, 4 EP $[z\vert y]$
	{1}	6.	$(\forall x)(\exists y)(Rxy)$	5 UG

11.	{1}	1.	$(\forall x)(\forall y)(Dx \land Ey \to Fxy)$	P
	{2}	2.	$(\forall x)(\forall y)(Dx \land Fxy \to Gy)$	P
	{3}	3.	$(\exists x)Dx$	P (for CP)
	{4}	4.	Dz	z P (for EP)
	{1}	5.	$Dz \land Ex \to Fzx$	1 US $[z\vert x][x\vert y]$
	{1}	6.	$Dz \to (Ex \to Fzx)$	5 T
	{1, 4}	7.	$Ex \to Fzx$	z, 4, 6 T
	{2}	8.	$Dz \land Fzx \to Gx$	2 US $[z\vert x][x\vert y]$
	{2}	9.	$Dz \to (Fzx \to Gx)$	8 T
	{2, 4}	10.	$Fzx \to Gx$	z, 4 9 T
	{1, 2, 4}	11.	$Ex \to Gx$	7, 10 T
	{1, 2, 3}	12.	$Ex \to Gx$	3, 4, 11 EP $[z\vert x]$
	{1, 2, 3}	13.	$(\forall x)(Ex \to Gx)$	12 UG
	{1, 2}	14.	$(\exists x)Dx \to (\forall x)(Ex \to Gx)$	3, 13 CP

14.	{1}	1.	$(\exists x)[Px \land (\forall y)(Py \land Rxy \to Qya)]$	P
	{2}	2.	$(\exists x)(Px \land \neg Qxa)$	P
	{3}	3.	$(\exists x)(\neg Px \land Qxa)$	P

{4}	4.	$Pu \land (\forall y)(Py \land Ruy \rightarrow Qya)$	u P (for EP)
{5}	5.	$Pv \land \neg Qva$	v P (for EP)
{4}	6.	$(\forall y)(Py \land Ruy \rightarrow Qya)$	u 4 T
{4}	7.	$Pv \land Ruv \rightarrow Qva$	u 6 US $[v \vert y]$
{4, 5}	8.	$\neg(Pv \land Ruv)$	u, v 5, 7 T
{4, 5}	9.	$\neg Pv \lor \neg Ruv$	u, v 8 T
{4, 5}	10.	$\neg Ruv$	u, v 5, 9 T
{4, 5}	11.	$Pu \land Pv \land \neg Ruv$	u, v 4, 5, 10 T
{4, 5}	12.	$(\exists x)(\exists y)(Px \land Py \land \neg Rxy)$	11 EG $[u \vert x][v \vert y]$
{1, 5}	13.	$(\exists x)(\exists y)(Px \land Py \land \neg Rxy)$	1, 4, 12 EP $[u \vert x]$
{1, 2}	14.	$(\exists x)(\exists y)(Px \land Py \land \neg Rxy)$	2, 5, 13 EP $[v \vert x]$

18.

{1}	1.	$(\forall x)(\forall y)(\forall x)(Rxy \land Ryz \rightarrow Rxz)$	P
{2}	2.	$(\forall x)(\forall y)(Rxy \rightarrow Ryx)$	P
{3}	3.	$(\forall x)(\exists y)Rxy$	P (for CP)
{3}	4.	$(\exists y)Rxy$	3 US $[x \vert x]$
{5}	5.	Rxz	x, z P (for $EP(z)$)
{2}	6.	$Rxz \rightarrow Rzx$	2 US $[x \vert x][z \vert y]$
{2, 5}	7.	Rzx	x, z 5, 6 T
{1}	8.	$Ruv \land Rvw \rightarrow Ruw$	1 US $[u \vert x][v \vert y][w \vert z]$
{1}	9.	$(\forall u)(\forall v)(\forall w)(Ruv \land Rvw \rightarrow Ruw)$	8 UG (3)
{1}	10.	$Rxz \land Rzx \rightarrow Rxx$	9 US $[x \vert u][z \vert v][x \vert w]$
{1, 2, 5}	11.	Rxx	x 5, 7, 10 T
{1, 2, 3}	12.	Rxx	4, 5, 11 EP $[z \vert y]$
{1, 2, 3}	13.	$(\forall x)Rxx$	12 UG
{1, 2}	14.	$(\forall x)(\exists y)Rxy \rightarrow (\forall x)Rxx$	3, 13 CP

23.

{1}	1.	$(\forall x)(\forall y)(Px \leftrightarrow Sxy)$	P
{2}	2.	$(\forall x)(\forall y)(Rxy \rightarrow Qy)$	P
{3}	3.	$(\forall x)(Qx \rightarrow Tx)$	P
{4}	4.	$(\exists x)Tx \rightarrow (\forall y)(\forall z)Ryz$	P
{5}	5.	$(\exists y)(Px \land Rxy)$	x P (for CP)
{6}	6.	$Px \land Rxw$	x, w P (for $EP(w)$)
{1}	7.	$Px \leftrightarrow Sxw$	1 US $[x \vert x][w \vert y]$
{1, 6}	8.	Sxw	x, w 6, 7 T
{2}	9.	$Rxw \rightarrow Qw$	2 US $[x \vert x][w \vert y]$
{2, 6}	10.	Qw	w 6, 9 T
{1, 2, 6}	11.	$Qw \land Sxw$	x, w 8, 10 T
{1, 2, 6}	12.	$(\exists z)(Qz \land Sxz)$	x, 11 EG $[w \vert z]$
{1, 2, 5}	13.	$(\exists z)(Qz \land Sxz)$	x, 5, 6, 12 EP $[w \vert y]$
{1, 2}	14.	$(\exists y)(Px \land Rxy) \rightarrow (\exists z)(Qz \land Sxz)$	x 5, 13 CP
{15}	15.	$(\exists z)(Qz \land Sxz)$	x P (for CP)
{16}	16.	$Qv \land Sxv$	x, v P (for $EP(v)$)

{1}	17. $Px \leftrightarrow Sxv$	1 *US* $[x\|x][v\|y]$
{1, 16}	18. Px	x 16, 17 *T*
{3}	19. $Qv \rightarrow Tv$	3 *US* $[v\|x]$
{3, 16}	20. Tv	v 16, 19 *T*
{3, 16}	21. $(\exists x)Tx$	20 *EG* $[v\|x]$
{3, 4, 16}	22. $(\forall y)(\forall z)Ryz$	4, 21 *T*
{3, 4, 16}	23. Rxz	22 *US* $[x\|y][z\|z]$
{1, 3, 4, 16}	24. $Px \wedge Rxz$	x 18, 23 *T*
{1, 3, 4, 16}	25. $(\exists y)(Px \wedge Rxy)$	x 24 *EG* $[z\|y]$
{1, 3, 4, 15}	26. $(\exists y)(Px \wedge Rxy)$	x 15, 16, 25 *EP* $[v\|z]$
{1, 3, 4}	27. $(\exists z)(Qz \wedge Sxz) \rightarrow (\exists y)(Px \wedge Rxy)$	15, 26 *CP*
{1, 2, 3, 4}	28. $(\exists z)(Qz \wedge Sxz) \leftrightarrow (\exists y)(Px \wedge Rxy)$	14, 27 *T*
{1, 2, 3, 4}	29. $(\forall x)[(\exists y)(Px \wedge Rxy) \leftrightarrow (\exists z)(Qz \wedge Sxz)]$	28 *UG*

B. 1.

{1}	1.	$(\forall x)(Cx \rightarrow Mx)$	*P*
{2}	2.	$(\exists x)(Cx \wedge \neg Tx)$	*P*
{3}	3.	$Cz \wedge \neg Tz$	z *P* (for *EP*)
{1}	4.	$Cz \rightarrow Mz$	1 *US* $[z\|x]$
{1, 3}	5.	Mz	z 3, 4 *T*
{1, 3}	6.	$Mz \wedge \neg Tz$	z, 3, 5 *T*
{1, 3}	7.	$(\exists x)(Mx \wedge \neg Tx)$	6 *EG* $[z\|x]$
{1, 2}	8.	$(\exists x)(Mx \wedge \neg Tx)$	2, 3, 7 *EP* $[z\|x]$

5.

{1}	1.	$(\forall x)(Dx \rightarrow Gx) \wedge (\forall x)(Sx \rightarrow Gx)$	*P*
{2}	2.	$(\forall x)(Gx \rightarrow Px)$	*P*
{3}	3.	$(\forall x)(Px \rightarrow \neg Ex)$	*P*
{2}	4.	$Gx \rightarrow Px$	2 *US* $[x\|x]$
{3}	5.	$Px \rightarrow \neg Ex$	3 *US* $[x\|x]$
{2, 3}	6.	$Gx \rightarrow \neg Ex$	4, 5 *T*
{1}	7.	$(\forall x)(Dx \rightarrow Gx)$	1 *T*
{1}	8.	$Dx \rightarrow Gx$	7 *US* $[x\|x]$
{1, 2, 3}	9.	$Dx \rightarrow \neg Ex$	6, 8 *T*
{1, 2, 3}	10.	$(\forall x)(Dx \rightarrow \neg Ex)$	9 *UG*

8.

{1}	1.	$(\forall x)(\forall y)(Sx \wedge Cy \rightarrow Axy)$	*P*
{2}	2.	$(\exists x)(\exists y)(Sx \wedge Ty \wedge Axy)$	*P*
{3}	3.	$(\exists x)(Px \wedge Cx \wedge Tx)$	*P*
{4}	4.	$(\exists y)(Su \wedge Ty \wedge Auy)$	u *P* (for *EP*)
{5}	5.	$Su \wedge Tv \wedge Auv$	u, v *P* (for *EP*(v))
{6}	6.	$Pw \wedge Cw \wedge Tw$	w *P* (for *EP*)
{1}	7.	$Su \wedge Cw \rightarrow Auw$	1 *US* $[u\|x][w\|y]$
{1, 5, 6}	8.	Auw	u, w 5, 6, 7 *T*
{1, 5, 6}	9.	$Su \wedge Pw \wedge Auw$	u, w 5, 6, 8 *T*

{1, 4, 6}	10.	$Su \wedge Pw \wedge Auw$	u, w 4, 5, 9 EP $[v\|y]$
{1, 4, 6}	11.	$(\exists y)(Su \wedge Py \wedge Auy)$	u 10 EG $[w\|y]$
{1, 3, 4}	12.	$(\exists y)(Su \wedge Py \wedge Auy)$	u 3, 6, 11 EP $[w\|x]$
{1, 3, 4}	13.	$(\exists x)(\exists y)(Sx \wedge Py \wedge Axy)$	12 EG $[u\|x]$
{1, 2, 3}	14.	$(\exists x)(\exists y)(Sx \wedge Py \wedge Axy)$	2, 4, 13 EP $[u\|x]$

11.

{1}	1.	$(\forall x)(Px \to Mx)$	P
{2}	2.	$(\forall x)[Ax \wedge (\exists y)$ $(My \wedge Txy) \to Gx]$	P
{3}	3.	$(\exists x)(\exists y)(Ax \wedge Ux \wedge$ $Py \wedge Txy)$	P
{4}	4.	$(\exists y)(Au \wedge Uu \wedge Py \wedge Tuy)$	u P (for EP)
{5}	5.	$Au \wedge Uu \wedge Pv \wedge Tuv$	u, v P (for $EP(v)$)
{2}	6.	$Au \wedge (\exists y)(My \wedge Tuy) \to Gu$	2 US $[u\|x]$
{1}	7.	$Pv \to Mv$	1 US $[v\|x]$
{1, 5}	8.	Mv	v 5, 7 T
{1, 5}	9.	$Mv \wedge Tuv$	u, v 5, 8 T
{1, 5}	10.	$(\exists y)(My \wedge Tuy)$	u 9 EG $[v\|y]$
{1, 5}	11.	$Au \wedge (\exists y)(My \wedge Tuy)$	u 5, 10 T
{1, 2, 5}	12.	Gu	u 6, 11 T
{1, 2, 5}	13.	$Au \wedge Uu \wedge Gu$	u 5, 12 T
{1, 2, 4}	14.	$Au \wedge Uu \wedge Gu$	u 4, 5, 14 EP $[v\|y]$
{1, 2, 4}	15.	$(\exists x)(Ax \wedge Ux \wedge Gx)$	14 EG $[u\|x]$
{1, 2, 3}	16.	$(\exists x)(Ax \wedge Ux \wedge Gx)$	3, 4, 15 EP $[u\|x]$

16.

{1}	1.	$(\forall x)[Sx \to (Ox \leftrightarrow (\forall y)[Pyx \to$ $(\exists z)(Nzy \wedge Txz)])]$	P
{2}	2.	$(\forall x)(Sx \to (Cx \leftrightarrow Of(x)))$	P
{3}	3.	$(\forall x)(Sx \to Sf(x))$	P
{4}	4.	$Sx \wedge (\forall y)[Pyf(x) \to$ $(\exists z)(Nzy \wedge Tf(x)z)]$	x P (for CP)
{3}	5.	$Sx \to Sf(x)$	3 US $[x\|x]$
{3, 4}	6.	$Sf(x)$	x 4, 5 T
{1}	7.	$Sf(x) \to (Of(x) \leftrightarrow (\forall y)(Pyf(x) \to$ $(\exists z)(Nzy \wedge Tf(x)z)))$	1 US $[f(x)\|x]$
{1, 3, 4}	8.	$Of(x) \leftrightarrow (\forall y)(Pyf(x) \to$ $(\exists x)(Nzy \wedge Tf(x)z))$	x 6, 7 T
{1, 3, 4}	9.	$Of(x)$	x 4, 8 T
{2}	10.	$Sx \to (Cx \leftrightarrow Of(x))$	2 US $[x\|x]$
{2, 4}	11.	$Cx \leftrightarrow Of(x)$	x 4, 10 T
{1, 2, 3, 4}	12.	Cx	x 9, 11 T
{1, 2, 3}	13.	$Sx \wedge (\forall y)[Pyf(x) \to$ $(\exists z)(Nzy \wedge Tf(x)z)] \to Cx$	4, 12 CP
{1, 2, 3}	14.	$(\forall x)(Sx \wedge (\forall y)[Pyf(x) \to$ $(\exists z)(Nzy \wedge Tf(x)z)] \to Cx)$	13 UG

C. 1.

{1}	1.	$(\forall x)(Dx \wedge Sx \rightarrow Px)$	*P*
{2}	2.	$(\exists x)(Dx \wedge \neg Px)$	*P*
{3}	3.	$Dy \wedge \neg Py$	*y P* (for *EP*)
{1}	4.	$Dy \wedge Sy \rightarrow Py$	1 *US* [y|x]
{1}	5.	$Dy \rightarrow (Sy \rightarrow Py)$	4 *T*
{1, 3}	6.	$Sy \rightarrow Py$	*y* 3, 5 *T*
{1, 3}	7.	$\neg Sy$	*y* 3, 6 *T*
{1, 3}	8.	$Dy \wedge \neg Sy$	*y* 3, 7 *T*
{1, 3}	9.	$(\exists x)(Dx \wedge \neg Sx)$	8 *EG* [y|x]
{1, 2}	10.	$(\exists x)(Dx \wedge \neg Sx)$	2, 3, 9 *EP* [y|x]

5.

{1}	1.	$(\exists x)(Ox \wedge (\forall y)(Ey \rightarrow Dxy))$	*P*
{2}	2.	Ex	*x P* (for *CP*)
{3}	3.	$Oz \wedge (\forall y)(Ey \rightarrow Dzy)$	*z P* (for *EP*)
{3}	4.	$(\forall y)(Ey \rightarrow Dzy)$	*z* 3 *T*
{3}	5.	$Ex \rightarrow Dzx$	*z* 4 *US* [x|y]
{2, 3}	6.	Dzx	*x, z* 2, 5 *T*
{2, 3}	7.	$Oz \wedge Dzx$	*x, z* 3, 6 *T*
{2, 3}	8.	$(\exists y)(Oy \wedge Dyx)$	*x* 7 *EG* [z|y]
{1, 2}	9.	$(\exists y)(Oy \wedge Dyx)$	*x* 1, 3, 8 *EP* [z|x]
{1}	10.	$Ex \rightarrow (\exists y)(Oy \wedge Dyx)$	2, 9 *CP*
{1}	11.	$(\forall x)(Ex \rightarrow (\exists y)(Oy \wedge Dyx))$	10 *UG*

9.

{1}	1.	$(\forall x)(\forall y)(Lx \wedge Ly \wedge Pxy \wedge \neg Exy \rightarrow \neg Ixy)$	*P*
{2}	2.	$(\forall x)(\forall y)(Lx \wedge Ly \wedge Pxy \rightarrow Nxy \vee Axy)$	*P*
{3}	3.	$(\forall x)(\forall y)(Lx \wedge Ly \wedge Axy \rightarrow Ixy)$	*P*
{4}	4.	$Lx \wedge Ly \wedge Pxy \wedge \neg Exy$	*x, y P* (for *CP*)
{1}	5.	$Lx \wedge Ly \wedge Pxy \wedge \neg Exy \rightarrow \neg Ixy$	1 *US* [x|x][y|y]
{1, 4}	6.	$\neg Ixy$	*x, y* 4, 5 *T*
{2}	7.	$Lx \wedge Ly \wedge Pxy \rightarrow Nxy \vee Axy$	2 *US* [x|x][y|y]
{2, 4}	8.	$Nxy \vee Axy$	*x, y* 4, 7 *T*
{3}	9.	$Lx \wedge Ly \wedge Axy \rightarrow Ixy$	3 *US* [x|x][y|y]
{3, 4}	10.	$Axy \rightarrow Ixy$	*x, y* 4, 9 *T*
{1, 3, 4}	11.	$\neg Axy$	*x, y* 6, 10 *T*
{1, 2, 3, 4}	12.	Nxy	*x, y* 8, 11 *T*
{1, 2, 3}	13.	$Lx \wedge Ly \wedge Pxy \wedge \neg Exy \rightarrow Nxy$	4, 12 *CP*
{1, 2, 3}	14.	$(\forall x)(\forall y)(Lx \wedge Ly \wedge Pxy \wedge \neg Exy \rightarrow Nxy)$	13 *UG* (2)

D. 1.

{1}	1.	$(\forall x)(Ax \wedge Bx \rightarrow Cx)$	*P*
{2}	2.	$(\exists x)(Ax \wedge \neg Cx)$	*P*

{3}	3.	$Ay \wedge \neg Cy$	y *P* (for *EP*)
{1}	4.	$Ay \wedge By \rightarrow Cy$	1 *US* $[y\vert x]$
{3}	5.	$\neg Cy$	y 3 *T*
{1, 3}	6.	$\neg(Ay \wedge By)$	y 4, 5 *T*
{1, 3}	7.	$\neg Ay \vee \neg By$	y 6 *T*
{3}	8.	Ay	y 3 *T*
{1, 3}	9.	$\neg By$	y 7, 8 *T*
{1, 3}	10.	$Ay \wedge \neg By$	y 8, 9 *T*
{1, 3}	11.	$(\exists x)(Ax \wedge \neg Bx)$	10 *EG* $[y\vert x]$
{1, 2}	12.	$(\exists x)(Ax \wedge \neg Bx)$	2, 3, 11 *EP* $[y\vert x]$

4.

{1}	1.	$(\forall x)[Px \rightarrow (\exists y)(Qy \wedge Rxy)]$	*P*
{2}	2.	$(\forall x)[Px \rightarrow (\forall y)(Rxy \rightarrow Sx)]$	*P*
{3}	3.	Px	x *P* (for *CP*)
{1}	4.	$Px \rightarrow (\exists y)(Qy \wedge Rxy)$	1 *US* $[x\vert x]$
{2}	5.	$Px \rightarrow (\forall y)(Rxy \rightarrow Sx)$	2 *US* $[x\vert x]$
{1, 3}	6.	$(\exists y)(Qy \wedge Rxy)$	x 3, 4 *T*
{2, 3}	7.	$(\forall y)(Rxy \rightarrow Sx)$	x 3, 5 *T*
{8}	8.	$Qz \wedge Rxz$	x, z *P* (for *EP*(z))
{2, 3}	9.	$Rxz \rightarrow Sx$	x 7 *US* $[z\vert y]$
{2, 3, 8}	10.	Sx	x 8, 9 *T*
{1, 2, 3}	11.	Sx	x 6, 8, 10 *EP* $[z\vert y]$
{1, 2}	12.	$Px \rightarrow Sx$	3, 11 *CP*
{1, 2}	13.	$(\forall x)(Px \rightarrow Sx)$	12 *UG*

E. **1.** z is flagged on line 3.
Line 4: Incorrect use of Rule *US*.

4. z is flagged on lines 4, 6, 8, and 9.
Line 9: Incorrect use of Rule *T*.
Line 12: Incorrect use of Rule *EP*.

10. x is flagged on lines 2 and 6; y is flagged on line 3.
Line 8: Incorrect use of Rule *EP*
Line 10: Incorrect use of Rule *EG*.

Chapter 9

A. **1.** $(\forall x)(Cx \rightarrow (\forall y)(Dy \rightarrow Lxy)), (\forall x)(Dx \rightarrow Fx) \therefore (\forall x)(Cx \rightarrow Fx)$.
Not valid: $\mathbf{S} = \langle \mathbf{U, C, D, F, L} \rangle$, $\mathbf{U} = \{\mathbf{a}\}$.

	C	**D**	**F**
a	*T*	*F*	*F*

4. $(\forall x)[Px \wedge Hx \rightarrow (\exists y)(Py \wedge Lxy)], Pd \wedge \neg Hd \wedge Lde \wedge Pe \wedge He \therefore Led$.
Not valid: $\mathbf{S} = \langle \mathbf{U, P, H, L, e, d} \rangle$, $\mathbf{U} = \{\mathbf{e, d}\}$.

	P	**H**
e	*T*	*T*
d	*T*	*F*

L	**e**	**d**
e	*T*	*F*
d	*T*	

7. $(\forall x)[Px \wedge Ox \rightarrow Vx \vee (\exists y)(Uy \wedge Axy)]$, $(\exists x)(Px \wedge Ox \wedge \neg Vx)$ $\therefore$ $(\exists x)Ux$.
Valid.

{1}	1.	$(\forall x)[Px \wedge Ox \rightarrow Vx \vee (\exists y)(Uy \wedge Axy)]$	P
{2}	2.	$(\exists x)(Px \wedge Ox \wedge \neg Vx)$	P
{3}	3.	$Pz \wedge Oz \wedge \neg Vz$	z P (for EP)
{1}	4.	$Pz \wedge Oz \rightarrow Vz \vee (\exists y)(Uy \wedge Azy)$	1 US $[z\|x]$
{1, 3}	5.	$Vz \vee (\exists y)(Uy \wedge Azy)$	z, 3, 4 T
{1, 3}	6.	$(\exists y)(Uy \wedge Azy)$	z 3, 5 T
{7}	7.	$Uw \wedge Azw$	z, w P (for $EP(w)$)
{7}	8.	Uw	w 7 T
{7}	9.	$(\exists x)Ux$	8 EG $[w\|x]$
{1, 3}	10.	$(\exists x)Ux$	6, 7, 9 EP $[w\|y]$
{1, 2}	11.	$(\exists x)Ux$	2, 3, 10 EP $[z\|x]$

11. $(\exists x)[Sx \wedge (\forall y)(Py \wedge Iyx \rightarrow By \vee Ay)]$, $(\forall x)[Sx \rightarrow (\exists y)(Py \wedge Iyx \wedge \neg By)]$.
(a) $\therefore$ $(\exists x)(Px \wedge Ax)$
Valid.

{1}	1.	$(\exists x)[Sx \wedge (\forall y)(Py \wedge Iyx \rightarrow By \vee Ay)]$	P
{2}	2.	$(\forall x)[Sx \rightarrow (\exists y)(Py \wedge Iyx \wedge \neg By)]$	P
{3}	3.	$Su \wedge (\forall y)(Py \wedge Iyu \rightarrow By \vee Ay)$	uP (for EP)
{2}	4.	$Su \rightarrow (\exists y)(Py \vee Iyu \wedge \neg By)$	2US $[u\|x]$
{2, 3}	5.	$(\exists y)(Py \wedge Iyu \wedge \neg By)$	u 3, 4T
{6}	6.	$Pv \wedge Ivu \wedge \neg Bv$	u, vP (for $EP(v)$)
{3}	7.	$(\forall y)(Py \wedge Iyu \rightarrow By \vee Ay)$	u 3T
{3}	8.	$Pv \wedge Ivw \rightarrow Bv \vee Av$	u 7US $[v\|y]$
{3, 6}	9.	$Bv \vee Av$	v 6, 8T
{3, 6}	10.	Av	v 6, 9T
{3, 6}	11.	$Pv \wedge Av$	v 6, 10T
{3, 6}	12.	$(\exists x)(Px \wedge Ax)$	11EG $[v\|x]$
{2, 3}	13.	$(\exists x)(Px \wedge Ax)$	5, 6, 12EP $[v\|y]$
{1, 2}	14.	$(\exists x)(Px \wedge Ax)$	1, 3, 13EP $[u\|x]$

(b) $\therefore$ $(\exists x)Sx \rightarrow (\exists x)(Px \wedge Ax)$.
Valid.
Use lines 1 through 14 in (a), followed by:

{1, 2}	15.	$(\exists x)Sx \rightarrow (\exists x)(Px \wedge Ax)$	14 T

(c) $\therefore$ $(\forall x)(\neg Sx) \rightarrow (\forall x)(Px \rightarrow \neg Ax)$.
Valid.

{1}	1.	$(\exists x)[Sx \wedge (\forall y)(Py \wedge Iyx \rightarrow By \vee Ay)]$	P
{2}	2.	$(\forall x)[Sx \rightarrow (\exists y)(Py \wedge Iyx \wedge \neg By)]$	P
{3}	3.	$(\forall x)(\neg Sx)$	P (for CP)
{4}	4.	$Su \wedge (\forall y)(Py \wedge Iyu \rightarrow By \vee Ay)$	u P (for EP)
{3}	5.	$\neg Su$	3 US $[u\|x]$
{3, 4}	6.	$Su \wedge \neg Su$	u 4, 5 T
{3, 4}	7.	$Q \wedge \neg Q$	6 T
{1, 3}	8.	$Q \wedge \neg Q$	1, 4, 7 EP $[u\|x]$
{1, 3}	9.	$(\forall x)(Px \rightarrow \neg Ax)$	8 T
{1}	10.	$(\forall x)(\neg Sx) \rightarrow (\forall x)(Px \rightarrow \neg Ax)$	3, 9 CP

15. $(\forall x)(\forall y)(\forall z)(Gx \wedge Py \wedge Ez \rightarrow Sxyz)$, $(\forall x)[(\exists y)(\exists z)(Py \wedge Ez \wedge Sxyz) \rightarrow Dx]$, $(\forall y)(\forall z)(Py \wedge Ez \rightarrow Smyz)$ $\therefore$ $Gm \rightarrow Dm$.
Not valid: $\mathbf{T} = \langle \mathbf{U, G, P, E, D, S, m} \rangle$, $\mathbf{U} = \{\mathbf{m}\}$.

	G	**D**	**P**
m	T	F	F

B. 1. Not valid: $\mathbf{S} = \langle \mathbf{U, A, B, C, D} \rangle$, $\mathbf{U} = \{\mathbf{a, b}\}$.

	A	**B**	**C**	**D**
a	T	T	T	T
b	F	F	F	F

7. Valid.

{1}	1.	$(\forall x)[Ax \rightarrow (\forall y)(\neg By \rightarrow Cxy)]$	P
{2}	2.	$(\exists x)[Ax \wedge (\forall y)(Dy \rightarrow \neg Cxy)]$	P
{3}	3.	Dx	x P (for CP)
{4}	4.	$Az \wedge (\forall y)(Dy \rightarrow \neg Czy)$	z P (for EP)
{4}	5.	$(\forall y)(Dy \rightarrow \neg Czy)$	z 4 T
{4}	6.	$Dx \rightarrow \neg Czx$	z 5 US $[x\|y]$
{3, 4}	7.	$\neg Czx$	x, z 3, 6 T
{1}	8.	$Az \rightarrow (\forall y)(\neg By \rightarrow Czy)$	1 US $[z\|x]$
{1, 4}	9.	$(\forall y)(\neg By \rightarrow Czy)$	z 4, 8 T
{1, 4}	10.	$\neg Bx \rightarrow Czx$	z 9 US $[x\|y]$
{1, 3, 4}	11.	Bx	x 7, 10 T
{1, 2, 3}	12.	Bx	x 2, 4, 11 EP $[z\|x]$
{1, 2}	13.	$Dx \rightarrow Bx$	3, 12 CP
{1, 2}	14.	$(\forall x)(Dx \rightarrow Bx)$	13 UG

10. Not valid: $\mathbf{S} = \langle \mathbf{U, A, B, C} \rangle$, $\mathbf{U} = \{\mathbf{a}\}$.

	A	**B**	**C**
a	T	F	T

12. Not valid: $\mathbf{S} = \langle \mathbf{U}, \mathbf{R} \rangle$, $\mathbf{U} = \{\mathbf{a}, \mathbf{b}\}$.

R	**a**	**b**
a	*T*	*T*
b	*F*	*T*

C. 3. Gx : x is a stag.
Ax : x is an animal.
Hx : x has horns.
$(\forall x)(Gx \rightarrow Hx)$, $(\forall x)(Gx \rightarrow Ax)$, $(\exists x)(Ax \wedge \neg Hx)$.

$\mathbf{S} = \langle \mathbf{U}, \mathbf{G}, \mathbf{H}, \mathbf{A} \rangle$, $\mathbf{U} = \{\mathbf{a}\}$.

	G	**A**	**H**
a	*F*	*T*	*F*

8. Lxy: x is looking at y.
$(\forall x)(\exists y)Lxy$, $(\exists x)(\exists y)\neg Lyx$.

$\mathbf{S} = \langle \mathbf{U}, \mathbf{L} \rangle$, $\mathbf{U} = \{\mathbf{a}, \mathbf{b}\}$.

L	**a**	**b**
a	*T*	*F*
b	*T*	

11. Px : x is a problem.
Mx : x is a mathematician.
Sxy : x solves y.
$(\exists x)[Px \wedge (\forall y)(Syx \rightarrow My)]$, $(\exists x)[Px \wedge (\forall y)(My \rightarrow \neg Syx)]$.

$\mathbf{T} = \langle \mathbf{U}, \mathbf{P}, \mathbf{M}, \mathbf{S} \rangle$, $\mathbf{U} = \{\mathbf{a}\}$.

	P	**M**
a	*T*	*F*

Saa is false.

15. Dx : x is a student.
$Axyz$: x was accompanied by y to z.
Fxy : x is y's father.
Mx : x is a meeting.
$(\forall x)(\forall y)(\forall z)(Dx \wedge Fyx \wedge Mz \rightarrow \neg Axyz)$,
$(\forall x)(\forall y)(\forall z)(Dx \wedge Mz \wedge Axyz \rightarrow Fyx)$,
$(\exists x)(\exists y)(Dx \wedge Fyx) \wedge (\exists x)(\forall y)(Dx \wedge \neg Fyx)$.

$\mathbf{S} = \langle \mathbf{U}, \mathbf{D}, \mathbf{A}, \mathbf{F}, \mathbf{M} \rangle$, $\mathbf{U} = \{\mathbf{a}, \mathbf{b}\}$.

	D	**M**
a	*T*	*F*
b	*T*	*F*

F	**a**	**b**
a	*T*	*F*
b		*F*

17. Dx : x is a student.
Ex : x went to the meeting.
Axy : x was accompanied by y.
Fxy : x is y's father.
Mxy : x is y's mother.
a : Alex, b : Betty.
$(\forall x)[Dx \wedge Ex \to (\exists y)(Fyx \wedge Axy)]$, $Da \wedge Ea \wedge (\exists x)(Mxa \wedge Aax)$, $Db \wedge (\forall x)(Fxb \to \neg Abx)$, $(\exists x)(Ex \wedge (\forall y)(\neg Fyx))$.

$$\mathbf{S} = \langle \mathbf{U}, \mathbf{D}, \mathbf{E}, \mathbf{F}, \mathbf{A}, \mathbf{M}, \mathbf{a}, \mathbf{b} \rangle, \qquad \mathbf{U} = \{\mathbf{a}, \mathbf{b}, \mathbf{c}\}.$$

	D	**E**
a	*T*	*T*
b	*T*	*F*
c	*F*	*T*

Maa, Faa, Aaa all true, **Fxb, Fxc** false for all **x**.

D. **3.** Consistent: $\mathbf{S} = \langle \mathbf{U}, \mathbf{P}, \mathbf{Q}, \mathbf{R} \rangle$, $\mathbf{U} = \{\mathbf{a}, \mathbf{b}\}$.

	P	**Q**	**R**
a	*T*	*F*	*F*
b	*F*	*T*	*F*

5. Inconsistent.

{1}	1.	$(\exists x)Px \to (\forall x)Qx$	*P*
{2}	2.	$Qa \wedge (\exists x)(Px \wedge \neg Qx)$	*P*
{2}	3.	$(\exists x)(Px \wedge \neg Qx)$	2 *T*
{4}	4.	$Py \wedge \neg Qy$	*y P* (for *EP*)
{4}	5.	Py	*y* 4 *T*
{4}	6.	$(\exists x)Px$	5 *EG* $[y\|x]$
{1, 4}	7.	$(\forall x)Qx$	1, 6 *T*
{1, 4}	8.	Qy	7 *US* $[y\|x]$
{1, 4}	9.	$Qy \wedge \neg Qy$	*y* 4, 8 *T*
{1, 4}	10.	$A \wedge \neg A$	9 *T*
{1, 2}	11.	$A \wedge \neg A$	3, 4 10 *EP* $[y\|x]$

9. Inconsistent.

{1}	1.	$(\forall x)[(\forall y)(Py \to Qx]$	*P*
{2}	2.	$(\forall x)Px$	*P*
{3}	3.	$(\exists y)\neg Qy$	*P*
{4}	4.	$\neg Qz$	*z P* (for *EP*)
{1}	5.	$(\forall y)Py \to Qz$	1 *US* $[z\|x]$
{2}	6.	Py	2 *US* $[y\|x]$
{2}	7.	$(\forall y)Py$	6 *UG*
{1, 4}	8.	$\neg(\forall y)Py$	4, 5 *T*
{1, 2, 4}	9.	$(\forall y)Py \wedge \neg(\forall y)Py$	7, 8 *T*
{1, 2, 3}	10.	$(\forall y)Py \wedge \neg(\forall y)Py$	3, 4, 9 *EP* $[z\|y]$

13. Consistent: $\mathbf{S} = \langle \mathbf{U}, \mathbf{P}, \mathbf{Q}, \mathbf{R} \rangle$, $\mathbf{U} = \{\mathbf{a}, \mathbf{b}\}$.

	P	**Q**
a	*T*	*T*
b	*T*	*T*

R	**a**	**b**
a	*T*	*F*
b	*F*	*T*

16. Consistent: $\mathbf{S} = \langle \mathbf{U}, \mathbf{R}, \mathbf{f} \rangle$, $\mathbf{U} = \mathbb{R}$, $\mathbf{f}(\mathbf{x}) = \mathbf{x}$, $\mathbf{Rxy} : \mathbf{x} = \mathbf{y}$.

E. 1. No, by the Soundness Theorem.
4. Apply Rule *UG*.
7. (a) All the formulas of Γ are true in the interpretation structure in which **P** is empty. Thus, Γ is consistent.

Chapter 10

A. 1.

{1}	1.	$(\forall x)(\varphi \wedge \psi)$	*P* (for *CP*)
{1}	2.	$\varphi \wedge \psi$	1 *US* $[x\|x]$
{1}	3.	ψ	2 *T*
{1}	4.	$(\forall x)\psi$	3 *UG*
{1}	5.	$\varphi \wedge (\forall x)\psi$	2, 4 *T*
	6.	$(\forall x)(\varphi \wedge \psi) \rightarrow \varphi \wedge (\forall x)\psi$	1, 5 *CP*
{7}	7.	$\varphi \wedge (\forall x)\psi$	*P* (for *CP*)
{7}	8.	$(\forall x)\psi$	7 *T*
{7}	9.	ψ	8 *US* $[x\|x]$
{7}	10.	$\varphi \wedge \psi$	7, 9 *T*
{7}	11.	$(\forall x)(\varphi \wedge \psi)$	10 *UG*
	12.	$\varphi \wedge (\forall x)\psi \rightarrow (\forall x)(\varphi \wedge \psi)$	7, 11 *CP*
	13.	$(\forall x)(\varphi \wedge \psi) \leftrightarrow \varphi \wedge (\forall x)\psi$	6, 12 *T*

B. 1. Logical Theorem.

{1}	1.	$(\exists x)(Px \wedge Qx)$	*P* (for *CP*)
{2}	2.	$Pu \wedge Qu$	*u P* (for *EP*)
{2}	3.	Pu	*u* 2 *T*
{2}	4.	$(\exists y)Py$	4 *EG* $[u\|y]$
{2}	5.	Qu	*u* 2 *T*
{2}	6.	$(\exists z)Qz$	5 *EG* $[u\|x]$
{2}	7.	$(\exists y)Py \wedge (\exists z)Qz$	4, 6 *T*
{1}	8.	$(\exists y)Py \wedge (\exists z)Qz$	1, 2, 7 *EP* $[u\|x]$
	9.	$(\exists x)(Px \wedge Qx) \rightarrow [(\exists y)Py \wedge (\exists z)Qz]$	1, 8 *CP*

3. Not a Logical Theorem: $\mathbf{S} = \langle \mathbf{U}, \mathbf{P}, \mathbf{Q} \rangle$, $\mathbf{U} = \{\mathbf{a}, \mathbf{b}\}$.

	P	**Q**
a	*T*	*F*
b	*F*	

8. Logical Theorem.

{1}	1.	$(\forall x)(\exists y)(Px \wedge Qy)$	P (for CP)
{1}	2.	$(\forall x)(Px \wedge (\exists y)Qy)$	1 LE (13)
{1}	3.	$(\forall x)Px \wedge (\exists y)Qy$	2 LT (12, T)
{1}	4.	$(\exists y)((\forall x)Px \wedge Qy)$	3 LT (13)
{1}	5.	$(\exists y)(\forall x)(Px \wedge Qy)$	4 LE (12, T)
	6.	$(\forall x)(\exists y)(Px \wedge Qy) \rightarrow (\exists y)(\forall x)(Px \wedge Qy)$	1, 5 CP

12. Logical Theorem.

{1}	1.	$(\forall x)Px \rightarrow (\exists y)\neg Qy$	P (for CP)
{1}	2.	$\neg(\exists y)\neg Qy \rightarrow \neg(\forall x)Px$	1 T
{1}	3.	$(\forall y)Qy \rightarrow (\exists x)\neg Px$	2 LE (2, 3)
{1}	4.	$(\forall x)Qx \rightarrow (\exists y)\neg Py$	3 LE (7, 8)
	5.	$[(\forall x)Px \rightarrow (\exists y)\neg Qy] \rightarrow$ $[(\forall x)Qx \rightarrow (\exists y)\neg Py]$	1, 4 CP

17. Not a Logical Theorem: **S** = ⟨**U, R**⟩, **U** = {**a, b**}.

R	**a**	**b**
a	*F*	*F*
b	*T*	*T*

22. Not a Logical Theorem: **S** = ⟨**U, P, Q**⟩, **U** = {**a, b**}.

	P	**Q**
a	*T*	*T*
b	*T*	*F*

C. We shall omit part (b) because it is the same as (a) with a negation sign in front.

1. (a) $(\exists x)(Ax \wedge Px)$.
   (c) $(\forall x)(\neg Ax \vee \neg Px)$ or $(\forall x)(Ax \rightarrow \neg Px)$.
   (d) No animals are pets.
5. (a) $(\forall x)(Qx \rightarrow Rx) \rightarrow (\forall y)(Iy \rightarrow Ry)$.
   (c) $(\forall x)(Qx \rightarrow Rx) \wedge (\exists y)(Iy \wedge \neg Ry)$.
   (d) All rational numbers are real numbers, and some integers are not real numbers.
9. (a) $(\exists x)(Px \wedge Oxe) \rightarrow (\forall x)(Px \rightarrow Oxe)$.
   (c) $(\exists x)(Px \wedge Oxe) \wedge (\exists x)(Px \wedge \neg Oxe)$.
   (d) Someone is older than George, and someone is not older than George.
13. (a) $(\forall x)(\forall y)(Sx \wedge Py \rightarrow Lxy)$.
   (c) $(\exists x)(\exists y)(Sx \wedge Py \wedge \neg Lxy)$.
   (d) Some student dislikes some professor.
17. (a) $(\exists x)(\exists y)(Ix \wedge Iy \wedge (\exists z)(Iz \wedge Bxzy))$.
   (c) $(\forall x)(\forall y)(Ix \wedge Iy \rightarrow (\forall z)(Iz \rightarrow \neg Bxzy))$.
   (d) No two integers have an integer between them.

20. (a) $(\forall x)(\forall y)(x \neq y \rightarrow [(Ex \wedge Oy) \vee (Ey \wedge Ox)])$.
    (c) $(\exists x)(\exists y)(x \neq y \wedge [(Ex \wedge Ey) \vee (Ox \wedge Oy)])$.
    (d) There are two distinct integers such that either both are even or both are odd.
23. (a) $(\forall x)(\forall y)(Ox \wedge Ey \rightarrow Ox+y)$.
    (c) $(\exists x)(\exists y)(Ox \wedge Ey \wedge \neg Ox+y)$.
    (d) There is an odd and an even integer such that their sum is not odd.

D. 1.

{1}	1.	$(\forall x)(Ax \rightarrow Bx)$	P
{2}	2.	$\neg(\exists x)(Ax \wedge Bx)$	P
{2}	3.	$(\forall x)(Ax \rightarrow \neg Bx)$	2 *LT* (6)
{1}	4.	$Ax \rightarrow Bx$	1 *US* $[x \vert x]$
{2}	5.	$Ax \rightarrow \neg Bx$	3 *US* $[x \vert x]$
{1, 2}	6.	$\neg Ax$	4, 5 *T*
{1, 2}	7.	$(\forall x)\neg Ax$	6 *UG*

5.

{1}	1.	$\neg(\exists x)(Px \vee Qx) \vee (\exists x)Rx$	P
{1}	2.	$(\exists x)(Px \vee Qx) \rightarrow (\exists x)Rx$	1 *T*
{1}	3.	$(\exists x)(Px \vee Qx) \rightarrow (\exists y)Ry$	2 *LE* (8)
{1}	4.	$(\forall x)(Px \vee Qx \rightarrow (\exists y)Ry)$	3 *LT* (18)
{1}	5.	$(\forall x)(\exists y)(Px \vee Qx \rightarrow Ry)$	4 *LE* (17)

9.

{1}	1.	$(\exists x)(Ax \wedge Bx \wedge Cx)$	P
{2}	2.	$\neg(\exists x)(Ax \wedge Bx \wedge Dx)$	P
{3}	3.	$Az \wedge Bz \wedge Cz$	z P (for *EP*)
{2}	4.	$(\forall x)(Ax \wedge Bx \rightarrow \neg Dx)$	2 *LT* (6)
{2}	5.	$Az \wedge Bz \rightarrow \neg Dz$	4 *US* $[z \vert x]$
{2, 3}	6.	$\neg Dz$	z, 3, 5 *T*
{2, 3}	7.	$Bz \wedge Cz \wedge \neg Dz$	z, 3, 6 *T*
{2, 3}	8.	$(\exists x)(Bx \wedge Cx \wedge \neg Dx)$	7 *EG* $[z \vert x]$
{1, 2}	9.	$(\exists x)(Bx \wedge Cx \wedge \neg Dx)$	1, 3, 8 *EP* $[z \vert x]$

12.

{1}	1.	$(\forall x)(Jx \rightarrow Kx) \rightarrow (\exists x)Lx \vee (\exists x)Mx$	P
{2}	2.	$\neg(\exists x)Mx$	P
{3}	3.	$\neg(\exists x)(Jx \wedge \neg Kx)$	P (for *CP*)
{3}	4.	$(\forall x)(Jx \rightarrow Kx)$	3 *LT* (6)
{1, 3}	5.	$(\exists x)Lx \vee (\exists x)Mx$	1, 4 *T*
{1, 2, 3}	6.	$(\exists x)Lx$	2, 5 *T*
{1, 2}	7.	$\neg(\exists x)(Jx \wedge \neg Kx) \rightarrow (\exists x)Lx$	3, 6 *CP*
{1, 2}	8.	$(\exists x)Lx \vee (\exists x)(Jx \wedge \neg Kx)$	7 *T*

16.

{1}	1.	$(\forall x)[Px \wedge (\exists y)(Qy \wedge Rxy) \rightarrow$ $(\exists z)(Sz \wedge Txz)]$	P
{2}	2.	$(\forall x)(\forall y)(Px \rightarrow Rxy)$	P
{3}	3.	$(\exists x)Px$	P
{4}	4.	$(\forall x)(\forall y)(Sy \rightarrow \neg Txy)$	P
{5}	5.	Pu	u P (for *EP*)
{4}	6.	$(\forall y)(Sy \rightarrow \neg Tuy)$	4 *US* $[u \vert x]$

{4}	7.	$\neg(\exists y)(Sy \wedge Tuy)$	6 *LT* (6)
{4}	8.	$\neg(\exists z)(Sz \wedge Tuz)$	7 *LE* (8)
{1}	9.	$Pu \wedge (\exists y)(Qy \wedge Ruz) \rightarrow (\exists z)(Sz \wedge Tuz)$	1 *US* [$u\|x$]
{1, 4}	10.	$\neg Pu \vee \neg(\exists y)(Qy \wedge Ruy)$	8, 9 *T*
{1, 4, 5}	11.	$\neg(\exists y)(Qy \wedge Ruy)$	*u* 5, 10 *T*
{1, 4, 5}	12.	$(\forall y)(Qy \rightarrow \neg Ruy)$	*u* 11 *LT* (6)
{1, 4, 5}	13.	$(\exists x)(\forall y)(Qy \rightarrow \neg Rxy)$	12 *EG* [$u\|x$]
{1, 3, 4}	14.	$(\exists x)(\forall y)(Qy \rightarrow \neg Rxy)$	3, 5, 13 *EP* [$u\|x$]

E. In this section, we shall omit the proofs of validity.

1. $(\forall x)(Gx \rightarrow Ix), \neg(\forall x)(Sx \rightarrow \neg Ix) \therefore (\exists x)(Sx \wedge Gx)$.
Invalid: $\mathbf{T} = \langle \mathbf{U, G, I, S} \rangle$, $\mathbf{U} = \{\mathbf{a}\}$.

	G	**I**	**S**
a	*F*	*T*	*T*

4. $\neg(\exists x)(Vx \wedge Fx), (\forall x)(Px \rightarrow Vx) \therefore \neg(\exists x)(Px \wedge Fx)$.
Valid.

8. $(\forall x)[Sx \wedge (\exists y)(Cy \wedge Axy) \rightarrow Hx], So \wedge \neg Ho \therefore \neg(\exists x)Cx$.
Invalid: $\mathbf{T} = \langle \mathbf{U, S, C, H, A, o} \rangle$, $\mathbf{U} = \{\mathbf{o}\}$.

	S	**C**	**H**
o	*T*	*T*	*F*

Aoo is false.

11. $(\forall x)[Sx \wedge (\exists y)(Cy \wedge Fxy) \rightarrow (\exists z)(Iz \wedge Axz)], So \wedge (\forall x)(Ix \rightarrow \neg Aox)$
$\therefore (\forall x)(Cx \rightarrow \neg Fox)$.
Valid.

F. We shall give the steps in finding a prenex normal form, but we shall omit the reasons.

1. $(\exists x)Px \rightarrow (\exists x)Qx$.
$(\exists x)Px \rightarrow (\exists y)Qy$.
$(\forall x)[Px \rightarrow (\exists y)Qy]$.
$(\forall x)(\exists y)[Px \rightarrow Qy]$.

4. $(\exists x)(Ax \wedge Bxy) \wedge (\exists y)(Cx \wedge Dxy)$.
$(\exists u)(Au \wedge Buy) \wedge (\exists y)(Cx \wedge Dxy)$.
$(\exists u)(Au \wedge Buy) \wedge (\exists v)(Cx \wedge Dxv)$.
$(\exists u)[(Au \wedge Buy) \wedge (\exists v)(Cx \wedge Dxv)]$.
$(\exists u)(\exists v)[(Au \wedge Buy) \wedge (Cx \wedge Dxv)]$.

8. $(\forall x)(\exists y)Rxy \rightarrow (\exists y)(\forall x)Rxy$.
$(\forall x)(\exists y)Rxy \rightarrow (\exists u)(\forall v)Rvu$.
$(\exists x)[(\exists y)Rxy \rightarrow (\exists u)(\forall v)Rvu]$.
$(\exists x)(\forall y)[Rxy \rightarrow (\exists u)(\forall v)Rvu]$.
$(\exists x)(\forall y)(\exists u)(\forall v)(Rxy \rightarrow Rvu)$.

11. $(\forall x)Jx \leftrightarrow (\exists x)Kx$
    $((\forall x)Jx \rightarrow (\exists x)Kx) \land ((\exists x)Kx \rightarrow (\forall x)Jx).$
    $((\forall x)Jx \rightarrow (\exists y)Ky) \land ((\exists z)Kz \rightarrow (\forall w)Jw).$
    $(\exists x)[Jx \rightarrow (\exists y)Ky] \land (\forall z)[Kz \rightarrow (\forall w)Jw].$
    $(\exists x)(\exists y)(Jx \rightarrow Ky) \land (\forall z)(\forall w)(Kz \rightarrow Jw).$
    $(\exists x)(\exists y)[(Jx \rightarrow Ky) \land (\forall z)(\forall w)(Kz \rightarrow Jw)].$
    $(\exists x)(\exists y)(\forall z)(\forall w)[(Jx \rightarrow Ky) \land (Kz \rightarrow Jw)].$

Chapter 11

A. 1. $Rs.$
   5. $Dd \lor Pd.$
   9. $(\forall x)(Bx \rightarrow \neg Mx).$
   13. $Hab \land (\forall x)(Hxb \rightarrow x = a).$
   18. $(\exists x)(Cx \land Lbx \land (\forall y)(Cy \land Lby \rightarrow x = y)).$
   24. $(\forall x)(\forall y)(Sx \land Fxcy \rightarrow Iy).$

C. 1. $I1 \land I2 \land 1 > 0 \land 2 > 0 \land 1 \neq 2.$
   4. $(\exists x)(\exists y)(Ix \land Iy \land x \neq y \land (\forall u)(\forall v)(Iu \land Iv \land Bxuy \land Bxvy \rightarrow u = v)).$
   7. $(\forall x)[Ix \land x > 1 \rightarrow Px \lor (\exists y)(Py \land (\exists z)(Iz \land x = y \cdot z))].$
   12. $(\forall x)(\forall y)(Px \land Py \land x \neq y \rightarrow (\exists z)(Lz \land Oxz \land Oyz)).$
   17. $(\exists x)(Px \land D2x \land (\exists y)(Ry \land Syx \land (\forall z)(Rz \land Szx \rightarrow z = y))).$
   22. $(\forall x)[Px \land D2x \rightarrow ((\exists w)(\exists y)(w \neq y \land Rw \land Ry \land Swx \land Syx) \leftrightarrow f(x) > 0)].$

D. 1. Consistent: $\mathbf{S} = \langle \mathbf{U}, \mathbf{A}, \mathbf{B}, \mathbf{a}, \mathbf{b} \rangle$, $\mathbf{U} = \{\mathbf{a}, \mathbf{b}, \mathbf{c}\}$, $\mathbf{a} = \mathbf{b}$.

	A	**B**
a	*T*	*T*
c	*T*	*F*

   5. Inconsistent.

{1}	1.	$Pa \lor (\forall x)Qx$	*P*
{2}	2.	$\neg Pb$	*P*
{3}	3.	$a = b \land \neg Qc$	*P*
{3}	4.	$a = b$	3 *T*
{2, 3}	5.	$\neg Pa$	2, 4 *I*
{1, 2, 3}	6.	$(\forall x)Qx$	1, 5 *T*
{1, 2, 3}	7.	Qc	6 *US* $[c\vert x]$
{1, 2, 3}	8.	$Qc \land \neg Qc$	3, 7 *T*

   9. Consistent: $\mathbf{S} = \langle \mathbf{U}, \mathbf{P}, \mathbf{R} \rangle$, $\mathbf{U} = \{\mathbf{a}, \mathbf{b}\}$.

	P
a	*T*
b	*F*

R	**a**	**b**
a	*F*	*T*
b	*T*	

12. Consistent: $\mathbf{S} = \langle \mathbf{U}, \mathbf{f} \rangle$, $\mathbf{U} = \mathbb{Z}$, $\mathbf{f}(\mathbf{x}) = \mathbf{x}$.

16. Inconsistent.

{1}	1.	$(\forall x)(\forall y)[x \circ y = a \rightarrow y = f(x)]$	P
{2}	2.	$(\forall x)(x \circ x = a)$	P
{3}	3.	$(\exists x)(x \neq f(x))$	P
{4}	4.	$z \neq f(z)$	z P (for EP)
{1}	5.	$z \circ z = a \rightarrow z = f(z)$	1 US $[z\|x][z\|y]$
{2}	6.	$z \circ z = a$	2 US $[z\|x]$
{1, 2}	7.	$z = f(z)$	5, 6 T
{1, 2, 4}	8.	$z = f(z) \wedge z \neq f(z)$	z 4, 7 T
{1, 2, 4}	9.	$A \wedge \neg A$	8 T
{1, 2, 3}	10.	$A \wedge \neg A$	3, 4, 9 EP $[z\|x]$

E. **1.** Invalid: $\mathbf{S} = \langle \mathbf{U}, \mathbf{J}, \mathbf{K}, \mathbf{a}, \mathbf{c} \rangle$, $\mathbf{U} = \{\mathbf{a}, \mathbf{c}\}$, $\mathbf{a} \neq \mathbf{c}$.

	J	**K**
a	*F*	
c	*T*	*T*

5. Valid.

{1}	1.	$(\forall x)(Rax \rightarrow a = x \vee a = b)$	P
{2}	2.	$(\exists x)Rax$	P
{3}	3.	$Sa \wedge \neg Sb$	P
{4}	4.	Ray	y P (for EP)
{1}	5.	$Ray \rightarrow a = y \vee a = b$	1 US $[y\|x]$
{1, 4}	6.	$a = y \vee a = b$	y 4, 5 T
{7}	7.	$a = b$	P (for IP)
{3, 7}	8.	$Sa \wedge \neg Sa$	3, 7 I
{3}	9.	$a \neq b$	7, 8 IP
{1, 3, 4}	10.	$a = y$	y 6, 9 T
{1, 3, 4}	11.	Raa	4, 10 I
{1, 2, 3}	12.	Raa	2, 4, 11 EP $[y\|x]$

8. Valid.

{1}	1.	$(\exists x)[Px \wedge (\forall y)(Py \rightarrow x = y)]$	P
{2}	2.	$Pz \wedge (\forall y)(Py \rightarrow z = y)$	z P (for EP)
{2}	3.	$(\forall y)(Py \rightarrow z = y)$	z 2 T
{2}	4.	$Py \rightarrow z = y$	z 3 US $[y\|y]$
{5}	5.	$z = y$	y, z P (for CP)
{2}	6.	Pz	z 2 T
{2, 5}	7.	Py	y 5, 6 I
{2}	8.	$z = y \rightarrow Py$	z 5, 7 CP
{2}	9.	$Py \leftrightarrow z = y$	z, 4, 8 T
{2}	10.	$(\forall y)(Py \leftrightarrow z = y)$	z 9 UG
{2}	11.	$(\exists x)(\forall y)(Py \leftrightarrow x = y)$	10 EG $[z\|x]$
{1}	12.	$(\exists x)(\forall y)(Py \leftrightarrow x = y)$	1, 2, 11 EP $[z\|x]$

12. Valid.

{1}	1.	$(\forall x)(\forall y)(Pxy \to Pyx)$	P
{2}	2.	$(\forall x)Pxf(x)$	P
{2}	3.	$Pf(x)f(f(x))$	2 US $[f(x)\vert x]$
{1}	4.	$Pf(x)f(f(x)) \to Pf(f(x))f(x)$	1 US $[f(x)\vert x][f(f(x))\vert y]$
{1, 2}	5.	$Pf(f(x))f(x)$	3, 4 T
{1, 2}	6.	$(\forall x)Pf(f(x))f(x)$	5 UG

16. Invalid: $\mathbf{S} = \langle \mathbf{U}, \mathbf{R} \rangle$, $\mathbf{U} = \mathbb{Z}$, $\mathbf{Rxy} : \mathbf{x} < \mathbf{y}$.

21. Valid.

{1}	1.	$(\forall x)(\forall y)(x \circ y = y \circ x)$	P
{2}	2.	$(\forall x)(\forall y)[f(x) \circ y = f(x \circ y)]$	P
{2}	3.	$f(x) \circ y = f(x \circ y)$	2 US $[x\vert x][y\vert y]$
{1}	4.	$x \circ y = y \circ x$	1 US $[x\vert x][y\vert y]$
{1, 2}	5.	$f(x) \circ y = f(y \circ x)$	3, 4 I
{2}	6.	$(\forall x)(\forall u)[f(x) \circ u = f(x \circ u)]$	2 LE (7)
{2}	7.	$f(y) \circ x = f(y \circ x)$	6 US $[y\vert x][x\vert u]$
{1}	8.	$(\forall x)(\forall u)(x \circ u = u \circ x)$	1 LE (7)
{1}	9.	$f(y) \circ x = x \circ f(y)$	8 US $[f(y)\vert x][x\vert u]$
{1, 2}	10.	$f(x) \circ y = x \circ f(y)$	5, 7, 9 I
{1, 2}	11.	$(\forall x)(\forall y)(f(x) \circ y = x \circ f(y))$	10 UG (2)

F. 1. Logical Theorem.

{1}	1.	Px	x P (for CP)
{1}	2.	$(\exists y)Py$	1 EG $[x\vert y]$
{1}	3.	$Px \wedge (\exists y)Py$	x, 1, 2 T
{1}	4.	$(\exists y)(Px \wedge Py)$	x 3 LT (13)
	5.	$Px \to (\exists y)(Px \wedge Py)$	1, 4 CP
{6}	6.	$(\exists y)(Px \wedge Py)$	x P (for CP)
{7}	7.	$Px \wedge Pz$	x, z P (for $EP(z)$)
{7}	8.	Px	x 7 T
{6}	9.	Px	x, 6, 7, 8 EP $[z\vert y]$
	10.	$(\exists y)(Px \wedge Py) \to Px$	6, 9 CP
	11.	$Px \leftrightarrow (\exists y)(Px \wedge Py)$	5, 10 T
	12.	$(\forall x)[Px \leftrightarrow (\exists y)(Px \wedge Py)]$	11 UG

5. Logical Theorem.

{1}	1.	$(\forall x)Rxx \wedge (\forall y)(g(f(y)) = f(y))$	P (for CP)
{1}	2.	$(\forall x)Rxx$	1 T
{1}	3.	$(\forall y)(g(f(y)) = f(y))$	1 T
{1}	4.	$g(f(y)) = f(y)$	3 US $[y\vert y]$
{1}	5.	$Rf(y)f(y)$	2 US $[f(y)\vert x]$
{1}	6.	$Rf(y)g(f(y))$	4, 5 I
{1}	7.	$(\exists z)Rzg(z)$	6 EG $[f(y)\vert z]$
	8.	$(\forall x)Rxx \wedge (\forall y)(g(f(y)) = f(y)) \to$ $(\exists z)Rzg(z)$	1, 7 CP

9. Logical Theorem.

{1}	1.	$(\forall x)(\forall y)(x \circ y = y \circ x) \wedge$ $(\forall x)(f(x) \circ x = a)$	P (for CP)
{1}	2.	$(\forall x)(f(x) \circ x = a)$	1 T
{1}	3.	$f(x) \circ x = a$	2 US $[x\|x]$
{1}	4.	$(\forall x)(\forall y)(x \circ y = y \circ x)$	1 T
{1}	5.	$f(x) \circ x = x \circ f(x)$	4 US $[f(x)\|x][x\|y]$
{1}	6.	$x \circ f(x) = a$	3, 5 I
{1}	7.	$(\exists y)(x \circ y = a)$	6 EG $[f(x)\|y]$
{1}	8.	$(\forall x)(\exists y)(x \circ y = a)$	7 UG
	9.	1. $\rightarrow$ 8.	1, 8 CP

G. Because most of the proofs are easy, we shall give the translations and add a hidden premise, H.

1. $(\exists x)(Bx \wedge Ax)$, $(\forall x)(Bx \rightarrow x = n \vee x = j)$, Sn $\therefore$ $Bj \wedge Aj$.
$H: (\forall x)(Sx \rightarrow \neg Ax)$.

4. $(\forall x)(Px \rightarrow Ex)$ $\therefore$ $(\exists x)(Sx \wedge Ex)$.
$H: (\forall x)(Sx \rightarrow Px) \wedge (\exists x)Sx$.

7. $(\exists x)[Rx \wedge (\forall y)(Ry \wedge x \neq y \rightarrow Fxy)]$ $\therefore$ $(\forall u)(\forall v)[Ru \wedge Rv \wedge (\forall y)(Ry \wedge y \neq u \rightarrow Fuy) \wedge (\forall z)(Rz \wedge z \neq v \rightarrow Fvz) \rightarrow u = v]$.
$H: (\forall x)(\forall y)(Fxy \rightarrow \neg Fyx)$.

11. $(\forall x)[(\forall y)(My \rightarrow \neg Oxys) \rightarrow Ux]$, $\neg(\exists x)(Cx \wedge Oaxs)$ $\therefore$ Ua.
$H: (\forall x)(Mx \rightarrow Cx)$.

15. $(\forall x)[Sx \wedge (g(x) = a \vee g(x) = b) \rightarrow Hx]$, $(\exists x)(Sx \wedge \neg Hx)$ $\therefore$ $(\exists x)(Sx \wedge Lg(x)b)$.
$H: (\forall x)(g(x) \neq a \wedge g(x) \neq b \rightarrow Lg(x)b)$.

Chapter 13

A. **3.**

{1}	1.	$(\forall x)(\forall y)(\forall z)(Rxy \wedge$ $Ryz \rightarrow Rxz)$	P
{2}	2.	$(\forall x)(\forall y)(Rxy \rightarrow Ryx)$	P
{3}	3.	$(\forall x)(\exists y)Rxy$	P
{3}	4.	$(\exists y)Rxy$	3 US $[x\|x]$
{5}	5.	Rxu	x, u P (for $EP(u)$)
{2}	6.	$Rxu \rightarrow Rux$	2 US $[x\|x][u\|y]$
{2, 5}	7.	Rux	x, u 5, 6 T
{1}	8.	$Rxu \wedge Rux \rightarrow Rxx$	1 US $[x\|x][u\|y][x\|z]$
{1, 2, 5}	9.	Rxx	x, 5, 7, 8 T
{1, 2, 3}	10.	Rxx	4, 5, 9 EP $[u\|y]$
{1, 2, 3}	11.	$(\forall x)Rxx$	10 UG

B. We shall only do part (a).

1. Theorem 4: $(\forall x)(x \circ x = e) \rightarrow (\forall x)(\forall y)(x \circ y = y \circ x)$.
Informal proof:
Suppose

(i) $x \circ x = e$, for all x.
In particular, it follows from (i) that

(ii) $(x \circ y) \circ (x \circ y) = e$.

(iii)
$$
\begin{aligned}
(y \circ x) \circ (x \circ y) &= y \circ (x \circ (x \circ y)) && \text{(by } A1) \\
&= y \circ ((x \circ x) \circ y) && \text{(by } A1) \\
&= y \circ (e \circ y) && \text{(by (i))} \\
&= y \circ (y \circ e) && \text{(by Theorem 2)} \\
&= (y \circ y) \circ e && \text{(by } A1) \\
&= e \circ e && \text{(by (i))} \\
&= e && \text{(by } A2)
\end{aligned}
$$

It follows from (ii) and (iii) that

(iv) $(x \circ y) \circ (x \circ y) = (y \circ x) \circ (x \circ y)$.
Consequently, the result follows from (iv) and Theorem 1.

5. Theorem 8: $(\forall x)[(x^*)^* = x]$.
Informal proof:
By Theorem 5 and $A3$, $x^* \circ x = e$.
Thus, by Theorem 7, $x = (x^*)^*$.

C. The groups we refer to are those given in the text in Section 13.3.
1. $(\exists x)(\forall y)(x = y)$ is false in Group (2) and true in Group (1).
5. $(\forall x)(\exists y)(x \circ y = y)$ is false in Group (2) and true in Group (1).
10. $(\forall x)(\forall y)(\exists z)(x = y \circ z \wedge x = z \circ y)$ is false in Group (5) (if $\mathbf{a}_4 = \mathbf{a}_1 \circ \mathbf{z}$, then $\mathbf{z} = \mathbf{a}_2$, and if $\mathbf{a}_4 = \mathbf{z} \circ \mathbf{a}_1$, then $\mathbf{z} = \mathbf{a}_3$) and is true in Group (1).

D. 1. Group.
7. Not a group because $2 \cdot 2 = 4 \notin \mathbf{U}$.
12. No. $A1$ is false because $0 \circ (0 \circ 1) \neq (0 \circ 0) \circ 1$.
15. Yes. $\mathbf{b} = \mathbf{e}$, $\mathbf{a}^* = \mathbf{c}$, $\mathbf{b}^* = \mathbf{b}$, $\mathbf{c}^* = \mathbf{a}$.

E. If inconsistent, we shall only give an informal proof of a contradiction.
1. Consistent: $\mathbf{U} = \{\mathbf{a}, \mathbf{b}\}$.

E	**a**	**b**
a	T	F
b	F	T

6. Inconsistent.
Assume that for some u, v, and w,

$$u \neq v \wedge Euw \wedge Evw \wedge \neg Euv.$$

However, Evw implies Ewv by $E2$. $Euw \wedge Ewv \wedge \neg Euv$ contradicts $E3$.

10. Consistent: **U** = {**a, b, c**}.

E	**a**	**b**	**c**
a	*T*	*F*	*F*
b	*F*	*T*	*F*
c	*F*	*F*	*T*

F. 1. (i) We shall only give informal proofs. Suppose

(1) $x \circ z = y \circ z$.

By $B3$, there is a u such that

(2) $z \circ u = e$.

Then, from (1),

(3) $(x \circ z) \circ u = (y \circ z) \circ u$

(4) $x \circ (z \circ u) = y \circ (z \circ u)$ (by (3) and $B1$)

(5) $x \circ e = y \circ e$ (by (4) and (2))

(6) $x = y$ (by (5) and $B2$)

4. (ii) Group (4), Section 13.3.
5. (i) Suppose

(1) $x \circ x = x$, for all x.

By $B2$.

(2) $x \circ e = x$, for all x.

Thus, by (1), (2), and 1(ii),

(3) $x = e$, for all x.

Therefore, $\neg(\exists x)(\exists y)(x \neq y)$, which is a contradiction.

H. 2. R is not an equivalence relation because R does not satisfy $E1$ in Group (2).
5. R is an equivalence relation.

Chapter 14

A. 1. $\sum_{i=1}^{n-1} i = \frac{(n-1)n}{2}, n \geq 2.$

Base Step: $n = 2$, $\sum_{i=1}^{2-1} i = 1, \frac{(2-1)2}{2} = 1.$

Induction Step: Suppose, for $m \geq 2$,

$$(IH) \quad \sum_{i=1}^{m-1} i = \frac{(m-1)m}{2}.$$

Then

$$\begin{aligned}
\sum_{i=1}^{m} i &= \sum_{i=1}^{m-1} i + m && \text{(Definition of } \Sigma) \\
&= \frac{(m-1)m}{2} + m && (IH) \\
&= \frac{m(m+1)}{2} && \text{(Algebra)}
\end{aligned}$$

Therefore, 1 follows by *MI*.

8. $\sum_{i=1}^{n} i(i+1) = \frac{1}{3}n(n+1)(n+2), n \geq 1.$

Base Step: $n = 1$, $\sum_{i=1}^{1} i(i+1) = 2, \frac{1}{3} \cdot 1(1+1)(1+2) = 2.$

Induction Step: Suppose, for $m \geq 1$,

$$(IH) \quad \sum_{i=1}^{m} i(i+1) = \frac{1}{3}m(m+1)(m+2).$$

Then

$$\begin{aligned}
\sum_{i=1}^{m+1} i(i+1) &= \sum_{i=1}^{m} i(i+1) + (m+1)(m+2) && \text{(Definition of } \Sigma) \\
&= \frac{1}{3}m(m+1)(m+2) + (m+1)(m+2) && (IH) \\
&= \frac{1}{3}(m+1)(m+2)(m+3) && \text{(Algebra)}
\end{aligned}$$

Therefore, 8 follows by *MI*.

12. $\sum_{i=1}^{n} i \cdot i! = (n+1)! - 1, n \geq 1.$

Base Step: $n = 1$, $\sum_{i=1}^{1} i \cdot i! = 1$, $(1+1)! - 1 = 1.$

Induction Step: Suppose, for $m \geq 1$,

$$(IH) \quad \sum_{i=1}^{m} i \cdot i! = (m+1)! - 1.$$

Then

$$\begin{aligned}\sum_{i=1}^{m+1} i\cdot i! &= \sum_{i=1}^{m} i\cdot i! + (m+1)\cdot(m+1)! && \text{(Definition of } \Sigma\text{)}\\ &= (m+1)! - 1 + (m+1)\cdot(m+1)! && (IH)\\ &= (m+2)\cdot(m+1)! - 1 && \text{(Algebra)}\\ &= (m+2)! - 1 && \text{(Definition of !)}\end{aligned}$$

Therefore, 12 follows by *MI*.

16. $n! > 3^n$, $n \geq 7$.
Base Step: $n = 7$, $7! = 5040$, $3^7 = 2187$.
Induction Step: Suppose, for $m \geq 7$,

$$(IH) \qquad m! > 3^m.$$

Then

$$\begin{aligned}(m+1)! &= (m+1)\cdot m! && \text{(Definition of !)}\\ &> (m+1)\cdot 3^m && (IH)\\ &> 3\cdot 3^m && \text{(Algebra, } m \geq 7\text{)}\\ &= 3^{m+1} && \text{(Algebra)}\end{aligned}$$

Therefore, 16 follows by *MI*.

21. $\displaystyle\sum_{i=1}^{n} \frac{1}{n+i} > \frac{13}{24}$, $n \geq 2$.
Base Step: $n = 2$, $\displaystyle\sum_{i=1}^{2} \frac{1}{2+i} = \frac{1}{3} + \frac{1}{4} = \frac{7}{12} > \frac{13}{24}$.
Induction Step: Suppose, for $m \geq 2$,

$$(IH) \qquad \sum_{i=1}^{m} \frac{1}{m+i} > \frac{13}{24}.$$

Then,

$$\begin{aligned}\sum_{i=1}^{m+1} \frac{1}{m+1+i} &= \sum_{i=1}^{m} \frac{1}{m+1+i} + \frac{1}{2m+2} && \text{(Definition of } \Sigma\text{)}\\ &= \sum_{i=1}^{m} \frac{1}{m+i} - \frac{1}{m+1} + \frac{1}{2m+1} + \frac{1}{2m+2} && \text{(Algebra)}\\ &> \frac{13}{24} - \frac{1}{m+1} + \frac{1}{2m+1} + \frac{1}{2m+2} && (IH)\\ &= \frac{13}{24} + \frac{1}{(2m+1)(2m+2)} && \text{(Algebra)}\\ &> \frac{13}{14} && (m \geq 2)\end{aligned}$$

Therefore, 21 follows by *MI.*

24. $2^{2n} - 1$ is divisible by 3, $n \geq 0$.
Base Step: $n = 0$, $2^{2 \cdot 0} - 1 = 0$, which is divisible by 3.
Induction Step: Suppose, for $m \geq 0$, there is an integer k such that

$$(IH) \qquad 2^{2m} - 1 = 3k.$$

Then

$$\begin{aligned} 2^{2m+2} - 1 &= 4 \cdot 2^{2m} - 1 && \text{(Algebra)} \\ &= 4(3k + 1) - 1 && (IH) \\ &= 3(4k + 1) && \text{(Algebra)} \end{aligned}$$

Therefore, 24 follows by *MI.*

30. If x is any positive real number, then $(1 + x)^n \geq 1 + nx$, $n \geq 0$.
Base Step: $n = 0$, $(1 + x)^0 = 1$, $1 + 0 \cdot x = 1$.
Induction Step: Suppose, for $m \geq 0$,

$$(IH) \qquad (1 + x)^m \geq 1 + mx.$$

Then

$$\begin{aligned} (1 + x)^{m+1} &= (1 + x)(1 + x)^m && \text{(Algebra)} \\ &\geq (1 + x)(1 + mx) && (IH) \\ &= 1 + (m + 1)x + mx^2 && \text{(Algebra)} \\ &\geq 1 + (m + 1)x && (\text{Algebra, } x,\ m \geq 0) \end{aligned}$$

Therefore, 30 follows by *MI.*

36. $\prod_{i=0}^{n} \cos(2^i x) = \dfrac{\sin(2^{n+1}x)}{2^{n+1}\sin x}$, $n \geq 0$.

Base Step: $n = 0$, $\prod_{i=0}^{0} \cos(2^i x) = \cos x$,

$$\frac{\sin(2^{0+1}x)}{2^{0+1}\sin x} = \frac{\sin 2x}{2\sin x} = \cos x \qquad \text{(Trigonometric Identity)}$$

Induction Step: Suppose, for $m \geq 0$,

$$(IH) \qquad \prod_{i=0}^{m} \cos(2^i x) = \frac{\sin(2^{m+1}x)}{2^{m+1}\sin x}.$$

Then

$$\begin{aligned} \prod_{i=0}^{m+1} \cos(2^i x) &= \cos(2^{m+1}x) \prod_{i=0}^{m} \cos(2^i x) && (\text{Definition of } \Pi) \\ &= \cos(2^{m+1}x)\frac{\sin(2^{m+1}x)}{2^{m+1}\sin x} && (IH) \\ &= \frac{\sin(2^{m+2}x)}{2^{m+2}\sin x} && \text{(Trigonometric Identity)} \end{aligned}$$

Therefore, 36 follows by *MI*.

For problems 43 through 58, we shall give the formula and omit the proof.

43. $\sum_{i=1}^{n} (2i - 1) = n^2, n \geq 1.$

47. $\prod_{i=2}^{n} \left(1 - \frac{1}{i^2}\right) = \frac{n+1}{2n}, n \geq 2.$

53. $n!5^n(4 - 5x)^{-(n+1)}, n \geq 1.$

57. (a) $180(n - 2), n \geq 3.$

58. (a) $4 \cdot 5^{n-1}, n \geq 1.$

B. 1. Theorem 7: $(\forall x)(\forall y)(x + y = 0 \rightarrow x = 0 \wedge y = 0)$.

Proof: The proof is by *MI*, where φ is the formula

$$(\forall y)(x + y = 0 \rightarrow x = 0 \wedge y = 0).$$

Base Step: $x = 0$. If $0 + y = 0$, then by Theorem 2, $y = 0$.
Induction Step: Suppose

$$(IH) \qquad z + y = 0 \rightarrow z = 0 \wedge y = 0, \text{for all } y.$$

Suppose $(z + 1) + y = 0$. Then, by *P3*, $z + (1 + y) = 0$ and, by *P1*, $z + (y + 1) = 0$. Using *P3* again, $(z + y) + 1 = 0$, but this contradicts *P8*. Therefore, $\varphi[z + 1|x]$ holds because its antecedent is false.

Theorem 7 follows by *MI*.

5. Theorem 11: $(\forall x)(\forall y)(\forall z)(z \neq 0 \wedge x \cdot z = y \cdot z \rightarrow x = y)$.

Proof: Suppose

(i) $z \neq 0 \wedge x \cdot z = y \cdot z.$

By Theorem 4, there is a u so that

(ii) $x = y + u \vee y = x + u.$

Suppose $x = y + u$. Then

(iii) $x \cdot z = (y + u) \cdot z$
$= y \cdot z + u \cdot z \qquad$ (Theorem 8)

Therefore, by (i),

$$y \cdot z + u \cdot z = y \cdot z + 0 \qquad (P6)$$

which implies, by Theorem 5, that $u \cdot z = 0$. By (i), $z \neq 0$. Thus, Theorem 9 implies $u = 0$. Therefore, $x = y + 0 = y$, by *P6*.

The proof is analogous if $y = x + u$.

10. Theorem 16: $(\forall x)(\forall y)(\forall z)(x < y \wedge y < z \rightarrow x < z)$.

Proof: If $x < y \wedge y < z$, then by *P10*, there exists $u \neq 0$ and $v \neq 0$ such that

$$y = x + u \wedge z = y + v.$$

Therefore,

$$\begin{aligned} z &= (x + u) + v \\ &= x + (u + v) \qquad (P3) \end{aligned}$$

Because $u \neq 0$ and $v \neq 0$, it follows from Theorem 7 that $u + v \neq 0$. Thus, by $P10$, $x < z$.

16. Theorem 22: $(\forall x)(\forall y)(\forall z)(x + z < y + z \rightarrow x < y)$.

Proof: Suppose $x + z < y + z$. Then, by $P10$, there is a $u \neq 0$ such that

$$\begin{aligned} y + z &= (x + z) + u \\ &= x + (z + u) \qquad (P3) \\ &= x + (u + z) \qquad (P1) \\ &= (x + u) + z \qquad (P3) \end{aligned}$$

Therefore, by Theorem 5, $y = x + u$ and, by $P10$, $x < y$.

19. Theorem 25: $(\forall x)(\forall y)(x < y + 1 \rightarrow x < y \lor x = y)$.

Proof: If $x \not< y$ and $x \neq y$, then by Theorem 19, $y < x$. However, $y < x$ and $x < y + 1$ contradicts Theorem 24.

C. 1. $(\forall x)(x|x)$.

Proof: By $P7$, $x = x \cdot 1$. Therefore, $P12$ implies $x|x$.

6. $(\forall x)(\forall y)(x|y \land y|x \rightarrow x = y)$.

Proof: Suppose $x|y \land y|x$. If $x = 0$, then $0|y$. By P_{12}, there is a z such that

$$\begin{aligned} y &= 0 \cdot z \\ &= z \cdot 0 \qquad (P2) \\ &= 0 \qquad (\text{Theorem } 6) \end{aligned}$$

Similarly, if $y = 0$, so does x.

Now assume $x \neq 0$ and $y \neq 0$. Then, by C5,

(i) $(x < y \lor x = y) \land (y < x \lor y = x)$.

It follows from (i) and Theorems 17 and 18 that $x = y$.

9. $(\forall x)(x|1 \rightarrow x = 1)$.

Proof: Suppose $x|1$. By C2, $1|x$. Therefore, by C6, $x = 1$.

Chapter 15

A. We shall give the formula and omit the proof.

1. $S_n = 2^{-n}$, $n \geq 0$.

5. $S_n = 2^{n+1} - 1$, $n \geq 0$.

10. $S_n = \frac{1}{3}(2 \cdot 4^n + 1)$, $n \geq 0$.

15. $S_n = 2 - 2^n$, $n \geq 0$.

20. $S_n = a \cdot 2^n + (b - 2a)n \cdot 2^{n-1}, n \geq 0.$

25. $S_n = 2 \cdot 3^{n-1}, n \geq 1.$

B. 1. $3, 3r - 7, 3r^2 - 7r + 2, 3r^3 - 7r^2 + 2r - 3, p(r).$

C. 1. Base Step: $n = 1, \quad f_1 + 2f_2 = 3, \quad f_4 = 3.$
$n = 2, \quad f_2 + 2f_3 = 5, \quad f_5 = 5.$
Induction Step: Suppose for $2 \leq k \leq m$,

$$(IH) \qquad f_k + 2f_{k+1} = f_{k+3}.$$

Then

$$\begin{aligned} f_{m+1} + 2f_{m+2} &= (f_m + f_{m-1}) + 2(f_{m+1} + f_m) \\ &= f_{m+3} + f_{m+2} && (IH) \\ &= f_{m+4} \end{aligned}$$

The theorem follows by *CI*.

5. Base Step: $n = 1, \quad f_{11} = 89, \quad 11f_6 + f_1 = 11 \cdot 8 + 1 = 89.$
$n = 2, \quad f_{12} = 144, \quad 11f_7 + f_2 = 11 \cdot 13 + 1 = 144.$
Induction Step: Suppose, for $2 \leq k \leq m$,

$$(IH) \qquad f_{k+10} = 11f_{k+5} + f_k.$$

Then

$$\begin{aligned} 11f_{m+6} + f_{m+1} &= 11(f_{m+5} + f_{m+4}) + (f_m + f_{m-1}) \\ &= f_{m+10} + f_{m+9} && (IH) \\ &= f_{m+11} \end{aligned}$$

The theorem follows by *CI*.

7. Base Step: $n = 1, \sum_{i=1}^{1} \ell_1 = 1, \ell_3 - 3 = 4 - 3 = 1.$

Induction Step: Suppose, for $m \geq 1$,

$$(IH) \qquad \sum_{i=1}^{m} \ell_i = \ell_{m+2} - 3.$$

Then

$$\begin{aligned} \sum_{i=1}^{m+1} \ell_i &= \sum_{i=1}^{m} \ell_i + \ell_{m+1} \\ &= \ell_{m+2} - 3 + \ell_{m+1} && (IH) \\ &= \ell_{m+3} - 3 \end{aligned}$$

The theorem follows by *CI*.

14. Base Step: $n = 1, \quad \ell_1 = 1, \quad 2^1 = 2.$
$n = 2, \quad \ell_2 = 3, \quad 2^2 = 4.$

Induction Step: Suppose, for $2 \leq k \leq m$,

$$(IH) \qquad \ell_k < 2^k.$$

Then

$$\begin{aligned} \ell_{m+1} &= \ell_m + \ell_{m-1} \\ &< 2^m + 2^{m-1} \qquad (IH) \\ &= 3 \cdot 2^{m-1} < 4 \cdot 2^{m-1} = 2^{m+1} \end{aligned}$$

The theorem follows by *CI*.

18. Base Step: $n = 7$, $f_8 = 21$, $f_7 = 13$, $\frac{21}{13} = 1.615$.
Induction Step: Suppose, for $m \geq 7$,

$$(IH) \qquad 1.614 < \frac{f_{m+1}}{f_m} < 1.620.$$

Then

$$\frac{f_{m+2}}{f_{m+1}} = \frac{f_{m+1} + f_m}{f_{m+1}} = 1 + \frac{f_m}{f_{m+1}}.$$

By (*IH*),

$$0.6196 > \frac{1}{1.614} > \frac{f_m}{f_{m+1}} > \frac{1}{1.620} > 0.617.$$

Therefore,

$$1.620 > 1.6196 > \frac{f_{m+2}}{f_{m+1}} > 1.617 > 1.614.$$

The theorem follows by *MI*.

D. **1.** If the word has length one, then it is either 0 or 1. If w is a word of length n in which all the 0's precede all the 1's, then a word of length $n + 1$ has the form $0w$ or $w1$.

5. Suppose w is of length one. If w is 0, then w^* is 1; if w is 1, then w^* is 0. Suppose w is a word of length $n + 1$. If w has the form $0v$, then $w^* = 1v^*$. If w has the form $1v$, then $w^* = 0v^*$.

8. $S_1 = 2$, $S_2 = 4$, $S_{n+1} = S_n + S_{n-1} + 1$, $n \geq 2$.

E. **1.** $S_1 = 1$, $S_2 = 2$, $S_{n+1} = 2S_n + S_{n-1}$, $n \geq 1$.
Prove $S_n \leq (\frac{5}{2})^n$, $n \geq 1$.

Base Step: $n = 1, \quad S_1 = 1, \quad \left(\frac{5}{2}\right)^1 = \frac{5}{2}.$

$n = 2, \quad S_2 = 2, \quad \left(\frac{5}{2}\right)^2 = \frac{25}{4}.$

Induction Step: Suppose, for $2 \leq k \leq m$,

$$(IH) \qquad S_k \leq \left(\frac{5}{2}\right)^k.$$

Then

$$\begin{aligned}
S_{m+1} &= 2S_m + S_{m-1} \\
&\leq 2\left(\frac{5}{2}\right)^m + \left(\frac{5}{2}\right)^{m-1} && (IH) \\
&= 6\left(\frac{5}{2}\right)^{m-1} && \text{(Algebra)} \\
&\leq \left(\frac{5}{2}\right)^2\left(\frac{5}{2}\right)^{m-1} && \left(6 < \left(\frac{5}{2}\right)^2\right) \\
&= \left(\frac{5}{2}\right)^{m+1} && \text{(Algebra)}
\end{aligned}$$

The theorem follows by *MI*.

5. $S_1 = 1;\ S_{n+1} = \dfrac{1}{1 + S_n},\ n \geq 1.$

(a) Prove $S_n = \dfrac{f_n}{f_{n+1}},\ n \geq 1.$
Base Step: $n = 1,\ S_1 = 1,\ f_1 = f_2 = 1.$
Induction Step: Suppose, for $m \geq 1$,

$$(IH) \qquad S_m = \frac{f_m}{f_{m+1}}.$$

$$\begin{aligned}
S_{m+1} &= \frac{1}{1 + S_m} \\
&= \frac{1}{1 + \dfrac{f_m}{f_{m+1}}} && (IH) \\
&= \frac{f_{m+1}}{f_{m+2}}
\end{aligned}$$

The theorem follows by *MI*.

(b) If S has a limit, prove $\lim_{n\to\infty} S_n = \dfrac{\sqrt{5} - 1}{2}.$

Proof: Suppose $\lim_{n\to\infty} S_n = L$. Then

$$L = \lim_{n\to\infty} S_{n+1} = \lim_{n\to\infty} \frac{1}{1 + S_n} = \frac{1}{1 + L}.$$

Solving for L,

$$L^2 + L - 1 = 0$$

$$L = \frac{-1 \pm \sqrt{5}}{2}.$$

However, because all terms of the sequence are positive, $L \geq 0$. Therefore, $L = \dfrac{\sqrt{5} - 1}{2}$.

7. $S_1 = 2$; $S_{n+1} = \frac{1}{4}[(S_n)^2 + 3]$.

(a) Prove $S_n > S_{n+1}$, $n \geq 1$. (Simple proof by *MI*).

(b) Prove S has a limit. Find the limit.

Proof: By part (a) and the definition of S, S is a decreasing sequence of positive terms, so it has a limit, L.

$$L = \lim_{n\to\infty} S_{n+1} = \lim_{n\to\infty} \frac{1}{4}[(S_n)^2 + 3] = \frac{1}{4}(L^2 + 3).$$

Solving for L, $L^2 - 4L + 3 = 0$, so we obtain $L = 3$ or $L = 1$. However, $L \neq 3$. Therefore, $L = 1$.

15. If $M = \begin{pmatrix} 0 & 1 \\ 1 & 1 \end{pmatrix}$, then $M^n = \begin{pmatrix} f_{n-1} & f_n \\ f_n & f_{n+1} \end{pmatrix}$, $n \geq 2$.

Base Step: $n = 2$.

$$M^2 = \begin{pmatrix} 1 & 1 \\ 1 & 2 \end{pmatrix}, f_1 = 1, f_2 = 1, f_3 = 2.$$

Induction Step: Suppose, for $m \geq 2$,

$$(IH) \qquad M^m = \begin{pmatrix} f_{m-1} & f_m \\ f_m & f_{m+1} \end{pmatrix}.$$

Then,

$$\begin{aligned} M^{m+1} &= \begin{pmatrix} 0 & 1 \\ 1 & 1 \end{pmatrix}\begin{pmatrix} f_{m-1} & f_m \\ f_m & f_{m+1} \end{pmatrix} \qquad (IH) \\ &= \begin{pmatrix} f_m & f_{m+1} \\ f_{m-1} + f_m & f_m + f_{m+1} \end{pmatrix} \\ &= \begin{pmatrix} f_m & f_{m+1} \\ f_{m+1} & f_{m+2} \end{pmatrix} \end{aligned}$$

The theorem follows by *MI*.

18. $(\forall y)[(\forall z)(z < y \rightarrow \varphi[z|x]) \rightarrow \varphi[y|x]] \rightarrow (\forall x)\varphi$.
Proof: Suppose, for all y,

(i) $(\forall z)(z < y \rightarrow \varphi[z|x]) \rightarrow \varphi[y|x]$.

If $y = 0$, then the antecedent of (i) holds. Thus, $\varphi[0|x]$ holds. Therefore, $(\forall x)\varphi$ follows by *CI*.

20. No positive formula is a tautology, and no positive formula is a contradiction. Prove that a positive formula is true when all its sentential variables are true and false when all its sentential variables are false by *CI* on the length of the formula.

24. (a) (i) $\rightarrow \vee \neg PQ\neg \wedge P\neg R$.
(c) (iii) Yes. $A \rightarrow (B \rightarrow (C \leftrightarrow D))$.

F. 1. $\neg Bx \vee Wx, \neg Fx \vee \neg Wx \therefore \neg Fx \vee \neg Bx$.
Proof: Use the two premises and R5.

7. $\neg Lx \vee Gx, \neg Gx \vee \neg Hx \vee \neg Ix, \neg Gx \vee \neg Hx \therefore \neg Lx \vee Ix \vee \neg Hx$.
Proof: From premises 1 and 3, derive $\neg Lx \vee \neg Hx$ using R5. Then derive the conclusion using R3.

11. $\neg(Fx \vee Mx) \vee \neg Cy \vee Lxy, \neg(\neg Fn \vee Lnp), \neg Mx \vee \neg Lxs, Mr$ $\therefore \neg(Cp \vee Cs)$.
Proof: Use an indirect proof. Assume

(i) $Cp \vee Cs$.

From premises 3 and 4, using R4, deduce

(ii) $\neg Lrs$.

From premise 1 and (ii), using R4, deduce

(iii) $\neg(Fr \vee Mr) \vee \neg Cs$.

From premise 2, using R3, deduce Fn and $\neg Lnp$.
From $\neg Lnp$ and premise 1, using R4, deduce

(iv) $\neg(Fn \vee Mn) \vee \neg Cp$.

From premise 4, using R3, deduce

(v) $Fr \vee Mr$,

and from Fn, using R3, deduce

(vi) $Fn \vee Mn$.

From (iii) and (v), using R5, deduce $\neg Cs$. From (i) and $\neg Cs$, using R5, deduce Cp. From (iv) and (vi), using R5, deduce $\neg Cp$.
Thus, we have a contradiction.

G. 1. Consistent, but not unique. *La, Pb, Tc, Vd* or *Lb, Pc, Ta, Vd*.
4. Inconsistent.

H. 1. The constants are $c_1, c_2, \ldots, c_9$ and $d_1 = \{c_1, c_2, c_3\}$, $d_2 = \{c_4, c_5, c_6\}$ and $d_3 = \{c_7, c_8, c_9\}$. The predicate symbols are *Bxy*: *x* is the same weight as *y*; *Lxy*: *x* is lighter than *y*; and *Hx*: *x* is the heaviest coin.

Assume $Bxy \vee Lxy \vee Lyx$, and then give an alternative proof.

Case 1. Bd_1d_2, $\neg Bc_7c_8 \vee Hc_9$, $\neg Lc_7c_8 \vee Hc_8$, $\neg Lc_8c_7 \vee Hc_7$.

Case 2. Ld_1d_2, $\neg Bc_4c_5 \vee Hc_6$, $\neg Lc_4c_5 \vee Hc_5$, $\neg Lc_5c_4 \vee Hc_4$.

Case 3. Ld_2d_1, $\neg Bc_1c_2 \vee Hc_3$, $\neg Lc_1c_2 \vee Hc_2$, $\neg Lc_2c_1 \vee Hc_1$.

Thus, the heavy coin can be determined in two weighings.

3. The constants are *a*: *U*; *b*: *V*; *c*: *W*. The predicate symbols are *Tx*: *x* is a Tu ($\neg Tx$: *x* is a Fa); *Sx*: *x*'s statement is true ($\neg Sx$: *x*'s statement is false). We are given (i) $\neg Tx \vee Sx$, (ii) $\neg Sx \vee Tx$.
(a) *U*: $\neg(Ta \vee Tb)$.
V: $Ta \vee Tb$, $\neg Ta \vee \neg Tb$.
Use an alternative proof.

Case 1. $\neg(Ta \vee Tb)$ and *Sa*. Then, by (ii), *Ta*. By $\neg(Ta \vee Tb)$ and R3, we deduce $\neg Tb$. Thus, *Sb*, which implies *Tb*. We have a contradiction.

Case 2. $Ta \vee Tb$ and $\neg Sa$. Then, by (i), $\neg Ta$. From $Ta \vee Tb$ and $\neg Ta$, we deduce *Tb*. Also, by R3, $\neg Ta$ and *Tb* imply *Sb*. Thus, we have $\neg Ta$ and *Tb*.

Chapter 16

A. 1. $(P \wedge Q) \Leftrightarrow \neg(\neg P \vee \neg Q)$.
4. $(\neg P) \Leftrightarrow (P \downarrow P)$; $(P \vee Q) \Leftrightarrow \neg(P \downarrow Q)$; use 1.
8. $(P \vee Q) \Leftrightarrow M^{\top}(\top, P, Q)$; use 1.

B. 1. Negation is not definable because any formula that only contains the connectives $\wedge$ and $\rightarrow$ and the sentential variable *P* is true when *P* is true.
3. Show that any formula that just contains the given connectives and two distinct sentential variables has an even number of *T*'s in its truth table. Thus, none of the connectives $\vee$, $\wedge$, or $\rightarrow$ are definable.

C. 1. a. $(\forall x)(\forall u)(Pxf(x) \wedge Qxg(x) \rightarrow Rf(x)g(x)u)$.
i. $(\forall x)(\forall u)(\forall z)[Fxf(x) \wedge Gf(x)x \rightarrow (Fug(x, y) \wedge Gg(x, u)z \rightarrow Hug(x, u)z)]$.
2. a. Pa, $(\forall x)(x \neq f(x))$; $\mathbf{U}_0 = \{\mathbf{a}\}$; $\mathbf{U}_1 = \{\mathbf{a}, \mathbf{f}(\mathbf{a})\}$; β_1 is $Pa \wedge a \neq f(a)$; $\mathbf{U}_2 = \{\mathbf{a}, \mathbf{f}(\mathbf{a}), \mathbf{f}^{(2)}(\mathbf{a})\}$; β_2 is $\beta_1 \wedge Pf(a) \wedge f(a) \neq f^{(2)}(a)$; $\mathbf{U}_3 = \{\mathbf{a}, \mathbf{f}(\mathbf{a}), \mathbf{f}^{(2)}(\mathbf{a}), f^{(3)}(\mathbf{a})\}$; β_3 is $\beta_2 \wedge Pf^{(2)}(a) \wedge f^{(2)}(a) \neq f^{(3)}(a)$. At each level, there is only one partial model. $\mathbf{S} = \langle \mathbf{U}, \mathbf{P}, \mathbf{f} \rangle$ is a model if $\mathbf{U} = \mathbb{N}$, $\mathbf{P} = \mathbf{U}$, and $\mathbf{f}(\mathbf{x}) = \mathbf{x} + \mathbf{1}$.
c. $(\forall x)(Rxf(x) \wedge x \neq f(x))$, $(\forall x)(\forall y)(Rxy \rightarrow Ryx)$, $(\forall x)\neg Rxx$; $\mathbf{U}_0 = \{\mathbf{a}\}$; $\mathbf{U}_1 = \{\mathbf{a}, \mathbf{f}(\mathbf{a})\}$; β_1 is $Raf(a) \wedge a \neq f(a) \wedge (Raa \rightarrow Raa) \wedge \neg Raa$. In

each partial model, $Raf(a)$ and $a \neq f(a)$ must be true and Raa false. It remains to choose the truth values of $Rf(a)f(a)$ and $Rf(a)a$. Thus, we have 2^2 partial models at level 1. $\mathbf{U}_2 = \{\mathbf{a}, \mathbf{f}(\mathbf{a}), \mathbf{f}^{(2)}(\mathbf{a})\}$; $\beta_2 = \beta_1 \wedge Rf(a)f^{(2)}(a) \wedge f(a) \neq f^{(2)}(a) \wedge (Raf(a) \rightarrow Rf(a)a) \wedge (Rf(a)f(a) \rightarrow Rf(a)f(a)) \wedge (Rf(a)a \rightarrow Raf(a)) \wedge \neg Rf(a)f(a)$. Continuing, we see that we get a model, **S**, if we take $\mathbf{x} \neq \mathbf{f}(\mathbf{x})$ and **Rxy** to be true, except if $\mathbf{x} = \mathbf{y}$. Take $\mathbf{S} = \{\mathbf{U}, \mathbf{R}, \mathbf{f}\}$ were $\mathbf{U} = \{\mathbf{a}, \mathbf{b}\}$, $\mathbf{f}(\mathbf{a}) = \mathbf{b}$, $\mathbf{f}(\mathbf{b}) = \mathbf{a}$,

R	**a**	**b**
a	*F*	*T*
b	*T*	*F*

3. a. $(\forall x)(\forall y)(Ax \wedge By \rightarrow \neg Lxy)$, $(\forall x)[(Bf(x) \rightarrow \neg Lxf(x)) \rightarrow Dx]$, $Aa \wedge \neg Da$; $\mathbf{U}_0 = \{\mathbf{a}\}$; $\mathbf{U}_1 = \{\mathbf{a}, \mathbf{f}(\mathbf{a})\}$; β_1 is $(Aa \wedge Ba \rightarrow \neg Laa) \wedge [(Bf(a) \rightarrow \neg Laf(a)) \rightarrow Da] \wedge (Aa \wedge \neg Da)$. One of the conjuncts in β_2 is $Aa \wedge Bf(a) \rightarrow \neg Laf(a)$. We can deduce the contradiction $Da \wedge \neg Da$ from β_1 and β_2.
   e. $(\forall x)(Ax \rightarrow Bx)$, $Ca \wedge (\forall y)(By \rightarrow Day)$, $(\forall x)(Cx \rightarrow Af(x) \wedge \neg Dxf(x))$; $\mathbf{U}_0 = \{\mathbf{a}\}$; $\mathbf{U}_1 = \{\mathbf{a}, \mathbf{f}(\mathbf{a})\}$; β_1 is $(Aa \rightarrow Ba) \wedge Ca \wedge (Ba \rightarrow Daa) \wedge (Ca \rightarrow Af(a) \wedge \neg Daf(a))$. We can deduce $Af(a) \wedge \neg Daf(a)$ from β_1. β_2 contains $Af(a) \rightarrow Bf(a)$ and $Bf(a) \rightarrow Daf(a)$ as conjuncts. Thus, we can deduce $Daf(a) \wedge \neg Daf(a)$.
4. a. $Aa \wedge (\forall y)(a = y \rightarrow \neg Ay)$; $\mathbf{U}_0 = \{\mathbf{a}\}$; $\mathbf{U}_1 = \{\mathbf{a}\}$; β_1 is $Aa \wedge (a = a \rightarrow \neg Aa)$ from which we derive the contradiction $Aa \wedge \neg Aa$.
   d. $(\forall x)(Px \wedge Qf(x)) \wedge (\forall y)(Pg(y) \rightarrow \neg Qy)$; $\mathbf{U}_0 = \{\mathbf{a}\}$; $\mathbf{U}_1 = \{\mathbf{a}, \mathbf{f}(\mathbf{a}), \mathbf{g}(\mathbf{a})\}$; β_1 is $Pa \wedge Qf(a) \wedge (Pg(a) \rightarrow \neg Qa)$. β_2 contains $Pg(f(a)) \rightarrow \neg Qf(a)$ as a conjunct, and β_3 contains $Pg(f(a))$ as a conjunct. Thus, we can deduce the contradiction $Qf(a) \wedge \neg Qf(a)$.

D. 1. Suppose that neither $\Gamma \cup \{\varphi\}$ nor $\Gamma \cup \{\neg\varphi\}$ is consistent, then there is a formula ψ such that $\Gamma \cup \{\varphi\} \vdash \psi \wedge \neg\psi$ and $\Gamma \cup \{\neg\varphi\} \vdash \psi \wedge \neg\psi$. Using *CP*, we obtain $\Gamma \vdash \varphi \rightarrow \psi \wedge \neg\psi$ and $\Gamma \vdash \neg\varphi \rightarrow \psi \wedge \neg\psi$. Using a tautology, we obtain $\Gamma \vdash \psi \wedge \neg\psi$. Contradiction.

3. a. It follows from the Soundness Theorem that if $\vdash \varphi$, then φ is true in every interpretation structure for φ that has $\leq n$ elements. Conversely, suppose φ is true in every interpretation structure for φ that has $\leq n$ elements. Then, if φ is not provable, by the Completeness Theorem and the hypothesis, there must be an interpretation structure for φ with more than n elements in which φ is false. That is, there are n elements, $\mathbf{b}_1, \mathbf{b}_2, \ldots, \mathbf{b}_n$ in the universe such that $\psi(\mathbf{b}_1, \mathbf{b}_2, \ldots, \mathbf{b}_n)$ is false. Therefore, φ is false in an interpretation structure whose universe is $\{\mathbf{b}_1, \mathbf{b}_2, \ldots, \mathbf{b}_n\}$. Contradiction.

5. a. $(\exists x)Px$ is neither provable nor is its negation provable.
   b. Each sentence φ is either true or false in **S**, and not both. If φ is true in **S**, then $\varphi \in \Gamma$. If φ is false in **S**, then $\neg\varphi$ is true in **S** so $\neg\varphi \in \Gamma$.

Index